European Climate and Clean Energy Law and Policy

European Climate and Clean Energy Law and Policy

Leonardo Massai

from Routledge

First published 2012 by Earthscan
4 Park Square, Milton Park, Abingdon, Oxon OX14 4RN
605 Third Avenue, New York, NY 10017

Earthscan is an imprint of the Taylor & Francis Group, an informa business

British Library Cataloguing in Publication Data
A catalogue record for this book is available from the British Library

Library of Congress Cataloging in Publication Data
Massai, L. (Leonardo)
European climate and clean energy law and policy / Leonardo Massai.
p. cm.
Includes bibliographical references and index.
1. Climatic changes-Law and legislation-European Union countries.
2. Greenhouse gas mitigation-Law and legislation-European Union countries.
3. Emissions trading-Law and legislation-European Union countries. I. Title.
KJE6246.M37 2011
344.2404'6342-dc22
2010047850

ISBN: 978−1−84971−203−3 (hbk)
ISBN: 978−1−84971−204−0 (pbk)
ISBN: 978−1−84977−544−1 (ebk)

Typeset in Bembo
by OKS Press Services

To Giulia and Ernesto

Contents

List of Acronyms and Abbreviations

AAU	assigned amount unit
AGBM	Ad Hoc Group on the Berlin Mandate
AOSIS	Alliance of Small Island States
APP	Asia-Pacific Partnership
ASEAN	Association of Southeast Asian Nations
AWG-LCA	Ad Hoc Working Group on Long-term Cooperative Action
AWG-KP	Ad Hoc Working Group on Further Commitments for Annex I Parties under the Kyoto Protocol
BAP	Bali Action Plan
BAPA	Buenos Aires Plan of Action
BSA	Burden Sharing Agreement
CACAM	Central Asia, Caucasus, Albania and Moldova
CBD	Convention on Biological Diversity
CCS	carbon capture and storage
CCPM	common and coordinated policy and measure
CDM	Clean Development Mechanism
CEEC	Central and Eastern European countries
CER	certified emission reduction
CFC	chlorofluorocarbons
CFI	Court of First Instance
CG11	Central Group 11
CHP	combined heat and power
CITL	Community Independent Transaction Log
CLRTAP	Convention on Long-Range Transboundary Air Pollution
CMP	Conference of the Parties/Meeting of the Parties
CO_2	carbon dioxide
COP	Conference of the Parties
COREPER	Committee of Permanent Representatives
DNA	Designated National Authority
DOE	Designated Operational Entity
EACI	Executive Agency for Competitiveness and Innovation
EAP	Environment Action Programme
EATD	European Allowance Trading Directive (2003/87/EC)
EC	European Community
ECCP	European Climate Change Programme
ECJ	European Court of Justice
EEA	European Environment Agency
EEC	European Economic Community
EEPR	European Energy Programme for Recovery
EEV	enhanced environmentally friendly vehicle
EEW	Energy Efficiency Watch

EFTA	European Free Trade Area
EIT	economy in transition
ELV	end-of-life vehicle
ENDS	Environmental Data Services
EPER	European Pollutant Emission Register
ERU	emission reduction unit
ET	emissions trading
ETS	emissions trading scheme
EU	European Union
EUA	European Union allowance
FAO	Food and Agriculture Organization
GEF	Global Environment Facility
GHG	greenhouse gas
GO	guarantee of origin
GWP	global warming potential
HFC	hydrofluorocarbon
IASG	Impact and Adaptation Steering Group
ICAO	International Civil Aviation Organization
ICT	information and communication technology
IET	international emissions trading
INC	Intergovernmental Negotiating Committee
IPCC	Intergovernmental Panel on Climate Change
IPPC	integrated pollution prevention and control
ISO	independent system operator
ITL	International Transaction Log
ITO	independent transmission operator
JAMA	Japanese Automobile Manufacturers Association
JI	joint implementation
JISC	Joint Implementation Supervisory Committee
JUSSCANNZ	Coalition of non-EU Annex I Parties, guided by Japan, the United States of America, Switzerland, Canada, Australia, Norway and New Zealand
KAMA	Korean Automobile Manufacturers Association
KP	Kyoto Protocol
LCV	light commercial vehicle
LDC	Least Developed Country
LPG	liquid petroleum gas
LULUCF	land use, land use change and forestry
MEA	multilateral environmental agreement
MEP	Member of European Parliament
MOP	Meeting of the Parties
MRV	monitoring, reporting and verification
NAMA	nationally appropriate mitigation action
NAP	National Allocation Plan
NAS	National Adaptation Strategy
NEEAP	National Energy Efficiency Action Plan

NGO	non-government organization
OECD	Organisation for Economic Co-operation and Development
PAM	policies and measures
PFC	perfluorocarbon
QELRC	quantified emission limitation and reduction commitment
POP	Persistent Organic Pollutants
R&D	research and development
REDD	reduce emissions from deforestation in developing countries
REEES	renewable energy and energy efficiency strategy
REIO	regional economic integration organization
RMU	removal unit
SAR	Second Assessment Report
SBI	Subsidiary Body for Implementation
SBSTA	Subsidiary Body for Scientific and Technological Advice
SCCF	Special Climate Change Fund
SEA	Single European Act
SET-PLAN	European Strategic Energy Technology Plan
SFM	sustainable forest management
SMEs	Small and Medium-sized enterprises
TEC	Treaty of European Community
TEN-T	trans-European transport network
TEU	Treaty on European Union
TFEU	Treaty on the Functioning of the European Union
TWh	Terawatt hour
UN	United Nations
UNCCD	United Nations Convention to Combat Desertification
UNCED	United Nations Conference on Environment and Development
UNECE	United Nations Economic Commission for Europe
UNEP	United Nations Environment Programme
UNFCCC	United Nations Framework Convention on Climate Change
VAT	value added tax
WCP	World Climate Programme
WMO	World Meteorological Organization
WSSD	World Summit on Sustainable Development
WTO	World Trade Organization

List of Figures and Tables

Figures

Tables

Preface

*T*he origin of this book can be traced back a number of years. It all started with the desire of the author to keep track of the increasing amount of legislation and activities of the European institutions, Member States and international community in the field of climate change and clean energy. As a practitioner and academic directly involved in the field, the need for a constant and frequent update on these matters was the main reason behind this book.

The official sources of information dealing with European Union (EU) climate and clean energy, mainly the website of the EU, soon proved to be too limited in their role and function to respond to the needs of experts working and researching in the field. Furthermore, requests for information from individuals with less experience have also grown substantially in number.

The collection of the material for this book started around 2003 when the author began to assemble legislation, policy and communication papers and official reports. One of the main features of the electronic book is the up-to-date linkages it gives to the original sources, representing a unique product in the panorama of EU and international literature. This work has grown rapidly in terms of volume and the importance of climate change and clean energy issues on the agenda of the international and European community, and it will contribute to further develop the agenda.

The book therefore presents an unprecedented compilation of information on climate and clean energy law and policy in Europe since the early 1990s, in terms of the themes considered and information provided. It aims to show the extent of the EU's commitment towards the development of an advanced environmental policy to fight global warming and to boost clean energy, independently from the developments in international law.

Due to the close relationships and interlinkages between the various issues addressed in this book, the ordering of the chapters, which cover different thematic areas, is solely based on the choice of the author, with a view to providing as much clarity as possible. Inevitably, chronological order is not always maintained and the same information can appear in different chapters.

Special recognition is due to Environmental Data Services (ENDS), Europe's major source of environmental news and information, which has often proved to be an excellent inspiration for the author.

CHAPTER I

Introduction to European Environmental Law

*T*he protection of the environment is one of the key pillars of EU law and policy with the European institutions fully engaged in the matter since the late 1970s, when the first regulations were adopted.[1] A specific title on environmental protection was first introduced in the system of European Community (EC) treaties by the Single European Act (SEA) of 1986 (Articles 130r to 130t) and further refined with the Treaty of Maastricht in 1992 (Articles 174 to 176) and the Treaty of Amsterdam of 1997. At the moment of writing, the Treaty of Lisbon is the last step of European integration and introduced several important reforms of EU institutions and law.[2] The changes initiated by the Treaty of Lisbon have a very limited impact on environmental protection, in particular as far as the decision-making procedures and the distribution of competences are concerned. They only provide for more clarity. In particular, the Treaty of Lisbon does not modify the requirement of unanimity in the Council for the adoption of legislation on environmental taxation and energy sources.

EU environmental policy is based on the principles of sustainable development and integration of environmental concerns in all other areas. Sustainable development is mentioned within the objectives of the EU (Article 3(3) Treaty on European Union (TEU)) after the incorporation of the objectives of the EC. The principle is mentioned in the same weak manner as in the previous treaties, namely along with the principles of economic growth and price stability, although with Lisbon reference is made to the sustainable development of Europe. Furthermore, Article 3(5) TEU refers to the role of the EU in the world and to the 'sustainable development of the Earth', and sustainable development is contained in the title on General Provisions on the Union's External Action (Article 21(2)d and f TEU). In this way, sustainable development is included in the list of objectives that guide the external action and the common foreign and security policy of the EU.

The Treaty of Lisbon introduces three references to the environmental integration principle, namely:

- Article 11 Treaty on the Functioning of the EU (TFEU) (ex Article 6 Treaty of EC (TEC));
- Article 13 TFEU: integration in the field of animal welfare;
- Article 194(2) TFEU: integration in the field of energy policy.

With the TFEU, the unique position occupied by the environmental integration principle in the TEC loses some relevance due to the fact that more references to the integration principle have been promoted to the section of provisions having general application, such as sex equality, employment and social policies, non-discrimination, etc. (Articles 7–13 TFEU).

Furthermore, objectives and principles of EU environmental protection are indicated under Article 191(1) and (2) of the TFEU. They are:

- Objectives:
  - preservation, protection and improvement of the quality of the environment;
  - protection of human health;
  - prudent and rational utilization of natural resources;
  - promotion of international measures to deal with environmental problems, in particular combating climate change;
- Principles:
  - high level of protection;
  - precautionary principle;
  - principle of prevention;
  - polluter pays principle.

The only change introduced by the TFEU in this section is the introduction of a specific reference to the fight against climate change among the objectives of European environmental policy.

Article 192 TFEU refers to the decision-making procedure in the field of environmental protection and mirrors ex Article 175 TEC, namely, ordinary legislative procedure under paragraph 1 (co-decision and qualified majority voting in the Council) and unanimity in the Council required for provisions of a fiscal nature, measures affecting town and country planning, management of water resources and land use, and national choices between different energy sources and the general structure of energy supply.

A new title on energy policy is introduced by the Treaty of Lisbon (Title XXI, Article 194 TFEU), which basically codifies the practice of the EU institutions and Member States on this matter. Article 194(1) indicates the objectives of EU energy policy:

- (a) ensure the functioning of the energy market;
- (b) ensure security of energy supply in the Union;

- (c) promote energy efficiency and energy saving and the development of new and renewable forms of energy;
- (d) promote the interconnection of energy networks.

The ordinary legislative procedure should be applied to achieve the objectives of paragraph 1 (Article 194(2)) without breaching the limits set under Article 192(2)(c) ('Member State's right to determine the conditions for exploiting its energy resources, its choice between different energy sources and the general structure of its energy supply'). Measures of a fiscal nature shall be adopted in accordance with a special legislative procedure (unanimity in the Council) as prescribed under Article 194(3).

EU environmental protection and clean energy policy and law are nowadays areas where the action of the EU and its Member States is most significant and advanced. The result of EU environmental policy is around 300 legal acts covering horizontal legislation, water and air pollution, climate change, management of waste and chemicals, biotechnology, nature protection, industrial pollution and risk management, noise and radiation protection. In terms of the status of environmental protection in the EU as of 2010, the 2009 Environmental Policy Reviews (Annual report on EU environment policy) released by the European Commission on 10 August 2010 shows that positive trends are observed in areas such as renewable energy, organic farming, the recycling of packaging waste and compliance with the Kyoto Protocol. On the contrary, negative trends were reported on energy use in the transport sector, nature protection, air pollution and waste generation.

One of the issues to be considered when addressing EU environmental law and policy, especially in relation with international law and multilateral environmental agreements (MEAs) is the division of internal and external competences between the EC and the Member States. Furthermore, with the reform of the EU treaties introduced by the Treaty of Lisbon, the EU has acquired an international legal personality and therefore replaced the EC.

The legislative competence of the EC and the Member States to act in a specific area of Community law was defined by the TEC and now by the TFEU. The competences in the different areas of Community law can be either shared between the EC and the Member States – in the majority of cases – or exclusively assigned to either of them. Before Lisbon, although in most cases the TEC explicitly identified the boundaries of the different Community policies and specified whether the competence was shared or exclusive, this distinction was sometimes open to different interpretations.

In this respect, the opinion of the European Court of Justice (ECJ) was necessary to clarify uncertainties generated by the lack of a clear demarcation between exclusive and shared competences in the EU. This is why, for instance, on several occasions, the European legislator considered Article 352 TFEU (ex Article 308 TEC, originally Article 235 TEC) as the legal basis for acting, particularly where the Treaty failed to give the Community specific and express legislative power in a certain area.[3] This was the case with early adopted European environmental legislation and policy until a specific title on environmental protection, which was not included in the Treaty of Rome,[4] was introduced by the SEA (1987). Before that, the Community institutions adopted several provisions aimed at the protection of the environment, opting, inter alia, for ex Article 308 TEC[5] as a legal basis. Ex Article 308 TEC conferred on the Council, acting unanimously on a proposal from the Commission and after consulting the Assembly,[6] the power to take the appropriate measures when 'the Treaty has not provided the necessary powers' in order to attain one of the Community objectives. A more frequently used legal basis for measures on environmental protection was Article 114 TFEU (ex Article 95 TEC) on the approximation of legislation aimed at the establishment and functioning of the internal market.

Since the amendments introduced by the SEA, the division of competences on environmental protection between the EU and the Member States is regulated under Articles 191 to 193 TFEU, title XX (ex Articles 174 to 176 TEC). Article 192 TFEU is the relevant legal basis, providing the EU general competence to adopt the legislative measures necessary to achieve the objectives stated in Article 191(1).[7]

The list of objectives and principles stated in title XX of the TFEU provides an exhaustive basis for the competences and actions of the EU in the field of environmental protection. Therefore, as defined by Epiney, it is 'difficult to imagine' a Community provision which is not covered by the objectives of Article 191(1).[8] In other words, Article 191(1) provides an indication of the boundaries within which the EU can operate to protect European and global environmental issues rather than a limit on the Union's scope of action. However, the extensive list of objectives and principles indicated under Article 191 does not confer the EU exclusive competence on the matter.

The Treaty of Lisbon introduces three categories of competences: exclusive (Article 3 TFEU), shared (Article 4 TFEU) and areas where the EU may only support, coordinate or supplement the actions of the Member States (Article 6 TFEU), excluding harmonization. The EU keeps exclusive competence for the conservation of marine biological resources under common fisheries policy, while the Treaty of Lisbon refers to environmental protection and energy policy as a shared competence (Article 4(2)(e) and (i) TFEU). This codifies the extensive jurisprudence of the ECJ on environmental protection. In the area of EU energy policy, this responds to the need to reduce EU dependency on imported energy supplies and to have a competitive advantage by initiating new low-carbon technologies.

The TFEU under Article 4 explicitly refers to the shared competences between the Union and the Member States:

> The Union shall share competence with the Member States where the Treaties confer on it a competence which does not relate to the areas referred to in Articles 3 and 6 [exclusive competence]. Shared competence between the Union and the Member States applies in the following principal areas:
>
> (d) agriculture and fisheries, excluding the conservation of marine biological resources;
> (e) environment;

(g) transport;
(i) energy; (Article 3(1) and (2) TFEU)

Even before the adoption of the Treaty of Lisbon, when such a clear indication of exclusive and shared competences was lacking, in practice environmental protection was always considered a matter of shared competence between the EC and the Member States. The first reason was that the TEC did not provide for exclusive competence on environmental protection to the Community, as it clearly did in other domains.[9] Second, the TEC, in Title XIX on Environment, left the Member States some room for action. Ex Article 174(2) provided the Member States with a safeguard clause, notably the power to adopt provisional measures of environmental protection subject to the control of the Community. Moreover, ex Article 176 TEC gave the Member States the possibility to introduce 'more stringent protective measures', provided that such measures are compatible with the EC Treaty and are notified to the European Commission.

If the confines of the distribution of internal competence within the Community were not always expressly defined by the EC Treaty, the identification of the EU's external competence certainly seemed no less complex. The term 'external power or competence' refers to the power of the Community to conclude an international agreement with third states and international organizations outside the EU. Although the Treaty of Rome did not mention the existence of clear external power for the EC, this gap has been filled by several amendments to the founding treaties and by the jurisprudence of the ECJ, which contributed clearly to the definition of the boundaries of the Community's external competence.[10] Explicit external competence of the Community was limited to commercial policy (ex Article 133 TEC), association agreements (ex Article 310 TEC), the maintenance of relations with international organizations (ex Articles 302 to 304 TEC), development policy (ex Article 181 TEC), environmental policy (ex Article 174 TEC), research and technology (ex Article 170 TEC), monetary and foreign exchange policy (ex Article 111 TEC) and economic, financial and technological cooperation with third countries (ex Article 181(a) TEC). In areas where the TEC was silent, the ECJ made clear since the beginning of the 1970s that even if no express external competence was granted to the Community by the treaties, implied external power could be applied in those cases. On several occasions, the Court confirmed the capacity of the Community to conclude an international agreement, either in areas falling within the scope of the objectives expressly established by the TEC, or in relation to areas where the Community order has created ad hoc internal powers to attain a specific objective.[11]

The TFEU introduced Article 216, which states that:

the Union may conclude an agreement with one or more third countries or international organisations where the Treaties so provide or where the conclusion of an agreement is necessary in order to achieve, within the framework of the Union's policies, one of the objectives referred to in the

Treaties, or is provided for in a legally binding Union act or is likely to affect common rules or alter their scope.

Furthermore, 'agreements concluded by the Union are binding upon the institutions of the Union and on its Member States'.

Furthermore, the Treaty of Lisbon could provide an added value to the external action of the EU in the field of environmental protection and clean energy. The changes related with the presidency of the European Council (Article 15(5) TEU) and the role of the High Representative for Foreign Affairs and Security Policy may contribute to boosting the EU's ability to negotiate on international environmental agreements (Article 18 TEU). The High Representative will be in charge of coordinating foreign issues while the current system foresees the leading of the Council Presidency based on a common position adopted by the 27 Member States. The High Representative could be better equipped to coordinate the interests of the Member States with the European Commission and able to present a more unified front at international negotiations.

The legal basis of the external competence of the Community in the field of environmental protection is provided by Article 191(4) TFEU, which confers on the Community and the Member States 'within their respective spheres of competence' the right to 'cooperate with third countries and with the competent international organisations'. This cooperation or shared competence can form the basis for an agreement to be concluded by the Community in accordance with Article 218 TFEU.[12] Furthermore, one of the objectives of the Community's environmental policy, referred to in Article 191(1) TFEU, is the promotion of 'measures at international level to deal with regional and worldwide environmental problems'.

MacLeod et al[13] correctly argue that the Community holds no exclusive external competence in those sectors, like environmental protection, where the Community shares its internal power to act with the Member States. As a consequence, ex Article 174(4) TEC is read as conferring on the Community and the Member States a shared competence in external matters. The practice of the Council and the Parliament concerning the conclusion of an international agreement by the EC and the Member States confirms this approach. EC legislation adopted to achieve a specific objective of environmental protection is nearly always the legal basis for justifying the external power of the EC and the Member States.

In theory, Member States can retain a certain power to independently conclude an international agreement with third parties in those areas where the Community has failed to provide an exhaustive set of legislation, in other words, in those sectors where the Community has not been able to attain the objectives of Article 191(1) TFEU. In practice, it is very unlikely that a conflict between the Community and the Member States will arise over the exercise of external power in the area of environmental protection. On the basis of the above considerations, it may be affirmed that, in principle, the external power of the Community and the Member States regarding environmental protection is not exclusive and is mainly based on a specific shared external competence provided for in Article 191(4) TFEU and on an implicit external power derived from the implementation of a clear internal competence regarding environmental protection.

Consequently, in the environmental sector it is common practice for the Community and the Member States to conclude so-called 'mixed agreements'.[14] While a definition of mixed agreement is lacking in both international and Community law, it can be concluded from Community and Member States' practice that a mixed agreement is an international treaty that can be ratified (a) by both the Community and the Member States, (b) only by the Member States, or (c) only by the Community. Some scholars have defined a mixed agreement as 'any treaty to which an international organization, some or all of its Member States and one or more third State are parties and for the execution of which neither the organization nor its Member States have full competence'.[15] Regarding environmental protection, all international agreements to which the European Community is a party are mixed agreements in the sense that one or more Member States are also parties to those treaties.[16]

A multilateral environmental agreement may take several forms: at the global level, it is usually an agreement, adopted within the framework of United Nations (UN) bodies;[17] at the regional level, it is an agreement concluded for instance, in the framework of the UN Economic Commission for Europe (UNECE) or the Council of Europe;[18] and, finally, at the intraregional level, it is an agreement entered into, for instance, in the field of management of seas or transboundary rivers.[19] The last typology of agreements normally applies to specific geographical areas and therefore not all 27 EU Member States participate. Below is an indicative list of relevant MEAs to which the EC and the Member States are parties:

- Geneva Convention on Long-Range Transboundary Air Pollution (CLRTAP; 1979);
- Cartagena Biosafety Protocol (2000) to the Rio Convention on Biological Diversity (CBD; 1992);
- Stockholm Convention on Persistent Organic Pollutants (POP; 2001);
- United Nations Framework Convention on Climate Change (UNFCCC; 1992) and the Kyoto Protocol (1997);
- Vienna Convention for the Protection of the Ozone Layer (1985) and the Montreal Protocol;
- Aarhus Convention on Access to Environmental Information (1998);
- UN Convention to Combat Desertification (UNCCD; 1994).

The UNFCCC and the Kyoto Protocol represent typical examples of mixed agreements where the EC and the Member States share the same status – Annex I parties – with the exception of Malta and Cyprus, and have the same rights and duties. Less clear are issues such as the hierarchy and balance between the Community and the Member States in the implementation and fulfilment of the obligations created by the Kyoto Protocol.

NOTES

1. Although the EC has been replaced by the EU since the entry into force of the Treaty of Lisbon on 1 December 2009, in this book the term EC is used as referring to the old first pillar of the TEU adopted in 1992 at Maastricht.

2. Consolidated versions of the TEU and the TFEU, Council of the EU, Brussels, 30 April 2008.

3. 'If action by the Community should prove necessary to attain, in the course of the operation of the common market, one of the objectives of the Community and this Treaty has not provided the necessary powers, the Council shall, acting unanimously on a proposal from the Commission and after consulting the European Parliament, take the appropriate measures' (ex Article 308 EC Treaty).

4. The Treaty of Rome established the European Economic Community (EEC) and was signed by France, West Germany, Italy, Belgium, the Netherlands and Luxembourg on 25 March 1957.

5. Council Directive 79/409/EEC of 2 April 1979 on the conservation of wild birds, *OJ* L 103, 25 April 1979, pp1–18; Council Decision 86/234/EEC of 10 June 1986 adopting multiannual research and development (R&D) programmes in the field of the environment (1986–1990), *OJ* L 159, 14 June 1986, pp31–35; Council Directive 82/884/EEC of 3 December 1982 on a limit value for lead in the air, *OJ* L 378, 31 December 1982, pp15–18; Council Recommendation 81/972/EEC of 3 December 1981 concerning the reuse of waste paper and the use of recycled paper, *OJ* L 355, 10 December 1981, pp56–57.

6. The body was renamed as the European Parliament in 1962.

7. 'The Council ... shall decide what action is to be taken by the Community in order to achieve the objectives referred to in Article 191' (Article 192(1) TFEU).

8. Epiney (2003), p46.

9. Only ex Articles 133 (commercial policy), 310 (association agreements), 302–304 (relation between the EC and international organizations) referred explicitly to the external power of the EC.

10. This is the case with the following EC policies: development, environment, research and technology, monetary and foreign exchange matters, education, culture, health, and common commercial policy.

11. ECJ, case 22/70 *Commission* v. *Council* [1971] *ECR* 263, cases 3, 4 and 6/76 *Kramer* [1976] *ECR* 1279, opinion 1/76 (Draft Agreement Establishing a Laying-up Fund for Inland Waterway Vessels) [1977] *ECR* 741, opinion 2/91 re *ILO Convention* 170 on Chemicals at Work [1993] *ECR* I-1061, opinion 2/94 (Accession of the Community to the European Human Rights Convention) [1996] *ECR* I-1579.

12. 'Within their respective spheres of competence, the Union and the Member States shall cooperate with third countries and with the competent international organisations. The arrangements for Union cooperation may be the subject of agreements between the Union and the third parties concerned. The previous subparagraph shall be without prejudice to Member States' competence to negotiate in international bodies and to conclude international agreements' (Article 191(4) TFEU).

13. MacLeod et al (1996), pp325–326.

14. On this matter, see O'Keeffe and Schermers (1983); MacLeod et al (1996), pp142–164; Rosas (2000) and Heliskoski (2001).

15. Schermers (1983), pp25–26.

16. MacLeod et al (1996), p329.

17. For instance, the UNFCCC (1994) or the United Nations CBD (1992).

18. For instance, the UNECE CLRTAP (1979) or the Convention on the Protection and Use of Transboundary Watercourses and International Lakes (1992).

19. For instance, the Barcelona Convention for the Protection of the Mediterranean (1978) or the Convention for the Protection of the Rhine (1976).

CHAPTER 2

The International Climate Regime

C limate change is one of the major environmental phenomena that the world has faced in recent times. Its coverage in the media as well as its importance in the development of international and European policies aimed at the protection of the environment have increased considerably in the recent past, reaching a peak with the Copenhagen summit in 2009, when all major world leaders convened to decide over the future of the international fight against global warming.

Carbon dioxide (CO_2) and other gases contribute to keeping the planet's surface warm and habitable in a natural way by trapping part of the solar heat within the atmosphere. The burning of fossil fuels, such as coal, gas and oil, and phenomena such as deforestation and the burning of forests, have artificially increased the amount of CO_2 in the earth's atmosphere, thus contributing to a rise in global temperature.[1] The release of greenhouse gases (GHGs) into the atmosphere is the main cause of the greenhouse effect, i.e. the excessive warming of earth's atmosphere caused by a disproportionate atmospheric concentration of water vapour, CO_2 and other GHGs. This so-called global warming directly and indirectly affects the ecosystems in many countries in the world. In particular, the alteration of the global temperature in the atmosphere contributes to several negative effects, notably an unbalanced increase in sea level, melting of glaciers, changes in plant habitat and animal migration, a high frequency of rain and changing rainfall at regional levels, as well as a more intense hydrological cycle.

What is important and relevant for this book is the origin of global warming. The international community is almost unanimous in identifying human-induced activities such as energy supply, transport and deforestation as the main cause of global warming. This is why the international community is, starting in the late 1980s, fully committed to tackling this problem from a legal and political perspective. The recognition of human influence on the global climate through the release of GHGs into the atmosphere was one of the key findings of the most recent studies and assessments of the Intergovernmental Panel on Climate Change (IPCC). The IPCC was established in 1988 by the World Meteorological Organization (WMO) and the UN Environment Programme (UNEP) in order to assist policy-makers in the identification and interpretation of scientific information on climate change and to provide important scientific inputs to the climate change negotiating process. The IPCC is composed of member countries and scientists that are divided into three working groups: (1) science of climate change; (2) impacts, vulnerability and adaptation; and (3) mitigation. In addition to these three working groups, the IPCC also includes a task force on national GHG inventories. In 1990, the First Assessment Report of the IPCC found that human-induced climate change was a global threat, and contributed to the launch of the negotiations on the UNFCCC. The Second Assessment Report (1995) confirmed 'a discernible human influence on global climate' and contributed to the intensification of the negotiations towards the adoption of the Kyoto Protocol. The Third Assessment Report (2001) provided new and stronger evidence of global warming and highlighted the importance of the integration of sustainable development policies with climate change mitigation measures. The Third Assessment Report emphasized the key role to be played by decision-makers at all levels, as well as the correct use of technologies to stop the growth of GHG emissions.[2] The IPCC findings are relevant to addressing questions such as whether global temperature is actually rising, to what extent GHG emissions affect changes in global temperature and, finally, to what extent human activity affects changes in global temperature. According to the IPCC, global warming is 'unlikely to be entirely natural in origin' and 'the balance of evidence suggests a discernible human influence on the global climate'. In 2007, the IPCC released its Fourth Assessment Report, concluding that climate change will continue in the coming centuries regardless of whether pollution continues. According to the IPCC, a rise of 1.8 up to 4.1°C in temperature by 2099 is highly probable, as well

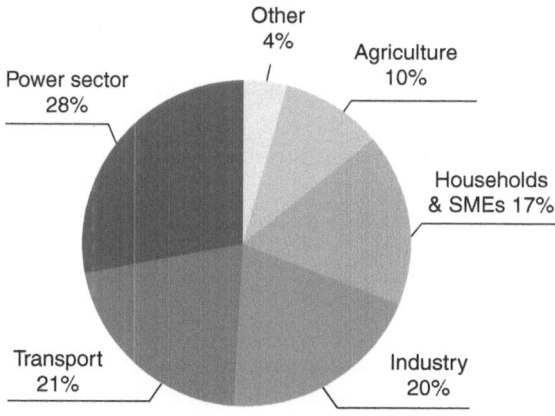

Figure 2.1 *Sources of GHG emissions in the EU*

Source: EEA

as an increase in sea level of 18 to 55cm by 2099. Finally, more scientific certainty is provided on the question of whether the increase in tropical storms such as hurricanes since 1970 is caused by human activity.

Figure 2.1 shows the major sources of GHG emissions in the EU, as identified by the European Environment Agency (EEA).

2.1   THE UNFCCC

The international climate change regime is based on the United Nations Framework Convention on Climate Change (UNFCCC),[3] signed in Rio de Janeiro in 1992 and entered into force in 1994. The Convention sets a series of general rules and objectives governing climate change mitigation, adaptation and capacity building, in particular with the aim of stabilizing the concentration of GHG emissions through cooperation between developed and developing countries – Annex I and non-Annex I parties.

The UNFCCC is an international treaty that was adopted at the UN headquarters in New York on 9 May 1992 and entered into force on 21 March 1994.[4] At the moment of writing there are 194 Parties (193 States and 1 regional economic integration organization) to the UNFCCC. According to Article 22(1) of the UNFCCC, 'the Convention shall be subject to ratification, acceptance, approval or accession by States and by regional economic integration organizations'. It is commonly accepted that MEAs allow for different procedures to become party to an international treaty in order to take account of the different constitutional systems of countries – i.e., in the case of accession, the state joins a treaty already negotiated and signed by other parties. The procedures indicated in Article 22(1) require the state's consent to be bound by an international treaty and all related obligations.

The UNFCCC sets the international legal framework for combating climate change and recognizes the vulnerability of the climate system affected by 'dangerous' anthropogenic emissions of CO_2 and other GHG emissions. The UNFCCC calls on governments to provide and make public relevant information on GHG emissions, as well as to develop national policies and best practices to combat global warming. The UNFCCC focuses on both mitigation of GHG emissions and adaptation to the adverse effects of climate change, in particular by requiring governments to cooperate and set up adequate national strategies accordingly. International cooperation is one of the major principles upon which the international climate regime is based, and particularly the UNFCCC and the Kyoto Protocol foster the transfer of technology, information and capacity from developed to developing countries.

The ultimate objective of the UNFCCC is the achievement of:

> stabilization of greenhouse gas concentrations in the atmosphere at a level that would prevent dangerous anthropogenic interference with the climate system. Such a level has to be achieved within a time-frame sufficient to allow ecosystems to adapt naturally to climate change, to ensure that food production is not threatened and to enable economic development to proceed in a sustainable manner. (Article 2)

The UNFCCC identifies the following principles as governing the climate change regime (Article 3):

- equity in the commitment of parties to protect the climate;
- recognition of developing countries' special needs and circumstances 'especially those that are particularly vulnerable to the adverse effects of climate change';
- precautionary principle according to which action should be taken to 'anticipate, prevent or minimise the causes of climate change and mitigate its adverse effects';
- promotion of sustainable development and growth.

Moreover, the UNFCCC recognizes the difference between developed and developing countries in terms of climate change, in particular refers to the 'common but differentiated responsibilities and respective capabilities' of parties and calls on the developed countries to 'take the lead in combating climate change and the adverse effects thereof' (Article 3(1)). The principle of common but differentiated responsibilities[5] is one of the key points upon which the international climate regime functions, as in the case of many other MEAs. In accordance with this principle, the industrialized countries should take the lead in the fight against climate change and in the promotion of adaptation measures (Articles 3(1) and 4(2)a UNFCCC). By the same token, the Convention clearly establishes, in Article 4(7), a link between the performance of Annex I parties in meeting 'their commitments under the Convention related to financial resources and transfer of technology' and the effectiveness of the implementation of the Convention commitments by developing countries.[6]

The distinction between developed and developing countries and the definition of different responsibilities and obligations among parties is the key milestone upon which the international climate regime created by the UNFCCC and the Kyoto Protocol is founded. The talks and negotiations at the international level on the future developments of the climate regime focus in particular on the redistribution of tasks, responsibilities and commitments among parties. The principle of common but differentiated responsibilities is also at the foundation of the establishment of different sets of obligations among parties, also because it provides the legal basis for the definition of a different status for parties that are granted a certain degree of flexibility in implementing the UNFCCC. In this respect, parties with economies in transition (EITs) to a market economy were given the possibility to negotiate and agree on a base year different from 1990 for the definition of the GHG reduction obligations under the Kyoto Protocol.

Annex I of the UNFCCC includes a list of industrialized countries that have committed themselves to reducing GHG emissions by the year 2000 to 1990 levels as prescribed under Article 4(2)b of the UNFCCC. These countries are the members of the Organisation for Economic Co-operation and Development (OECD) in 1992, excluding Mexico, together with designated countries undergoing the transition to a market economy – EITs[7] –, the EC and Turkey. Non-Annex I parties are those not listed in Annex I (see Table 2.1), notably developing countries such as least developed countries, most vulnerable countries, oil-exporting countries and other parties.

Annex II of the UNFCCC includes the list of parties – Annex I parties excluding the EITs – that are required to assist the developing countries in the implementation of the Convention by providing new and additional financial resources necessary to meet the expenses of the preparation and submission of national inventories of GHG emissions and the costs of adaptation to the adverse effects of climate change. Furthermore, Annex II parties are required to promote the transfer of technology and know-how (Article 4(3), 4(4) and 4(5) of the UNFCCC).

Articles 4 and 12 of the UNFCCC identify and list the essential commitments applying indistinctly to all parties:

- Preparation of national inventories of GHG emissions and removals by sinks (Article 4(1)a);[8]
- Preparation of national and regional programmes containing measures regarding climate change (Article 4(1)b):[9]
  – mitigation by addressing GHG emissions sources and sinks;
  – adaptation;
- Transfer and promotion of technologies, practices and processes aimed at the reduction of GHG emissions (Article 4(1)c);[10]
- Promotion of sustainable management and cooperation in the management of sinks (Article 4(1)d);[11]
- Cooperation in the preparation for adaptation to climate change (Article 4(1)e),[12] consideration of climate change in economic and environmental policy and actions (Article 4(1)f),[13] promotion of research, systematic observation and development of data archives to reduce uncertainties about the causes and effects of climate change (Article 4(1)g),[14] cooperation in the exchange of information (Article 4(1)h),[15] promotion of education, training and public awareness regarding climate change (Article 4(1)i);[16]
- Communication of information in accordance with Article 12 of the UNFCCC (Article 4(1)j), notably a national inventory of anthropogenic emissions by sources and removals, a description of steps undertaken to implement the Convention and any other relevant information).[17]

Table 2.1 *Annex I parties to the UNFCCC*

Australia	Austria	Belarus	Belgium
Bulgaria	Canada	Croatia	Czech Republic
Denmark	EC	Estonia	Finland
France	Germany	Greece	Hungary
Iceland	Ireland	Italy	Japan
Latvia	Liechtenstein	Lithuania	Luxembourg
Monaco	Netherlands	New Zealand	Norway
Poland	Portugal	Romania	Russian Fed.
Slovak Republic	Slovenia	Spain	Sweden
Switzerland	Turkey	Ukraine	UK
US			

Furthermore, on the grounds of the principle of common but differentiated responsibilities, Article 4 of the UNFCCC lists the differentiated commitments that apply only to Annex I and Annex II parties separately:

- For Annex I parties:
  - adoption of national policies and measures (PAMs) aimed at the mitigation of climate change (Article 4(2)a);[18]
  - submission of relevant information on the above mentioned PAMs as well as on projected GHG emissions 'with the aim of returning individually or jointly to their 1990 levels' (Article 4(2)b);[19]
- For Annex II parties:[20]
  - financial and technical assistance to developing countries to enable them to comply with the obligations of information under Article 12 UNFCCC (Articles 4(3) and 4(5));[21]
  - assistance to developing countries in meeting the costs of adaptation to the adverse effects of climate change (Article 4(4)).[22]

Finally, Article 4 of the UNFCCC provides for specific entitlements of certain parties, notably a certain degree of flexibility allowed to EITs in order to address climate change and negotiate an appropriate base year for the calculation of the historical level of GHG emissions (Articles 4(6)),[23] as well as for 'the specific needs and special situations of the least developed countries' (Article 4(9)) and of developing countries with economies that are vulnerable to the implementation of mitigation measures (Article 4(10)).[24]

Mention has already been made of the role of Annex II parties, whose main obligation is to assist and provide financial support to developing countries in the implementation of the Convention. As regards non-Annex I parties, Article 4(7) makes the effective implementation of their commitments dependent on the effective implementation by developed countries of their obligations to provide financial support and transfer technology to those countries. Furthermore, Article 4(8) urges parties to undertake 'actions related to funding, insurance and transfer of technology' to address the specific needs of developing countries, especially with regard to adaptation measures.

Common to several existing multilateral environmental agreements, the institutional framework upon which the international climate regime is based relies on the Conference of the Parties (COP), which is the supreme body of the UNFCCC and was established to keep under regular review the implementation of the Convention and to adopt the decisions necessary to promote the implementation of the Convention (Article 7 of the UNFCCC). The COP is supported by two subsidiary bodies that are relevant for the implementation of the UNFCCC: the Subsidiary Body for Scientific and Technological Advice (SBSTA) (Article 9), responsible for providing information and advice on scientific and technological matters, and the Subsidiary Body for Implementation (SBI) (Article 10) that assists the COP in the assessment and the review of the effective implementation of the Convention. The COP meets in plenary where parties are requested to adopt decisions usually forwarded by the subsidiary bodies. In practice, the different issues on the agenda are separated into negotiating items that are discussed individually by the parties. The subsidiary bodies are responsible for the adoption of draft decisions on each of the negotiating items. All decisions within the framework of the UNFCCC are adopted by consensus, in line with UN practice. This 'default' procedure aims at seeking the agreement of most parties. No consensus is required for decisions on the rules of procedure.[25] The COP can take 'decisions necessary to promote the effective implementation' of the UNFCCC – formally non-binding – and can adopt additional protocols by consensus that are subject to ratification, and amendments to the UNFCCC can be adopted by a three-quarters majority vote, but subject to subsequent acceptance by parties.

The Secretariat of the UNFCCC (Article 8) is located in Bonn and acts as a coordination and administrative office in charge of, amongst others, collecting and preparing reports on the basis of the relevant information on GHG emission sources and sinks as well as national policies and measures submitted by parties. The UNFCCC has established a mechanism for the provision of financial resources and technology transfer administered by its own Council but subject to COP guidance (Article 11). This financial mechanism is the Global Environment Facility (GEF), which, like the IPCC, is not a formal institution of the UNFCCC.

2.2 THE KYOTO PROTOCOL

In 1997 during COP3 the Kyoto Protocol[26] was adopted as an international binding agreement with a target of a 5 per cent overall reduction of GHG emissions between 2008 and 2012 from 1990 levels for Annex B parties (Annex I states that have agreed quantified emission limitation and reduction commitments (QELRCs) within the Protocol).

Right after the entry into force of the UNFCCC, pressing and increasing scientific evidence of climate change urged the international community to initiate a discussion on a stricter and more efficient system for the reduction of GHG emissions at the global level. To this end, in 1995, the first COP to the UNFCCC (COP1) endorsed the creation of the so-called Ad Hoc Group on the Berlin Mandate (AGBM). The Berlin Mandate was initiated with the aim of establishing quantifiable limitation and reduction targets for GHG emissions within a defined and clear commitment period. The Berlin Mandate included an agreement on the establishment of a pilot phase for projects between Annex I parties and non-Annex I parties aimed at the reduction of GHG emissions and the achievement of the UNFCCC objective. COP2 endorsed the IPCC Second Assessment Report (SAR) in which the human influence on the global climate was confirmed and the preparation of a new text in the form of a 'protocol or another legal instrument' was initiated. In 1997, at COP3, the details of the Berlin Mandate were translated into the Kyoto Protocol, a new international agreement in which the industrialized world accepted binding targets for the reduction of GHG emissions by 2012 compared to 1990 levels.[27]

The Kyoto Protocol is a legal instrument 'related' to the UNFCCC that was opened for signature at the UN headquarters in 1998. The Kyoto Protocol is a legally binding instrument to reduce GHG emission concentrations in the atmosphere in the commitment period of 2008–2012. These reductions are calculated in relation to a base year, which for the majority of Annex I parties is 1990. The idea to establish a multi-year target of a more flexible nature than a single-year target was first presented by the US delegation in 1997, supported by Iceland, New Zealand, Norway and the Russian Federation. The main reasoning behind the adoption of a multi-year approach is twofold: on the one hand, the intention to reduce the risk that annual fluctuations in the level of GHG emissions could be provoked by different natural factors; on the other hand, the necessity to ensure the success of emissions trading and the borrowing of assigned amounts as a way out for Annex I parties in non-compliance with their reduction obligations.[28]

The rules for entry into force of the Protocol (Article 25)[29] require 55 Parties to the Convention to ratify (or approve, accept or accede to) the Protocol, including Annex I Parties accounting for 55 per cent of that group's CO_2 emissions in 1990. The Kyoto Protocol entered into force on 16 February 2005, following the submission of the instrument of ratification by the Russian Federation to the secretariat 90 days earlier. On Friday 22 October 2004, the Russian Parliament voted in favour of ratification and the official instrument of ratification was submitted to the UNFCCC Secretariat on 18 November 2004.[30] Ratification by the Russian Federation became essential for the entry into force of the Kyoto Protocol according to the complex procedure foreseen under Article 25(1),[31] particularly after the refusal of the US and Australia to accept this treaty. This was due to the fact that the US and Australia accounted for 36.1 and 2.1 per cent, respectively, of the global level of GHG emissions, while the share of the Russian Federation was 17.4 per cent. The Kyoto Protocol followed and reinforced the basic idea of the UNFCCC, namely the need for international action to mitigate GHG emissions and to adapt to the adverse effects of climate change. Both EU environment ministers within the EU Environment Council of 14 October 2004[32] and the European Commission welcomed Russian ratification.

At the moment of writing there are 192 Parties (191 states and 1 regional economic integration organization) to the Kyoto Protocol to the UNFCCC, which amounts to 63.7 per cent of the total percentage of Annex I Parties' GHG emissions. The status of ratification of the Kyoto Protocol appears in Table 2.2.

The Kyoto Protocol builds upon the UNFCCC; both treaties share the same principles and ultimate objectives, namely the reduction of global-level GHG emissions and the distinction between developed and developing countries, as well as the same institutional structure. Article 2 of the Kyoto Protocol contains the basic principles that Annex I parties shall follow in order to meet their QELRCs. These are in line with Article 3 of the UNFCCC: promotion of sustainable development, cooperation among parties in the implementation of policies and measures for the purposes of Article 4(2) and minimization of the adverse effects of climate change.

The COP serving as the Meeting of the Parties (MOP) to the Kyoto Protocol, has similar functions, roles, duties and responsibilities to those of the COP for the Convention. The COP/MOP convenes at the same place and in the same period as the COP, but meetings are held separately and parties that are members of the

Table 2.2 *Status of ratification of the Kyoto Protocol*

Participant	Signature	Ratification Acceptance (A) Accession (a) Approval (AA)	Entry into force	% of emissions
ALBANIA		1 Apr 2005 a	30 Jun 2005	
ALGERIA		16 Feb 2005 a	17 May 2005	
ANGOLA		8 May 2007 a	6 Aug 2007	
ANTIGUA AND BARBUDA	16 Mar 1998	3 Nov 1998	16 Feb 2005	
ARGENTINA	16 Mar 1998	28 Sep 2001	16 Feb 2005	
ARMENIA		25 Apr 2003 a	16 Feb 2005	
AUSTRALIA*	29 Apr 1998	12 Dec 2007	11 Mar 2008	2.1%
AUSTRIA *	29 Apr 1998	31 May 2002	16 Feb 2005	0.4%
AZERBAIJAN		28 Sep 2000 a	16 Feb 2005	
BAHAMAS		9 Apr 1999 a	16 Feb 2005	
BAHRAIN		31 Jan 2006 a	1 May 2006	
BANGLADESH		22 Oct 2001 a	16 Feb 2005	
BARBADOS		7 Aug 2000 a	16 Feb 2005	
BELARUS*		26 Aug 2005 a	24 Nov 2005	
BELGIUM*	29 Apr 1998	31 May 2002	16 Feb 2005	0.8%
BELIZE		26 Sep 2003 a	16 Feb 2005	
BENIN		25 Feb 2002 a	16 Feb 2005	
BHUTAN		26 Aug 2002 a	16 Feb 2005	
BOLIVIA	9 Jul 1998	30 Nov 1999	16 Feb 2005	
BOSNIA AND HERZEGOVINA		16 Apr 2007 a	15 Jul 2007	
BOTSWANA		8 Aug 2003 a	16 Feb 2005	
BRAZIL	29 Apr 1998	23 Aug 2002	16 Feb 2005	
BRUNEI DARUSSALAM		20 Aug 2009 a	18 Nov 2009	
BULGARIA*	18 Sep 1998	15 Aug 2002	16 Feb 2005	0.6%
BURKINA FASO		31 Mar 2005 a	29 Jun 2005	
BURUNDI		18 Oct 2001 a	16 Feb 2005	
CAMBODIA		22 Aug 2002 a	16 Feb 2005	
CAMEROON		28 Aug 2002 a	16 Feb 2005	
CANADA*	29 Apr 1998	17 Dec 2002	16 Feb 2005	3.3%
CAPE VERDE		10 Feb 2006 a	11 May 2006	
CENTRAL AFRICAN REPUBLIC		18 Mar 2008 a	16 Jun 2008	
CHAD		18 Aug 2009 a	17 Nov 2009	
CHILE	17 Jun 1998	26 Aug 2002	16 Feb 2005	
CHINA	29 May 1998	30 Aug 2002 AA (1)	16 Feb 2005	
COLOMBIA		30 Nov 2001 a	16 Feb 2005	
COMOROS		10 Apr 2008 a	9 Jul 2008	
CONGO		12 Feb 2007 a	13 May 2007	
COOK ISLANDS	16 Sep 1998	27 Aug 2001	16 Feb 2005	
COSTA RICA	27 Apr 1998	9 Aug 2002	16 Feb 2005	
COTE D'IVOIRE		23 Apr 2007 a	22 Jul 2007	
CROATIA*	11 Mar 1999	30 May 2007	28 Aug 2007	
CUBA	15 Mar 1999	30 Apr 2002	16 Feb 2005	
CYPRUS		16 Jul 1999 a	16 Feb 2005	
CZECH REPUBLIC*	23 Nov 1998	15 Nov 2001 AA	16 Feb 2005	1.2%
DEMOCRATIC PEOPLE'S REPUBLIC OF KOREA		27 Apr 2005 a	26 Jul 2005	
DEMOCRATIC REPUBLIC OF CONGO		23 Mar 2005 a	21 Jun 2005	

DENMARK*	29 Apr 1998	31 May 2002 (2)	16 Feb 2005	0.4%
DJIBOUTI		12 Mar 2002 a	16 Feb 2005	
DOMINICA		25 Jan 2005 a	25 Apr 2005	
DOMINICAN REPUBLIC		12 Feb 2002 a	16 Feb 2005	
ECUADOR	15 Jan 1999	13 Jan 2000	16 Feb 2005	
EGYPT	15 Mar 1999	12 Jan 2005	12 Apr 2005	
EL SALVADOR	8 Jun 1998	30 Nov 1998	16 Feb 2005	
EQUATORIAL GUINEA		16 Aug 2000 a	16 Feb 2005	
ERITREA		28 Jul 2005 a	26 Oct 2005	
ESTONIA*	3 Dec 1998	14 Oct 2002	16 Feb 2005	
ETHIOPIA		14 Apr 2005 a	13 Jul 2005	
EUROPEAN UNION	29 Apr 1998	31 May 2002 AA	16 Feb 2005	
FIJI	17 Sep 1998	17 Sep 1998	16 Feb 2005	
FINLAND*	29 Apr 1998	31 May 2002	16 Feb 2005	0.4%
FRANCE	29 Apr 1998	31 May 2002 AA	16 Feb 2005	2.7%
GABON		12 Dec 2006 a	12 Mar 2007	
GAMBIA		1 Jun 2001 a	16 Feb 2005	
GEORGIA		16 Jun 1999 a	16 Feb 2005	
GERMANY*	29 Apr 1998	31 May 2002	16 Feb 2005	7.4%
GHANA		30 May 2003 a	16 Feb 2005	
GREECE*	29 Apr 1998	31 May 2002	16 Feb 2005	0.6%
GRENADA		6 Aug 2002 a	16 Feb 2005	
GUATEMALA	10 Jul 1998	5 Oct 1999	16 Feb 2005	
GUINEA		7 Sep 2000 a	16 Feb 2005	
GUINEA-BISSAU		18 Nov 2005 a	16 Feb 2005	
GUYANA		5 Aug 2003 a	16 Feb 2005	
HAITI		6 Jul 2005 a	4 Oct 2005	
HONDURAS	25 Feb 1999	19 Jul 2000	16 Feb 2005	
HUNGARY*		21 Aug 2002 a	16 Feb 2005	0.5%
ICELAND *		23 May 2002 a	16 Feb 2005	0.0%
INDIA		26 Aug 2002 a	16 Feb 2005	
INDONESIA	13 Jul 1998	3 Dec 2004	3 Mar 2005	
IRAN (ISLAMIC REPUBLIC OF)		22 Aug 2005 a	20 Dec 2005	
IRAQ		28 Jul 2009 a	26 Oct 2009	
IRELAND*	29 Apr 1998	31 May 2002	16 Feb 2005	0.2%
ISRAEL	16 Dec 1998	15 Mar 2004	16 Feb 2005	
ITALY *	29 Apr 1998	31 May 2002	16 Feb 2005	3.1%
JAMAICA		28 Jun 1999 a	16 Feb 2005	
JAPAN *	28 Apr 1998	4 Jun 2002 A	16 Feb 2005	8.5%
JORDAN		17 Jan 2003 a	16 Feb 2005	
KAZAKHSTAN	12 Mar 1999	19 Jun 2009	17 Sep 2009	
KENYA		25 Feb 2005 a	26 May 2005	
KIRIBATI		7 Sep 2000 a	16 Feb 2005	
KUWAIT		11 Mar 2005 a	9 Jun 2005	
KYRGYZSTAN		13 May 2003 a	16 Feb 2005	
LAO PEOPLE'S DEMOCRATIC REPUBLIC		6 Feb 2003 a	16 Feb 2005	
LATVIA*	14 Dec 1998	5 Jul 2002	16 Feb 2005	0.2%
LEBANON		13 Nov 2006 a	11 Feb 2007	
LESOTHO		6 Sep 2000 a	16 Feb 2005	
LIBERIA		5 Nov 2002 a	16 Feb 2005	

(continued)

Table 2.2 – *continued*

Participant	Signature	Ratification Acceptance (A) Accession (a) Approval (AA)	Entry into force	% of emissions
LIBYAN ARAB JAMAHIRIYA		24 Aug 2006 a	22 Nov 2006	
LIECHTENSTEIN *	29 Jun 1998	3 Dec 2004	3 Mar 2005	
LITHUANIA *	21 Sep 1998	3 Jan 2003	16 Feb 2005	
LUXEMBOURG *	29 Apr 1998	31 May 2002	16 Feb 2005	0.1%
MADAGASCAR		24 Sep 2003 a	16 Feb 2005	
MALAWI		26 Oct 2001 a	16 Feb 2005	
MALAYSIA	12 Mar 1999	4 Sep 2002	16 Feb 2005	
MALDIVES	16 Mar 1998	30 Dec 1998	16 Feb 2005	
MALI	27 Jan 1999	28 Mar 2002	16 Feb 2005	
MALTA	17 Apr 1998	11 Nov 2001	16 Feb 2005	
MARSHALL ISLANDS	17 Mar 1998	11 Aug 2003	16 Feb 2005	
MAURITANIA		22 Jul 2005 a	20 Oct 2005	
MAURITIUS		9 May 2001 a	16 Feb 2005	
MEXICO	9 Jun 1998	7 Sep 2000	16 Feb 2005	
MICRONESIA (FEDERATED STATES OF)	17 Mar 1998	21 Jun 1999	16 Feb 2005	
MONACO	29 Apr 1998	27 Feb 2006	28 May 2006	0.0%
MONGOLIA		15 Dec 1999 a	16 Feb 2005	
MONTENEGRO		4 Jun 2007 a	2 Sep 2007	
MOROCCO		25 Jan 2002 a	16 Feb 2005	
MOZAMBIQUE		18 Jan 2005 a	18 Apr 2005	
MYANMAR		13 Aug 2003 a	16 Feb 2005	
NAMIBIA		4 Sep 2003 a	16 Feb 2005	
NAURU		16 Aug 2001 a	16 Feb 2005	
NEPAL		16 Sep 2005 a	15 Dec 2005	
NETHERLANDS *	29 Apr 1998	31 May 2002 A (3)	16 Feb 2005	1.2%
NEW ZEALAND	22 May 1998	19 Dec 2002 (4)	16 Feb 2005	0.2%
NICARAGUA	7 Jul 1998	18 Nov 1999	16 Feb 2005	
NIGER	23 Oct 1998	30 Sep 2004	16 Feb 2005	
NIGERIA		10 Dec 2004 a	10 Mar 2005	
NIUE	8 Dec 1998	6 May 1999	16 Feb 2005	
NORWAY *	29 Apr 1998	30 May 2002	16 Feb 2005	0.3%
OMAN		19 Jan 2005 a	19 Apr 2005	
PAKISTAN		11 Jan 2005 a	11 Apr 2005	
PALAU		10 Dec 1999 a	16 Feb 2005	
PANAMA	8 Jun 1998	5 Mar 1999	16 Feb 2005	
PAPUA NEW GUINEA	2 Mar 1999	28 Mar 2002	16 Feb 2005	
PARAGUAY	25 Aug 1998	27 Aug 1999	16 Feb 2005	
PERU	13 Nov 1998	12 Sep 2002	16 Feb 2005	
PHILIPPINES	15 Apr 1998	20 Nov 2003	16 Feb 2005	
POLAND *	15 Jul 1998	13 Dec 2002	16 Feb 2005	3.0%
PORTUGAL *	29 Apr 1998	31 May 2002 AA	16 Feb 2005	0.3%
QATAR		11 Jan 2005 a	11 Apr 2005	
REPUBLIC OF KOREA	25 Sep 1998	8 Nov 2002	16 Feb 2005	
REPUBLIC OF MOLDOVA		22 Apr 2003 a	16 Feb 2005	
ROMANIA *	5 Jan 1999	19 Mar 2001	16 Feb 2005	1.2%
RUSSIAN FEDERATION *	11 Mar 1999	18 Nov 2004	16 Feb 2005	14.4%

RWANDA		22 Jul 2004 a	16 Feb 2005	
SAINT KITTS AND NEVIS		8 Apr 2008 a	7 Jul 2008	
SAINT LUCIA	16 Mar 1998	20 Aug 2003	16 Feb 2005	
SAINT VINCENT AND THE GRENADINES	19 Mar 1998	31 Dec 2004	31 Mar 2005	
SAMOA	16 Mar 1998	27 Nov 2000	16 Feb 2005	
SAN MARINO		28 April 2010	27 Jul 2010	
SAO TOME AND PRINCIPE		25 Apr 2008 a	24 Jul 2008	
SAUDI ARABIA		31 Jan 2005 a	1 May 2005	
SENEGAL		20 Jul 2001 a	16 Feb 2005	
SERBIA		19 Oct 2007 a	17 Jan 2008	
SEYCHELLES	20 Mar 1998	22 Jul 2002	16 Feb 2005	
SIERRA LEONE		10 Nov 2006 a	8 Feb 2007	
SINGAPORE		12 Apr 2006 a	11 Jul 2006	
SLOVAKIA *	26 Feb 1999	31 May 2002	16 Feb 2005	0.4%
SLOVENIA *	21 Oct 1998	2 Aug 2002	16 Feb 2005	
SOLOMON ISLANDS	29 Sep 1998	13 Mar 2003	16 Feb 2005	
SOMALIA		26 July 2010	24 Oct 2010	
SOUTH AFRICA		31 Jul 2002 a	16 Feb 2005	
SPAIN *	29 Apr 1998	31 May 2002	16 Feb 2005	1.9%
SRI LANKA		3 Sep 2002 a	16 Feb 2005	
SUDAN		2 Nov 2004 a	16 Feb 2005	
SURINAME		25 Sep 2006 a	24 Dec 2006	
SWAZILAND		13 Jan 2006 a	13 Apr 2006	
SWEDEN *	29 Apr 1998	31 May 2002	16 Feb 2005	0.4%
SWITZERLAND *	16 Mar 1998	9 Jul 2003	16 Feb 2005	0.3%
SYRIAN ARAB REPUBLIC		27 Jan 2006 a	27 Apr 2006	
TAJIKISTAN		29 Dec 2008 a	29 Mar 2009	
THAILAND	2 Feb 1999	28 Aug 2002	16 Feb 2005	
THE FORMER YUGOSLAV REPUBLIC OF MACEDONIA		18 Nov 2004 a	16 Feb 2005	
TIMOR-LESTE		14 Oct 2008 a	12 Jan 2009	
TOGO		2 Jul 2004 a	16 Feb 2005	
TONGA		14 Jan 2008 a	13 Apr 2008	
TRINIDAD AND TOBAGO	7 Jan 1999	28 Jan 1999	16 Feb 2005	
TUNISIA		22 Jan 2003 a	16 Feb 2005	
TURKEY *		28 May 2009 a	26 Aug 2009	
TURKMENISTAN	28 Sep 1998	11 Jan 1999	16 Feb 2005	
TUVALU	16 Nov 1998	16 Nov 1998	16 Feb 2005	
UGANDA		25 Mar 2002 a	16 Feb 2005	
UKRAINE *	15 Mar 1999	12 Apr 2004	16 Feb 2005	
UNITED ARAB EMIRATES		26 Jan 2005 a	26 Apr 2005	
UNITED KINGDOM*	29 Apr 1998	31 May 2002 (5) (6)	16 Feb 2005	
UNITED REPUBLIC OF TANZANIA		26 Aug 2002 a	16 Feb 2005	
UNITED STATES OF AMERICA *	12 Nov 1998			4.3%
URUGUAY	29 Jul 1998	5 Feb 2001	16 Feb 2005	
UZBEKISTAN	20 Nov 1998	12 Oct 1999	16 Feb 2005	
VANUATU		17 Jul 2001 a	16 Feb 2005	
VENEZUELA		18 Feb 2005 a	19 May 2005	
VIET NAM	3 Dec 1998	25 Sep 2002	16 Feb 2005	
YEMEN		15 Sep 2004 a	16 Feb 2005	

(continued)

Table 2.2 — *continued*

Participant	Signature	Ratification Acceptance (A) Accession (a) Approval (AA)	Entry into force	% of emissions
ZAMBIA	5 Aug 1998	7 Jul 2006	5 Oct 2006	
ZIMBABWE		30 Jun 2009 a	28 Sep 2009	

Source: www.unfccc.org

Notes: * indicates an Annex I party to the UNFCCC.

(1) In a communication received on 30 August 2002, the Government of the People's Republic of China informed the Secretary-General of the following: 'In accordance with article 153 of the Basic Law of the Hong Kong Special Administrative Region of the People's Republic of China of 1990 and article 138 of the Basic Law of the Macao Special Administrative Region of the People's Republic of China of 1993, the Government of the People's Republic of China decides that the Kyoto Protocol to the United Nations Framework Convention on Climate Change shall provisionally not apply to the Hong Kong Special Administrative Region and the Macao Special Administrative Region of the People's Republic of China.'
Further, in a communication received on 8 April 2003, the Government of the People's Republic of China notified the Secretary-General of the following: 'In accordance with the provisions of Article 153 of the Basic Law of the Hong Kong Special Administrative Region of the People's Republic of China of 1990, the Government of the People's Republic of China decides that the United Nations Framework Convention on Climate Change and the Kyoto Protocol to the United Nations Framework Convention on Climate Change shall apply to the Hong Kong Special Administrative Region of the People's Republic of China.
The United Nations Framework Convention on Climate Change continues to be implemented in the Macao Special Administrative Region of the People's Republic of China. The Kyoto Protocol to the United Nations Framework Convention on Climate Change shall not apply to the Macao Special Administrative Region of the People's Republic of China until the Government of China notifies otherwise.'
In a communication received on 14 January 2008, the Government of the People's Republic of China notified the Secretary-General of the following: 'In accordance with Article 138 of the Basic Law of the Macao Special Administrative Region of the People's Republic of China, the Government of the People's Republic of China decides that the Kyoto Protocol to the United Nations Framework Convention on Climate Change shall apply to the Macao Special Administrative Region of the People's Republic of China.'
(2) With a territorial exclusion to the Faroe Islands.
(3) For the Kingdom in Europe.
(4) With the following declaration: '... consistent with the constitutional status of Tokelau and taking into account the commitment of the Government of New Zealand to the development of self-government for Tokelau through an act of self-determination under the Charter of the United Nations, this ratification shall not extend to Tokelau unless and until a Declaration to this effect is lodged by the Government of New Zealand with the Depositary on the basis of appropriate consultation with that territory.'
(5) By a communication received on 27 March 2007, the Government of Argentina notified the Secretary-General of the following: 'The Argentine Republic objects to the extension of the territorial application to the Kyoto Protocol to the United Nations Framework Convention on Climate Change of 11 December 1997 with respect to the Malvinas Islands, which was notified by the United Kingdom of Great Britain and Northern Ireland to the Depositary of the Convention on 7 March 2007.
The Argentine Republic reaffirms its sovereignty over the Malvinas Islands, the South Georgia and South Sandwich Islands and the surrounding maritime spaces, which are an integral part of its national territory, and recalls that the General Assembly of the United Nations adopted resolutions 2065 (XX), 360 (XXVIII), 31/49, 37/9, 38/12, 39/6, 40/21, 41/40, 42/19 and 43/25, which recognize the existence of a dispute over sovereignty and request the Governments of the Argentine Republic and the United Kingdom of Great Britain and Northern Ireland to initiate negotiations with a view to finding the means to resolve peacefully and definitively the pending problems between both countries, including all aspects on the future of the Malvinas Islands, in accordance with the Charter of the United Nations.'
(6) On 4 April 2006, the Government of the United Kingdom informed the Secretary-General that the Protocol shall apply to the Bailiwick of Guernsey and the Isle of Man. On 2 January 2007: in respect of Gibraltar. On 7 March 2007: in respect of Bermuda, Cayman Islands, Falkland Islands (Malvinas) and the Bailiwick of Jersey.

Convention but have not ratified the Kyoto Protocol are allowed to participate in the COP/MOP as observers but have no rights in the decision-making process. COP/MOP1 was held together with COP11 in December 2005 in Montreal, Canada, following the entry into force of the Kyoto Protocol in February 2005. The main task of COP/MOP1 was to consider for approval all COP decisions adopted before 2005, in particular the Marrakech Accords of 2001. Prior to COP/MOP1, negotiations and talks over the rules of the Kyoto Protocol were conducted within the framework of the COP, thus also including countries parties to the Convention but not having ratified the Kyoto Protocol. Before the entry into force of the Kyoto Protocol, the COPs adopted several decisions that were forwarded to COP/MOP1 for consideration and adoption.

The two subsidiary bodies created by the Convention, SBSTA and SBI, as well as the bureau of the COP, also serve and assist the COP/MOP in its work and decisions. Parties to the UNFCCC and to the Kyoto Protocol participate in the meetings of the convention bodies through a national delegation that negotiates on the various negotiating issues on behalf of the national government. Based on UN practice, parties take part in the bodies' meetings via negotiating groups, which allows them to better meet their national and regional interests and to form common negotiating positions accordingly. In respect of countries with limited staff resources and expertise on climate change issues, the participation of delegation members in the negotiating groups is a way to fill the gaps of knowledge and lack of personnel that they face as regards the many complex items that countries have to deal with in the implementation of the international climate regime. These groups are usually defined on the basis of their regional location and reflect the basic North–South division proper of all UN institutions. However, within the international climate regime, the differentiation of groups is not only based on geographical considerations; there are also many differences in how they aim to reduce GHG emissions and in the efficiency of their energy systems, as well as in the role of technology and domestic fuel resources. The proliferation of negotiating items is due to the complexity of the issues discussed and to the different positions of the parties in this respect. The negotiating groups participating in the COPs and COP/MOPs are:

- Group 77 and China, including more than 130 parties and comprising all developing countries. In this group, countries often take different negotiating positions, which is why usually smaller groupings are established, such as the African UN Regional Group, the Alliance of Small Island States (AOSIS) and the group of Least Developed Countries (LDCs);
- EU (27 Member States);
- Umbrella Group – evolution of the JUSSCANNZ (Coalition of non-EU Annex I Parties, guided by Japan, the United States of America, Switzerland, Canada, Australia, Norway and New Zealand) group during the Kyoto Protocol negotiations – including the following non-EU developed countries: Australia, Canada, Iceland, Japan, New Zealand, Norway, the Russian Federation, Ukraine and the US;
- Central Group 11 (CG11) (dissolved in 2001) including most EITs listed in Annex I of the UNFCCC;

- Central Asia, Caucasus, Albania and Moldova (CACAM), including non-Annex I parties from Asia and Central and Eastern Europe.

The Kyoto Protocol contains two Annexes: Annex A (see Table 2.3), which includes the list of GHGs regulated by the Kyoto Protocol as well as an indicative list of sectors and source categories responsible for GHG emissions. Annex B (see Table 2.4) contains the same list of countries included in Annex I of the UNFCCC – with the exception of Turkey and Belarus, which were not parties to the UNFCCC when the text of the Kyoto Protocol was agreed – that have assumed legally binding commitments for the period 2008–2012 pursuant to Article 3(1) of the Kyoto Protocol.

According to Article 3 of the Kyoto Protocol, Annex I parties shall ensure 'individually or jointly' that GHG emissions do not exceed 'their assigned amounts calculated pursuant to their quantified emission limitation and reduction commitments inscribed in Annex B' in the commitment period 2008–2012 with a view to reducing the global level of GHG emissions by 'at least 5 per cent below 1990 levels' in that period. The QELRCs listed in Annex B are expressed in percentages of 1990 emissions of GHGs and the assigned amounts correspond to the total amount of GHG emissions in CO_2 equivalent that a country is allowed to emit during the 2008–2012 period (Article 3(7)). Article 3(5) of the Kyoto Protocol allows the EITs to use a base year different from 1990 in accordance with Decision 9/CP.2 of COP2. Article 3(6) allows such countries 'a certain degree of flexibility' in the implementation of the Kyoto Protocol. As mentioned above, the word 'shall' included in Article 3(1) contributes to giving these commitments a legally binding nature. This binding nature is reinforced by the rules of the non-compliance procedure established on the basis of Article 18 of the Kyoto Protocol. Explicit reference to the binding character of QELRCs was made in the AGBM4 conclusions and obtained the support of several parties also in future years, for instance, the US at COP2, the Geneva Ministerial Declaration, France, Germany and Poland.[33]

The Kyoto Protocol does not foresee any binding commitment for non-Annex I parties in the first commitment period. According to Article 3(9) of the Kyoto Protocol, commitments for Annex I parties in the period post-2012 shall take the form of an amendment to Annex B, and the international negotiations should start at the latest in 2005 (seven years before the end of the first commitment period).

Article 4 of the Kyoto Protocol establishes the possibility for Annex I parties to reach an agreement on the joint fulfilment of GHG emission reduction

Table 2.3 *Annex A of the Kyoto Protocol*

GHGs	Sectors/source categories
Carbon dioxide	Energy (fuel combustion, fugitive emissions from fuels)
Methane	Industrial processes (mineral products, chemical industry, metal production,
Nitrous oxide	other)
Hydrofluorocarbons	Solvent and other product use (agriculture, enteric fermentation, manure
Perfluorocarbons	management, rice cultivation, other)
Sulphur hexafluoride	Waste (solid waste disposal on land, wastewater handling, waste incineration, other)

Table 2.4 *Annex B of the Kyoto Protocol*

Party	Quantified emission limitation and reduction commitment (percentage of base year or period)
Australia	108
Austria	92
Belgium	92
Bulgaria*	92
Canada	94
Croatia*	95
Czech Republic*	92
Denmark	92
Estonia*	92
EU	92
Finland	92
France	92
Germany	92
Greece	92
Hungary	94
Iceland	110
Ireland	92
Italy	92
Japan	94
Latvia*	92
Liechtenstein	92
Lithuania	92
Luxembourg	92
Monaco	92
Netherlands	92
New Zealand	100
Norway	101
Poland	94
Portugal	92
Romania*	92
Russian Federation*	100
Slovak Republic*	92
Slovenia*	92
Spain	92
Sweden	92
Switzerland	92
Ukraine*	100
UK and Northern Ireland	92
US	93

Note: * Countries that are undergoing the process of transition to a market economy.

commitments. This Article was introduced in the final text of the Kyoto Protocol at the request of the European Community, which decided to fulfil its commitments jointly with the Member States (EU15).[34] Consequently, the global target of the EC has been redistributed internally among the Member States in different percentages following an agreement by the ministers of the environment at Council level. This topic is addressed in Chapter 3.

2.3 THE KYOTO PROTOCOL: OPERATIONAL MEASURES

At the COP following the adoption of the Kyoto Protocol – COP4 of 1998 in Buenos Aires – the parties and delegates agreed on a preliminary agenda for the development of the operational details of the Kyoto Protocol mechanisms and for the definition of rules and details regarding the main issues left unresolved in the previous negotiation phase. The first document in this regard is the Buenos Aires Plan of Action (BAPA),[35] including the work programme for the definition of the operational rules of the Kyoto Protocol with regard to the flexible mechanisms and the compliance regime, as well as to modalities, rules and guidelines, in particular for verification, reporting and accountability regarding GHG emissions.

The negotiations on the details of the Kyoto Protocol rules and modalities started in Buenos Aires in 1998 and continued until 2001, when the parties to the UNFCCC, extraordinarily convened in Bonn from 16 to 27 June 2001 at COP6, part 2, following the political failure of COP6 in The Hague, agreed on the final text. This agreement was translated into legal texts after long and exhausting negotiations in Marrakech in November 2001 and the Marrakech Accords were finally adopted at COP7. In particular, parties in Marrakech agreed on the definition of the rules governing the participation and functioning of the flexible mechanisms, on the composition and principles of the non-compliance regime and on the details of the role that activities related to forestry shall assume in the framework of the Kyoto Protocol rules.

The Marrakech Accords define the framework for the implementation of the Kyoto Protocol and include, amongst others, a list of draft decisions on the details of the flexible mechanisms, reporting and methodologies, land use, land use change and forestry (LULUCF) and the non-compliance regime to be adopted by COP/MOP1. In respect of the flexible mechanisms, the Marrakech Accords provided for the prompt start of the Clean Development Mechanism (CDM), identified the eligibility criteria to participate in the flexible mechanisms, and finally included the rules for the implementation of the flexible mechanisms. Other issues mentioned in the Marrakech Accords were the support for developing countries, capacity building needs, the transfer of technology, the answer to the adverse effects of climate change, as well as the establishment of three funds, namely the LDC Fund, the Special Climate Change Fund (SCCF) and the Adaptation Fund.

Furthermore, the Marrakech Accords identified the rules for the establishment of additional bodies for the implementation of the Kyoto Protocol, namely:

- The CDM Executive Board in charge of supervising the CDM under the Kyoto Protocol and responsible for the accreditation of operational entities, the approval of baseline methodologies, the approval and registration of CDM projects, the maintenance of the CDM registry, etc.;
- The Joint Implementation Supervisory Committee (JISC), in charge of supervising, registering and verifying joint implementation (JI) projects;
- The Compliance Committee governing the non-compliance regime.

In respect of the non-compliance regime, the Marrakech Accords included a draft decision identifying the structure, composition and rules of this system.

COP/MOP1 held in December 2005 in Montreal adopted all the package decisions included in the Marrakech Accords. Furthermore, COP11 and COP/MOP1 in Montreal adopted two important decisions on the future of the international climate regime, thus opening the way for further commitments of the developed and developing countries in respect of the fight against global warming in the period after 2012.[36]

2.3.1 Flexible mechanisms

First, the Kyoto Protocol established the so-called flexible or market-based mechanisms, a set of instruments to reduce GHG emissions in a cost-effective way, notably by providing Annex I parties with the possibility to implement projects and to invest in those countries where the marginal cost of the abatement of 1 tonne of CO_2 equivalent is lower than at the domestic level. The flexible mechanisms are instruments designed to combat global warming by combining the protection of the environment, namely the reduction of GHG emissions worldwide – no matter where – with the necessity to lower the costs of the actions required to comply with those reduction obligations. The flexible mechanisms were included in the text of the Kyoto Protocol, with the support, in particular, of the Umbrella Group, whose representatives proposed a complementary alternative to green domestic actions for Annex I parties. In opposition to this, several other parties expressed concerns over the extent to which the flexible mechanisms could be used by Annex I parties to meet the reduction targets, in order to guarantee the environmental integrity of the system created by the Kyoto Protocol as well as the full respect of the principle of common but differentiated responsibilities. To this end, the principle of supplementarity was introduced in the text of the Kyoto Protocol (Articles 6 and 17): flexible mechanisms 'shall be supplemental to domestic actions'. The flexible mechanisms are separated into two types: the emissions trading system and the project-based mechanisms. They are:

- JI (Article 6 Kyoto Protocol (KP)): providing for Annex I parties to implement projects that reduce emissions, or remove carbon from the air, in other Annex I parties, in return for emission reduction units (ERUs);
- CDM (Article 12 KP): providing for Annex I parties to implement projects that reduce emissions in non-Annex I parties, in return for certified emission reductions (CERs), and assist the host parties in achieving sustainable development and contributing to the ultimate objective of the Convention;
- International emissions trading (IET) (Article 17 KP): providing for Annex I parties to acquire units from other Annex I parties. These units may be in the form of assigned amount units (AAUs), removal units (RMUs), ERUs and CERs.

2.3.2 The non-compliance regime

Article 18 of the Kyoto Protocol requires the COP/MOP to adopt, at its first session, 'appropriate and effective procedures and mechanisms to determine and to address cases of non-compliance with the provisions of this Protocol'. In accordance with Article 18, this non-compliance regime shall be adopted as an amendment to the Protocol if including binding consequences for the parties. The text of the Kyoto Protocol therefore provides for the establishment of a non-compliance regime and leaves it to the parties, under the guidance of the COP, to define the details and procedures regarding the functioning of such a mechanism. The rules and details of the non-compliance regime were agreed on at COP7 in Marrakech and further adopted by COP/MOP1. The establishment of an ad hoc system designed to address compliance by parties with the Kyoto Protocol obligations is important and innovative, particularly in comparison with the existing rules for the enforcement of international environmental law and the non-compliance regimes established under the existing MEAs. Unlike the non-compliance regime of the Kyoto Protocol and in line with a more classic approach, both the Convention and the Kyoto Protocol refer to the system for the settlement of disputes under Articles 14 and 19 respectively. These articles rely on a more general approach that recommends that parties 'seek a settlement of the dispute through negotiation or any other peaceful means of their own choice'. More precisely, the Convention, in Article 14(2), requires parties to notify their own preferred method for the settlement of disputes, namely through submission of the dispute to the International Criminal Court and/or arbitration.

One of the main issues related to the compliance system, left unresolved first by the Kyoto Protocol negotiations within the COP and COP/MOP and then by the Marrakech Accords, concerns the binding force of this regime. Although Article 18 of the Kyoto Protocol requires parties to adopt 'any procedures and mechanisms under this Article entailing binding consequences ... by means of an amendment to this Protocol', the rules of the non-compliance regime of the Protocol have been adopted as a COP decision. However, the fact that parties have agreed on the adoption of such rules under the COP can be considered as an implicit expression of their willingness to be bound by the consequences of non-compliance mentioned in that decision. Furthermore, no parties have so far formally objected to the binding value of the decisions of the Compliance Committee.

The non-compliance regime of the Kyoto Protocol consists of the Compliance Committee, which is divided into a facilitative branch and an enforcement branch. The facilitative branch assists all parties in the Kyoto Protocol implementation, while the enforcement branch is a quasi-judicial body that can propose decisions on parties in non-compliance, such as non-eligibility to participate in the flexible mechanisms. The main task of the Compliance Committee is to consider the 'questions of implementation' referring to the compliance by Annex I parties with the different obligations established by the Kyoto Protocol. For clarity, these obligations are grouped in this book in three categories:

- Monitoring, reporting and verification (MRV) obligations under Articles 5(1)(2) and 7(1)(4) of the Kyoto Protocol;
- Eligibility requirements as defined under Articles 6, 12 and 17 of the Kyoto Protocol and the Marrakech Accords;
- QELRCs under Article 3(1) of the Kyoto Protocol.[37]

The enforcement branch is also responsible for determining the consequences of non-compliance by Annex I parties. These consequences are:

- MRV obligations: submission of a plan addressing the reasons for non-compliance as well as detailed measures and a timetable to reinstate compliance;[38]
- Eligibility requirements: suspension of eligibility to participate in one or more of the flexible mechanisms[39] (eligibility is suspended whenever the branch determines non-compliance by an Annex I party with the eligibility criteria and can be reinstated at the moment the party demonstrates its compliance with the specific criterion);[40]
- Limitation and reduction commitments:[41]
  - each tonne of emissions in excess multiplied by 1.3 will be deducted from the party's assigned amount for the second commitment period;
  - preparation of a detailed compliance action plan;
  - suspension and eventual reinstatement of the party's eligibility to transfer carbon units under Article 17 of the Kyoto Protocol.

The expedited procedure to reinstate eligibility as mentioned above was introduced in the Marrakech Accords at the specific request of Japan and consists in the party's right to request, either through an expert review team or directly to the enforcement branch, the restoration of eligibility in case the problem has been rectified and the relevant criteria are met.

2.3.3 Land use, land use change and forestry

The second important feature introduced by the Kyoto Protocol is the significant role assigned, at the international level, to the so-called carbon sinks. More precisely, the Kyoto Protocol provides Annex I parties with the possibility to offset their emissions by increasing the amount of GHGs removed from the atmosphere through activities relating to LULUCF. Article 2(1) of the Kyoto Protocol includes, among the duties of Annex I parties, the 'protection and enhancement of sinks and reservoirs of greenhouse gases ... the promotion of sustainable forest management practices, afforestation and reforestation ... [and] ... the promotion of sustainable forms of agriculture'. Annex I parties are given the possibility to account for the removals from the atmosphere of GHG emissions resulting from afforestation, reforestation and deforestation activities (Article 3(3)), as well as forest management, cropland management, grazing land management and revegetation activities (Article 3(4)) that have taken place since 1990, in their assigned amounts listed in Annex B of the Kyoto Protocol. These removals are eligible to generate emission reduction credits called RMUs and the Marrakech Accords set a limit for each party in respect of the amount of credits which can be claimed to offset GHG emissions. Moreover, the Marrakech Accords and the subsequent COP/MOP decisions deal with various pending issues concerning the scientific and methodological uncertainties that relate to LULUCF activities and that have delayed the full utilization of these activities by Annex I parties – for example, the definition of forest baselines for the calculation of the carbon stock potential and net changes.

2.4   POST-2012

The Kyoto Protocol set quantified emission limitation and reduction commitments for Annex I parties for the first commitment period (2008–2012) in accordance with Article 3(1). According to Article 3(9), commitments for subsequent periods for Annex I parties shall result in amendments to Annex B to the Kyoto Protocol and the COP serving as the MOP shall start the negotiating process by 2005.

The G8 Gleneagles Summit of 6–8 July 2005 tackled the issue of climate change and produced the following documents:

- Signed Version of Gleneagles Communiqué on Africa, Climate Change, Energy and Sustainable Development;
- Climate Change, Clean Energy and Sustainable Development;
- Climate Change: Plan of Action.

The Asia-Pacific Partnership on Clean Development and Climate (APP) was launched on 28 July 2005 at the 12th ASEAN (Association of Southeast Asian Nations) regional forum in Vientiane (Laos). The APP is composed of the US, Australia, Japan, China, India and Korea and is aimed at the established of a partnership in the field of climate change and technological development among the members. The founding documents are the Charter (2007) and the Vision Statement (2005).

International negotiations on the post-2012 process, expected to begin formally at COP/MOP1 in Canada in December 2005 as prescribed under Article 3(9) of the Kyoto Protocol, started early in May 2005 in Bonn at the UNFCCC Seminar of Governmental Experts to discuss global climate change policies after the Kyoto Protocol first commitment period and convened by the UNFCCC Secretariat. Although the meeting did not produce any official document, many delegates defined the seminar as a 'minor diplomatic breakthrough' on the path toward a post-2012 strategy to tackle global warming.

Parties to the UNFCCC and to the Kyoto Protocol respectively met for COP11 and COP/MOP1 in Montreal from 28 November to 9 December 2005 and one of the main issues at stake was the future of the Kyoto Protocol and the establishment of new GHG emissions reductions binding targets for industrialized countries as well as a few developing countries such as China, India and Brazil, whose per capita GHG emissions are rising rapidly. The main outcomes of the Montreal international negotiations were:

- Five-year programme of work on adaptation, impacts and vulnerability (COP11);
- Dialogue on long-term cooperative action to address climate change by enhancing implementation of the Convention (COP11);
- Several decisions on budgetary issues (COP11);
- Adoption of the Marrakech Accords (COP/MOP1);
- Consideration of commitments for subsequent periods for Annex I Parties to the Convention under Article 3(9) of the Kyoto Protocol (COP/MOP1);

- Procedures and mechanisms relating to compliance under the Kyoto Protocol (COP/MOP1);
- Several decisions on the implementation of flexible mechanisms (COP/MOP1).

The Dialogue on long-term cooperative action to address climate change by enhancing implementation of the Convention launched by COP11 was designed to identify a long-term cooperative action to address climate change focusing on, inter alia: sustainable development; adaptation; technology transfer; and market-based opportunities.

The 'Convention Dialogue' consisted of a programme of four workshops leading up to COP13 aimed at enhancing long-term cooperation by all Parties under the Convention. The workshops were:

- First workshop (15–16 May 2006);
- Second workshop (15–16 November 2006);
- Third workshop (16–17 May 2007);
- Fourth workshop (27–31 August 2007, Vienna, Austria).

In Montreal, the Ad Hoc Working Group on Further Commitments for Annex I Parties under the Kyoto Protoco (AWG-KP) was established by COP/MOP 1 with a view to amend Annex B to the Kyoto Protocol and establish further commitments by Annex I parties on the basis of Article 3(9) of the Protocol.

The 24th sessions of the Subsidiary Bodies – SBSTA and SBI of the UNFCCC – were held in Bonn from 17–26 May 2006 together with the first session of the AWG-KP. These sessions were preceded by the First Workshop on the dialogue on long-term cooperative action to address climate change by enhancing implementation of the Convention, which took place on 15–16 May 2006 in Bonn and where parties discussed undertaking future actions on development goals in a sustainable way, adaptation, technology and market-based opportunities. The first session of the AWG-KP ended with an agreement on the Planning of future work.

COP12 and COP/MOP2 were hosted from 6–17 November 2006 in Nairobi, Kenya. Among other issues, parties discussed the future of the climate regime and confirmed the multi-track approach based on the AWG-KP and the Dialogue (Convention).

On 16 February 2007 the Legislators Forum on Climate Change, composed of parliamentarians of the G8 countries as well as representatives of China, Brazil, Mexico and South Africa, adopted a statement calling for world leaders to start negotiations on a post-2012 international agreement on climate change at the July 2007 G8 summit of Heiligendamm, Germany.

The 26th sessions of the Subsidiary Bodies were held in Bonn from 7–18 May 2007 and contributed to move the 'world closer to comprehensive negotiations on post-2012 climate change regime' and to open the way for the Bali Conference of December 2007 (COP13).

The G8 summit convened in June 2007 in Heiligendamm, Germany, contributed to the international discussion on the post-2012 process. The Summit Declaration of 7 June 2007 included the agreement on 'strong and early action' to fight climate change from the major developed countries within the framework of the UNFCCC. The final communiqué referred to the halving of global GHG emissions by 2050 by the major emitters but no legally binding commitments were considered. The joint statement by the German presidency of G8 and the heads of state and/or governments of Brazil, China, India, Mexico and South Africa released on 8 June 2007 referred to the formula that developed and developing countries shall act in tackling climate change in respect of the principle of 'common but differentiated responsibilities'. During the summit, the five developing countries confirmed their intention of not engaging in any long-term GHG reduction commitment.

On 1 August 2007, for the first time ever, the UN General Assembly held an informal debate on climate change as a global challenge to discuss the future of climate policy. Government representatives, UN officials, scientists and stakeholders from all over the world met to pave the way for a Kyoto-like agreement to be agreed by 2009 within the UNFCCC system.

The fourth session of the Ad Hoc Working Group on Further Commitments for Annex I Parties under the Kyoto Protocol (AWG-KP4) and the fourth workshop under the dialogue on long-term cooperative action to address climate change by enhancing implementation of the Convention took place in Vienna, Austria, on 27–31 August 2007. The Vienna Climate Change Talks 2007 concluded on 31 August 2007 with a general agreement by industrialized countries on the urgent need of a significant reduction of GHGs by 2020 in order to tackle global warming. The meeting was held under the auspices of the UNFCCC and considered issues related to the implementation of the Convention as well as the future of the international climate regime.

On 24 September 2007 the event The Future in our Hands: Addressing the Leadership Challenge of Climate Change was held in the UN Headquarters, where the UN Secretary General Ban Ki-moon met with more than 150 countries. At the meeting, world leaders showed political will and commitment towards a post-2012 new climate change regime.

On 27 and 28 September 2007 the Major Economies Meeting on Energy Security and Climate Change (17 largest CO_2 emitters) was held in Washington, DC, to discuss the future of the international efforts to combat global warming. The US government recognized that climate change is 'a real problem', urged the international community to 'address climate change' but still opposed legally binding emission reduction obligations. The US position was considered 'isolated' by other participants.

From 3 to 14 December 2007 COP13 and COP/MOP13 convened in Bali, Indonesia, to discuss, among other issues, the future of the international climate regime. The 27th sessions of the SBI and SBSTA as well as the resumed fourth session AWG-KP took place in Bali too. The final agreement reached by parties in Bali, labelled by COP president Witoelar in its closing statement as a 'breakthrough', may not represent what the EU was aiming at, namely a precise and concrete commitment to reduce anthropogenic GHG emissions by 25–40 per cent by 2020, but still was considered significant as it signed the return of the US in the negotiating process for the first time after the withdrawal from the Kyoto

Protocol track of March 2001. Open and controversial was the issue of how the requests of an extremely fragmented international community will be combined in the near future.

At COP13, parties agreed on the Bali Action Plan (BAP), which established the Ad Hoc Working Group on Long-term Cooperative Action (AWG-LCA), a subsidiary body under the Convention mandated to elaborate the main issues identified in the Convention Dialogue.[42] The main goal of the BAP and consequently the AWG-LCA is the enhancement of the implementation of the Convention in order to achieve its ultimate objective in accordance with its principles and commitments.[43] More precisely, the BAP identified the following five areas of negotiations: shared vision for long-term cooperative action; mitigation; adaptation; finance; and technology and capacity building.[44] The BAP was accompanied by a set of decisions adopted by COP/MOP3 and established a two-track process (Convention and Kyoto Protocol) aiming at the identification of a post-2012 global climate regime to be adopted by COP15 and COP/MOP5 in Copenhagen at the end of 2009. The AWG-LCA was required to provide its conclusions on the 'full, effective and sustained implementation of the Convention' while the Kyoto Protocol track signed the continuation of the work of the AWG-KP, which was required to provide recommendations to COP/MOP5 for adoption of new binding GHG emission reduction and limitation commitments for Annex I parties under the Kyoto Protocol.

The main difference between the AWG-LCA and AWG-KP are their mandates and objectives, respectively the enhancement of the implementation of the UNFCCC and the amendment of the Kyoto Protocol, and the fact that the US was not directly engaged in the AWG-KP track.[45] Furthermore, parties at the high-level segment of COP13 agreed on the Bali Roadmap[46] setting the stage for a two-year programme of negotiations aimed at the identification of the most adequate solution to the continuation of the international climate regime under the form of a new legally binding agreement within the frameworks of the Kyoto Protocol and UNFCCC. This negotiation programme was based on eight separate sessions of the AWG-LCA and AWG-KP, to take place in Bangkok, Bonn, Accra and Poznan in 2008 and Bonn (I and II), Bangkok and Copenhagen in 2009.

At the core of the Bali Roadmap, important decisions adopted were: the launch of the Adaptation Fund (COP/MOP3), a commitment to reduce emissions from deforestation in developing countries – REDD – (COP13), as well as the agreement on a new programme for technology transfer (COP13 SBSTA and SBI). The full set of COP13 and COP/MOP3 decisions is provided in Table 2.5.

On 15 December 2007, the European Commission and the European Parliament welcomed the result of the Bali conference. The EU expressed its satisfaction that the Bali Climate talks gave birth to an agreement launching formal negotiations on a global climate regime for post-2012.

The second major economies meeting was held on 1 February 2008 in Honolulu, Hawaii, and gathered together representatives of the 16 biggest carbon-emitting countries. The conclusions of the meeting did not mention any binding reduction targets for the participating countries but progress was noticed in respect of the international talks towards a post-2012 agreement.

Table 2.5 *COP13 and COP/MOP3 decisions*

COP13	COP/MOP3
Decision 1/CP.13 Bali Action Plan	Decision 1/CMP.3 Adaptation Fund
Decision 2/CP.13 Reducing emissions from deforestation in developing countries: approaches to stimulate action	Decision 2/CMP.3 Further guidance relating to the CDM
Decision 3/CP.13 Development and transfer of technologies under the SBSTA	Decision 3/CMP.3 Guidance on the implementation of Article 6 of the Kyoto Protocol
Decision 4/CP.13 Development and transfer of technologies under the SBI	Decision 4/CMP.3 Scope and content of the second review of the Kyoto Protocol pursuant to its Article 9
Decision 5/CP.13 Fourth Assessment Report of the IPCC	Decision 5/CMP.3 Compliance under the Kyoto Protocol
Decision 6/CP.13 Fourth review of the financial mechanism	Decision 6/CMP.3 Good practice guidance for land-use, land-use change, and forestry activities under Article 3, paragraphs 3 and 4, of the Kyoto Protocol
Decision 7/CP.13 Additional guidance to the Global Environment Facility	
Decision 8/CP.13 Extension of the mandate of the Least Developed Countries Expert Group	Decision 7/CMP.3 Demonstration of progress in achieving commitments under the Kyoto Protocol by Parties included in Annex I to the Convention
Decision 9/CP.13 Amended New Delhi work programme on Article 6 of the Convention	Decision 8/CMP.3 Compilation and synthesis of supplementary information incorporated in fourth national communications submitted in accordance with Article 7, paragraph 2, of the Kyoto Protocol
Decision 10/CP.13 Compilation and synthesis of fourth national communications	
Decision 11/CP.13 Reporting on global observing systems for climate	
Decision 12/CP.13 Budget performance and the functions and operations of the secretariat	Decision 9/CMP.3 Implications of possible changes to the limit for small-scale afforestation and reforestation clean development mechanism project activities
Decision 13/CP.13 Programme budget for the biennium 2008–2009	Decision 10/CMP.3 Budget performance for the biennium 2006–2007
Decision 14/CP.13 Date and venue of COP14 and COP15	Decision 11/CMP.3 Programme budget for the biennium 2008–2009

At the fourth meeting of the Gleneagles dialogue, energy and environment ministers from 20 developed and developing countries failed to reach an agreement on a common approach for the future of the international climate regime. The former British prime minister invited the world leaders to bridge the differences and proposed a new climate change initiative aimed at the shaping of a new global agreement.

The Bangkok Climate Change Talks held in the period 31 March to 4 April 2008 concluded with an agreement on a work plan for the negotiations on the post-2012 international climate change with the view to agree on a comprehensive deal by 2009 in Copenhagen. Delegates from 162 countries agreed on the continuation of the flexible mechanisms – emissions trading (ET), JI and CDM – as supplementary measures to reduce GHG emissions. Parties also decided to advance the discussion on issues such as the sectoral approach to tackle climate change or advancing adaptation through finance and technology.

The third meeting of world's biggest carbon-emitting countries (16) was held on 9 April 2008 in Paris (L'élaboration d'un régime multilatéral sur le climat au-delà de 2012) and dealt with sectoral agreements, technological cooperation, adaptation to climate and forests. At the meeting, the EU urged once again industrialized countries to engage in ambitious commitments to reduce GHG emissions and did not close the door to a sector-specific agreement. President of France Sarkozy invited the major GHG-emitting countries, including China and India, to stop the defensive approach to the creation of the post-2012 regime and highlighted the danger created by climate change to food supply worldwide.

The themes discussed at the G8 Environment Ministers meeting held on 26 May 2008 in Kobe City, Japan, were climate change, biodiversity and the 3Rs (reduce, reuse and recycle). The results of the meeting were the Kobe Call for Action for Biodiversity and the Kobe 3R Action Plan. On climate change, the G8 ministers agreed on the Chair's Summary, which included a call for the adoption of a long-term GHG reduction goal for industrialized countries and a 'strong political will' to 'go beyond' the agreement reached in Heiligendamm in 2007 mentioning the halving of global GHG emissions by 2050.

G8 leaders in Japan agreed on 8 July 2008 on a joint statement that included a generic and global long-term GHG emission reduction target of at least 50 per cent by 2050. The joint statement did not mention mid-term targets, nor national plans or base year, but recognized the leadership role of G8 countries in the fight against climate change. On 9 July 2008, the Leaders of Major Economies (G8 countries together with Brazil, China, India, Indonesia, Republic of Korea, Mexico and South Africa) adopted a declaration on energy security and climate change that did not include clear and strict commitments to reduce GHG emissions, nor actions to be undertaken as part of the post-2012 strategy. The joint statement only vaguely referred to long-term reduction targets and reiterated the principle of common but differentiated responsibilities to be adopted in the definition of mitigation measures.

COP14 and COP/MOP4 were held in Poznan from 1 to 12 December 2008 and focused on the structure and possible contents of the future draft negotiating text for the agreement to be concluded in 2009 in Copenhagen. Other issues considered at the meetings were: improvements to the CDM, an adaptation fund for developing countries and long-term action on climate change. Although the conference ended with no real progress in the definition of new GHG emission reduction targets, the final press release of 12 December 2008 of the secretariat reported 'a clear commitment from governments to shift into full negotiating mode next year' in Copenhagen. In Poznan parties agreed that a first draft of the negotiating text for Copenhagen would be presented in Bonn in June 2009. Furthermore, parties reached an agreement on the following issues:

- endorsement of the GEF's 'Poznan Strategic Programme on Technology Transfer' aimed at boosting foreign investments in developing countries in order to raise funding for adaptation and mitigation activities;
- launching of the Adaptation Fund as a legal entity providing direct access to developing countries (no agreement was reached on the increase of the 2 per cent levy on the transactions under the CDM);
- CDM: inclusion of carbon capture and storage (CCS) projects, enhancing regional distribution, extension of the eligibility criteria for afforestation and

reforestation projects (no agreement on the possibility for project developers to appeal on the decisions of the Executive Board);

Finally, on 12 December 2008 a joint ministerial declaration was issued by a group of developed and developing countries on a shared vision on long-term cooperative action on climate change. This recognized, amongst other issues, the urgent need to address climate change, the right to sustainable development, the importance of adaptation and the need to move towards a society with low carbon consumption. Table 2.6 shows the decisions adopted by COP14 and COP/MOP4.

In Poznan, parties under the AWG-KP agreed on the Work Programme for 2009, which according to paragraph 49 'will focus on agreeing on further commitments for Annex I Parties under the Kyoto Protocol'.[47] The Work Programme also indicated the main issues to be addressed in 2009 under the AWG-KP, namely:

- the scale of emission reductions to be achieved by Annex I parties in aggregate;
- the contribution of Annex I parties, individually or jointly, consistent with Article 4 of the Kyoto Protocol;
- other issues such as the duration of the commitment period(s), how quantified emission limitation and reduction objectives could be expressed, base year, improvements to emissions trading and the project-based mechanisms; definitions, modalities, rules and guidelines for the treatment of LULUCF in the second commitment period; coverage of GHGs, sectors and source categories; potential environmental, economic and social consequences, and possible approaches targeting sectoral emissions.

In 2009, the ad hoc working groups convened four times for parallel sessions: April, June and August in Bonn, Germany; October in Bangkok, Thailand; November in Barcelona, Spain; and December in Copenhagen, Denmark.

On 2 April 2009 the Leaders of the G20 met in London and in the final communiqué reaffirmed the commitment to address climate change and to reach an agreement in the Copenhagen summit scheduled for December 2009.

AWG-KP7 and AWG-LCA5 were held in Bonn in the period 29 March to 8 April 2010. For the AWG-LCA three workshops were held on: economic and social consequences of response measures; mitigation by developed and developing countries; and mitigation in the agricultural sector. Progress was achieved in many areas of the text, in particular on nationally appropriate mitigation actions (NAMAs) and adaptation. AWG-KP7 addressed the following issues on the basis of four notes prepared by the Chair:

- emission reductions by Annex I parties under the Protocol focusing on:
  - scale of emission reduction of GHG emissions for developed countries by 2050 in aggregate;
  - scale of emission reduction of GHG emissions for developed countries by 2020 at the individual level;
  - actions and commitments for the reduction of GHG emissions by major developing countries;

Table 2.6 *COP14 and COP/MOP4 decisions*

COP14	COP/MOP4
Advancing the Bali Action Plan	Report of the Adaptation Fund Board
Development and transfer of technologies	Further guidance relating to the clean
Financial mechanism of the Convention: fourth	development mechanism
review of the financial mechanism	Advancing the work of the Ad
Additional guidance to the Global Environment	Hoc Working Group on Further Commitments
Facility	for Annex I Parties under the Kyoto Protocol
Further guidance for the operation of	Compliance Committee
the Least Developed Countries Fund	Guidance on the implementation of Article
Capacity-building for developing countries under the	6 of the Kyoto Protocol
Convention	Capacity-building for developing countries under
Continuation of activities implemented jointly under	the Kyoto Protocol
the pilot phase	Privileges and immunities for individuals serving
Administrative, financial and institutional matters	on constituted bodies established under the
	Kyoto Protocol
	Administrative, financial and institutional matters

Source: www.unfccc.org

- legal form of the agreed outcome;
- potential environmental, economic and social consequences, including spillover effects, of tools, policies, measures and methodologies available to Annex I parties;
- other issues of the Poznan Work Programme, in particular flexible mechanisms and LULUCF.

At Bonn I, it appeared already quite clear that the two negotiating tracks were progressing at different speeds, especially in consideration of the fact that the AWG-LCA was able to identify a compact text to form the basis for future negotiations. Under the AWG-KP the gap between the positions of developing and developed Parties was not new, in particular on issues such as the mandate and scope of the AWG-KP and its relation with the works of the AWG-LCA.[48]

This difference in the position of Annex I and non–Annex I parties concerned mainly two issues. The first issue was the form of the negotiations and the difficulties created with the establishment of the two-track process (Kyoto Protocol and LCA), which was highlighted by the problem of the different status of the US in each of the two tracks, i.e. as an Annex I party that had not ratified the Kyoto Protocol. Another difficulty was that under the Kyoto Protocol, parties agreed to legally binding commitments subject to the assessment and consequences of the Compliance Committee, while under the Convention, parties are simply required to comply with commitments, which are not subject to a review mechanism as in the Protocol. The second issue of great concern was related to 'numbers', namely the establishment of new binding GHG reduction and limitation commitments for Annex I parties under Annex B to the Protocol. This showed clearly how the two tracks were creating ambiguities. Under the AWG-KP, non-Annex I parties aimed at the fulfilment of the mandate of the AWG-KP, namely the amendment of Annex B of the Protocol listing the QELRCs in accordance with Article 3(9) of the Protocol.

Meanwhile, many Annex I parties preferred to have a more 'comprehensive and coherent'[49] discussion about the efforts to mitigate climate change to be undertaken by Annex I parties that have not ratified the Kyoto Protocol (i.e. the US) and by major developing countries with the merging of the two tracks together.

AWG-LCA6 and AWG-KP8 were held in conjunction with the annual meeting of the UNFCCC's subsidiary bodies, notably the 30th sessions of SBSTA and SBI, from 1 to 12 June 2009. The focus of AWG-LCA6 was on the elaboration of the Chair's first draft negotiating text presented by the Chair only shortly before the opening of the session through a first and second reading, which allowed parties to provide comments and clarification on the different proposals in the text.[50] The result of AWG-LCA6 was a revised negotiating text of 199 pages, very complex and full of proposals, brackets and options.[51]

At AWG-KP8, three contact groups were established, notably on Annex I parties' emissions reductions in aggregate and individually, other issues and legal matters. In the final plenary, the Chair recognized the progress achieved in this session, in particular on reviewing the various proposals, streamlining the text and clarifying the various options. The major focus of AWG-KP8 was on further commitments for Annex I parties and amendments of the Kyoto Protocol pursuant to Article 3(9).

The Bonn session of June 2009 was also characterized by concerns regarding compliance with the so-called 'six months rule'. This referred to the provisions of the Convention and the Protocol that request parties to communicate to the UNFCCC Secretariat any proposal for new protocols under the Convention (Article 17 of the Convention) or amendments to the Kyoto Protocol (Articles 20(2) and 21(3) of the Protocol) at least six months before the next COP and COP/MOP meeting. Proposals for a new protocol under the Convention were submitted by five parties,[52] while 12 proposals for amendments of the Kyoto Protocol were received by the secretariat.[53]

In June, parties also agreed on the establishment of two additional meetings originally not foreseen by the Bali Roadmap, but required at this stage of negotiations in consideration of the complexity of the issues and options to be defined before Copenhagen. The additional meetings were one week of informal consultations among Parties from 11–14 August 2009 and an additional week of negotiations to be held from 2–6 November 2009 in Barcelona as second parts to AWG-LCA5 and AWG-KP9.

The informal session of 10–14 August 2010 produced no official documentation given its informal character. However, some progress was made by parties on issues in the programme of both of the ad hoc working groups. Progress made in Bonn was captured by the documentation to facilitate negotiations among parties and presented before Bangkok. At the conclusion of the August Bonn session, strong divisions remained among the parties concerning:

- convergence of the two tracks;
- Annex I emission reductions commitments, in particular the lack of progress by Annex I parties in showing their proposals for individual commitments;

- the role of major developing countries, in particular the reluctance by these countries in engaging in any type of discussion concerning future commitments of GHG emission reductions;
- the rules for accounting of LULUCF;
- mitigation actions: role of offsets and NAMAs;
- adaptation: finance and classification of most vulnerable countries;
- finance, in particular the presence of too many different proposals by the parties on the governance, structure and implementation of a new financial mechanism;
- REDD Plus: parties requested to keep all ideas reflected in the text.

The first part of the AWG-KP9 and AWG-LCA7 took place in Bangkok between Monday 28 September and Friday 9 October 2009. For the AWG-LCA the secretariat issued additional guidance material, namely document Inf.2 of 15 September 2009,[54] containing reordered and consolidated sections of the AWG-LCA revised negotiating text prepared by the facilitators. Document Inf.2 summarized the work of sub-groups undertaken in the August informal consultations. It was a collection of tools and ideas at the disposal of parties and served as a basis for further work on Inf.1. It did not replace document Inf.1. At AWG-LCA7, five groups and six sub-groups (on mitigation) met with the main task to move from the revised negotiation text to non-papers to be forwarded to Barcelona.[55] In addition, an updated list of proposed new institutional arrangements and a list of numerical proposals made by parties both in the context of the long-term global goal of emission reductions (shared vision) and proposals of mitigation goals (mitigation chapter) were prepared. For the AWG-KP, after the August session, documentation was revised by the Chair on the basis of the non-papers forwarded by the facilitators of the various contact groups with the aim of facilitating negotiations among parties. The document on other proposed amendments to the Kyoto Protocol, including several proposals submitted by parties to change the Protocol, was neither modified in August nor in the other sessions before Copenhagen. In Bangkok, three contact groups met under the AWG-KP: (1) Annex I parties, emission reductions; (2) other issues; and (3) response measures.

The results of the work undertaken by the contact groups established by the AWG-KP in Bangkok are:

- Contact group on potential consequences, 7 October 2009;
- Contact group on the scale of emission reduction by Annex I Parties, 9 October 2009;
- Contact group on other issues identified at the resumed sixth session of the AWG-KP:
  - Land use, land use change and forestry, 8 October 2009;
  - Emissions trading and the project-based mechanisms, 9 October 2009;
  - Greenhouse gases, sectors and source categories (baskets), 9 October 2009.

There were two main positive outcomes of the Bangkok session. In some parts, texts were refined and many informal meetings were held that contributed to reduce the gaps in the positions among parties. On the negative side, parties were still far from agreement on the legal form of the agreed outcome, although this issue did not constitute a formal agenda item but was more the object of informal consultations among parties. In this respect, several options were still considered. In the AWG-KP some of the options discussed were:

- an amendment to the Protocol on the section of Annex I parties' GHG emission reductions and a set of COP and COP/MOP decisions covering operational items;
- an amendment to the Protocol allowing new Annex I party emission reductions together with other significant changes to the Protocol such as the flexible mechanisms;
- the substitution of the Kyoto Protocol with a new agreement (possibly another Protocol to the UNFCCC) based on the five building blocks of the AWG-LCA and only some parts of the Kyoto Protocol.

The second part of the AWG-KP9 and AWG-LCA7 was held in Barcelona from 2–6 November 2009. In Barcelona, parties had only five days' time to prepare the final package of texts to be forwarded to Copenhagen. The main issues addressed related to clarity on emission reduction commitments for Annex I parties under the Kyoto Protocol, on NAMAs by developing countries, and on long- and short-term financing under the LCA framework.

Under the resumed seventh session of the AWG-LCA, parties managed to make some progress on the text and finally produced a series of new non-papers to be forwarded to Copenhagen as an annex to the meeting report. Negotiations were held in the same contact group settings as in Bangkok.

In Barcelona, the first official words, after a lot of speculation in the media, on the form of the final outcome were given. In informal consultations, parties heard for the first time that the conclusion of a new legally binding agreement in Copenhagen was very unlikely to be achieved. Officials referred to a strong decision with 'political commitment to immediate action on all the building blocks of the Bali Action Plan'.[56] Such a decision should probably contain a commitment to work further on a legally ratifiable instrument. The core decision would be comprehensive and could be supported by annexes listing numbers related to mitigation and finance, or annexes and complementary decisions detailing how the different mechanisms would function (adaptation, support for NAMAs and MRV actions, finance mechanism, technology transfer mechanism). Non-Annex I parties questioned the nature and legal force of a political commitment and expressed disappointment about the lack of ambition related to the adoption of a decision as final outcome.

Under the AWG-KP, the following four contact groups continued with their work: (1) Annex I parties' emission reductions (aggregate/individual); (2) other issues identified by the AWG-KP at its resumed sixth session (emissions trading and the project-based mechanisms, LULUCF, methodological issues); (3) potential consequences; and (4) legal matters. These groups considered reviewed documentation, prepared by the Chair on the basis of work at AWG-KP9, first part, and produced the following updated documentation:

- Contact group on potential consequences, 6 November 2009;
- Contact group on other issues identified at the resumed sixth session of the AWG-KP:
  - Land use, land use change and forestry, 6 November 2009;
  - Emissions trading and the project-based mechanisms, 6 November 2009;
  - Greenhouse gases, sectors and source categories (baskets), 6 November 2009.

At the end of the Barcelona talks, it emerged clearly that the two weeks of negotiations foreseen in Copenhagen would be insufficient to conclude a new overarching legal architecture for the climate change regime after 2012. Too little progress on the texts under consideration in the AWG-KP and AWG-LCA was made and too many unresolved issues remained for the decision of political leaders. The clear reference to the adoption of a 'political' COP decision under the AWG-LCA was a direct message that the job could not be finished in 2009.

The Copenhagen conference on climate change was held from 7–19 December 2009 and hosted several official meetings at the same time: the annual COP to the UNFCCC and COP serving as the MOP to the Kyoto Protocol, namely the fifteenth session of the COP (COP15) and the fifth session of the COP serving as the MOP to the Kyoto Protocol (CMP5 or COP/MOP5); the 31st sessions of the UNFCCC's SBSTA (SBSTA31) and SBI (SBI31) as well as the tenth session of the AWG-KP (AWG-KP10) and the eighth session of the AWG-LCA under the UNFCCC (AWG-LCA8).

The AWG-KP10 and AWG-LCA8 meetings were supposed to be the final act of the negotiation process for the post-2012 phase launched by COP13 in December 2007 in Indonesia. The deadline created by the BAP was not respected since parties were not able to agree on the conclusions of the work of the two ad hoc working groups. The Copenhagen Accord (Decision 2/CP.15)[57] resulted as a political decision agreed by the majority of the parties but not formally adopted by COP15 because of the lack of consensus. At the moment of writing, 114 parties have agreed to the Accord and have been listed in its chapeau accordingly. The Copenhagen Accord did not mention any legally binding GHG emission reduction commitment, nor did it provide clarity on the future of the Kyoto Protocol. It did extend the mandate of the ad hoc working groups. The major points addressed by the Copenhagen Accord are:

- keeping global temperature below 2°C;
- review of the Accord by 2015 with the possibility to consider a limit of 1.5°C;
- establishment of a High Level Panel, the Copenhagen Green Climate Fund and the Technology Mechanism;
- Annex I parties required to implement individually or jointly the quantified economy-wide emissions targets for 2020 (Appendix I);
- Non-Annex I parties to the Convention required to implement mitigation actions (Appendix II);

- Funding increases (collective commitment by developed countries to provide new and additional resources approaching US$30 billion for the period 2010–2012 and developed countries commitment to mobilize jointly $100 billion a year by 2020 to address the needs of developing countries).

Several other decisions were adopted by COP15 and CMP5 as indicated in Table 2.7.

After Copenhagen, AWG-KP11 and AWG-LCA9 resumed in Bonn in the period 9–11 April 2010 and focused on organizational issues. The AWG-KP and AWG-LCA requested their Chairs to prepare updated documentation and a text

Table 2.7 *COP15 and COP/MOP5 decisions*

COP15	COP/MOP5
Decision 1/CP.15 Outcome of the work of the Ad Hoc Working Group on Long-term Cooperative Action under the Convention	Decision 1/CMP.5 English Outcome of the work of the Ad Hoc Working Group on Further Commitments for Annex I Parties under the Kyoto Protocol
Decision 2/CP.15 English Copenhagen Accord	
Decision 3/CP.15 English Amendment to Annex I to the Convention	Decision 2/CMP.5 English Further guidance relating to the clean development mechanism
Decision 4/CP.15 English Methodological guidance for activities relating to reducing emissions from deforestation and forest degradation and the role of conservation, sustainable management of forests and enhancement of forest carbon stocks in developing countries	Decision 3/CMP.5 English Guidance on the implementation of Article 6 of the Kyoto Protocol
	Decision 4/CMP.5 English Report of the Adaptation Fund Board
	Decision 5/CMP.5 English Review of the Adaptation Fund
Decision 5/CP.15 English Work of the Consultative Group of Experts on National Communications from Parties not included in Annex I to the Convention	Decision 6/CMP.5 English Compliance Committee
	Decision 7/CMP.5 English Capacity-building under the Kyoto Protocol
Decision 6/CP.15 English Fourth review of the financial mechanism	Decision 8/CMP.5 English Updated training programme for members of expert review teams participating in annual reviews under Article 8 of the Kyoto Protocol
Decision 7/CP.15 English Additional guidance to the Global Environment Facility	
Decision 8/CP.15 English Capacity-building under the Convention	Decision 9/CMP.5 English Administrative, financial and institutional matters
Decision 9/CP.15 English Systematic climate observations	Decision 10/CMP.5 English Programme budget for the biennium 2010–2011
Decision 10/CP.15 English Updated training programme for greenhouse gas inventory review experts for the technical review of greenhouse gas inventories from Parties included in Annex I to the Convention	
Decision 11/CP.15 English Administrative, financial and institutional matters	
Decision 12/CP.15 English Programme budget for the biennium 2010-2011	
Decision 13/CP.15 English Dates and venues of future sessions	

Source: www.unfccc.org

to facilitate negotiations among parties and agreed that two sessions be held between AWG-KP12 and AWG-LCA10 and COP16 in Cancun, Mexico.

AWG-KP12 and AWG-LCA10 were held in conjunction with the 32nd session of the subsidiary bodies (SBI and SBSTA). For the AWG-LCA, a revised negotiating text of 17 May 2010[58] was presented, including many of the draft decision texts negotiated in Copenhagen in 2009. The AWG-KP requested the secretariat to organize 'an in-session workshop at its thirteenth session on the scale of emission reductions to be achieved by Annex I Parties in aggregate and the contribution of Annex I Parties, individually or jointly, to this scale', to update relevant documentation, and to organize, before its 13th session, 'a pre-sessional workshop on forest management accounting, including any new available information, taking into account progress made during its twelfth session regarding the use of reference levels'. Furthermore, the AWG-KP requested the Secretariat to prepare a paper that 'identifies and explores all the legal options available, aiming at ensuring that there is no gap between the first and subsequent commitment periods' and 'identifies the legal consequences and implications of a possible gap between the first and subsequent commitment periods'.[59]

The thirteenth session of the AWG-KP and the eleventh session of the AWG-LCA took place from 2 to 6 August 2010 in Bonn, Germany. Negotiations continued with the view to COP16 and CMP6 scheduled for the end of the year in Cancun. Progress was achieved in the AWG-KP track, especially on the issue of accounting rules for LULUCF, while the works under the AWG-LCA did not advance but rather moved backwards in some particular areas, for example, REDD Plus. For the first time since the launch of the post-2012 climate change talks in Bali in 2007, the AWG-KP was able to produce a single negotiating text (draft proposal by the Chair of 6 August 2010).[60] The text was a compilation of draft decisions, notably:

- Chapter I: Draft decision -/CMP.X Amendments to the Kyoto Protocol pursuant to its Article 3, paragraph 9;
- Chapter II: Draft decision -/CMP.X LULUCF;
- Chapter III: Draft decision -/CMP.X ET and the project-based mechanisms;
- Chapter IV: Draft decision -/CMP.X GHGs, sectors and source categories, common metrics to calculate the CO_2 equivalence of anthropogenic emissions by sources and removals by sinks, and other methodological issues;
- Chapter V: Draft decision -/CMP.X Consideration of information on potential environmental, economic and social consequences, including spillover effects, of tools, policies, measures and methodologies available to Annex I parties.

The draft text under the AWG-LCA track was significantly altered by several submissions and additions by parties so that at the end of the week the number of pages increased greatly (tripled).[61] Amongst others, new wording on MRV was introduced and the draft decision on REDD Plus reopened significantly.

On the legal form of the agreed outcome, informal consultations were held in Bonn where a non-paper designed to facilitate the exchange of views among parties on the matter was distributed by the Mexican government, hosting COP16 and CMP6 in Cancun. Three possible outcomes of COP16 and CMP6 were outlined in

Bonn: a legally binding agreement, a set of COP decisions or a combination of both. Also, for the first time since the start of the post-2012 climate change talks, three separate solutions were identified as possible outcomes: one treaty covering industrialized countries currently committed under the Kyoto Protocol, a second agreement dealing with US emissions and a third agreement for developing countries.[62]

NOTES

1. Other human-induced activities contributing to global warming are the emissions of methane from landfills, rice paddies and livestock as well as the release in the atmosphere of chlorofluorocarbons (CFCs) from refrigerators and air conditioners.

2. IPCC (2001).

3. GE.05-62220 UNFCCC of May 2007 established a scheme to protect the climate system for present and future generations and amended the original text drawn up in New York on 9 May 1992.

4. The UNFCCC was adopted on 9 May 1992 at the Intergovernmental Negotiating Committee's (INC) resumed fifth session, within the framework of the negotiations on a climate change agreement launched by the UN General Assembly in December 1990. The UNFCCC was then opened for signature from 4 to 14 June 1992 at the Earth Summit (UN Conference on Environment and Development – UNCED) in Rio de Janeiro, Brazil. All relevant information, documents and details of the UNFCCC are available at: www.unfccc.int. For a better understanding of the UNFCCC, see UNEP and UNFCCC (2002a) (2002b) and UNFCCC (2006).

5. Articles 3(1) and 4(1) of the UNFCCC and Article 10 of the Kyoto Protocol.

6. 'The extent to which developing country Parties will effectively implement their commitments under the Convention will depend on the effective implementation by developed country Parties of their commitments under the Convention related to financial resources and transfer of technology and will take fully into account that economic and social development and poverty eradication are the first and overriding priorities of the developing country Parties'.

7. Within the UNFCCC and the Kyoto Protocol, EITs are the ten new EU Member States without Malta and Cyprus and including Croatia, Belarus, the Russian Federation and Ukraine. In the context of this book, the term 'EITs' refers only to ten EU accession countries, without Malta and Cyprus, unless otherwise specified.

8. 'Develop, periodically update, publish and make available to the Conference of the Parties, in accordance with Article 12, national inventories of anthropogenic emissions by sources and removals by sinks of all greenhouse gases not controlled by the Montreal Protocol, using comparable methodologies to be agreed upon by the Conference of the Parties' (Article 4(1)a of the UNFCCC).

9. 'Formulate, implement, publish and regularly update national and, where appropriate, regional programmes containing measures to mitigate climate change by addressing anthropogenic emissions by sources and removals by sinks of all greenhouse gases not controlled by the Montreal Protocol, and measures to facilitate adequate adaptation to climate change' (Article 4(1)b of the UNFCCC).

10. 'Promote and cooperate in the development, application and diffusion, including transfer, of technologies, practices and processes that control, reduce or prevent anthropogenic emissions of greenhouse gases not controlled by the Montreal Protocol in all relevant sectors, including the energy, transport, industry, agriculture, forestry and waste management sectors' (Article 4(1)c of the UNFCCC).

11. 'Promote sustainable management, and promote and cooperate in the conservation and enhancement, as appropriate, of sinks and reservoirs of all greenhouse gases not controlled by the Montreal Protocol, including biomass, forests and oceans as well as other terrestrial, coastal and marine ecosystems' (Article 4(1)d of the UNFCCC).

12. 'Cooperate in preparing for adaptation to the impacts of climate change; develop and elaborate appropriate and integrated plans for coastal zone management, water resources and agriculture, and for the protection and rehabilitation of areas, particularly in Africa, affected by drought and desertification, as well as floods' (Article 4(1)e of the UNFCCC).

13. 'Take climate change considerations into account, to the extent feasible, in their relevant social, economic and environmental policies and actions, and employ appropriate methods, for example impact assessments, formulated and determined nationally, with a view to minimizing adverse effects on the economy, on public health and on the quality of the environment, of projects or measures undertaken by them to mitigate or adapt to climate change' (Article 4(1)f of the UNFCCC).

14. 'Promote and cooperate in scientific, technological, technical, socio-economic and other research, systematic observation and development of data archives related to the climate system and intended to further the understanding and to reduce or eliminate the remaining uncertainties regarding the causes, effects, magnitude and timing of climate change and the economic and social consequences of various response strategies' (Article 4(1)g of the UNFCCC).

15. 'Promote and cooperate in the full, open and prompt exchange of relevant scientific, technological, technical, socio-economic and legal information related to the climate system and climate change, and to the economic and social consequences of various response strategies' (Article 4(1)h of the UNFCCC).

16. 'Promote and cooperate in education, training and public awareness related to climate change and encourage the widest participation in this process, including that of non-governmental organizations' (Article 4(1)i of the UNFCCC).

17. 'Communicate to the Conference of the Parties information related to implementation, in accordance with Article 12' (Article 4(1)j of the UNFCCC).

18. 'Each of these Parties shall adopt national policies and take corresponding measures on the mitigation of climate change, by limiting its anthropogenic emissions of greenhouse gases and protecting and enhancing its greenhouse gas sinks and reservoirs. These policies and measures will demonstrate that developed countries are taking the lead in modifying longer-term trends in anthropogenic emissions consistent with the objective of the Convention, recognizing that the return by the end of the present decade to earlier levels of anthropogenic emissions of carbon dioxide and other greenhouse gases not controlled by the Montreal Protocol would contribute to such modification, and taking into account the differences in these Parties' starting points and approaches, economic structures and resource bases, the need to maintain strong and sustainable economic growth, available technologies and other individual circumstances, as well as the need for equitable and appropriate contributions by each of these Parties to the global effort regarding that objective. These Parties may implement such policies and measures jointly with other Parties and may assist other Parties in contributing to the achievement of the objective of the Convention and, in particular, that of this subparagraph' (Article 4(2)a of the UNFCCC).

19. 'In order to promote progress to this end, each of these Parties shall communicate, within six months of the entry into force of the Convention for it and periodically thereafter, and in accordance with Article 12, detailed information on its policies and measures referred to in subparagraph (a) above, as well as on its resulting projected anthropogenic emissions by sources and removals by sinks of greenhouse gases not controlled by the Montreal Protocol for the period referred to in subparagraph (a), with the aim of returning individually or jointly to their

1990 levels these anthropogenic emissions of carbon dioxide and other greenhouse gases not controlled by the Montreal Protocol. This information will be reviewed by the Conference of the Parties, at its first session and periodically thereafter, in accordance with Article 7' (Article 4(2)b of the UNFCCC).

20. Annex II of the UNFCCC lists the same countries as included in Annex I, with the exception of parties with EITs.

21. 'The developed country Parties and other developed Parties included in Annex II shall provide new and additional financial resources to meet the agreed full costs incurred by developing country Parties in complying with their obligations under Article 12, paragraph 1. They shall also provide such financial resources, including for the transfer of technology, needed by the developing country Parties to meet the agreed full incremental costs of implementing measures that are covered by paragraph 1 of this Article and that are agreed between a developing country Party and the international entity or entities referred to in Article 11, in accordance with that Article. The implementation of these commitments shall take into account the need for adequacy and predictability in the flow of funds and the importance of appropriate burden sharing among the developed country Parties' and 'The developed country Parties and other developed Parties included in Annex II shall take all practicable steps to promote, facilitate and finance, as appropriate, the transfer of, or access to, environmentally sound technologies and know-how to other Parties, particularly developing country Parties, to enable them to implement the provisions of the Convention. In this process, the developed country Parties shall support the development and enhancement of endogenous capacities and technologies of developing country Parties. Other Parties and organizations in a position to do so may also assist in facilitating the transfer of such technologies' (Articles 4(3) and (5) of the UNFCCC).

22. 'The developed country Parties and other developed Parties included in Annex II shall also assist the developing country Parties that are particularly vulnerable to the adverse effects of climate change in meeting costs of adaptation to those adverse effects' (Article 4(4) of the UNFCCC).

23. 'In the implementation of their commitments under paragraph 2 above, a certain degree of flexibility shall be allowed by the Conference of the Parties to the Parties included in Annex I undergoing the process of transition to a market economy, in order to enhance the ability of these Parties to address climate change, including with regard to the historical level of anthropogenic emissions of greenhouse gases not controlled by the Montreal Protocol chosen as a reference' (Article 4(6) of the UNFCCC).

24. 'The Parties shall take full account of the specific needs and special situations of the least developed countries in their actions with regard to funding and transfer of technology' and 'The Parties shall, in accordance with Article 10, take into consideration in the implementation of the commitments of the Convention the situation of Parties, particularly developing country Parties, with economies that are vulnerable to the adverse effects of the implementation of measures to respond to climate change. This applies notably to Parties with economies that are highly dependent on income generated from the production, processing and export, and/or consumption of fossil fuels and associated energy-intensive products and/or the use of fossil fuels for which such Parties have serious difficulties in switching to alternatives' (Articles 4(9) and (10) of the UNFCCC).

25. UNFCCC, Draft Rules of Procedure of the Conference of the Parties and its Subsidiary Bodies FCCC/CP/1996/2, 22 May 1996.

26. Kyoto Protocol to the UNFCCC of 11 December 1997 at Kyoto within the framework of the Convention negotiation COP3 establishing an international binding agreement for achieving the quantified emission limitation and reduction commitments in order to promote sustainable development in pursuit of the ultimate objective of the UNFCCC.

27. The Kyoto Protocol was adopted by the Parties to the UNFCCC on 11 December 1997 in Kyoto, Japan, at the Third Conference of the Parties to the UNFCCC (COP3). For additional information, documents and details on the Kyoto Protocol and subsequent decisions, see note 4 above.

28. UNFCCC (2000), p36.

29. Kyoto Protocol to the UNFCCC of 11 December 1997 at Kyoto within the framework of the Convention negotiation COP3 establishing an international binding agreement for achieving the quantified emission limitation and reduction commitments in order to promote sustainable development in pursuit of the ultimate objective of the UNFCCC, p19.

30. Statement of 18 November 2004 attributable to the UN Secretary General Kofi Annan upon receiving Russian Federation ratification for the 1997 Kyoto Protocol to the UNFCCC from the Russian Permanent Representative Andrey Denisov.

31. 'This Protocol shall enter into force on the ninetieth day after the date on which not less than 55 Parties to the Convention, incorporating Parties included in Annex I which accounted in total for at least 55 per cent of the total carbon dioxide emissions for 1990 of the Parties included in Annex I, have deposited their instruments of ratification, acceptance, approval or accession' (Article 25(1) of the Kyoto Protocol).

32. Press release of 14 October 2004 of the Council of the EU regarding the main results of the 2610th Council Meeting on environment at Luxembourg.

33. UNFCCC (2000), p32.

34. At the time of ratification there were 15 EU Member States.

35. The BAPA, Decision 1/CP.4, FCCC/CP/1998/16/Add.1, 25 January 1999.

36. Dialogue on long-term cooperative action to address climate change by enhancing implementation of the Convention, Decision 1/CP.11, FCCC/CP/2005/5/Add.1 of 30 March 2006, and consideration of commitments for subsequent periods for Annex I Parties to the Convention under Article 3(9) of the Kyoto Protocol (KP), Decision 1/CMP.1, FCCC/KP/CMP/2005/8/Add.1 of 30 March 2006.

37. Procedures and mechanisms relating to compliance under the Kyoto Protocol, Decision 27/CMP.1, section V, paragraph 4, FCCC/KP/CMP/2005/8/Add.3, 30 March 2006.

38. Decision 27/CMP.1, section XV, paragraphs 1, 2 and 3.

39. The suspension to participate in a flexible mechanism depends on which of the specific eligibility requirements are not met, and, consequently, there may well arise a situation where a party is allowed to participate in one but not all flexible mechanisms. Obviously, the suspension to participate in IET may affect JI and CDM, namely the exchange of ERUs and/or CERs.

40. Decision 27/CMP.1, section XV, paragraph 4. For the expedited procedures for the enforcement branch see Decision 27/CMP.1, section X.

41. Decision 27/CMP.1, section XV, paragraph 5.

42. BAP, Decision 1/CP.13, FCCC/CP/2007/6/Add.1, 14 March 2008.

43. See UNFCCC, Article 2, which states: 'The ultimate objective of this Convention and any related legal instruments that the Conference of the Parties may adopt is to achieve, in accordance with the relevant provisions of the Convention, stabilization of greenhouse gas concentrations in the atmosphere at a level that would prevent dangerous anthropogenic interference with the climate system. Such a level should be achieved within a time-frame sufficient to allow ecosystems to adapt naturally to climate change, to ensure that food production is not threatened and to enable economic development to proceed in a sustainable manner.'

44. See BAP, note 42.

45. The US has not ratified the Kyoto Protocol and therefore it participates in the meetings of the COP/MOP and AWG-KP with an observer status.

46. The Bali Roadmap, Address to closing plenary by his Excellency Mr. Rachmat Witoelar, President, UN Climate Change Conference, Closing of Joint High-Level Segment, Bali, 15 December 2007.

47. Report of the Ad Hoc Working Group on Further Commitments for Annex I Parties under the Kyoto Protocol on its resumed sixth session, held in Poznan from 1 to 10 December 2008, FCCC/KP/AWG/2008/8, 4 February 2009.

48. Developed and developing countries are often referred to respectively as Annex I parties to the UNFCCC and non-Annex I parties. Parties inscribed in Annex I to the UNFCCC are considered industrialized countries by the UNFCCC.

49. The terms convergence and coherence were used often by Annex I parties during the negotiations, in particular by members of the Umbrella Group, such as Japan, Canada, the US and Australia. This is evidenced, for instance, in the Report of the AWG-KP on its seventh session, held in Bonn from 29 March to 8 April 2009 (FCCC/KP/AWG/2009/5, 13 May 2009), at 13.

50. Negotiating text: note by the Chair (FCCC/AWGLCA/2009/8, 19 May 2009).

51. Revised negotiating text: note by the secretariat (FCCC/AWGLCA/2009/INF.1, 22 June 2009).

52. These were the US, Tuvalu, Japan, Australia and Costa Rica.

53. These were from the EU, Tuvalu (2), Philippines, New Zealand, Colombia, Algeria and other Group of 77 and China countries, Belarus, Australia, Japan, Bolivia on behalf of Malaysia, Paraguay and Venezuela, and Papua New Guinea.

54. AWG-LCA, reordering and consolidation of text in the revised negotiating text, note by the secretariat, FCCC/AWGLCA/2009/INF.2, 15 September 2009.

55. The work of the five groups is outlined below. The six sub-groups focused on adaptation, finance, technology, mitigation, capacity building and a shared vision for long-term cooperative action.

56. Resumed seventh session, Barcelona, 2–6 November 2009, informal consultations on legal form of the agreed outcome as reported by the author.

57. Decision 2/CP.15, Copenhagen Accord, FCCC/CP/2009/11/Add.1, 30 March 2010, pp4–9.

58. AWG-LCA, Text to facilitate negotiations among Parties, Note by the secretariat, FCCC/AWGLCA/2009/INF.2, 15 September 2009, FCCC/AWGLCA/2010/6, 17 May 2010.

59. AWG-KP, Report of the AWG-KP on its 12th session, held in Bonn from 1–11 June 2010, FCCC/KP/AWG/2010/7, 28 June 2010.

60. AWG-KP, 13th session Bonn, 2–6 August 2010, Agenda item 3, Consideration of further commitments for Annex I Parties under the Kyoto Protocol, Draft proposal by the Chair, FCCC/KP/AWG/2010/CRP.2, 6 August 2010.

61. AWG-LAC under the Convention, 11th session, Bonn, 2–6 August 2010, Item 3 of the provisional agenda, Preparation of an outcome to be presented to the COP for adoption at its 16th session to enable full, effective and sustained implementation of the Convention through long-term cooperative action now, up to and beyond 2012, Text to facilitate negotiations among parties, FCCC/AWGLCA/2010/8, 9 July 2010.

62. At the moment of writing, it seems quite premature that a final agreement or a set of agreements can be reached at COP16, while it appears more realistic to consider the conclusion of the climate change talks for the post-2012 regime in South Africa at the end of 2011.

CHAPTER 3

The Foundations of European Climate Policy

3.1 EARLY PROVISIONS

In general, the basic goals of the European climate and clean energy policy that have been developed over years can be summarized as:

- Promotion of energy efficiency;
- Promotion of electricity generated from renewable sources;
- Research on and attention to cost efficiency;
- Cooperation with the industrial sector, non-government organizations (NGOs) and other stakeholders.

In the period 1988–1990, the need for legislation to tackle climate change was for the first time officially recognized by the European Commission with Communication to the Council – The greenhouse effect and the Community Commission work programme concerning the evaluation of policy options to deal with the greenhouse effect COM(1988)656.[1] This document reflected the requests of the northern EU Member States to act at the international level for the reduction of CO_2, even if already in 1986 a resolution of the European Parliament had already raised the same issue. Communication COM(1988)656 introduced the greenhouse issue and its effects, impacts and consequences in Europe, the existing international framework and perspectives (World Conference on 'The changing atmosphere' of 1988) and possible actions at the European level (research, preventive action, planned adaptation and cooperation with developing countries).

At the level of the EU Council of Ministers, the first steps towards the establishment of a solid European climate policy date back to the Dublin meeting of June 1990, where the European Council urged the adoption of targets and strategies for limiting the emissions of GHGs and to the Environment Council of

29 October 1990 where European ministers agreed to take actions to stabilize, by 2000, CO_2 emissions at the 1990 level in the EC. The Environment Council of 13 December 1991 invited the Commission to take the lead in formulating proposals for concrete policies and measures. The first package of legislative proposals responding to climate change was therefore prepared by the European Commission as early as 1991 and was intended to be adopted by the Environment Council prior to the Earth Summit of the UN Conference on Environment and Development (UNCED) held from 3–14 June 1992. The Council failed to reach an agreement on the harmonization of energy taxation and the entire package was dropped.

Although several provisions indirectly contributing to fight global warming were adopted by European institutions during the 1980s, it is only since the beginning of the 1990s that the EC and its Member States have started to develop an advanced environmental policy exclusively aiming at the reduction of GHG emissions and directly aiming at the development of clean energy. This was mainly due to the obligations resulting from the international climate regime created by the UNFCCC and the Kyoto Protocol.

The first Community legislation expressly designed to be adopted within the framework of what in the future would be called the European climate policy was Council Decision 93/389/EEC which established a monitoring mechanism for anthropogenic CO_2 and other greenhouse gas emissions not controlled by the Montreal Protocol in the Member States, and which required the Member States to 'publish and implement national programmes for limiting' their CO_2 emissions.[2] In 1993, the EU and the Member States were already required to initiate legislation for the monitoring of GHG emissions in line with the international climate regime. It was therefore the entry into force of the UNFCCC first and subsequently the Kyoto Protocol that contributed to the boosting of EU climate and clean energy policy, as a consequence of what was required by international and EC law.

Under international law the EU and the Member States were driven by the following requirements:

- Article 4(1)b UNFCCC 'All Parties, taking into account their common but differentiated responsibilities and their specific national and regional development priorities, objectives and circumstances, shall:
  - ...
  - b) Formulate, implement, publish and regularly update national and, where appropriate, regional programmes containing measures to mitigate climate change by addressing anthropogenic emissions by sources and removals by sinks of all greenhouse gases not controlled by the Montreal Protocol, and measures to facilitate adequate adaptation to climate change';
- Article 2(11) of the Kyoto Protocol 'Each Party included in Annex I, in achieving its quantified emission limitation and reduction commitments under Article 3, in order to promote sustainable development, shall:
  - (a) Implement and/or further elaborate policies and measures in accordance with its national circumstances'.

Under Community law, the EU institutions and the Member States were guided by the EC Treaty:

- Institutions:
    - Article 174 TEC (Objectives and principles of EU environmental policy);
    - Article 175(1) TEC 'The Council, acting in accordance with the procedure referred to in Article 251 and after consulting the Economic and Social Committee and the Committee of the Regions, shall decide what action is to be taken by the Community in order to achieve the objectives referred to in Article 174';
- Member States
    - Article 10 TEC 'Member States shall take all appropriate measures, whether general or particular, to ensure fulfilment of the obligations arising out of this Treaty or resulting from action taken by the institutions of the Community. They shall facilitate the achievement of the Community's tasks';
    - Article 176 TEC 'The protective measures adopted pursuant to Article 175 shall not prevent any Member State from maintaining or introducing more stringent protective measures. Such measures must be compatible with this Treaty. They shall be notified to the Commission'.

By accepting rules of the international climate regime, the EC and the Member States are therefore committed to contributing to the fight against global warming by developing and implementing national policies aimed at the mitigation of, and adaptation to, climate change.

Since the early 1990s, the EC's position on the fight against global warming has focused on the following objectives:

- Ensuring the environmental integrity of the international climate regime designed under the UNFCCC and the Kyoto Protocol;
- Ensuring compliance with the quantified emission limitation and reduction commitments included in Annex B of the Kyoto Protocol;
- Ensuring that the reduction of GHG emissions is achieved through the introduction of national policies and measures and only in a way supplemental to the implementation of the flexible mechanisms;
- Maintaining a multilateral approach in the fight against global warming;
- Promoting sustainable development.

The main EU laws on energy, air, water, waste and industrial pollution, as well as horizontal legislation such as the directive on environmental impact assessment and the directive on public access to environmental information, represented in 2004 the core of the relevant environmental *acquis communautaire* in relation to climate policy. The following list covers the main pieces of legislation relevant to the fight against global warming at the time of the 2004 EU enlargement:

- Council Decision 93/389/EEC as amended by Decision 99/296/EC for a monitoring mechanism of EC CO_2 and other GHG emissions;[3]

- Council Decision 94/69/EEC concerning the conclusion of the UNFCCC;[4]
- Council Directive 96/62/EC on ambient air quality assessment and management;[5]
- Commission Recommendations 1999/125/EC, 2000/303/EC and 2000/304/EC on the reduction of CO_2 emissions from passenger cars;[6]
- Council Directive 1999/13/EC on the limitation of emissions of volatile organic compounds due to the use of organic solvents in certain activities and installations;[7]
- Council Directive 1999/32/EC on the reduction of sulphur content of certain liquid fuels;[8]
- Council Directive 2001/81/EC on national emissions ceilings for certain atmospheric pollutants;[9]
- Council Directive 2002/3/EC relating to ozone in ambient air;[10]
- Council Directive 2001/100/EC of the European Parliament and of the Council of 7 December 2001 amending Council Directive 70/220/EEC on the approximation of the laws of the Member States on measures to be taken against air pollution by emissions from motor vehicles;[11]
- Council Directive 2001/1/EC of the European Parliament and of the Council of 22 January 2001 amending Council Directive 70/220/EEC concerning measures to be taken against air pollution by emissions from motor vehicles;[12]
- Council Directive 2003/17/EC of the European Parliament and of the Council of 3 March 2003 amending Directive 98/70/EC relating to the quality of petrol and diesel fuels;[13]
- Council Directive 96/61/EC concerning integrated pollution prevention and control (IPPC);[14]
- Council Directive 2001/80/EC on the limitation of emissions of certain pollutants into the air from large combustion plants.[15]

3.2   THE EU IN THE INTERNATIONAL CLIMATE REGIME

The UNFCCC adopted at the Earth Summit was the result of long and intense negotiations. Since the beginning of the negotiations, the position of the EC delegation favoured the introduction of stringent provisions to combat global warming. Such an approach is reflected in the final text agreed:

- 'The ultimate objective of this Convention and any related legal instruments that the Conference of the Parties may adopt is to achieve, in accordance with the relevant provisions of the Convention, stabilization of greenhouse gas concentrations in the atmosphere at a level that would prevent dangerous anthropogenic interference with the climate system', Article 2 UNFCCC;
- A wide range of commitments for developed countries to mitigate climate change is inscribed under Article 4(2) UNFCCC.

The EC and the Member States approved the UNFCCC through Council Decision 94/69/EC of 15 December 1993.[16] With Decision 94/69/EC, the president of the Council of the EU is given mandate to deposit the instrument of approval to the Secretariat of the UNFCCC in accordance with Article 22(1) of the Convention as well as the declaration of competence as set out in Annex B to that decision in accordance with Article 22(3) UNFCCC.

Both the UNFCCC and the Kyoto Protocol follow the practice of agreements adopted within the framework of the UN, i.e. they do not refer specifically to the EC but are open to the participation of 'regional economic integration organizations' (REIOs).[17] Article 1 of the UNFCCC recognizes a REIO as:

> an organization constituted by sovereign States of a given region which has competence in respect of matters governed by this Convention or its protocols and has been duly authorized, in accordance with its internal procedures, to sign, ratify, accept, approve or accede to the instruments concerned.

Therefore the EC participates in the international climate regime as a REIO and is included, together with the Member States, in the list of Annex I parties to the UNFCCC.

Under the UNFCCC, all parties are required to report periodically on the steps they are taking or envisage undertaking to implement the Convention. In accordance with the principles of equity and common but differentiated responsibilities and respective capabilities (Article 3(1) of the UNFCCC), Annex I parties 'should take the lead in combating climate change and the adverse effects thereof'. Accordingly, Annex I and non-Annex I parties have a different timetable for the submission of national communications and national inventories of GHG emissions (Articles 4(1) and 12 of the UNFCCC). In accordance with Article 4(2) of the UNFCCC, the EC and its Member States are jointly committed to limit anthropogenic CO_2 emissions.

As required under Article 12 of the UNFCCC, the EC has so far submitted to the secretariat the following information:

- First Communication from the EC under the UNFCCC (available in hard copy only);[18]
- Second Communication from the EC under the UNFCCC, Brussels, 26 June 1998, SEC(1998)1770;[19]
- Third Communication from the EC under the UNFCCC, Brussels, 20 December 2001, SEC(2001)253;[20]
- Fourth Communication from the EC under the UNFCCC, Communication from the Commission, Brussels, 8 February 2006, COM(2006)40.[21]

After the signature of the Kyoto Protocol, the EC started to draft its medium- and long-term strategy vis-à-vis climate change. A series of Community documents starting from 1998 explain the first set of EU policies towards climate change:

- Commission communication to the Parliament and the Council, Climate Change – Towards an EU Post-Kyoto Strategy COM(1998)353;[22]
- Commission communication to the Parliament and the Council, Preparing for Implementation of the Kyoto Protocol COM(1999)230;[23]

- Proposal for a Council Decision concerning ratification of the Kyoto Protocol by the European Community COM(2001)579.[24]

2001 was an important turning point for EU policy on climate change, in particular since the public announcement of the US administration to withdraw from any commitment related to the Kyoto Protocol. Since then, the EU has tried to assume a leading role on climate policy at the international level, as confirmed over the years by the very strong position of the EU at COP negotiations aimed at ensuring the environmental integrity of the Kyoto Protocol and ambitious GHG emission reduction targets, and by the quantity and quality of legislation at the Community level aimed at combating global warming.

Proposal COM(2001)579 included the draft council decision required for the ratification of the Kyoto Protocol by the EC and the Member States. In particular, the proposal provided for general remarks on the importance of climate change and its recognition at the international level, clarity on the legal basis for the ratification of the Kyoto Protocol (Article 174(4) in conjunction with Article 300(2) and (3)), reference to the shared competence between the Community and the Member States as well as to the Burden Sharing Agreement (BSA), the consistency with other Community policies, the financial implication of the ratification, and a full explanation of the six articles of the decision.

The EC and its Member States agreed in the Council Conclusion of 4 March 2002[25] to opt for a coordinated ratification of the Kyoto Protocol by the Community and the Member States. In order to ensure the simultaneous deposit of the instrument of ratification by the Community and its Member States and to enable the Kyoto Protocol to enter into force before the Johannesburg World Summit on Sustainable Development (WSSD)[26] in September 2002, the Member States committed 'to make every effort to deposit their instruments of ratification or approval at the same time as those of the Community and, as far as possible, no later than 1 June 2002'.[27]

The EC and the Member States ratified the Kyoto Protocol with the adoption of Council Decision 2002/358/EC of 25 April 2002 concerning the approval, on behalf of the European Community, of the Kyoto Protocol to the United Nations Framework Convention on Climate Change and the joint fulfilment of commitments thereunder.[28] The EC decision to ratify the Kyoto Protocol in April 2002 went along with the intention to give a strong political input to the global process towards the fight against global warming before the Johannesburg WSSD, which gathered together representatives of governments from all over the world from 26 August to 4 September 2002.

The EC and the EU15 are Annex I and Annex II parties to the UNFCCC, and are included in Annex B of the Kyoto Protocol.[29] With regard to the new Member States (the EU12 or EU10 if excluding Malta and Cyprus), they are also included in both the Annex I to the UNFCCC and Annex B to the Kyoto Protocol, with the exception of Malta and Cyprus, but are still considered as non-Annex I parties under the UNFCCC. In this respect, in the international climate regime the EC is considered, as regards some of the obligations created by the Kyoto Protocol, as a

REIO of 15 and not 27 Member States (after the two latest enlargements in 2004 and 2007).

Council decision 2002/358/EC concerning the approval, on behalf of the European Community, of the Kyoto Protocol to the UNFCCC and the joint fulfilment of commitments thereunder gave effect to an international treaty under Community law and provides for the declaration on the division of competences between the Community and the Member States. With Council Decision 2002/358/EC, the Kyoto Protocol and its obligations are fully integrated in the *acquis communautaire* and the Community as well as all the Member States (old, EU15, and new, EU27)[30] are bound by those provisions from the international and the Community law perspective. Furthermore, this decision formalized, at the EC level, the EU BSA, which creates individual GHG emission reduction obligations for the Community and the Member States.

With regard to the legal basis chosen for the adoption of Decision 2002/358/EC, although the initial proposal COM(2001)579[31] referred to Article 174(4) EC Treaty as legal basis, the EC decided to change the legal basis to Article 175(1) as a consequence of the ECJ opinion on the legal basis necessary for the conclusion of a mixed agreement.[32] Finally, while the procedure foreseen by Article 175(1) TEC required qualified majority voting, Decision 2002/358/EC was adopted with unanimity. After, a few Member States required the adoption of this instrument on the basis of Article 175(2) TEC (the procedure requiring unanimity), in consideration of the substantial impact that the ratification of the Kyoto Protocol could have on the national energy policies of the Member States. Council Decision 2002/358/EC also authorized the European Commission to adopt further decisions through the comitology procedure,[33] as, for instance, Decision 2006/944/EC determining the emission levels respectively allocated to the Community and to the Member States in terms of CO_2 equivalent, as required under Article 4(2) of the Kyoto Protocol.[34]

The distribution of powers and competences between the EC and the Member States to conclude and participate in the Kyoto Protocol is indicated in Annex III of Council Decision 2002/358/EC, where:

> the European Community declares that, in accordance with the Treaty establishing the European Community, and in particular Article 175(1) thereof, it is competent to enter into international agreements, and to implement the obligations resulting therefrom.

The UNFCCC and the Kyoto Protocol required both the Community and the Member States to deposit their own instruments of ratification, approval or accession with the secretariat under Articles 22(1) and 24 respectively. The EC and the Member States deposited their instruments of ratification, approval or accession with the Depositary on 31 June 2002. On that same day, the EU15 notified the UNFCCC Secretariat of their decision to jointly commit to the Kyoto Protocol in accordance with Article 4. In the Member States, ratification of the Kyoto Protocol took place in accordance with the national procedures established by the different constitutional systems. Since 31 May 2002, the Kyoto Protocol has been part of the *acquis communautaire* and therefore all obligations arising from this treaty are binding on all Member States.

3.3 THE EU BURDEN SHARING AGREEMENT

Article 3(1) of the Kyoto Protocol sets the QELRCs for Annex I parties listed in Annex B to the Kyoto Protocol. Developed countries shall reduce overall GHG emissions 'by at least 5 per cent below 1990 levels in the commitment period 2008 to 2012'. Article 3(1) of the Kyoto Protocol refers to the right of Annex I parties to ensure 'individually or jointly, that their aggregate anthropogenic carbon dioxide equivalent emissions of the greenhouse gases listed in Annex A do not exceed their assigned amounts', determined on the basis of the percentages inscribed in Annex B.

The text of Article 3 of the Kyoto Protocol was subject to long and difficult negotiations at the meetings of the Ad Hoc Group of the Berlin Mandate, with the two most controversial issues being the legal nature of the QELRCs and the provisions related to the joint commitment. In this context, the EU position was aimed at the introduction of the alternative 'individually or jointly' with the view to distribute the overall target among the 15 Member States. Support for this position was provided by, among others, the Czech Republic, Hungary and the Russian Federation. Opposition came from several other parties, in particular the JUSSCANNZ group headed by the US delegation.[35] At the eighth meeting of the AGBM held in Bonn from 22 to 31 October 1997, three text options were presented by parties as to the joint commitment issue: (1) the EU position including the words 'individually or jointly'; (2) the texts of JUSSCANNZ; and (3) the G77 and China group, which did not mention this concept. Concerns about including the reference to 'individually or jointly', mainly from Australia and Japan, were related to the lack of legal clarity regarding the distribution of competences between the EU and the Member States. Some Annex I parties were afraid that the EU reduction commitment would be softened by including the GHG emissions surplus of the new Member States in the calculation of the EU15 targets. The Community's proposal was accepted by COP3 in Japan only after the EC had made a few concessions to the JUSSCANNZ requests that the EU enlargement would not affect the joint agreement in the Kyoto Protocol's first commitment period.[36]

While Article 3(1) provides the option for Annex I parties to fulfil their QELRCs individually or jointly, Article 4 of the Kyoto Protocol sets out the rules of such a joint commitment. Article 4(1) of the Kyoto Protocol provides all Annex I parties, whether nation states or REIOs, with the opportunity to fulfil their limitation and reduction commitments jointly.[37] In practice, although Article 4(1) refers generically to Annex I parties, the joint commitment option is referred to the EC and the Member States, who supported the introduction of this article in the negotiations for the adoption of the Kyoto Protocol.

In accordance with Article 4 of the Kyoto Protocol, the EC and the Member States are jointly committed to a quantified GHG emission reduction of 8 per cent below 1990 levels by the period 2008–2012. Due to the inability of EU15 to agree on a common European-wide GHG reduction target to be applied by every Member State, this obligation has been distributed internally at the European level in the form of differentiated GHG emission limits for each of the pre-2004 Member States. This agreement is known as the Burden Sharing Agreement – EU bubble.[38] These limits are expressed in terms of percentages by which the Member States must reduce or in some cases stabilize or increase their emissions in comparison with their levels in the base year (1990). The EU BSA was notified to the UNFCCC with the instrument of ratification of the Kyoto Protocol by the Community, namely Annex II to Council Decision 2002/358/EC.[39]

In Geneva in 1990, on the occasion of the Second World Climate Conference aimed at a review of the UNEP/WMO World Climate Programme (WCP) and at the identification of policy recommendations and actions at the global level, the European Commission made the first official attempt to establish a European-wide BSA, indicating individual GHG emissions reduction targets for the Member States to be introduced internally as early as 1991. Subsequently, the issue of the EU BSA was set aside, especially because of the reluctance of several Member States (France, the UK and the southern European states). In the meantime, in 1994, the UNFCCC was ratified by the Community. A new proposal for a general reduction target to be distributed among the Member States was prepared by the European Commission in 1996, but the most serious input to the definition of the EU Bubble came from the Dutch presidency in the first half of 1997 and was finally formulated as a proposal in March 1997, only a few months before the Kyoto Protocol was adopted in Japan. The first draft of the EU BSA covered only CO_2 emissions and a revision of this agreement was needed soon after it had first been drawn up. In fact, in the meantime, the Kyoto Protocol was adopted in 1997 at COP3, including six GHGs in the agreement and identifying differentiated targets for Annex I parties. The first round of EU internal negotiations on the distribution of the common target in 1997 was characterized by the Dutch proposal, which aimed at introducing stringent reduction targets for the Member States.

This proposal was objected to by several European capitals and especially by the members of the Cohesion Fund mentioned above, which were clearly in favour of less stringent targets. The deal was finalized in 1998 thanks to a UK proposal that took into account a number of changes included in the Kyoto Protocol, such as the coverage of six instead of three GHGs, as agreed previously. This proposal softened the Dutch proposal by accepting concessions to the 'cohesion' countries in terms of less rigid GHG emissions reduction targets. The final agreement of the Environment Council of June 1998[40] represents the existing framework of the EU Bubble and established strong reduction obligations for nearly all Member States. Furthermore, the individual GHG emissions reduction targets agreed under the EU bubble were obviously in line with the UNFCCC's main objective, namely stabilization of CO_2 emissions at the 1990 level by 2000. However, the importance of the EU BSA in terms of its effective contribution to the fight against global warming is reduced considerably by the fact that this agreement is not based on a top-down approach whereby the individual targets would be set by EC law; rather, the power to negotiate on the basis of national circumstances was left to the national governments.[41]

The concept of the BSA adopted under the Kyoto Protocol is not new in Community law. Under Directive 88/609/EC on the limitation of emissions of certain pollutants from large combustion plants into the air, a set of common sulphur dioxide emission ceilings were agreed by the Member States.[42] The Triptych approach is at the foundation of the differentiation of commitments under the international climate regime among Member States.[43] The Triptych approach is based on the identification of an overall target to be distributed among parties on the basis of fairness. The initial identification of the different national targets was based on the level of GHG emissions generated by three major sectors: electricity generation, energy-intensive industries and households.[44] Moreover, several other criteria such as national energy structure and consumption rate, economic stability, difference in marginal abatement costs and political considerations contributed to the identification of Member States' individual contributions.

As mentioned above, under the request of a number of members of the Umbrella Group, conditions or 'safeguards' were inserted in Article 4 of the Kyoto Protocol, with a view to limiting the possibility of the EC and the EU15 getting support from Central and Eastern European countries (CEECs) in the fulfilment of their QERLCs. To this aim, Article 4(3) and (4) of the Kyoto Protocol limits the EC joint commitment, both in geographical and temporal terms.[45] Again, the reasoning behind this limit can be found in the requests from parties to avoid any advantage for those parties who would have adopted a joint commitment. In accordance with Article 4(3), the joint commitment of the EC and the Member States shall not be modified until the end of the first commitment period (2012), thus de facto limiting such a commitment to the EU15.[46] Following Article 4(4), the joint commitment shall not be affected by a change in the composition of the REIO in the first commitment period of the Kyoto Protocol.[47] In accordance with Article 4(3) and (4) of the Kyoto Protocol, the joint GHG emission and reduction limitation commitment of the EC and the Member States is fixed until 2012 and covers the EU15 that originally negotiated it, but excludes the new members after the enlargement of 1 May 2004 and 1 January 2007. Consequently, from an international and Community law perspective, the limits created by Article 4 of the Kyoto Protocol and its application to the joint commitment of the EC and the Member States are of significant relevance due to the EU enlargements of 2004 and 2007. In particular, the consequences for the EC and the Member States in the event of non-compliance with the Kyoto Protocol obligations are to be studied under international and European Community law.[48] Since the enlargement occurred after the adoption of Council Decision 2002/358/EC by the EC, the ratification and implementation of the Kyoto Protocol are part of the *acquis communautaire* and are therefore a compulsory obligation for the new Member States from a Community law viewpoint. Under Community law, international agreements ratified by the EC and the Member States form an integral part of the Community legal system on the basis of ex Article 300(7) TEC and of the jurisprudence of the ECJ.[49] The new Member States who acceded to the EU in 2004 and 2007 are therefore required to accept the obligations created by international treaties concluded by the Community, unless otherwise provided in the Treaty of Accession to the EU. This was specified under Article 6 of the Act of Accession of 2003.[50]

Finally, Article 4(5) and (6) of the Kyoto Protocol deals with the issue of responsibility and compliance by the REIO and its parties with the QELRCs.[51] In the event that the Community is not in compliance with the agreed GHG emission limitation and reduction commitments, the Member States are individually responsible under international law for the entire failure on the basis of their 'level of emissions' indicated in the BSA and therefore communicated to the UNFCCC. If, for instance, at the end of the first commitment period (2008–2012) the EC is not in compliance with the 8 per cent GHG emissions reduction obligation and Italy and Spain are in non-compliance with the level of QELRCs agreed under the BSA – a reduction of 6.5 per cent and an increase of 15 per cent by 2012, respectively, compared to 1990 – Article 4(5) considers these two countries in breach of the Kyoto Protocol GHG emissions reduction obligations and responsible for non-compliance under international law. The interpretation of Article 4(6) is more complex when considering the issue of responsibility for and consequences of non-compliance with the agreed levels of GHG emissions reductions. In particular, as regards the responsibility of parties for failure to achieve a total combined level of emission reductions, Article 4(6) refers to the individual responsibility of each Member State of the REIO and introduces this type of responsibility. The Member States are responsible for their own level of GHG emissions. In respect of the example mentioned above where Italy and Spain may be in a situation of non-compliance with their GHG emissions reduction targets, not only those countries are held liable under international law; the Community as Annex I party is also responsible for the non-compliance with its reduction commitments.

The final details of the EU BSA are shown in Table 3.1.

3.4   INTERNAL STRUCTURE AND CLIMATE CHANGE NEGOTIATIONS

As mentioned above, environmental protection and climate change are matters falling under the competence of both the Community and the Member States. It is therefore commonly accepted that in this case the responsibility for negotiating, concluding and implementing the UNFCCC and the Kyoto Protocol is the result of close cooperation between the Member States and the EU institutions, in particular the European Commission. The duty of close cooperation applied to the external relations of the Community was introduced by the ECJ in ruling 1/78[52] and opinion 2/91,[53] later reformulated in opinion 1/94[54] and in the FAO case.[55]

At least before the entry into force of the Treaty of Lisbon, which introduced several institutional changes such as the elimination of the six months rotating presidency of the Council, in the international climate regime the EU delegation is

Table 3.1 *Quantified emission limitation and reduction commitments for the purpose of determining the respective emission levels allocated to the European Community and its Member States in accordance with Article 4 of the Kyoto Protocol*

Party	Quantified emission reduction commitment under the Kyoto Protocol (percentage 2008–2012 compared to base year)
European Community	− 8
Austria	− 13
Belgium	− 7.5
Denmark	− 21
Finland	0
France	0
Germany	− 21
Greece	25
Ireland	13
Italy	− 6.5
Luxembourg	− 28
Netherlands	− 6
Portugal	27
Spain	15
Sweden	4
UK	− 12.5

formally represented by the so-called troika: a representative of the Commission and representatives of both the present and next EU presidencies. In reality, the negotiations on the different agenda items discussed at the COP and COP/MOP meetings are left to the experts of the European Commission and the Member States that represent the Community and all the Member States and act under the flag of the designated EU Presidency.[56] Behind the scenes, the principle of close cooperation is effectively applied through regular coordination meetings between the representatives of the Member States and the EU institutions aimed at discussing and seeking a common position on the different agenda items. The EU coordination meetings are held either in Brussels or at the location of the international talks in which the EC and the Member States participate.

The international position of the EU on climate change is adopted prior to any COP of the UNFCCC by the Council, which usually brings along the inputs and concerns of the European Commission and the European Parliament, as well as the Member States. Such a decision is taken by qualified majority voting in the Council. The responsibility for drafting the EU position to be presented at the international level falls under the competence of the EU presidency in cooperation with the Commission and the Parliament (consultative role). The draft position is scrutinized by the Council working group on climate change, COREPER I and, finally, the environment ministers of the Member States. The EU position for the various COP under the Convention are indicated below:

- EU position for COP3 in Kyoto (Japan) − 16 October 1997;[57]
- EU position for COP4 in Buenos Aires (Argentina) − 6 October 1998;[58]

- EU position for COP5 in Bonn (Germany) − 12 October 1999;[59]
- EU position for COP6 in The Hague (Netherlands) − 7 November 2000;[60]
- Community strategy on climate change − Council Conclusions − EU position for COP6bis in Bonn (Germany) − 6 June 2001;[61]
- EU position for COP7 in Marrakech (Morocco) − 29 October 2001;[62]
- COP7 − Marrakech EU Final Report;
- Follow-up to COP7 Marrakech − 12 December 2001;[63]
- EU position for COP8 in New Delhi (India) − 17 October 2002;
- European Commission welcomes outcome of climate change conference (COP8);
- EU position for COP9 in Milan (Italy) − 27 October 2003;
- EU position for COP10 in Buenos Aires (Argentina) − 3 December 2004;
- EU position for COP11 and COP/MOP1 in Montreal (Canada) − 17 October 2005;[64]
- EU position for COP12 and COP/MOP2 in Nairobi (Kenya) − 23 October 2006;[65]
- EU position for COP13 and COP/MOP3 in Bali (Indonesia) − 30 October 2007;[66]
- UNFCCC: preparation for the Poznan negotiations − Council conclusions − 20 October 2008;[67]
- Council Conclusions on EU position for the Copenhagen Climate Conference (7−18 December 2009).[68]

Since the beginning of the 1990s, so-called European climate diplomacy aimed at the curbing of global warming, has been characterized by the adoption of advanced secondary legislation to comply with the international obligations at the European level and by the definition of strategies and documents to lead and to influence the international debate on the climate regime.

The general keystones of the position of the EC in the international negotiations under the UNFCCC and the Kyoto Protocol are summarized as follows:

- Support for the environmental integrity of the international climate regime designed under the UNFCCC and the Kyoto Protocol;
- Definition of ambitious quantified emission limitation and reduction commitments for Annex I parties;
- Introduction of the supplementarity principle in the use of the flexible mechanisms established by the Kyoto Protocol (JI, CDM and IET);
- Safeguard and enhance the multilevel nature of the international climate regime;
- Promotion of sustainable development, especially in developing countries.

With regard to the Community legislative procedure, since the Treaty changes introduced by the Treaty of Maastricht and the Treaty of Amsterdam, the Council and the Parliament have seen their power increased under the co-decision

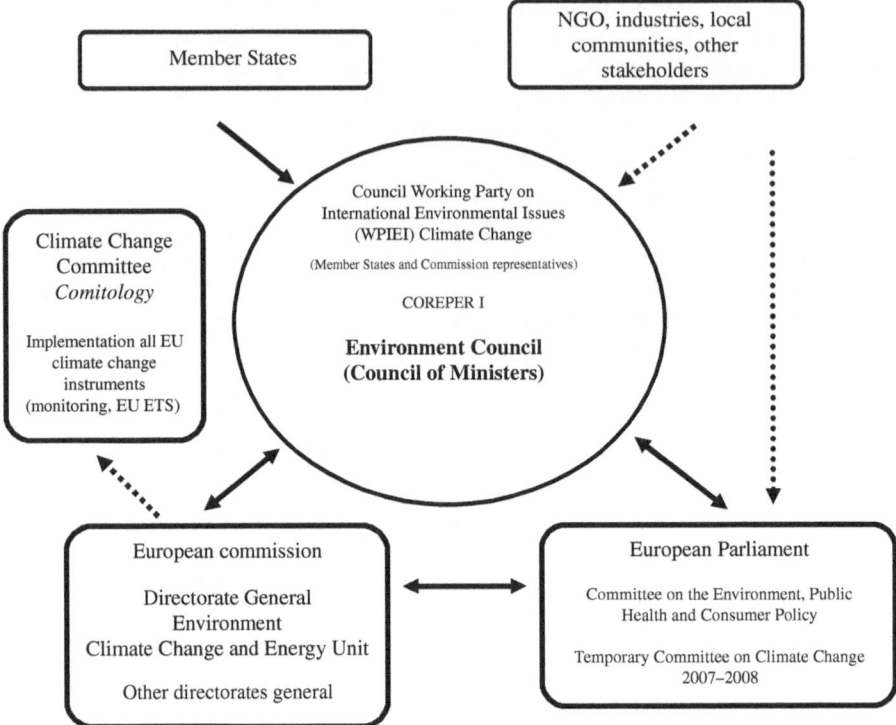

Figure 3.1 *The distribution of the responsibilities and tasks of European climate policy (pre-Lisbon Treaty)*

procedure (ex Article 251 TEC), which was the usual procedure required by ex Article 175 TEC for the definition of the Community environmental policy.

Figure 3.1 provides an idea of the distribution of responsibilities and tasks of climate policy – before the entry into force of the Treaty of Lisbon – within the European institutions.[69]

On 17 February 2010, after the entry into force of the Treaty of Lisbon, two new directorates general were created by the European Commission as part of the organizational reshuffle due to the allocation of portfolios to Commissioners. These are Directorate General Energy (ENER) and Directorate General Climate Action (CLIM). Directorate General Energy is composed of the departments of the former Transport and Energy Directorates dealing with energy issues and of the Task Force Energy transferred from the External Relations Directorate General. Directorate General Climate Action is composed of activities from Directorates of Environment, External Relations related to international negotiations on climate change, and Enterprise and Industry related to climate change.

NOTES

1. Communication from the Commission to the Council of 16 November 1988 on the Greenhouse Effect and the Community, establishing the Commission work programme concerning the evaluation of policy evaluation to deal with the green house effect and a draft resolution on the greenhouse effect and the Community, COM(1988)656.

2. Council Decision 93/389/EEC of 24 June 1993 for a monitoring mechanism of Community CO_2 and other GHG emissions, *OJ* L 167, 9 July 1993, pp31–33; further details about this decision and its consequent amendments are provided in Chapter 9.

3. Council Decision 93/389/EEC as amended by Decision 99/296/EC for a monitoring mechanism of Community CO_2 and other GHG emissions, *OJ* L 117, 5 May 1999.

4. Council Decision 94/69/EEC concerning the conclusion of the United Nations Framework Convention on Climate Change, *OJ* L 33, 7 February 1994.

5. Council Directive 96/62/EC on ambient air quality assessment and management, *OJ* L 296, 21 November 1996, pp55–63.

6. Commission Recommendations 1999/125/EC, 2000/303/EC and 2000/304/EC on the reduction of CO_2 emissions from passenger cars, *OJ* L 40, 13 February 1999, pp49–50, and *OJ* L 100, 20 April 2000, pp55–58, respectively.

7. Council Directive 1999/13/EC on the limitation of emissions of volatile organic compounds due to the use of organic solvents in certain activities and installations, *OJ* L 85, 29 March 1999, pp1–26.

8. Council Directive 1999/32/EC on the reduction of sulphur content of certain liquid fuels, *OJ* L 121, 11 May 1999, pp13–18.

9. Council Directive 2001/81/EC on national emissions ceilings for certain atmospheric pollutants, *OJ* L 309, 27 November 2001, pp22–30.

10. Council Directive 2002/3/EC relating to ozone in ambient air, *OJ* L 67, 9 March 2002, pp14–30.

11. Council Directive 2001/100/EC of the European Parliament and of the Council of 7 December 2001 amending Council Directive 70/220/EEC on the approximation of the laws of the Member States on measures to be taken against air pollution by emissions from motor vehicles, *OJ* L 16, 18 January 2002, pp32–34.

12. Council Directive 2001/1/EC of the European Parliament and of the Council of 22 January 2001 amending Council Directive 70/220/EEC concerning measures to be taken against air pollution by emissions from motor vehicles, *OJ* L 035, 6 February 2001, pp34–35.

13. Council Directive 2003/17/EC of the European Parliament and of the Council of 3 March 2003 amending Directive 98/70/EC relating to the quality of petrol and diesel fuels, *OJ* L 76, 22 March 2003, pp10–19.

14. Council Directive 96/61/EC concerning integrated pollution prevention and control (IPPC), *OJ* L 257, 10 October 1996, pp26–48.

15. Council Directive 2001/80/EC on the limitation of emissions of certain pollutants into the air from large combustion plants, *OJ* L 309, 27 November 2001, pp1–27.

16. Council Decision 94/69/EEC of 15 December 1993 concerning the conclusion of the United Nations Framework Convention on Climate Change, *OJ* L 33, 7 February 1994, p11.

17. For instance, the Convention on the Control of Transboundary Movements of Hazardous Waste and Their Disposal (Basel Convention), Articles 2(20) and 22(1).

18. Communication from the Commission under the United Nations Framework Convention on Climate Change of 11 June 1996, Brussels, dealing with measures taken and adopted in the period 1990–August 1995, COM (96) 217 final.

19. Second communication from the European Community under the United Nations Framework Convention on Climate Change of 26 June 1998 dealing with measures taken and adopted in the period September 1995 to October/November 1997, and updates the state of implementation of measures to address climate change that are or will be pursued at Community level, SEC(1998)1770.

20. Commission staff working paper on the Third Communication from the European Community under the United Nations Framework Convention on Climate Change of 30 November 2001, Brussels, providing an overview of climate change related policies and measures at European level since 1998, SEC(2001)2053.

21. Communication from the Commission on the Fourth National Communication from the European Community under the United Nations Framework Convention on Climate Change COM(2006)40, Brussels, 8 February 2006.

22. Commission communication to the Parliament and the Council, Climate Change – Towards an EU Post-Kyoto Strategy establishing a first analysis on the shaping of a post Kyoto strategy COM(1998)353, Brussels, 3 June 1998.

23. Communication from the Commission to the Council and the European Parliament – Preparing for implementation of the Kyoto Protocol COM(1999)230, Brussels, 19 May 1999.

24. Proposal for a Council Decision concerning the approval, on behalf of the European Community, of the Kyoto Protocol to the United Nations Framework Convention on Climate Change and the joint fulfilment of commitments thereunder submitted by the Commission COM(2001)579, Brussels, 23 October 2001.

25. 2413rd Council meeting on the environment of 4 March 2002, 6592/02 Presse 47 – G p4.

26. Johannesburg Summit 2002 – the World Summit on Sustainable Development of 26 August to 4 September 2002, which gathered heads of state and government, national delegates and leaders from NGOs, businesses and other major groups to focus the world's attention and direct action toward meeting difficult challenges concerning sustainable development.

27. Finally, the Kyoto Protocol did not enter into force by the Johannesburg Summit of 2002 because the procedure foreseen under Article 25 of the Kyoto Protocol was not fulfilled at that time.

28. Council Decision 2002/358/EC of 25 April 2002 concerning the approval, on behalf of the European Community, of the Kyoto Protocol to the United Nations Framework Convention on Climate Change and the joint fulfilment of commitments thereunder, OJ L 130 of 15 May 2002.

29. In this book, EU15 refers to the 15 Member States member of the EU at the moment of the ratification of the Kyoto Protocol by the European Community (2002); the term EU12 refers to the Member States who joined the EU on 1 May 2004 and 1 January 2007.

30. With the term EU27 we refer to Member States of the EU after the two enlargements of 1 May 2004 and 1 January 2007.

31. Recital 11, COM (2001)579, 'The subject matter of the Kyoto Protocol comes under the heading of Community environment policy. This proposal is based on Article 174(4) of the Treaty establishing the European Community, in conjunction with the first sentence of Article 300(2) and the first subparagraph of Article 300(3). Article 174(4) confers express competence on the Community to conclude the Kyoto Protocol, while Article 300 lays down the procedural requirements. The Commission's proposal is subject to approval by a qualified majority in the Council after consultation of the European Parliament.'

32. Opinion 2/00 on the Cartagena Protocol on Biosafety, 6 December 2001, *ECR* I-09713.

33. Articles 5 and 7 of Council Decision 1999/468/EC of 28 June 1999 laying down the procedures for the exercise of implementing powers conferred on the Commission, *OJ* L 184,17 July 1999, p23.

34. Commission Decision 2006/944/EC of 14 December 2006 determining the respective emission levels allocated to the Community and each of its Member States under the Kyoto Protocol pursuant to Council Decision 2002/358/EC.

35. See UNFCCC (2002), pp30–32.

36. See UNFCCC (2000), pp58–59.

37. 'Any Parties included in Annex I that have reached an agreement to fulfil their commitments under Article 3 jointly, shall be deemed to have met those commitments provided that their total combined aggregate anthropogenic carbon dioxide equivalent emissions of the greenhouse gases listed in Annex A do not exceed their assigned amounts calculated pursuant to their quantified emission limitation and reduction commitments inscribed in Annex B and in accordance with the provisions of Article 3. The respective emission level allocated to each of the Parties to the agreement shall be set out in that agreement', Article 4(1) of the Kyoto Protocol.

38. Press release 9402/98 of 19 June 1998 of the Council of the EU, Conclusions of the community strategy on climate change, Council Environment 16–17 June 1998, establishing Member States' commitments in accordance with Article 4 of the Kyoto Protocol, Appendix I.

39. See note 28 above.

40. See note 38 above.

41. See Haigh (1996).

42. Council Directive 88/609/EEC of 24 November 1988 on the limitation of emissions of certain pollutants into the air from large combustion plants, *OJ* L 336, 7 December 1988, pp1–13.

43. The Triptych approach considered the differentiation of commitments within the EU Bubble and was developed in the Netherlands during the EU Dutch Presidency (January–June 1997), see Grönenberg (2002).

44. Torvanger and Ringius (2000).

45. Article 4(3) and (4) can be considered as 'blocking clause', see Massai (2010a).

46. 'Any such agreement shall remain in operation for the duration of the commitment period specified in Article 3, paragraph 7', Article 4(3) of the Kyoto Protocol.

47. 'If Parties acting jointly do so in the framework of, and together with, a regional economic integration organization, any alteration in the composition of the organization after adoption of this Protocol shall not affect existing commitments under this Protocol. Any alteration in the composition of the organization shall only apply for the purposes of those commitments under Article 3 that are adopted subsequent to that alteration', Article 4(4) of the Kyoto Protocol.

48. For an extensive consideration of the legal implications of the EU enlargements of 2004 and 2007 on the joint commitment of the European Community and the Member States under the Kyoto Protocol, see Massai (2010a).

49. 'The provisions of the agreement, from the coming into force thereof, form an integral part of Community law', ECJ, case 181/73 *Haegeman v. Belgian State* [1974] *ECR* 449.

50. Act concerning the conditions of accession of the Czech Republic, the Republic of Estonia, the Republic of Cyprus, the Republic of Latvia, the Republic of Lithuania, the Republic of Hungary, the Republic of Malta, the Republic of Poland, the Republic of Slovenia and the Slovak Republic and the adjustments to the Treaties on which the European Union is founded, *OJ* L 236, 23 September 2003, pp33–50.

51. 'In the event of failure by the Parties to such an agreement to achieve their total combined level of emission reductions, each Party to that agreement shall be responsible for its own level of emissions set out in the agreement', Article 4(5) of the Kyoto Protocol. 'If Parties acting jointly do so in the framework of, and together with, a regional economic integration organization which is itself a Party to this Protocol, each member State of that regional economic integration organization individually, and together with the regional economic integration organization acting in accordance with Article 24, shall, in the event of failure to achieve the total combined level of emission reductions, be responsible for its level of emissions as notified in accordance with this Article', Article 4(6) of the Kyoto Protocol.

52. ECJ, ruling 1/78 (Draft Convention on the Physical Protection of Nuclear Materials) [1978] *ECR* 2151, paras. 34–36.

53. ECJ, opinion 2/91 (ILO Convention) [1993] *ECR* I-1061, para. 36.

54. ECJ, opinion 1/94 (WTO Agreement) [1994] *ECR* I-5267, para. 19.

55. ECJ, case C-25/94, *FAO* [1996] *ECR* I-1469, paras. 48–49.

56. For instance, in 2009, the climate change negotiations on the adoption of the post-2012 regime were conducted under the flag of the Czech Republic in the first semester and under the flag of Sweden in the second semester.

57. Council conclusions of 3 March, 19 June and 16 October 1997 on community strategy on climate change constituting the Community's negotiating mandate for the Third Conference of the Contracting Parties to the UN Framework Convention on Climate Change to take place in Kyoto in December 1997.

58. Council conclusions of 6 October 1998 on community strategy on climate change under the Convention and under the Kyoto Protocol at the Fourth Conference of the Parties (COP4) to the Framework Convention on Climate Change (UNFCCC) to take place in Buenos Aires in November 1998.

59. Council conclusions of the 2207th Council meeting of 12 October 1999 Luxembourg on community strategy on climate change, establishing commitment to making substantial progress at the fifth session of the Conference of the Parties (COP5) to the United Nations Framework Convention on Climate Change (UNFCCC), in order to implement the Buenos Aires Plan of Action at the sixth session of the Conference of the Parties (COP6).

60. Council conclusions of the 2302nd Council meeting on 7 November 2000, Brussels, on community strategy on climate change, focusing on the essential questions to be raised at the 6th Conference of Parties to the United Nations Framework Convention on Climate Change (COP6) in The Hague, 13–24 November 2000.

61. Press release of the 2355th Council Meeting Environment of 7 June 2001 Luxembourg.

62. Council conclusions of the 2378th Council meeting on 29 October 2001, Luxembourg, on community strategy on climate change with a view to the 7th Conference of the Parties (COP7) to the United Nations Framework Convention on Climate Change (UNFCCC) to be held in Marrakech from 29 October to 9 November 2001.

63. 2399th Council meeting Environment, Brussels, 12 December 2001.

64. Press release on the 2684th Council Meeting of 17 October 2005 Luxembourg, establishing council conclusions on climate change with the decision to place climate change high on the agenda of the summit with Canada, to prepare the 11th Conference of the Parties to the United Nations Framework Convention on Climate Change (COP11) and the first Conference of the Parties serving as the meeting of the Parties to the Kyoto Protocol (COP/MOP1) in Montreal in December 2005.

65. Press release on the 2757th Council Meeting of 23 October 2006 Luxembourg, establishing council conclusions on climate change with view to the 12th Conference of the Parties to the United Nations Framework Convention on Climate Change (COP12) and the

2nd Conference of the Parties serving as the meeting of the Parties to the Kyoto Protocol (COP/MOP2) in sub-Saharan Africa.

66. Press release on the 2826th Council Meeting of 30 October 2007 Luxembourg, establishing council conclusions reflecting the EU's position in preparation for the 13th session of the Conference of the Parties to the United Nations Framework Convention on Climate Change (COP13) to be held in Bali in December with the aim of launching the negotiation process for a global and comprehensive post-Kyoto agreement.

67. Press release on the 2398th Council Meeting Environment of 20 October 2008, Luxembourg.

68. Council conclusions on EU position for the Copenhagen Climate Conference (7–18 December 2009), the Council of the EU, 2968th Environment Council meeting, Luxembourg, 21 October 2009.

69. On the issue, see van Schaik and Egenhofer (2003).

CHAPTER 4

The Environment Council

*T*he Environment Council is composed of all the environment ministers of the Member States and is convened about four times a year. Decisions are adopted by qualified majority in co-decision with the European Parliament. The Environment Council deals with all issues related with the protection of the environment in Europe, as well as with fostering the harmonious, balanced and sustainable development of economic activities in respect of a high level of environmental quality. The action of the Environment Council in these fields is based, amongst other issues, on the precautionary principle and preventive action, as well as the stance that environmental damage should as a priority be rectified at source and that the polluter should pay.

This chapter includes references to the latest meetings of the Environment Council with specific information on decisions adopted concerning climate change, clean energy and related issues:

- EU Council Environment of 21 December 2004, Brussels, confirmed the EU commitment towards future additional cuts of the GHG emissions of the Community and the Member States.
- EU Council Environment of 10 March 2005, Brussels, adopted a resolution on new global commitments for GHG reductions in the post-Kyoto phase.
- EU Council Environment of 24 June 2005, Luxembourg.
- Informal meeting of environment and agriculture ministers of 9–12 September 2005 (UK presidency) discussed the 'significant challenges and opportunities that climate change presents to European agriculture' and 'examine agriculture's role in reducing greenhouse emissions'.
- EU Council Environment of 17 October 2005, Brussels – Background note and EU Council Environment of 17 October 2005 – results: preparation for

COP11 and development of the EU strategy on the post-2012 climate change talks.

- EU Council Environment of 2 December 2005, Brussels, strong support to the Commission plans to include the aviation sector in the European GHG emission trading system.
- EU Council Environment of 9 March 2006, Brussels, supported the EURO 5 regulation proposal of the Commission on CO_2 emissions from passenger cars and agreed on the directive on end-use energy efficiency and energy services.
- EU Council Environment of 27 June 2006, Luxembourg, agreed on new air quality limits within the framework of the air quality framework directive (Directive 96/62/EC).
- EU Council Environment of 23 October 2006, Luxembourg, agreed on the air quality legislation, expressed the first concerns over the second national allocation plans presented by the Member States within the framework of the EU ETS and adopted the EU positions for COP12 and COP/MOP2.
- EU Council Environment of 18 December 2006, Brussels, welcomed the progress of the Nairobi summit and supported a global commitment of a reduction of GHG emissions of 30 per cent by 2020.
- EU Council Environment of 20 February 2007, Brussels, confirmed the European position on the post-2012 targets (20 per cent cut of GHG emissions by 2020 with the option to move to 30 per cent provided that other industrialized countries agree on similar commitments) and decided almost unanimously to include aviation in the EU emissions trading system, as well as to support binding legislation to reduce CO_2 for new cars and vans.
- European Council of 8–9 March 2007, Brussels, Presidency Conclusions, agreed on binding targets to reduce GHG emissions by 20 per cent in 2020 compared to 1990 levels and to increase the share of electricity from renewable sources of 20 per cent by 2020.
- EU Council Environment of 30 October 2007, Luxembourg, adopted the common position for COP13 in Bali in December 2007.
- EU Council Environment of 20 December 2007, Brussels, reached political agreement on the proposal to include the aviation sector in the EU ETS as well as progress in the discussion over the revision of the EU fuel quality directive.
- EU Council Environment of 3 March 2008, Brussels, discussed the integrated climate and energy package and the reduction of CO_2 from passenger cars.
- EU Council Environment of 5 June 2008, Luxembourg, discussed the 2008 package on integrated climate and energy measures and the proposal for the setting of emission performance standards for new passenger cars. On the climate and energy package, disagreement was noted in respect of the base year for the calculation of the Member States' obligations to reduce GHG emissions, on flexibility in burden sharing legislation (Sweden) and on the trading of emissions credits between EU ETS and non-EU ETS sectors (finally rejected).
- EU Council Environment of 20 October 2008, Luxembourg, discussed the integrated climate and energy package but no consensus was reached on several key issues (disagreement on carbon leakage and industrial sectors at risk and on

a solidarity mechanism recognizing the reductions of GHG emissions achieved in the Eastern European countries since 1990). The Council adopted conclusions highlighting the EU position for negotiations under COP14 and COP/MOP4 in Poznan in December 2008 calling for, among others things, significant efforts to be undertaken by developing countries.

- European Council of 11 and 12 December 2008 Brussels, Presidency Conclusions, agreed on the European Economic Recovery Plan and on the integrated climate and energy package that should contribute to the compliance by the EU with the climate and energy targets agreed in 2007 and 2008 (20 + 20 + 20).
- 2928th EU Council Environment of 2 March 2009, Brussels.
- 2953rd EU Council Environment of 25 June 2009, Luxembourg.
- 2968th EU Council Environment of 21 October 2009, Luxembourg.
- 2977th EU Council Environment of 23 November 2009, Brussels.
- 2988th EU Council Environment of 22 December 2009, Brussels.
- 3002nd EU Council Environment of 15 March 2010, Brussels.
- 3021st EU Council Environment of 11 June 2010, Luxembourg.

CHAPTER 5

Clean Energy

E nergy policy is one of the main pillars of the European climate and clean energy strategy. Since the early 1990s, European institutions and Member States have based EU policy in this area on the promotion of renewable energy and energy efficiency as well as on the introduction of European-wide energy taxation. Unfortunately, one of the main obstacles for the adoption of advanced regulations on energy is the text of the EC Treaty, recently consolidated under the TFEU: while the co-decision procedure (Article 294 TFEU) is the basis upon which 'action is to be taken by the Community in order to achieve the objectives referred to in Article 191' (Article 192(1) TFEU), 'provisions primarily of a fiscal nature' as well as 'measures significantly affecting a Member State's choice between different energy sources and the general structure of its energy supply' shall be adopted by 'the Council acting unanimously', in derogation of the co-decision procedure mentioned above (Article 192(2) TFEU).

Unfortunately, neither the amendments to the EC Treaty included in the Treaty of Amsterdam (1997) or the Treaty of Nice (2001), nor the Constitution for Europe[1] (signed in June 2004 but never entered into force) or the Lisbon Treaty (TFEU), have provided changes to this paragraph. Under these terms, it remains very challenging for the EC to adopt legislation on energy issues, especially following the last EU enlargement and the increase of the number of Member States in the Council.

The introduction of a new title on energy in the Treaty, as provided for by the Treaty of Lisbon, has produced no substantial changes to the previous situation. In this area, the lack of an explicit legal basis in the EC Treaty regarding the adoption of legislation in the field of energy and environment has not limited the Community's activity in this area. The EU institutions, in close cooperation with the Member States, have adopted significant legislation in the area, and this is due in part to the use of Treaty articles covering other areas of Community law such as environmental protection[2] and taxation[3] and, in part due to the doctrine of implied powers, that is to say an extensive interpretation of the EC Treaty, in particular ex Articles 95 and 308. An example is provided by the explanatory memorandum attached to the Proposal for a Directive on the promotion of the use of energy from renewable sources COM(2008)19, which led to the adoption of Directive 2009/28/EC on the promotion of the use of energy from renewable sources and amending and subsequently repealing Directives 2001/77/EC and 2003/30/EC.[4] This Directive was adopted by the European Parliament and the Council with regard to ex Articles 175(1) and 95 of the EC Treaty. The explanatory memorandum refers to the shared competence between the Community and the Member States in the energy field by highlighting the need for concerted action in order to achieve an effective green energy policy and an equal distribution of efforts among the Community.[5]

The new title on energy policy is introduced by the Treaty of Lisbon[6] (Title XXI, Article 194 TFEU) and this is the codification of current practice and procedures, being the only innovation provided by the incorporation of energy efficiency and renewable forms of energy in the objectives of EU energy policy (Article 194(1)c TFEU). Before Lisbon EC energy law is adopted on the basis of internal market provisions (ex Article 95 TEC), competition provisions (ex Articles 81–88 TEC) and environmental protection (ex Article 175 TEC).

Article 194 TFEU clearly gives EU energy policy a strong environmental angle, although this is restricted within the framework of the internal market (free movement of goods and fair competition and therefore maybe only referring to imports and exports of energy). Article 194(1) reiterates the connection between the internal market and environmental protection and refers to energy as shared competence (solidarity between the EU and the Member States).[7]

Article 194(2) provides for the legal basis for adoption of legislation and confirms what is indicated above. The ordinary legislative procedure (former co-decision procedure) is the default rule. Unanimity is required in the cases indicated by Article 192(2)c, namely in the field of exploiting energy resources, choices between different sources of energy and general structure of energy supply. A special legislative procedure is introduced for measures primarily of a fiscal nature.[8]

Before examining the different aspects of EU energy policy aimed at the protection of the environment, we should briefly consider the general approach of the EU towards a sustainable and environmentally friendly consideration of energy needs in Europe.

First, the integration of environmental protection concerns into Community energy policy was the direct consequence of several amendments to the EC Treaty, in particular the reference to the integration principle under Article 6 TEC after the adoption of the Treaty of Amsterdam. To this end, the Commission released Communication of 14 October 1998: Strengthening environmental integration within Community energy policy COM(1998)571[9] directed to the incorporation of environmental protection concerns into the definition of other EU policies such as transport, energy and industrial policy.

As pointed out in Commission Communication COM(1997)167 An overall view of energy policy and actions[10] released on 23 April 1997, the production, transport and use of energy have a considerable impact on the environment. Therefore, one of the main challenges facing Community energy policy was the incorporation of the environmental dimension into its objectives and actions while developing a sustainable energy policy. Communication COM(1997)167 proposed several specific measures aimed at the development of a sustainable energy policy.

Since the beginning of 2000, the focus of the European institutions on energy-related issues moved even further towards the general principle of sustainable development. To this aim, the EU adopted in May 2001 a strategy for the promotion of sustainable development, namely the Commission's proposal to the Gothenburg European Council, Communication from the Commission of 15 May 2001, A sustainable Europe for a better world: a European Union strategy for sustainable development COM(2001)264.[11] Such a strategy designed the long-term perspective of EU policies, including energy policy, on the basis of economic, social and sustainable development parameters.

The contribution of the EU to global sustainable development was stressed by the Commission with Communication COM(2002)82, Towards a global partnership for sustainable development.[12] The EU strategy on sustainable development of 2001 was updated in 2005 by the Commission Communication of 13 December 2005 on the review of the Sustainable Development Strategy, A platform for action COM(2005)658.[13]

On 8 March 2006 the European Commission released the Green Paper, A European strategy for sustainable, competitive and secure energy COM(2006)105.[14] The Green Paper aimed at the development of a more coordinated EU policy on energy-related issues, although a few energy issues were left to the Member States' competence. The Green Paper called for improvements in energy efficiency, renewable energy and CCS and it aimed at achieving more than half of the EU energy production from low carbon-intensive sources by 2025.

The public hearing on the Strategic EU Energy Review[15] of 22 September 2006 gathered together several stakeholders to discuss European energy policy. Attention primarily fell on the need to focus on energy efficiency and renewable energy. Other issues considered were subsidiarity, the role of nuclear power and adaptation to environmental disasters.

On 14 December 2006 the European Parliament adopted a non-legislative resolution[16] responding to the Commission energy strategy proposal. The resolution called for the increase of carbon-free sources to cover energy demands by 2050, such as energy saving, efficiency and renewable energies.

On 10 January 2007 the European Commission presented the Communication, an energy policy for Europe,[17] which set the new European strategy on energy policy and was based, in part, on the EU Energy Policy Data.[18] The package included:

- Renewable Energy Road Map;[19]
- Progress in Renewable Electricity;[20]

- Progress in Biofuels;[21]
- Internal Market Gas and Electricity;[22]
- Gas and Electricity Infrastructure;[23]
- Nuclear Energy;
- Sustainable Power Generation from Fossil Fuels;[24]
- Strategic Energy Technology Plan.[25]

The main points of the Energy Policy for Europe were:

- The reduction of GHG emissions by 20 per cent by 2020 in order to ensure a limit to global warming to no more than 2°C above pre-industrial temperatures;
- The promotion of CCS;
- The improvement of energy efficiency in vehicles, appliances, homes and factories;
- The establishment of a minimum target of 20 per cent of renewable energy share;
- The investment of €1 billion over the following six years for research into renewable energies;
- A focus on nuclear power;
- The improvement of the European energy network connections.

On 19 September 2007, the European Commission released the 3rd package of measures in the energy sector[26] aiming at a further liberalization of the energy markets in the EU. According to the Commission, the new legislative measures will provide new space for investments in renewable energy and energy efficiency, as well as liberalization contributing to the effectiveness of the EU emissions trading scheme (ETS).

The package included the following proposals:

- Proposal for a Directive amending Directive 2003/54/EC concerning common rules for the internal market in electricity;[27]
- Proposal for a Regulation establishing an Agency for the Cooperation of Energy Regulators COM(2007)530;[28]
- Proposal for a regulation amending Regulation (EC) 1228/2003 on the conditions for access to the network for cross-border exchanges in the electricity COM(2007)531;[29]
- Proposal for a Directive amending Directive 2003/55/EC concerning common rules for the internal market in natural gas COM(2007)529;[30]
- Regulation amending Regulation (EC) 1775/2005 on conditions for access to the natural gas transmission networks COM(2007)532.[31]

On 22 November 2007 the European Commission released Communication COM(2007)723 on the European Strategic Energy Technology Plan (SET-PLAN)[32] aimed at the enhancement and development of cost-effective low-carbon technologies in Europe. Furthermore, the SET-PLAN aimed also at the reduction of cost of clean energy through the development of new technologies.

On 23 January 2008 the European Commission presented the EU climate and energy package including, amongst others, the following provisions directly concerning the issue of clean energy:

- Proposal for a Directive on the promotion of the use of energy from renewable sources COM(2008)19;[33]
- Communication on a first assessment of national energy efficiency action plans as required by Directive 2006/32/EC on energy end-use efficiency and energy services COM(2008)11;[34]
- Communication on supporting early demonstration of sustainable power generation from fossil fuels COM(2008)13 (CO_2 CCS).[35]

The EU Portuguese presidency revealed on 3 December 2007 a Vision Paper for EU Strategic Energy Technology Plan[36] indicating the road to achieve the EU objective to reduce GHG emissions by 60–80 per cent by 2050. The plan called for a 40 per cent reduction of CO_2 emissions in the transport sector and stronger energy efficiency targets, especially in the housing and power sectors.

On 28 February 2008 the EU energy ministers agreed on a resolution[37] approving the SET-PLAN with the removal of the reference to project support for research on nuclear fission. Divisions among the Member States remained on the further liberalization of the EU energy markets.

On 13 November 2008 the European Commission presented the second Strategic Energy Review package titled 'An EU energy security and solidarity action plan' COM(2008)781[38] aimed at increasing energy security and supporting the 20-20-20 climate package (Memo).[39] The energy review package aimed at a 15 per cent reduction in oil use by 2020 and identified five key areas for the promotion of sustainable energy supply (infrastructure, external relations, oil and gas stocks, energy efficiency, indigenous energy resources). Furthermore, the package:

- highlighted a new strategy to stimulate investments in low-carbon energy networks;
- included an Energy Security and Solidarity Action Plan looking at the period 2020–2050;
- proposed key legislation on energy efficiency of buildings and energy-using products.

According to the EEA report of June 2008, Energy and Environment 2008[40] released on 21 November 2008 and investigating the impact on the environment of energy consumption in the EU, the increase of primary energy consumption in Europe up to 2030 could only be reversed by 'more stringent climate and energy policies'. According to the report, the business-as-usual scenario would bring a rise in energy consumption of 26 per cent by 2030. Measures identified in the report were the enhancement of energy efficiency and the use of renewable energy.

On 24 March 2009 a compromise deal[41] was reached between the European Parliament and the Council on a set of legislation on the further liberalization of the internal energy market. The focus of the package is on the separation of production from the supply phase. The package provides for three options to reach this goal: (1) unbundling of energy companies; (2) involvement of an independent system operator (ISO); and (3) involvement of an independent transmission operator (ITO). The new legislation would strengthen the rights of consumers and would enhance energy efficiency, renewable energy and the implementation of the EU ETS. Finally, the new law would give the Member States the option to require distribution network operators to prioritize grid access to renewable energy, waste-to-energy and combined heat and power (CHP) installations. The deal was endorsed by the European Parliament on 22 April 2009 when a series of amendments to the Council common position on the Proposal for a Directive concerning common rules for the internal market in electricity and repealing Directive 2003/54/EC was adopted.

On 13 July 2009 Regulation (EC) No 663/2009 establishing a programme to aid economic recovery by granting Community financial assistance to projects in the field of energy was adopted by the European Parliament and the Council.[42] The European Energy Programme for Recovery (EEPR) aims at increasing investment in infrastructure and technology projects in the energy sector in Europe, improving the security of supply of the Member States and contributing to the implementation of the 20-20-20 objectives for 2020.

On 16 July 2009 the European Commission presented the following two legislative proposals to improve security of gas supplies in the framework of the internal gas market: Proposal for a Council Regulation concerning the notification to the Commission of investment projects in energy infrastructure within the European Community and repealing Regulation (EC) No 736/96, COM(2009)361;[43] and Proposal for a regulation concerning measures to safeguard security of gas supply and repealing Directive 2004/67/EC, COM(2009)363.[44]

The main goal of the two proposals is the strengthening of the existing system and the promotion of preventive measures to be adopted by the Member States and their gas market participants to face the consequences of potential disruptions to gas supplies. In particular, Proposal COM(2009)363 provides the players in the energy market, namely gas suppliers and transmission system operators, with the right to deal with the disruption for as long as possible before any state intervention is taken. Furthermore, this proposal promotes coordination at the regional and EU level to ensure the right measures are adopted by the competent authorities.

Proposal COM(2009)361 ensures more transparency on the evolution of energy infrastructure in the main energy sectors such as oil (including biofuels), electricity and gas, but also in related areas such as the transport and storage of carbon related to energy production. Increased transparency in planned and ongoing investment projects would reduce the risks of infrastructure gaps in the future and would provide more stability to potential investors.

On 27 April 2010 the European Commission presented to the Council and the European Parliament the Report on the implementation of the European Energy

Programme for Recovery COM(2010)191.[45] The Annex to the report provided detailed information on the major energy projects under the EEPR, which should contribute to the economic recovery in the EU by increasing the security of energy supply and establishing cross-border infrastructure. The Report showed that almost the entire EEPR financial envelope (€3.98 billion) will be committed in the course of spring 2010. The Report indicated that the amount of uncommitted funds will be known by the end of 2010 and noted that a few projects may fail the eligibility test to get funding.

5.1   ENERGY TAXATION

The impossibility of finding common ground between the European institutions and the Member States on the introduction of any kind of taxation or fiscal regulation related to energy products is not only due to the limitation of the EC Treaty provisions indicated above, but also due to the Member States having shown on many occasions their inability to find common agreement on these themes.[46] The first big failure in the Council in the negotiations for the establishment of a European-wide energy tax dates to 1994, when Member States were unable to find a compromise on the Communication of the Commission for an amended proposal for a Council Directive on excise duties on motor fuels from agricultural sources COM(1994)147 (obsolete).[47]

Council Directive 95/60/EC of 27 November 1995 on fiscal marking of gas oils and kerosene was adopted with the view to prevent tax evasion and contribute to the correct functioning of the internal market.[48] To this end, the directive established a common marking system to identify gas oil and kerosene in Europe. Directive 95/60/EC was followed by Commission Decision 2006/428/EC of 22 June 2006 repealing Commission Decision 2001/574/EC and establishing the solvent Yellow 124 as the common fiscal marker for gas oils and kerosene and fixing the marking level of at least 6mg and not more than 9mg of marker per litre of mineral oil.[49]

On 6 May 1997, the European Commission released a proposal for a Council Directive restructuring the Community framework for the taxation of energy products COM(1997)30[50] pursuing the following three main objectives:

• Restructuring of the European system for the taxation of energy products;
• Fulfilling of the legislative gap for energy products taxation, so far applied only to mineral oils as indicated in Council Directive 92/12/EEC of 25 February 1992 on the general arrangements for products subject to excise duty and on the holding, movement and monitoring of such products;[51]
• Promoting energy efficiency through the reduction of energy imports and CO_2 emissions.

Communication from the Commission on Tax policy in the European Union – Priorities for the years ahead COM(2001)260 of 23 May 2001 contained a chapter on 'Energy and environmental taxation', where the system of environmental taxation was addressed.[52] According to this communication, EU energy taxation in 2001 was building three main pillars: '(i) excise duties, (ii) VAT (value added tax) and (iii) specific levies'. In particular, at that time in the EU there was a common system of taxation as far as excise duties on mineral oils and VAT were concerned while this was not applicable to energy products other than mineral oils.

On 20 March 2003, the EU Ministers of Economics and Finance gave their political agreement to the proposed Directive on a Community framework for the taxation of energy products. Consequentially, on 27 October 2003, after many years of negotiations, the Council adopted Directive 2003/96/EC restructuring the Community framework for the taxation of energy products and electricity,[53] which entered into force on 1 January 2004. Directive 2003/96/EC modified the European minimum rate system for energy products, previously covering mineral oils, coal, natural gas and electricity. In particular, the main goal of the new system introduced by Directive 2003/96/EC was the reduction of distortion of competition in the Community energy market, specifically through the harmonization of the rates of tax on energy products, as well as tax legislation on mineral oils and other energy products (previously not covered by Community legislation). Furthermore, Directive 2003/96/EC contributed to the increase of incentives for energy efficiency and to allow Member States to grant tax incentives to companies investing in the reduction of emissions.

Directive 2003/96/EC contained several amendments to the original Commission proposal mentioned above. In particular, the possibility for Member States 'to apply under fiscal control total or partial exemptions or reductions in the level of taxation to' a long list of different electricity and energy products (Article 15), as well as transitional arrangements for several sectors and Member States. On this issue, on 1 May 2004 Council Directive 2004/74/EC and Council Directive 2004/75/EC were adopted and introduced transitional arrangements for the new Member States.[54]

On 28 January 2004 the European Commission presented a Proposal for a Council Directive amending Directive 2003/96/EC as regards the possibility for certain Member States to apply, in respect of energy products and electricity, temporary exemptions or reductions in the levels of taxation COM(2004)42.[55] The proposal referred to those States who acceded the EU in 2004 and their right to obtain transitional arrangements in terms of temporary derogation with regard to specific pieces of legislation was explicitly stated under Article 55, Title II, Part 5 of the Treaty of Accession.

On 30 June 2006 the European Commission adopted Communication COM(2006)342 Review of the derogations in Annexes II and III of Council Directive 2003/96/EC that expire by the end of 2006 establishing minimum tax rates on all energy products and introducing an expiring date by the end of 2006.[56] As a follow-up of a review undertaken by the Commission over the 111 derogations to the 2003 energy tax directive, the Communication announced that almost all exemptions should not be renewed by the end of 2006. Among the few exceptions to be renewed were the tax exemptions for local public transport and renewable energies.

On 5 October 2006 the ECJ affirmed in Case C-368/04[57] that the permission of the European Commission is required in the event the EU Member States want to grant national firms rebates on energy taxes. The Court considered unlawful the state aid scheme introduced by the Austrian government in 1996 to provide firms with tax reductions on electricity and natural gas consumption.

On 13 March 2007 the European Commission released Proposal for a Council Directive amending Directive 2003/96/EC as regards the adjustment of special tax arrangements for gas oil used as motor fuel for commercial purposes and the coordination of taxation of unleaded petrol and gas oil used as motor fuel COM(2007)52.[58]

On 28 March 2007 the European Commission released a Green Paper on market-based instruments for environment and related policy purposes COM(2007)140.[59] The Commission paper promotes a greater use of the so-called market-based instruments such as green taxation for the achievement of the EU climate and energy goals.

On 10 December 2009 Ireland approved a tax on CO_2 emissions for sectors not covered by the EU ETS. The tax will apply to transport fuels such as petrol and diesel, as well as for non-transport fuels such as kerosene, marketed gas oil, fuel oil, liquid petroleum gas (LPG) and natural gas. It is still unclear whether or not the measure will apply to solid fuels such as coal.

On 4 January 2010 the French Constitutional Court (Conseil Constitutionnel) rejected the proposal of the French government to introduce a carbon tax on fossil fuel use. The High Court considered the proposed measure as inefficient and unfair due to the fact that 93 per cent of emissions from industrial installations would have been exempted from the tax since they were already covered by the EU ETS. Also agriculture, road hauliers and maritime transport sectors would have been exempted. Consequently, the levy would have been applied mostly to the household sector. According to the Court the tax would have not contributed to the fight against climate change and furthermore would have breached the principle of tax equality.

On 12 May 2010 opposition to the French idea to introduce a CO_2 border tax emerged from several sources inside the European Commission. An early version of a policy paper of the European Commission on the issue of moving to a 30 per cent CO_2 emission reduction target referred to carbon leakage and in particular to the carbon border tax. The policy paper described that such a tax 'could at best only be envisaged for a very limited number of standardised commodities, such as steel or cement'. Furthermore, diverging views on the issue were registered among the different commissioners. In support of the introduction of a carbon border tax, the WTO/UNEP report 'Trade and Climate Change' released on 26 June 2009 affirmed that such a tax would be feasible from a technical point of view if calculated per production process instead of per single product.[60]

On 23 June 2010 the EU Commissioner for Taxation, Customs, Anti-fraud and Audit Algirdas Šemeta announced a review of the Impact Assessment on the Proposal for a Council Directive amending Directive 2003/96/EC COM(2007)52.[61] The review should consider the introduction of a duty on CO_2 emissions in energy taxes consisting in a minimum flat rate of €0.02 per

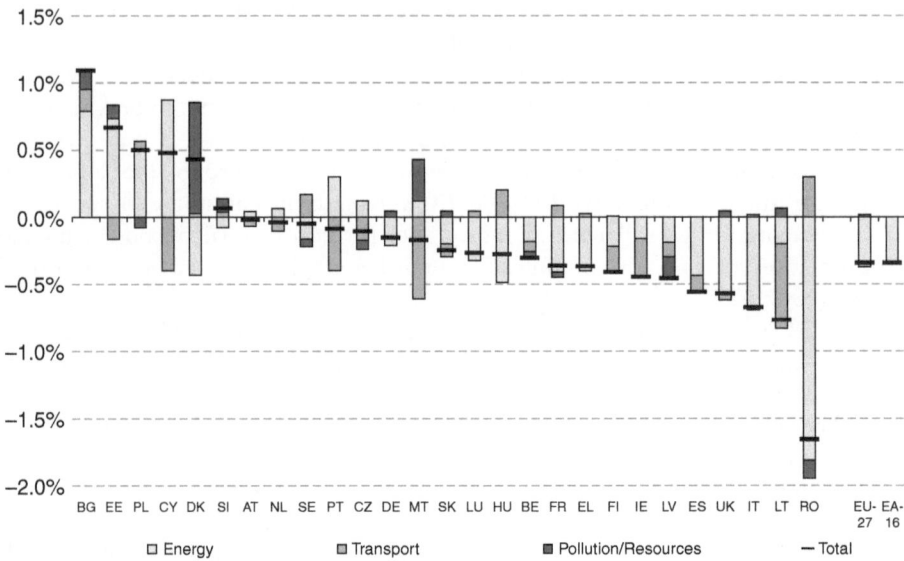

Figure 5.1 *The evolution of the structure of environmental taxes in the period 2000–2008 (percentage of gross domestic product)*
Source: Eurostat (2010)

kilogram of CO_2 (€20 per tonne). Concerns have been raised inside the European Commission on the impact of such a tax on sectors such as agriculture and on households and whether derogations for these sectors could be envisaged.

The report 'Taxation trends in the European Union' released by Eurostat on 28 June 2010 indicates that the level of environmental taxes in the EU decreased in 2008 to 2.4 per cent (from 2.9 per cent in 1999) as a percentage of the overall taxation. The reason behind such a reduction is mainly lower energy use. Among the Member States, Denmark has the highest level of energy taxes (€286 per tonne of oil equivalent), while Romania the lowest charge at €26.

Figure 5.1 depicts the evolution of the structure of environmental taxes in the period 2000–2008.

5.2 SAVE, ALTENER AND RESEARCH

5.2.1 SAVE

The basic foundations of the European programme SAVE, the only European-wide programme aiming at the promotion of energy efficiency and energy saving practices in several different sectors, are:

- Improvement of energy efficiency in all sectors;
- Rational use of energy, in particular in the building and industry sectors, but also in commerce and the domestic sector as well as transport;

- Stabilization of GHG emissions;
- Production of information, studies and pilot actions, as well as the establishment of local and regional energy management agencies.

The SAVE programme required Member States to revise their national energy strategies and was divided in two phases:

- 1st phase 1991–1995, Council Decision 91/565/EEC of 29 October 1991 concerning the promotion of energy efficiency in the Community (SAVE programme):[62]
  - Improvement and integration of existing national energy programmes with new technologies;
  - No quantified objectives;
  - Improvement of energy efficiency and cost efficiency;
  - Improvement of technical infrastructure and environmental impact;
  - Pilot projects to foster energy efficiency in different sectors;
  - Improvement of energy consumption monitoring.
- 2nd phase 1996–2000, Council Decision 96/737/EC of 16 December 1996 concerning a multiannual programme for the promotion of energy efficiency in the Community – SAVE II:[63]
  - Introduction of new technologies;
  - Sector pilot actions aiming at accelerating investments and/or improving energy use patterns;
  - Monitoring of progress on energy efficiency;
  - Exchange of experiences among stakeholders.

Following the adoption of Council Decision 1999/21/EC, Euratom of 14 December 1998 adopting a multiannual framework programme for actions in the energy sector (1998–2002) and connected measures,[64] the European Parliament and the Council agreed in February 2000 on a new SAVE programme whose details were defined under Decision 647/2000/EC of the European Parliament and of the Council of 28 February 2000 adopting a multiannual programme for the promotion of energy efficiency (SAVE).[65]

The following directives have been adopted within the framework of the SAVE programme:

- Directive 96/57/EC of the European Parliament and of the Council of 3 September 1996 on energy efficiency requirements for household electric refrigerators, freezers and combinations thereof;[66]
- Commission Directive 95/12/EC of 23 May 1995 implementing Council Directive 92/75/EEC with regard to energy labelling of household washing machines;[67]
- Commission Directive 96/60/EC of 19 September 1996 implementing Council Directive 92/75/EEC with regard to energy labelling of household combined washer-driers.[68]

5.2.2 ALTENER

The Community ALTENER programme focused exclusively on the promotion of new and renewable energy sources for centralized and decentralized production of electricity and heat and their integration into local environment and energy systems.

The first phase of ALTENER ran for five years and ended in 1997: Council Decision 93/500/EEC of 13 September 1993 concerning the promotion of renewable energy sources in the Community (ALTENER programme).[69]

ALTENER II (1998–2002) was initiated in 1998 through the adoption of Council Decision 98/352/EC of 18 May 1998 concerning a multiannual programme for the promotion of renewable energy sources in the Community (ALTENER II)[70] and Council Decision 646/2000/EC of the European Parliament and of the Council of 28 February 2000 adopting a multiannual programme for the promotion of renewable energy sources in the Community (ALTENER) (1998–2002).[71] ALTENER II aimed at the promotion of activities in the renewable energy field and the encouragement of both private and public investment in the sector. It also set the Community Strategy and Action Plan outlined in Communication of the Commission to the Council and the European Parliament, Energy for the future – renewable sources of energy: White Paper for a Community Strategy and Action Plan, COM(1997)599.[72]

Within the framework of the ALTENER programme several projects were implemented in Europe aiming at increases in the use of new and renewable energy sources, in particular regarding electricity production, heat production, alternative fuels and small-scale applications.

5.2.3 Research (JOULE-THERMIE and ENERGY)

The EU Programme JOULE-THERMIE, aimed at the promotion of energy technologies, was a project included in the fourth framework programme of the Non-Nuclear RTDD (Research, Technological Development and Demonstration) Programme and it ran for four years (1995–1998). JOULE-THERMIE was an extension of a previous programme called THERMIE (1990–1994) and was adopted through Council Regulation (EEC) 2008/90 of 29 June 1990 concerning the promotion of energy technology in Europe.[73] Under the framework of THERMIE, 726 projects were implemented in the field of technology development in the period 1990–1994.

The THERMIE programme (1995–1998) was adopted with Council Decision 94/806/EC of 23 November 1994 adopting a specific programme for research and technological development, including demonstration, in the field of non-nuclear energy (1994 to 1998),[74] with a focus on the promotion of research and technology development in the energy sector and aimed at:

• The reduction of energy consumption;
• The improvement of the use of energy;

- The diffusion of demonstration activities, dissemination and support measures, as well as concerted actions and technology stimulation.

In 2003 a new programme was adopted through Decision 1230/2003/EC of the European Parliament and of the Council of 26 June 2003 adopting a multiannual programme for action in the field of energy: 'Intelligent Energy – Europe' (2003–2006).[75]

The Intelligent Energy programme (2003–2006) 'will support sustainable development in the energy context' contributing to the 'security of energy supply, competitiveness, and environmental protection' and to the promotion of 'transparency, coherence and the complementarity of all the actions and other related measures in the field of energy' (Article 2 of Decision 1230/2003/EC). The programme is implemented by the Executive Agency for Competitiveness and Innovation – EACI (former Energy Executive Agency) and supported several projects, the establishment of local/regional energy agencies and the organization of European events on the subject.

NOTES

1. The final text of the draft Treaty establishing a Constitution for Europe as adopted by consensus by the European Convention on 13 June and 10 July 2003 and submitted to the president of the European Council in Rome on 18 July 2003, presented by the Secretariat to the Members of the Convention on 18 July 2003, Brussels.

2. This is the case with Community legislation aimed at the promotion of renewable energy sources and energy efficiency adopted on the basis of ex Article 175(1) TEC on the protection of the environment.

3. This is the case, for instance, with Council Directive 2003/96/EC of 27 October 2003 restructuring the Community framework for the taxation of energy products and electricity, which was adopted having regard to Article 93 of the EC Treaty and which is based on the harmonization of taxation legislation and on the necessity to ensure the proper functioning of the internal market and the achievement of the objectives of other Community policies.

4. Directive 2009/28/EC of the European Parliament and of the Council of 23 April 2009 on the promotion of the use of energy from renewable sources and amending and subsequently repealing Directives 2001/77/EC and 2003/30/EC, OJ L 140, 5 June 2009, pp16–62.

5. Proposal for a Directive of the European Parliament and of the Council on the promotion of the use of energy from renewable sources, COM(2008)19, Brussels, 23 January 2008, p9.

6. See note 3 in Chapter 1.

7. 'In the context of the establishment and functioning of the internal market and with regard for the need to preserve and improve the environment, Union policy on energy shall aim, in a spirit of solidarity between Member States, to: (a) ensure the functioning of the energy market; (b) ensure security of energy supply in the Union; (c) promote energy efficiency and energy saving and the development of new and renewable forms of energy; and (d) promote the interconnection of energy networks' Article 194(1) TFEU.

8. 'Without prejudice to the application of other provisions of the Treaties, the European Parliament and the Council, acting in accordance with the ordinary legislative procedure, shall establish the measures necessary to achieve the objectives in paragraph 1. Such measures shall be

adopted after consultation of the Economic and Social Committee and the Committee of the Regions. Such measures shall not affect a Member State's right to determine the conditions for exploiting its energy resources, its choice between different energy sources and the general structure of its energy supply, without prejudice to Article 192(2)(c)' Article 194(2) TFEU. 'By way of derogation from paragraph 2, the Council, acting in accordance with a special legislative procedure, shall unanimously and after consulting the European Parliament, establish the measures referred to therein when they are primarily of a fiscal nature' Article 194(3) TFEU.

9. Communication from the Commission strengthening environmental integration within Community energy policy, establishing a set of measures for further integration of environmental considerations in energy policy and reviewing progress made so far COM(1998)571, Brussels, 14 October 1998.

10. Communication from the Commission establishing an overall overview of energy policy and actions, COM(1997)167, Brussels, 23 April 1997.

11. Communication from the Commission, A Sustainable Europe for a Better World: A European Union Strategy for Sustainable Development concerning Commission's proposal to the Gothenburg European Council, COM(2001)264, Brussels, 15 May 2001.

12. Communication from the Commission to the European Parliament, the Council, The Economic and Social Committee and the Committee of the Regions, Towards a global partnership for sustainable development COM(2002)82, Brussels, 13 February 2002.

13. Communication from the Commission to the Council and the European Parliament on the review of the Sustainable Development Strategy, A platform for action COM(2005)658, Brussels, 13 December 2005.

14. Communication from the Commission establishing Green Paper: A European Strategy for Sustainable, Competitive and Secure Energy, COM(2006)105, Brussels, 8 March 2006.

15. Public hearing organized in Brussels on Friday 22 September 2006 for the purpose of the public to express directly their views as to the content of the Strategic EU Energy Review to be adopted by the Commission on 10 January 2007.

16. Press release of 14 December 2006 on the adoption of the Commission's Green Paper on a European strategy for sustainable, competitive and secure energy by the European Parliament.

17. Communication from the Commission to the European Council and the European Parliament, An Energy Policy for Europe, COM(2007)1, Brussels, 10 January 2007.

18. Commission staff working document of 10 October 2007 accompanying several documents on the EU Energy Policy Data.

19. Communication from the Commission to the European Council and the European Parliament establishing the Renewable Energy Road Map; renewable energies in the 21st century; building a more sustainable future, COM(2006)848, Brussels, 10 January 2007.

20. Communication from the Commission to the European Council and the European Parliament establishing the Green Paper follow-up action; report on progress in renewable electricity, COM(2006)849, Brussels, 10 January 2007.

21. Communication from the Commission to the European Council and the European Parliament establishing the Biofuels Progress Report; report on the progress made in the use of biofuels and other renewable fuels in the Member States of the European Union, COM(2006)845, Brussels, 10 January 2007.

22. Communication from the Commission to the European Council and the European Parliament establishing the document Prospects for the internal gas and electricity, COM(2006)841, Brussels, 10 January 2007.

23. Communication from the Commission to the European Council and the European Parliament establishing the Priority Interconnection Plan, COM(2006)846, Brussels, 10 January 2007.

24. Communication from the Commission to the European Council and the European Parliament establishing the document, Sustainable power generation from fossil fuels: aiming for near-zero emissions from coal after 2020, COM(2006)843, Brussels, 10 January 2007.

25. Communication from the Commission to the European Council, the European Parliament, the European economic and social committee and the committee of the regions establishing the document, Towards a European Strategic Energy Technology Plan, COM(2006)847, Brussels, 10 January 2007.

26. Explanatory memorandum to the Commission draft Proposal for a Directive of the European parliament and of the council amending directive 2003/54/EC and 2003/55/EC and for a regulation of the European parliament and of the council amending regulation (EC) No 1228/2003 and (EC) No 1775/2005 and establishing an Agency for the Cooperation of Energy Regulators, COM(2007) Draft, Brussels, 19 September 2007.

27. Commission Proposal for a Directive of the European parliament and of the council amending Directive 2003/54/EC of the European Parliament and of the Council of 26 June 2003 concerning common rules for the internal market in electricity, COM(2007)528, Brussels, 19 September 2007.

28. Commission Proposal for a Regulation of the European parliament and of the council establishing an Agency for the Cooperation of Energy Regulators, COM(2007)530, Brussels, 19 September 2007.

29. Commission Proposal of 19 September 2007 for a Regulation of the European Parliament and of the Council amending regulation EC No 1228/2003 on conditions for access to the network for cross-border exchanges in electricity, COM(2007)531, Brussels, 19 September 2007.

30. Commission Proposal for a Directive of the European Parliament and of the Council amending Directive 2003/55/EC concerning common riles for the internal market in natural gas COM(2007)529, Brussels, 19 September 2007.

31. Commission Proposal for a Regulation of the European Parliament and of the Council amending regulation (EC) No 1775/2005 on conditions for access to the natural gas transmission networks, COM(2007)532, Brussels, 19 September 2007.

32. Communication from the Commission to the Council, the European Parliament, the European economic and social committee and the committee of the regions of 11 November 2007 presenting the European strategic energy technology SET-PLAN 'Towards a low carbon future', COM(2007)723, Brussels, 22 November 2007.

33. Commission Proposal for a Directive of the European Parliament and of the Council on the promotion of the use of energy from renewable sources, COM(2008)19, Brussels, 23 January 2008.

34. Communication from the Commission to the Council and the European Parliament establishing a first assessment of national energy efficiency action plans, as required by Directive 2006/32/EC on energy end-use efficiency and energy services, entitled Moving Forward Together on Energy Efficiency, COM(2008)11, Brussels, 23 January 2008.

35. Communication from the Commission to the European Parliament, the Council, The European Economic and Social Committee and the Committee of the Regions entitled Supporting Early Demonstration of Sustainable Power Generation from Fossil Fuels, COM(2008)13, Brussels, 23 January 2008.

36. Vision Paper for EU Strategic Energy Technology Plan prepared by the Presidency of the European Union of 15 November 2007.

37. Press release of 28 February 2008 of the Council of the European Union on the 2854th Council meeting concerning transport, telecommunications and energy, establishing the Council's adoption of the European strategic energy technology plan.

38. Communication from the Commission to the European Parliament, the Council, the European Economic and Social Committee and the Committee of the Regions presenting the Second Strategic Energy Review entitled an EU Energy Security and Solidarity Action Plan, COM(2008)781, Brussels, 13 November 2008.

39. Memo prepared by the European Commission on the EU energy security and solidarity action plan: 2nd strategic energy review.

40. Report of the European Environment Agency of June 2008 on energy and environment specifically the impact of energy consumption on the environment in the EU.

41. Press release of the European Parliament of June 2009 on the compromise reached between the Members of the Parliament and the Council Presidency concerning the electricity and gas markets.

42. Regulation (EC) No 663/2009 of the European Parliament and of the Council of 13 July 2009 establishing a programme to aid economic recovery by granting Community financial assistance to projects in the field of energy, OJ L 200, 31 July 2009, pp31–45.

43. Proposal for a Council Regulation concerning the notification to the Commission of investment projects in energy infrastructure within the European Community and repealing Regulation (EC) No 736/96, COM(2009)361, Brussels, 16 July 2009.

44. Proposal for a Regulation of the European Parliament and of the Council concerning measures to safeguard security of gas supply and repealing Directive 2004/67/EC, COM(2009)363, Brussels, 16 July 2009.

45. Report from the Commission to the Council and the European Parliament on the implementation of the European Energy Programme for Recovery, COM(2010)191, Brussels, 27 April 2010.

46. On EU energy regulation, see Roggenkamp et al (2007).

47. Communication of the Commission for an amended proposal for a Council Directive on excise duties on motor fuels from agricultural sources, COM(1994)147, OJ C 209, 29 July 1994, p9.

48. Council Directive 95/60/EC of 27 November 1995 on fiscal marking of gas oils and kerosene, OJ L 291, 6 December 1995, pp46–47.

49. Commission Decision 2006/428/EC of 22 June 2006 establishing a common fiscal marker for gas oils and kerosene (notified under document number C(2006) 2383), OJ L 172, 24 June 2006, pp15–16.

50. Proposal for a Council Directive restructuring the Community framework for the taxation of energy products, COM(1997)30, OJ C 139, 6 May 1997, p14.

51. Council Directive 92/12/EEC of 25 February 1992 on the general arrangements for products subject to excise duty and on the holding, movement and monitoring of such products, OJ L 76, 23 March 1992, pp1–13.

52. Communication from the Commission to the Council, the European Parliament and the Economic and Social Committee on Tax policy in the European Union – Priorities for the years ahead, COM(2001)260, Brussels, 23 May 2001.

53. Directive 2003/96/EC of 27 October 2003 restructuring the Community framework for the taxation of energy products and electricity, OJ L 283, 31 October 2003, pp51–70.

54. Council Directive 2004/74/EC of 29 April 2004 amending Directive 2003/96/EC as regards the possibility for certain Member States to apply, in respect of energy products and electricity, temporary exemptions or reductions in the levels of taxation, OJ L 195, Brussels, 2 June 2004, pp26–31 and Council Directive 2004/75/EC of 29 April 2004 amending Directive 2003/96/EC as regards the possibility for Cyprus to apply, in respect of energy products and electricity, temporary exemptions or reductions in the levels of taxation, OJ L 195, Brussels, 2 June 2004, pp31–32.

55. Proposal for a Council Directive amending Directive 2003/96/EC as regards the possibility for certain Member States to apply, in respect of energy products and electricity, temporary exemptions or reductions in the levels of taxation COM(2004)42, Brussels, 28 January 2004.

56. Communication from the Commission to the Council, Review of the derogations in Annexes II and III of Council Directive 2003/96/EC that expire by the end of 2006, COM(2006)342, Brussels, 30 June 2006.

57. Judgment of the European Court of Justice (Third Chamber) of 5 October 2006 following a reference for a preliminary ruling by the Verwaltungsgerichtshof Austria concerning the interpretation of Article 88(3) EC.

58. Proposal for a Council Directive amending Directive 2003/96/EC as regards the adjustment of special tax arrangements for gas oil used as motor fuel for commercial purposes and the coordination of taxation of unleaded petrol and gas oil used as motor fuel COM(2007)52, Brussels, 13 March 2007.

59. European Commission Green Paper on market-based instruments for environment and related policy purposes, COM(2007)140, Brussels, 28 March 2007.

60. WTO/UNEP (2009).

61. Commission Staff Working Document of 2007 accompanying the proposal for a Council Directive amending Directive 2003/96/EC as regards the adjustment of special tax arrangements for gas oil used as motor fuel for commercial purposes and the coordination of taxation of unleaded petrol and gas oil used as motor fuel, Impact Assessment, SEC(2007)171/2.

62. Council Decision 91/565/EEC of 29 October 1991 concerning the promotion of energy efficiency in the Community (SAVE programme), OJ L 307, 8 November 1991 pp34–36.

63. Council Decision 96/737/EC of 16 December 1996 concerning a multiannual programme for the promotion of energy efficiency in the Community – SAVE II, OJ L 335, 24 December 1996, pp50–53.

64. Council Decision 1999/21/EC, Euratom of 14 December 1998 adopting a multiannual framework programme for actions in the energy sector (1998–2002) and connected measures, OJ L 7, 13 January 1999, pp16–19.

65. Decision 647/2000/EC of the European Parliament and of the Council of 28 February 2000 adopting a multiannual programme for the promotion of energy efficiency (SAVE) (1998 to 2002), OJ L 79, 30 March 2000, pp6–9.

66. Directive 96/57/EC of the European Parliament and of the Council of 3 September 1996 on energy efficiency requirements for household electric refrigerators, freezers and combinations thereof, OJ L 236, 18 September 1996, pp36–43.

67. Commission Directive 95/12/EC of 23 May 1995 implementing Council Directive 92/75/EEC with regard to energy labelling of household washing machines, OJ L 136, 21 June 1995, pp1–27.

68. Commission Directive 96/60/EC of 19 September 1996 implementing Council Directive 92/75/EEC with regard to energy labelling of household combined washer-driers, OJ L 266, 18 October 1996, pp1–27.

69. Council Decision 93/500/EEC of 13 September 1993 concerning the promotion of renewable energy sources in the Community (ALTENER programme), OJ L 235, 18 September 1993, pp41–44.

70. Council Decision 98/352/EC of 18 May 1998 concerning a multiannual programme for the promotion of renewable energy sources in the Community (ALTENER II), OJ L 159, 3 June 1998, pp53–57.

71. Council Decision 646/2000/EC of the European Parliament and of the Council of 28 February 2000 adopting a multiannual programme for the promotion of renewable energy sources in the Community (ALTENER) (1998–2002) *OJ* L 79, 30 March 2000, pp1–5.

72. Communication of the Commission to the Council and the European Parliament, Energy for the future – renewable sources of energy: White Paper COM(1997)599, Brussels, 26 November 1997.

73. Council Regulation (EEC) 2008/90 of 29 June 1990 concerning the promotion of energy technology in Europe, *OJ* L 185, 17 July 1990, p1.

74. Council Decision 94/806/EC of 23 November 1994 adopting a specific programme for research and technological development, including demonstration, in the field of non-nuclear energy (1994 to 1998), *OJ* L 334, 22 December 1994, pp87–108.

75. Decision 1230/2003/EC of the European Parliament and of the Council of 26 June 2003 adopting a multiannual programme for action in the field of energy: 'Intelligent Energy – Europe' (2003–2006), *OJ* L 176, 15 July 2003, pp29–36.

CHAPTER 6

Energy Efficiency

B efore addressing the details of energy efficiency law and policy in this chapter, as well as renewable energy in the next chapter, it is important to highlight a few key areas that are at the foundations of European policy and law in those areas.[1]

The Resolution of the Council and the Representatives of the Governments of the Member States, meeting within the Council of 1 February 1993 on a Community programme of policy and action in relation to the environment and sustainable development explicitly mentions energy policy as:

> a key factor in the achievement of sustainable development. While the Community's energy sector is making steady progress in dealing with local and regional environmental problems such as acidification, global issues are daily growing in importance. The challenge of the future will be to ensure that economic growth, efficient and secure energy supplies and a clean environment are compatible objectives. The achievement of this balance requires a strategic perspective well beyond the period covered by this Programme. The key elements of the strategy up to 2000 will be improvement in energy efficiency and the development of strategic technology programmes moving towards a less carbon-intensive energy structure including, in particular, renewable energy options.[2]

On 24 September 1998 the European Parliament and Council adopted Decision No 2179/98/EC on the review of the European Community programme of policy and action in relation to the environment and sustainable development 'Towards sustainability'.[3] The 5th Environment Action Programme (EAP) of the EC covered the period 1992–2000 and its main objective was the promotion of sustainable development in the growth of the EC. The 5th EAP considered environmental issues such as climate change, water pollution and waste management and introduced new concepts such as shared responsibility, new environmental instruments and more engagement by all actors at different levels. It established performance levels to be achieved by 2000. Among its priority areas, the 5th EAP included 'reducing the consumption of energy from non-renewable sources; improving the management of mobility by developing efficient and clean modes of transport'. Energy was one of the target sectors, and in particular the 5th EAP stated that 'action in this field is indispensable if sustainable development is to be achieved; this will require an improvement in energy efficiency, a reduction in the consumption of fossil fuels and the promotion of renewable energy sources'.

On 29 May 2001 the European Commission presented Communication COM(2001)31 on the Sixth Environment Action Programme of the European Community, 'Environment 2010: Our future, Our choice'.[4] The 6th EAP of the European Community covers the period 22 July 2002 to 21 July 2012 and identifies four priority areas for action: climate change; biodiversity; environment and health; and sustainable management of resources and wastes. On climate change, the 6th EAP aims at the reduction of GHG emissions by 8 per cent by 2008–2012, compared with 1990 levels, and by 20–40 per cent by means of an effective international agreement in the longer term (by 2020). In order to achieve these objectives the 6th EAP indicated 'the reduction of greenhouse gases by means of specific measures to improve energy efficiency, to make increased use of renewable energy sources, to promote agreements with industry and to make energy savings'.

Energy saving measures aimed at the reduction of energy consumption and waste can contribute significantly, amongst others, to the decrease of GHG emissions and to the achievement of sustainable development. In particular, energy efficiency policies and measures can have a direct positive effect in the reduction of GHG emissions in sectors such as primary energy (fuel combustion from energy industries, manufacturing and construction, transport and others as well as fugitive emissions from solid fuels and oil and natural gas) and industrial processes (mineral, chemical, metal, others). Energy efficiency is usually achieved through policies and technologies development in buildings, appliances, transport, industry and end-use applications such as lighting.

Since the 1990s the promotion of energy efficiency has been one of the core strategies of the EU and the Member States in the fight against climate change and the reduction of GHG emissions.[5] In particular, EU policy and law on energy efficiency covers:

- End-use and services;
- Voluntary agreements;
- Industry;
- Buildings;
- Cogeneration;
- Eco-design;
- Labelling.

A decisive input to the development of specific European legislation aimed at the promotion of energy efficiency is related with the implementation of the second phase of the European Climate Change Programme. More precisely, the European

Commission issued on 10 December 2003 a Proposal for a Directive of the European Parliament and the Council on the Promotion of End-use efficiency and Energy Services COM(2003)739,[6] aiming at the improvement of end-use energy efficiency and the reduction of GHG emissions in the Member States. This proposal identified several operational measures, including the development of the market for energy services covering:

> the retail supply and distribution of extensive net-bound energy carriers, such as electricity and natural gas, together with other important energy types, such as district heating, heating fuel, coal and lignite, forestry and agricultural energy products and transport fuels.

The proposal indicated a savings target of an annual amount of energy equal to 1 per cent of energy distributed or sold to final consumers in six years for the Member States in terms of energy efficiency improvements and reaching sufficient market demand for energy services. It also included a savings target for the public sector equivalent to 1.5 per cent level per year.

On 29 November 2004 European energy ministers met to discuss proposal COM(2003)739 and failed to agree on the mandatory aspects of such a legislative proposal. They decided to subordinate the national targets of energy efficiency in the Member States, and the consequent adoption of energy efficiency measures, to the principle of cost-effectiveness, with benefits higher than costs (see revised text of proposal COM(2003)739 prepared by the EU Council presidency, 24 September 2004 and Council Transport, Telecommunications and Energy, Conclusions of 29 November 2004). Moreover, on 30 November 2004 the Committee on the Environment, Public Health and Consumer Policy of the European Parliament discussed the Member of European Parliament Korhol's report on the same proposal and confirmed the Member States' approach of introducing more flexibility on such an issue.

The European Parliament agreed at first reading on the text of the draft end-use energy efficiency and energy services directive to introduce more rigorous objectives for the Member States. The Members of European Parliament (MEPs) proposed setting binding targets every three years (3 per cent by 2006–2009, 4 per cent by 2009–2012 and 4.5 per cent by 2012–2015), with an overall energy saving of at least 11.5 per cent in the period 2006–2015. As to the public sector, MEPs proposed making energy efficiency one of the criteria for awarding public service contracts.

On 27 June 2005 the Council Transport, Telecommunications and Energy adopted a common position softening the proposal of the European Parliament and rejecting the idea of differentiated and higher targets on energy efficiency for the public sector (only a non-binding 'indicative' target to increase efficiency by 6 per cent over eight years) and not requiring energy firms to meet quantitative targets to provide energy services or energy audits.

On 22 June 2005, the Energy Commissioner Piebalgs presented the EU Green Paper on Energy Efficiency COM(2005)265,[7] which focused on:

- Competitiveness and the Lisbon agenda: the EU could save 20 per cent of energy consumption in a cost-effective manner and create millions of new jobs;

- Environmental protection and the Kyoto Protocol: energy saving as the most effective and cost-effective way to reduce GHG emissions and to contribute to the post-2012 fight against climate change;
- Security of supply: capping and reducing energy demand to promote a balanced policy for energy security.

By the same token, the 'Intelligent Energy initiative', establishing a target of saving 380 million tonnes of oil equivalent by 2020, was adopted by a group of European parliamentarians representing a substantial part of the EU Member States.

Directive 2005/32/EC of 6 July 2005 establishing a framework for the setting of eco-design requirements for energy-using products and amending Council Directive 92/42/EEC and Directives 96/57/EC and 2000/55/EC of the European Parliament and of the Council[8] identified a first scheme of environmental design requirements for common appliances (energy-using products) likely to have a significant impact on the environment. Directive 2005/32/EC aimed at ensuring the free movement of those products in the internal market and provided the requirements that such products must fulfil in order to be placed in the market. The directive applied not only to the 'means of transport for persons or goods'. The directive defined 'eco-design' as the 'integration of environmental aspects into product design with the aim of improving the environmental performance of the EuP [energy-using product throughout its whole life cycle' and set eco-design parameters for energy-using products.

On 18 July 2005 the European Commission launched a four-year campaign to increase the share of renewable energy in the EU25 to 12 per cent by 2010. The Sustainable Energy Europe Campaign 2005–2008 was aimed at raising the awareness of decision-makers at local, regional, national and European level, and to spread best practices in the sector of sustainable energy technologies.

The draft text of the energy end-use efficiency and energy services Directive was agreed on 30 November 2005 by the European Parliament and the Council. The directive obliged the Member States in the period 2008–2017 to save at least an additional 1 per cent of their annual final energy consumption. The directive referred also to local and regional energy actors and it covered areas such as energy-efficient and cost-effective lighting, heating, hot water, ventilation and transportation.

On 24 November 2005, the European Parliament's Industry Committee voted to reopen the discussion on a series of amendments to the draft directive on energy end-use efficiency and energy services previously rejected by the Council at first reading. The resolution of the Industry Committee urged the introduction of tougher obligations for the Member States in terms of energy efficiency. The committee called for Member States to increase energy saving by 11.5 per cent in six years, while governments had proposed cuts in energy consumption of 6 per cent in six years.

The Energy Council of 2 December 2005 adopted a resolution urging the European Commission to develop as soon as possible an action plan on energy efficiency to implement the indications of the Green Paper released in June 2005.

EU energy ministers also called for the inclusion of energy efficiency principles into the revision of the EU's sustainable development strategy to be undertaken by the European Commission.

The Proposal for a Directive on end-use energy efficiency and energy services COM(2003)739 was discussed by the European Parliament on 6 December 2005. The inclusion into the new directive of a requirement for the Member States to submit national energy efficiency action plans was proposed by the MEPs with the support of the Council.

A new initiative to boost energy efficiency measures and public support was launched by the European Parliament on 25 January 2006. Energy Efficiency Watch was an initiative of MEPs coordinated by the European renewable energy forum Eufores and aimed at the monitoring of European and national performance on energy efficiency.

The EU Council Environment of 9 March 2006 agreed on the final text of the end-use energy efficiency and energy services directive proposal and no amendments were proposed. On 5 April 2006 Directive 2006/32/EC on energy end-use efficiency and energy services and repealing Council Directive 93/76/EEC was adopted by the European Parliament and the Council.[9]

The response of the European Parliament to the European Commission's Green Paper on energy efficiency was a non-legislative resolution adopted on 1 June 2006, which called for higher targets for energy savings in the EU by 2020 (the European Commission proposed an increase of 20 per cent). Furthermore, the Parliament requested stricter minimum standard requirements, improvements in information campaigns and energy labelling, as well as improvements in the performance of office equipment, consumer electronics and industrial engines.

On 17 May 2006 Directive 2006/32/EC on energy end-use efficiency and energy service entered into force. The aim of the directive was the promotion of energy end-use efficiency in Member States. The correct implementation of the directive should foster a market for energy services and for the delivery of other energy efficiency measures to final consumers. The directive established an indicative target of a 9 per cent cut in energy use in respect of the business-as-usual scenario in the period 2008–2017. Member States were required to submit National Energy Efficiency Action Plans (NEEAPs) including the details of Member States' strategies to meet the 9 per cent energy savings target. NEEAPs shall be submitted in accordance with the following time plan: first plan by 30 June 2007, second plan by 30 June 2011 and third plan by 30 June 2014.

On 19 October 2006, the European Commission released Communication COM(2006)545 'Action Plan for energy efficiency: realising the potential',[10] setting out a detailed plan for policies and measures to be adopted in the next six years in order to boost energy efficiency in Europe. The energy efficiency plan aims at a 20 per cent reduction in energy consumption by 2020 and includes a list of targets for energy saving to be introduced in the energy end-use efficiency directive and new targets for transport and other economic sectors. The plan also recommends the adoption of policies and measures for the following six years in different sectors aiming at the above mentioned objectives, in particular more than

70 initiatives aiming at the improvement of energy efficiency in the following seven main sectors are recommended:

- Products;
- Building and energy services;
- Energy transformation;
- Transport;
- Financing and economic incentives;
- Changes in behaviour;
- International partnerships.

The draft Council conclusions on the Commission's Action Plan on Energy Efficiency of 1 November 2006 showed a disagreement among the Member States on two important issues. First, the text did not contain any reference to the Commission's proposal 'to facilitate a more targeted and coherent use of energy taxation'. Second, no agreement was reached over the way to reduce CO_2 emissions of cars in the transport sector.

The European Commission Communication COM(2007)1 'An Energy Policy for Europe' of 10 January 2007 focused on the improvement of energy efficiency, especially in vehicles, appliances, homes and factories.[11] An Energy Policy for Europe identified three main challenges of EU energy policy:

- Sustainability;
- Security of supply;
- Competitiveness.

In this Communication, the European Commission proposed that European energy policy should aim at:

- The EU objective in international negotiations of 30 per cent reduction in GHG emissions by developed countries by 2020 compared to 1990. In addition, global GHG emissions reduction of up to 50 per cent by 2050 compared to 1990, implying reductions in industrialized countries of 60–80 per cent by 2050;
- The EU commitment to achieve, in any event, at least a 20 per cent reduction of GHGs by 2020 compared to 1990.

The action plan of an Energy Policy for Europe was based on the transformation of Europe into a highly energy-efficient and low-CO_2 energy economy, fostering a new industrial revolution, based on low-carbon growth and increasing the amount of local, low-emission energy that we produce and use. To this aim, the following measures were identified:

- Creation of a real internal energy market to be achieved through unbundling, effective regulation, transparency, infrastructure, network security, adequacy of electricity generation and gas supply capacity, and energy as a public service;

- Fostering of solidarity between Member States and security of supply for oil, gas and electricity;
- Creation of a long-term commitment to the reduction of GHG emissions;
- Establishment of an ambitious programme of energy efficiency measures;
- Setting of a long-term target for renewable energy;
- Establishment of a SET-PLAN;
- Building a low-CO_2 fossil fuel future;
- Addressing the future of nuclear energy policy;
- Development of an international energy policy addressing Europe's energy and development.

On 6 February 2007 the Commission adopted Decision 2007/74/EC of 21 December 2006 establishing harmonized efficiency reference values for separate production of electricity and heat in application of Directive 2004/8/EC of the European Parliament and of the Council.[12] The Commission document was aimed at fostering the correct implementation of the CHP directive ensuring a high level of energy efficiency in the combined production of electricity and heat. In particular, cogeneration plants shall provide energy savings of at least 10 per cent according to the established benchmarks.

The European Council of 8–9 March 2007 Brussels, Presidency Conclusions, welcomed the EU energy action plan for 2007–2009, which included:

- A non-binding commitment to reduce energy consumption in the EU by 20 per cent in comparison with 2020 projections to be achieved through an increase in energy efficiency;
- A proposal for an energy technology plan aiming at the improvement of research in renewable energy, low carbon and energy efficiency;
- The separation of supply and production activities by the electricity network operators.

The list of NEEAPs to be submitted by the Member States to the Commission by 30 June 2007 (and every three years until 2016) in accordance with Article 14(2) of Directive 2006/32/EC showed the level of commitment of the Member States to reach the 9 per cent indicative energy savings target by 2016.

Communication COM(2007)380 on VAT rates other than standard VAT rates released by the European Commission on 5 July 2007, introduced the idea of a lower VAT rate for energy-efficient appliances. In the preparation of a legislative proposal expected by 2009, the Commission asked the Member States for inputs and opinions on the possibility of applying a lower tax regime instead of energy efficiency improvements in appliances.[13]

In December 2007 the European Commission adopted a Consultation Document on the revision of the Energy Labelling Directive 92/75/EEC of 22 September 1992 on the indication by labelling and standard product information of the consumption of energy and other resources by household appliances aimed at the extension of the energy labelling scheme to boilers, televisions and electric motors.[14] The revision of the Energy Labelling Directive was foreseen by the EU energy plan adopted in 2007, aiming at the reduction of energy consumption by 20 per cent by 2020 through the improvement of energy efficiency.

On 23 January 2008, the European Commission tabled Communication COM(2008)11 on a first assessment of national energy efficiency action plans as required by Directive 2006/32/EC on energy end-use efficiency and energy services, which considered 17 national action plans submitted by the Member States to the European Commission by October 2007.[15] The assessment showed that Member States should do more to increase energy efficiency measures at the domestic level. In Communication COM(2008)11, the European Commission provided the details of the measures to be adopted in 2008 in order to boost energy efficiency, namely the revision of existing EU legislation on the energy performance of buildings, energy labelling and energy taxation, as well as the preparation of new reports on sustainable consumption and production and on the improvement of energy efficiency via information and communication technologies.

On 3 April 2008 the European Commission sent reasoned opinions to Greece and Latvia for their failure to submit NEEAPs, as required under the end-use energy efficiency and energy services directive. The action plans were due to be submitted by 30 June 2007.

On 13 May 2008 the European Commission released the policy paper addressing the challenge of energy efficiency through information and communication technologies COM(2008)241.[16] The paper highlighted the strong potential of information and communication technologies (ICTs) to cut EU CO_2 emissions, in particular in the building, lighting and power sectors. The main goal of the paper was to explain the extent to which the promotion of ICTs can contribute to the EU commitments to the enhancement of energy efficiency, for instance through the establishment of voluntary agreements with the industry on green procurement. The development of pilot projects and the promotion of research and cooperation in the sector were identified as key tools to achieve the paper's aims.

The first results of the screening of the NEEAPs submitted by the Member States in accordance with the Energy Efficiency and Energy Services Directive were released on 27 June 2008 by Energy Efficiency Watch (EEW). The EEW report showed that EU Member States had not seriously committed to the adoption of energy saving measures and in particular the NEEAPs did not show how the Member States intended to comply with the reduction target in end energy use of 9 per cent by 2016. According to the report, Member States failed to communicate sufficient details on the planning of energy efficiency measures, as well as on financing.

On 10 September 2008 the French presidency drafted a ministerial non-legislative resolution on the implementation of the framework directives on the energy labelling and eco-design of energy-using products. The main goal of the resolution was putting political pressure on the adoption of energy efficiency standards for a wide range of appliances. In particular, the resolution called for the introduction of a ban on energy-intensive and inefficient light bulbs such as incandescent lamps for domestic use from 2010. The measure had to be formally adopted by the European Commission through a committee of experts from the Member States.

On 6 October 2008 the French presidency presented a draft report to the European Council on energy security – Presidency Briefing 14090/08 aimed at enhancing energy efficiency in the EU in order to combat climate change. The report emphasized the importance of the energy efficiency action plan adopted in 2006 and urged the adoption of the proposals on directives on energy labelling and the energy performance of buildings by the end of 2009.

On 9 October 2008 the energy department of the Commission drafted a legislative proposal on a Commission directive for a recast of the energy performance of buildings directive (2002/91/EC). The early draft of the proposal extended the energy efficiency standards set at the national level to all buildings regardless of the 1000 square metre limit included in Directive 2002/91/EC.

On 10 October 2008 the EU energy ministers in the Council adopted a non-legislative resolution backing the call of the European Commission for the introduction of a ban on inefficient incandescent light bulbs.

On 13 November 2008 the European Commission presented two legislative proposals aimed at improving energy efficiency in the EU as part of the second Strategic Energy Review package 'An EU energy security and solidarity action plan'. The Proposal for a Directive on the energy performance of buildings (recast) COM(2008)780 introduced the obligation for all buildings (no limitation on the size) undergoing a major renovation to comply with energy efficiency standards set at the national level.[17] Furthermore, the proposal required the Member States to define specific penalties for non-compliance obliged the public sector to comply with even more stringent targets and required that energy performance certificates have to be considered in sales and renting activities. The second legislative plan was the Proposal for a Directive on energy labelling on the indication by labelling and standard product information of the consumption of energy and other resources by energy-related products (recast) COM(2008)778,[18] extending the energy labelling requirement to all energy-related products with a significant impact on energy consumption, notably enlarging the scope of the 1992 energy labelling directive from domestic appliances to the industrial and commercial sectors. The information to be included in the labelling would have to be decided on a product-by-product basis and public authorities would have an obligation to purchase products under a certain efficiency class. Finally, the proposal introduced minimum efficiency standards for green public procurement.

Furthermore, on 13 November 2008 the Commission presented a Proposal for a Directive on labelling of tyres with respect to fuel efficiency and other essential parameters, which would oblige tyre manufacturers in the EU to display from 2012 information on the environmental performance of their products, notably on fuel efficiency, wet grip and noise pollution.[19] The requirements set in the proposal would apply to passenger cars and light and heavy duty vehicles.

On 26 November 2008 the European Commission presented the Communication to the European Council COM(2008)800 A European Economic Recovery Plan, designed to assist the European economy to overcome the world financial crisis.[20] The plan included a focus on energy efficiency, greener products and green cars. On energy efficiency, the Commission invited the Member States to adopt the directive on the revision of energy efficiency and buildings, proposed cuts on property taxes for efficient buildings and envisaged a partnership between public and private sectors to reduce energy use in buildings. On greener products, the Commission announced the intention to introduce lower EU sales taxes for greener products and services, as well as new environmental standards for energy-using appliances within the eco-design directive. Finally, the Commission proposed another public–private partnership to urge the car industry to develop greener cars.

On 8 December 2008 the Member States at the Eco-design Regulatory Committee endorsed the proposal of the Commission to phase out incandescent bulbs in the period 2009–2012. The measure was adopted within the framework of the energy-using products directive and targeted major lamps used in households, such as incandescent lamps, halogen lamps and compact fluorescent lamps.

On 8 and 9 December 2008 the Transport, Communications and Energy Council agreed informally on a general approach to extend the scope of the eco-design directive (2005/32/EC) to further categories of products, notably those products with high rates of energy consumption in the design phase. These 'energy-related products' are windows, piping, shower heads and insulation.

On 17 December 2008 the European Commission adopted a draft regulation establishing a mandatory limit on standby electricity consumption, which was the first measure adopted within the framework of the Eco-design Regulatory Committee established by the eco-design directive (2005/32/EC). By 2020 the standby regulation would reduce by three-quarters (from current levels) electricity consumption of the standby mode of several consumer appliances.

On 31 March 2009 Member States representatives meeting in the Eco-design and Energy Labelling Regulatory Committee agreed on the adoption of new energy labelling and energy efficiency requirements for televisions, fridges and freezers, washing machines and dishwashers. According to the Member States' experts, new classes of efficiency beyond A level will be progressively introduced for the mentioned categories of products.

On 1 April 2009 the Committee on Industry, Research and Energy of the European Parliament adopted a series of amendments to the proposal for a revision of the directive on energy performance of buildings. MEPs proposed the creation of an EU fund by 2014 aimed at financing improvements in energy efficiency in buildings and agreed on the introduction of minimum national energy efficiency standards for all buildings under major renovation. Finally MEPs proposed that by 2019 all new buildings in the EU should have zero energy consumption.

On 22 April 2009 the European Parliament adopted a Legislative Resolution on the Proposal for a Directive on labelling of tyres with respect to fuel efficiency and other essential parameters COM(2008)779. In the resolution, MEPs adopted almost all recommendations proposed by the Industry Committee of the Parliament and strengthened the original proposal of the Commission for the introduction of a mandatory tyre efficiency requirement from 2012. Under the adopted resolution, Member States would be allowed to promote tyres with at least a C level of efficiency and the labelling system would apply to all tyres produced after July 2012.

On 23 April 2009 the European Parliament voted on the amendments of the 2002 Energy performance of buldings directive, which established that by 31 December 2018 the Member States must ensure that all newly constructed buildings produce as much energy as they consume on-site, for instance through solar panels or heat pumps. Furthermore, the amended directive would require Member States to establish intermediate national targets for existing buildings that should amount to minimum percentages of energy consumption by 2015 and 'zero energy' by 2020. The definition of zero energy buildings would be clarified and established by the Commission by 2010. Finally, Member States were required to set national action plans by mid-2011, aimed at the development of financial instruments for the improvement of energy efficiency in buildings (i.e. low-interest loans and fiscal rebates).

On 24 April 2009 the European Parliament approved a deal reached with the Council on the Proposal for a Directive on eco-design requirements for energy-related products COM(2008)399, which will regulate products with an impact on energy consumption during use, such as construction and water-using products. The list of products covered by the directive will be provided by the European Commission by 2011. In 2012 the Commission will review the implementation of the directive with a view to extending its coverage.

At a meeting in Sweden on 23 July 2009 the energy ministers of the EU were not able to agree on the establishment of a binding target on energy efficiency as part of a consultation launched by the European Commission on the review of the EU's energy efficiency action plan. The extent of such a target and its timeframe were the two main issues of concern among the Member States. The existing proposal referred to 20 per cent by 2020 but at the moment of writing the target was still not tabled officially.

On 22 July 2009 the European Commission adopted the following four new eco-design regulations aimed at the improvement of energy efficiency of industrial electric motors, water circulators, televisions, and refrigerators and freezers:

- Commission Regulation (EC) No 641/2009 of 22 July 2009 implementing Directive 2005/32/EC of the European Parliament and of the Council with regard to eco-design requirements for glandless standalone circulators and glandless circulators integrated in products;[21]
- Commission Regulation (EC) No 640/2009 of 22 July 2009 implementing Directive 2005/32/EC of the European Parliament and of the Council with regard to eco-design requirements for electric motors;[22]
- Commission Regulation (EC) No 643/2009 of 22 July 2009 implementing Directive 2005/32/EC of the European Parliament and of the Council with regard to eco-design requirements for household refrigerating appliances;[23]
- Commission Regulation (EC) No 642/2009 of 22 July 2009 implementing Directive 2005/32/EC of the European Parliament and of the Council with regard to eco-design requirements for televisions.[24]

The laws will help to save 190 terawatt hours (TWh) of electricity per year by 2020. Industrial electric motors will have to meet minimum efficiency standards from 2011 and 'variable speed drives' will be promoted. Low- and standard-efficiency water circulators will be phased out by 2013 and substituted by high-efficiency circulators as from 2015. As of 2010 all televisions for sale will be more efficient and refrigerator and freezer models with a current efficiency rating of class B and below removed from the market.

NOTES

1. Chapter 6 on Energy Efficiency and Chapter 7 on Renewable Energy are intended to provide a non-exhaustive overall picture of the activities of the EU in these areas. Given the close relation between energy efficiency, renewable energy and climate policy in general, the reader will find reference to the first two areas of legislation in many other parts of this book.

2. Resolution of the Council and the Representatives of the Governments of the Member States, meeting within the Council of 1 February 1993 on a Community programme of policy and action in relation to the environment and sustainable development – A European Community programme of policy and action in relation to the environment and sustainable development, OJ C 138, 17 May 1993, pp1–4, points 21 and 22.

3. Decision No 2179/98/EC of the European Parliament and of the Council of 24 September 1998 on the review of the European Community programme of policy and action in relation to the environment and sustainable development 'Towards sustainability', OJ L 275, 10 October 1998, pp1–13.

4. Communication from the Commission to the Council, the European Parliament, the Economic and Social Committee and the Committee of the Regions on the Sixth Environment Action Programme of the European Community, 'Environment 2010: Our future, Our choice', including Proposal for a Decision of the European Parliament and of the Council Laying down the Community Environment Action Programme 2001–2010 COM(2001)31, OJ C 154E, 29 May 2001, pp218–225.

5. This book predominantly considers measures and policies on energy efficiency introduced as of 2000.

6. Proposal for a Directive of the European Parliament and the Council on the Promotion of End-use Efficiency and Energy Services COM(2003)739, Brussels, 10 December 2003.

7. Communication of the European Commission, EU Green Paper on Energy Efficiency COM(2005)265 Brussels, 22 June 2005.

8. Directive 2005/32/EC of 6 July 2005 establishing a framework for the setting of eco-design requirements for energy-using products and amending Council Directive 92/42/EEC and Directives 96/57/EC and 2000/55/EC of the European Parliament and of the Council, OJ L 121, 22 July 2005, pp29–58.

9. Directive 2006/32/EC of the European Parliament and of the Council of 5 April 2006 on energy end-use efficiency and energy services and repealing Council Directive 93/76/EEC, OJ L 114, 27 April 2006, pp64–85.

10. Communication from the Commission to the European Council and the European Parliament COM(2006)545, Action Plan for energy efficiency: realising the potential, Brussels, 19 October 2006.

11. Communication from the Commission to the European Council and the European Parliament COM(2007)1, An Energy Policy for Europe, Brussels, 10 January 2007.

12. Decision of 21 December 2006 establishing harmonized efficiency reference values for separate production of electricity and heat in application of Directive 2004/8/EC of the European Parliament and of the Council 2007/74/EC, *OJ* L 32 of 6 February 2007, pp183–188.

13. Communication of the European Communities on VAT rates other than standard VAT rates, COM(2007)380, Brussels, 5 July 2007.

14. Consultation Document on the revision of the Energy Labelling Directive 92/75/EEC on the indication by labelling and standard product information of the consumption of energy and other resources by household appliances, 22 September 1992.

15. Communication on a first assessment of national energy efficiency action plans as required by Directive 2006/32/EC on energy end-use efficiency and energy services, COM(2008)11, Brussels, 23 January 2008.

16. European Commission, policy paper 'Addressing the challenge of energy efficiency through information and communication technologies', COM(2008)241, Brussels, 13 May 2008.

17. European Commission Proposal for a Directive on the energy performance of buildings (recast) COM(2008)780, Brussels, 13 November 2008.

18. European Commission Proposal for a Directive on energy labelling on the indication by labelling and standard product information of the consumption of energy and other resources by energy-related products (recast) COM(2008)778, Brussels, 13 November 2008.

19. European Commission Proposal for a Directive on labelling of tyres with respect to fuel efficiency and other essential parameters, Brussels, COM(2008)779, Brussels, 13 November 2008.

20. Communication of the Commission to the European Council, A European Economic Recovery Plan, COM(2008)800, Brussels, 26 November 2008.

21. Commission Regulation (EC) No 641/2009 of 22 July 2009 implementing Directive 2005/32/EC of the European Parliament and of the Council with regard to eco-design requirements for glandless standalone circulators and glandless circulators integrated in products, *OJ* L 191, 23 July 2009, pp35–41.

22. Commission Regulation (EC) No 640/2009 of 22 July 2009 implementing Directive 2005/32/EC of the European Parliament and of the Council with regard to eco-design requirements for electric motors, *OJ* L 191, 23 July 2009, pp26–34.

23. Commission Regulation (EC) No 643/2009 of 22 July 2009 implementing Directive 2005/32/EC of the European Parliament and of the Council with regard to eco-design requirements for household refrigerating appliances, *OJ* L 191, 23 July 2009, pp53–68.

24. Commission Regulation (EC) No 642/2009 of 22 July 2009 implementing Directive 2005/32/EC of the European Parliament and of the Council with regard to eco-design requirements for televisions, *OJ* L 191, 23 July 2009, pp42–52.

CHAPTER 7

Renewable Energy

F ollowing the 5th and 6th EAPs of the EC, where the promotion of and
increase in the share of renewable energy in the production and consumption
of electricity was one of the key milestones, the first provision at the EC level
directly addressing renewable energy was Directive 2001/77/EC of the European
Parliament and of the Council of 27 September 2001 on the promotion of
electricity produced from renewable energy sources in the internal electricity
market.[1] Directive 2001/77/EC was designed to increase 'the contribution of
renewable energy sources to electricity production in the internal market for
electricity and to create a basis for a future Community framework thereof'
(Article 1). Among others, Directive 2001/77/EC set indicative targets for
Member States in terms of the contribution of electricity produced from
renewable energy sources to gross electricity consumption by 2010. These were
indicated in the Annex to the directive, as depicted in Table 7.1.

Furthermore, Directive 2001/77/EC established reporting obligations for
Member States as to their domestic actions and progress towards the compliance
with the above mentioned targets. In particular, in accordance with Article 3(3)
and Article 6(2) of the directive, Member States were required to:

> publish, for the first time not later than 27 October 2003 and thereafter
> every two years, a report which includes an analysis of success in meeting the
> national indicative targets taking account, in particular, of climatic factors
> likely to affect the achievement of those targets and which indicates to what
> extent the measures taken are consistent with the national climate change
> commitment

and:

Table 7.1 *Reference values for Member States' national indicative targets for the quantity of electricity produced from renewable energy sources to gross electricity consumption by 2010*

	RES-E TWh 1997	RES-E% 1997	RES-E% 2010
Belgium	0.86	1.1	6.0
Denmark	3.21	8.7	29.0
Germany	24.91	4.5	12.5
Greece	3.94	8.6	20.1
Spain	37.15	19.9	29.4
France	66.00	15.0	21.0
Ireland	0.84	3.6	13.2
Italy	46.46	16.0	25.0
Luxembourg	0.14	2.1	5.7
Netherlands	3.45	3.5	9.0
Austria	39.05	70.0	78.1
Portugal	14.30	38.5	39.0
Finland	19.03	24.7	31.5
Sweden	72.03	49.1	60.0
United Kingdom	7.04	1.7	10.0
Community	338.41	13.9%	22%

to publish, not later than 27 October 2003, a report on the evaluation existing legislative and regulatory framework with regard to authorisation procedures or the other procedures laid down in Article 4 of Directive 96/92/EC, which are applicable to production plants for electricity produced from renewable energy sources.

According to Article 8 of the directive, on the basis of the reports submitted by the Member States, the European Commission was required to present to the European Parliament and the Council, no later than 31 December 2005 and thereafter every five years, a summary report on the implementation of this directive.

A first assessment of the Member States' performance in terms of promotion of renewable energy was provided by Communication of the European Commission COM(2004)366 – The share of renewable energy in the EU Commission, Report in accordance with Article 3 of Directive 2001/77/EC, evaluation of the effect of legislative instruments and other Community policies on the development of the contribution of renewable energy sources in the EU and proposals for concrete actions released on 26 May 2004.[2] The report indicated that in order to overcome obstacles to growth of renewable energy, such as technical limits and the need to respect the principle of cost-effectiveness, new long-term EU renewable energy targets (2020) would be proposed by the Commission early in 2007.

The Energy Council on 9 September 2004 confirmed the approach of the Commission in postponing renewable energy goals for electricity generation in Europe. While Communication COM(2004)366 was unclear on the possibility to introduce future legally binding targets in terms of renewable energy shares of the

Member States, the Council draft declaration of 8 September 2004 appeared even less stringent, making reference to indicative targets for 2020.

The results of the Council Transport, Telecommunications and Energy of 29 November 2004 backed the proposal of the Commission and agreed on postponing any decision on post-2010 targets for EU renewable electricity generation until after 2007. Energy ministers also decided to include in the post-2010 strategy the considerations related to the latest developments of international negotiations on climate change and the EU ETS.

On 25 April 2005, the Environment Committee of the European Parliament adopted a draft opinion on the share of renewable energy in the EU and proposals for concrete actions focusing on the introduction of legislation on heating and cooling from renewable energy sources and calling for a 20 per cent target of renewable energy share by 2020. On such a basis, the Industry and Energy Committee of the Parliament adopted on 24 June 2005 a report urging the EU to aim at a 25 per cent target in terms of energy coming from renewable sources by 2020. On 29 September 2005, the European Parliament adopted such a report, namely the report on the share of renewable energy in the EU and proposals for concrete actions (2004/2153(INI)), with a reference to a 20 per cent mandatory target for renewable energy in total EU energy consumption by 2020.

With the Declaration of Edinburgh (2005), the MEPs of more than 20 European countries called on the EU to develop a combined renewable energy and energy efficiency strategy (REEES) and stressed the need for the adoption of stronger and binding targets for the Member States on those matters. The declaration was adopted within the framework of the latest biannual inter-parliamentary meeting on renewables and energy efficiency.

On 7 December 2005 the European Commission released Communication on the biomass action plan COM(2005)628,[3] including a proposal to draw up new legislation on renewable heating and cooling in 2006.

On 8 February 2006 the European Commission presented Communication on the EU strategy for biofuels COM(2006)34.[4] Communication COM(2006)34 contained an EU plan to increase the production and the use of transport biofuels based on actions to be taken in areas such as: stimulation of biofuel demand, environmental protection, development of production, feedstock supplies, external trade, improvement of international development and research. Moreover, the plan urged the EU to ensure sustainability was respected in the cultivation for biofuels, as well as ensure the reduction of CO_2 emissions through the removal of limits on blending ethanol into petrol, as well as biodiesel into diesel.

The Council Agriculture and Fisheries of 23 January 2006 welcomed the Commission proposal on the promotion of biomass energy COM(2005)628, which was considered by the European ministers a key step to reduce GHG emissions in Europe.

The European Commission annual report on the Lisbon strategy – Time to move up a gear – released on 25 January 2006 recognized the question of energy supply as one of the top EU priorities and stressed the need to boost energy efficiency, renewable energy and green technologies.

The Committee on Industry, Research and Energy of the European Parliament adopted on 26 January 2006 a draft report with recommendations to the Commission on heating and cooling from renewable sources of energy, expressing firm support for binding targets to promote renewable energy in heating and cooling and for doubling the share of renewables by 2020.

The EU energy ministers meeting in Brussels on 14 March 2006 expressed strong support for the implementation of a biomass energy action plan issued in December and the promotion and growth in renewable energies for the period post-2010. It also backed the intention of the Commission to issue an energy efficiency action plan by the end of 2006.

The European heads of government met for the 2006 spring European Council meeting in Brussels and agreed on a call for European energy policy focusing on the balance between security of supply, competitiveness and environmental sustainability. The Presidency Conclusions of the Brussels European Council of 23 and 24 March 2006 provided support to the conclusions of the Council Transport, Telecommunications and Energy mentioned above, which referred to the increase of renewable energy production and suggested, in respect of long-term targets for renewable energy sources, a share of 15 per cent for renewable energy and 8 per cent for biofuels by 2015 (targets in place at that time were 10 per cent for renewable energy and 5.75 per cent for transport biofuels, both by 2010).

The Council of the EU response to the Commission Communication on a biomass action plan COM(2005)628 and to the Commission Communication on an EU strategy for biofuels COM(2006)34 was a position statement adopted by the EU ministers on 8–9 June 2006, calling for a balanced and integrated increase in biofuel energy use.

On the promotion of biofuels, the results of the public consultation on the directive proposal on biofuels launched by the Commission revealed that the target of 5.75 per cent biofuels share by 2010 was considered difficult to meet and called for the introduction of binding targets in order to promote biofuels also by 2015 and 2020.

On 4 October 2006 the Industry Committee of the European Parliament amended in several parts the Commission proposal on the establishment of the EU transport biofuels strategy. The Parliament response to the Commission included, inter alia, a request to introduce a ban to the use of biofuels derived from palm oil, mainly produced in Southeast Asia, as well as a call for a 'mandatory and comprehensive' certification scheme to ensure sustainable production of biofuels.

The following three documents prepared by the European Commission accompanied the above mentioned Energy Policy for Europe presented on 10 January 2007:

• Communication Renewable Energy Road Map; Renewable energies in the 21st century: building a more sustainable future COM(2006)848 identifying a goal of a 20 per cent share of renewable in the EU energy mix by 2020. The renewable energy road map aimed at combating climate change, the increase of dependence on oil and fossil fuel, as well as the rise in energy prices;[5]

- Biofuels Progress Report, Report on the progress made in the use of biofuels and other renewable fuels in the Member States of the European Union identifying a 10 per cent binding target for biofuels use as vehicle fuel by 2020;[6]
- Green Paper follow-up action, Report on progress in renewable electricity COM(2006)849, which showed that the overall share of renewable electricity by 2010 will fall just short of the original target agreed in 2004 (21 per cent), reaching 19 per cent by 2010.[7]

On 14 February 2007 the European Parliament adopted a resolution on climate change, which included a proposition for a target of 25 per cent of renewable energy in the share of EU energy use by 2020.

The Council Transport, Telecommunications and Energy of 15 February 2007 supported the proposal of the Commission – An Energy Policy for Europe – for the introduction of a mandatory 10 per cent target for biofuels by 2020, but failed to agree on a 20 per cent target for renewable energy share in the EU.

The European Council of 8–9 March 2007, Brussels, Presidency Conclusions, welcomed the EU energy action plan for 2007–2009 including a binding target to increase the use of renewables of 20 per cent by 2020 to be divided into binding national targets among the 27 Member States (Commission Proposal for a Directive expected by the end of 2007); plus a binding target for the share of biofuels in EU transport fuel consumption of 10 per cent by 2020.

The Commission released in April 2007 a public consultation document on Biofuel issues in the new legislation on the promotion of renewable energy. The document provided for initial details of the legislative proposal of the Commission over the targets for the promotion of biofuels. According to the document, in order to receive market benefits, biofuel products would have to prove their positive carbon balance.

The Directorate General for Agriculture and Rural Development of the European Commission released on 25 July 2007 the Impact Assessment on the Renewable Energy Roadmap, which ensured that the impact on land use of the 10 per cent target for biofuels in the EU27 agreed in March 2007 by the Council would not harm environmental sustainability, or other sectors such as food, feed, pulp and paper.

On 25 September 2007 the European Parliament adopted a resolution on the Road Map for Renewable Energy in Europe to be tabled by the European Commission and supported the introduction of a target of 20 per cent and 10 per cent shares respectively for renewables and biofuels by 2020. Furthermore, the Parliament called for a strong legal framework to guarantee that the 20 per cent goal is met by the Member States, as well as the full integration of renewable energy sources in the EU internal energy market.

According to a draft version of the Proposal for a Directive on the promotion of the use of renewable energy sources circulated on 5 December 2007, with the new regime the European producers of electricity from renewable energy sources would be allowed to sell tradable certificates of renewable energy (guarantees of origin – GOs) to operators outside of the EU. Such trading could be used by the Member States to comply with the 2020 targets for renewable energy. The proposal identified different mechanisms and rules for the trading of renewable energy certificates, in particular for green electricity produced after 2010.

On 23 January 2008, the European Commission tabled the Proposal for a Directive on the promotion of the use of energy from renewable sources COM(2008)19 establishing different mandatory targets for the Member States in relation to the increase of the share of renewable energy in the EU level of gross final consumption of energy. In the transport sector this was equivalent to 20 per cent by 2020 and a 10 per cent binding minimum target for biofuels in transport by 2020. Furthermore, the proposal ensured that full control of the renewable energy support instruments would remain under the Member States.[8] It further introduced a new trading system for renewable energy certificates (GOs). According to the proposal, only Member States in compliance with their interim targets ahead of the 20 per cent target by 2020 would be allowed to sell the GOs of their domestic producers to other Member States. The interim targets were defined according to an 'indicative trajectory' based on the following increases in the use of renewable sources in the period 2005–2020:

- 25 per cent by 2012;
- 35 per cent by 2014;
- 45 per cent by 2016;
- 65 per cent by 2018.

Under proposal COM(2008)19, Member States were required to submit national renewables action plans to the European Commission by March 2010. These plans have to include the details of targets and measures for the share of renewables in the electricity, transport, and heating and cooling sectors at the national level.

The mandatory targets of the Member States on the share of energy from renewable sources in final consumption of energy included in Annex I of proposal COM(2008)19 are shown in Table 7.2.

On 29 January 2008 the Agriculture Committee of the European Parliament adopted a non-binding resolution urging the European Commission to propose a new directive on the promotion of the use of gas produced from animal and plant wastes (biogas) for electricity production, heating and cooling, and road transport. According to the committee, the proposed directive should include production targets, a system for the incentive of new installations as well as quality standards for biowaste.

On 11 March 2008 the European Commission submitted to the EU governments a paper including Technical clarifications on guarantees of origin (GOs) and transfer schemes, which highlighted the details of the new trading system designed for the promotion of renewable energy sources in the EU included in the integrated climate and energy package adopted in January 2008. The paper of the Commission clarified that GO trading could not start before 2013, thus leaving Member States free to give priority to feed-in tariff systems in the meantime.

On 12 March 2008 the Committee on Agriculture and Rural Development of the European Parliament adopted a non-legislative resolution on sustainable

Table 7.2 *National overall targets for the share of energy from renewable sources in final energy consumption in 2020*

	Share of energy from renewable sources in final consumption of energy, 2005 (S_{2005})	Target for share of energy from renewable sources in final consumption of energy, 2020 (S_{2020})
Belgium	2.2%	13%
Bulgaria	9.4%	16%
The Czech Republic	6.1%	13%
Denmark	17.0%	30%
Gemany	5.8%	18%
Estonia	18.0%	25%
Ireland	3.1%	16%
Greece	6.9%	18%
Spain	8.7%	20%
France	10.3%	23%
Italy	5.2%	17%
Cyprus	2.9%	13%
Latvia	34.9%	42%
Lithuania	15.0%	23%
Luxembourg	0.9%	11%
Hungary	4.3%	13%
Malta	0.0%	10%
The Netherlands	2.4%	14%
Austria	23.3%	34%
Poland	7.2%	15%
Portugal	20.5%	31%
Romania	17.8%	24%
Slovenia	16.0%	25%
The Slovak Republic	6.7%	14%
Finland	28.5%	38%
Sweden	39.8%	49%
United Kingdom	1.3%	15%

agriculture and biogas: a need for review of EU legislation urging the European Commission to propose a directive on the exploitation of gas from agricultural waste (biogas and biowaste) and to include the promotion of biogas installations in the Proposal for a Directive on renewable energy.

The ad hoc working party, established on 22 February 2008 by COREPER (Committee of Permanent Representatives) in order to draft a common sustainability scheme for biofuels production, released its final criteria on 8 May 2008. Only biofuels produced in accordance with those sustainability criteria would count towards the EU target of 10 per cent share for biofuels in road transport fuels by 2020. The final version of these criteria would have to be decided by the Council of Ministers and included in both proposals for a directive on fuel quality COM(2007)18 and for a directive on the promotion of the use of energy from renewable sources COM(2008)19. The sustainability criteria include, among others, minimum levels of CO_2 emissions for the production of biofuels; environmental and social factors that biofuels imported into the EU shall comply with; and a methodology to calculate CO_2 savings.

The Committee on Industry, Research and Energy of the European Parliament was divided on the obligation for the Member States to ensure that at least 10 per cent of the share of renewable energy in transport fuel comes from biofuels (COM(2008)19). At a meeting on 29 May 2008, MEPs in the Committee could eventually reach a compromise for an obligation at 8 per cent.

On 10 July 2008 the French presidency presented draft revisions to proposal COM(2008)19 on the promotion of the use of energy from renewable sources, including the support for a mechanism to assess the compliance of Member States with their national targets for renewable energy use. Furthermore, the French presidency proposal provided for a new definition of renewable energy generation certificates (GOs) thus confirming that GOs would not be tradable but only used to verify Member States' progress towards compliance with their renewables targets.

EU energy ministers met for a two-day informal meeting in Paris on 4 and 5 July 2008 and announced their opposition to the 10 per cent biofuels target in road transport fuels by 2020, as proposed by the Commission in January 2008. Ministers confirmed their concerns over the need for clear sustainability criteria for the production and imports of biofuels as well as the potential negative effects of these measures on land use, biodiversity and food prices, especially in developing countries.

Similar conclusions were reached on 7 July 2008 in the Environment Committee of the European Parliament, where MEPs approved a few amendments to the draft renewables directive and voted against the introduction of the 10 per cent EU target to increase the share of biofuels. Among the compromise amendments adopted, the most relevant was the obligation to provide a 4 per cent share of renewable energy sources in transport fuels by 2015. Up to half of this target would have to be provided by first generation biofuels and a fifth by green electricity, hydrogen and second generation biofuels.

On 11 September 2008 the Industry Committee of the European Parliament voted a set of compromise amendments to the draft renewable energy directive including binding interim targets for Member States before 2020, the introduction of a penalty system for countries in non-compliance with the reduction targets and the modification of the 10 per cent target for biofuels. In respect of biofuels, the Parliament's amendments introduced the possibility to review the 10 per cent target by introducing the obligation to evaluate the impact of the EU biofuels targets on food security and biodiversity by 2014. Furthermore, the committee adopted a GHG savings threshold of 45 per cent for biofuels, rising to 60 per cent from 2015 and also introduced the principle of financial penalties to be imposed on Member States failing to comply with the targets set by the directive.

As a follow up of the COREPER meeting of 17 September 2008, on 23 September 2008 the French presidency presented a few compromise amendments on the proposal for the new renewable energy directive. The amendments attenuated the binding targets on the share of renewables in the energy consumption of the Member States by 2020, notably by introducing a review clause for 2014 that would give a certain degree of flexibility to the Member States in the compliance with their targets. The changes proposed by the French presidency were reflected in a draft compromise text agreed on 29 October 2008 with a broad consensus by the Member States. The review clause in 2014 was confirmed and furthermore the Member States agreed on the sustainability criteria for biofuels (35 per cent GHG emissions savings compared to conventional fuels, increasing to 50 per cent by 2017) and on the default values for biofuels (Annex I) necessary to prove compliance by the producers with the emissions savings targets.

Informal agreement on the directive on the promotion of the use of energy from renewable sources was reached between the European Parliament and the Council on 9 December 2008 after a number of Member States dropped their opposition to the new law. On 12 December 2008 the European Council reached a unanimous agreement on the integrated climate and energy package and urged the Council and the Parliament to adopt the draft legislation including the details of the final compromise to which paragraph 20 of the European Council conclusions referred. On 17 December 2008, the European Parliament approved with a big majority the package including a Legislative Resolution on the Proposal for a Directive of the European Parliament and of the Council on the promotion of the use of energy from renewable sources COM(2008)19. As mentioned above, COM(2008)19 proposed to establish mandatory targets for the Member States in relation to the share of renewable energy in the level of gross final consumption of energy and in the transport sector.

Directive 2009/28/EC on the promotion of the use of energy from renewable sources and amending and subsequently repealing Directives 2001/77/EC and 2003/30/EC was adopted by the European Parliament and the Council on 23 April 2009.[9]

Directive 2009/28/EC regulates the cooperation between the Member States and third countries in this sector (statistical transfers, joint projects, GOs, information and training on access to electricity grid) and establishes sustainability criteria for biofuels and bioliquids. National targets have been defined in consistency with the EU overall target of 20 per cent share of energy from renewable sources by 2020 (Article 3). The level of the targets is the same indicated in the proposal of January 2008, with the only exception being Latvia whose baseline (2005) and target (2020) have been reduced to 32.6 per cent and 40 per cent, respectively. Directive 2009/28/EC imposes a minimum level of 10 per cent of energy from renewable sources in all forms of transport by 2020. Under Article 4 the Member States are required to submit by 30 June 2010 a national action plan in which adequate measures designed to comply with the national target shall be identified. The Commission will be responsible for the evaluation of the national plans and may issue a recommendation addressed to the Member States. Under Article 17, Directive 2009/28/EC sets out specific sustainability criteria for biofuels and other bioliquids cultivated inside or outside the Community. These are:

- GHG emissions savings from the use of biofuels and other bioliquids: 35 per cent until 2017, 50 per cent from 2017 and 60 per cent after 2017;
- biofuels and other bioliquids shall not be made from raw materials obtained from land with high biodiversity value;
- biofuels and other bioliquids shall not be made from raw materials obtained from land with high carbon stock;

- biofuels and other bioliquids shall not be made from raw materials obtained from land that was peatland in January 2008.

Member States are required to submit a national progress report to the Commission by 31 December 2011 and every two years thereafter. The reports shall include information on national progress towards the targets, information about the support scheme and the system of GOs (Article 22).

Directive 2009/28/EC will be reviewed by the Commission by 2014, which will have no consequences on the national renewable energy targets or on the support mechanisms (Article 23(8)). This condition is merely symbolic, however, since the European Commission has the power to review any EU legislation and propose changes at any time.

On 24 April 2009 the European Commission released the Renewable Energy Progress Report COM(2009)192[10] in accordance with Article 3 of Directive 2001/77/EC, Article 4(2) of Directive 2003/30/EC and the EU Biomass Action Plan COM(2005)628. The report confirmed that the non-binding renewable energy target set under Directive 2001/77/EC, notably the increase of 12 per cent of the share of renewables in the EU final energy consumption by 2010, respectively 21 per cent and 5.75 per cent for the electricity and transport sectors, will not be achieved. The prediction of the Commission considered an increase in the share for the electricity and transport sectors of 19 per cent and 4 per cent. The slow progress of the Member States in the increase of renewable energy is monitored by the Commission, which has initiated since 2004 61 legal proceedings against Member States for their failure to take action in accordance with the obligations set by Directive 2001/77/EC.

On 30 June 2009 the European Commission released a Decision establishing a template for national renewable energy action plans under Directive 2009/25/EC that Member States were required to submit by 30 June 2010. The national plans include information and details on the Member States' plans to achieve the 2020 targets for renewable energy in transport, electricity production, heating and cooling, revised building codes, and biomass and biofuel sustainability criteria. The reason behind the introduction of the template is to ensure completeness and comparability among Member States on the implementation of the new renewables directive.

NOTES

1. Directive 2001/77/EC of the European Parliament and of the Council of 27 September 2001 on the promotion of electricity produced from renewable energy sources in the internal electricity market, *OJ* L 283, 27 October 2001, pp33–40.

2. Communication of the Commission to the Council and the European Parliament, The share of renewable energy in the EU Commission Report in accordance with Article 3 of Directive 2001/77/EC, evaluation of the effect of legislative instruments and other Community policies on the development of the contribution of renewable energy sources in the EU and proposals for concrete actions COM(2004)366, Brussels, 26 May 2004.

3. European Commission, Communication on the biomass action plan COM(2005)628, Brussels, 7 December 2005.

4. European Commission, Communication on the EU strategy for biofuels COM(2006)34, Brussels, 8 February 2006.

5. European Commission Communication Renewable Energy Road Map; Renewable energies in the 21st century: building a more sustainable future COM(2006)848, Brussels, 10 January 2007.

6. European Commission, Biofuels Progress Report, Report on the progress made in the use of biofuels and other renewable fuels in the Member States of the European Union, Brussels, 10 January 2007.

7. Communication from the Commission to the Council and the European Parliament, Green Paper follow-up action, Report on progress in renewable electricity COM(2006)849, Brussels, 10 January 2007.

8. European Commission, Proposal for a Directive on the promotion of the use of energy from renewable sources COM(2008)19, Brussels, 23 January 2008.

9. Directive 2009/28/EC of the European Parliament and of the Council of 23 April 2009 on the promotion of the use of energy from renewable sources and amending and subsequently repealing Directives 2001/77/EC and 2003/30/EC, OJ L 140, 5 June 2009, pp16–62.

10. European Commission, Renewable Energy Progress Report COM(2009)192, Brussels, 24 April 2009.

CHAPTER 8

Transport Sector and GHG Emissions

C arbon dioxide emissions from the transport sector represent a large portion of GHG emissions in the EU. In general terms the following points were considered the priorities of the transport policy of the EC in relation to the fight against climate change in the early 1990s:

- Improvement of the rail system;
- Investments in trans-European networks to reduce road traffic;
- Harmonization of professional commercial fuels;
- Improvement of public transport.

Among the first measures adopted by the EC in this field Communication from the Commission to the Council and the European Parliament on a community strategy to reduce CO_2 emissions from passenger cars and improve fuel economy COM(1995)689[1] and the conclusions of the Council Environment of 25 June 1996 can be highlighted.

On 12 September 2001 the European Commission issued the White Paper European transport policy for 2010: Time to decide COM(2001)370,[2] establishing a long-term strategy on transport that included recommendations and ideas on a large range of policies and measures necessary to foster European sustainable transport. Amongst others, measures indicated in the White Paper consider the boosting of rail and maritime connections for long-distance freight transport. The main points of the White Paper were:

- Regulated competition: improving quality in the road sector, revitalizing the railways, controlling the growth in air transport;
- Linking up the modes of transport: linking up sea, inland waterways and rail,

helping to start up intermodal services: the new Marco Polo programme, creating favourable technical conditions;

• Unblocking major routes: towards multimodal corridors giving priority to freight, a high-speed passenger network, improving traffic conditions, major infrastructure projects;
• The headache of funding: unsafe roads, the facts behind costs to users, transport with a human face, rationalizing urban transport;
• Managing the globalization of transport in view of Community enlargement.

On 22 June 2006 the Commission released the Communication Keep Europe moving – Sustainable mobility for our continent, Mid-term review of the European Commission's 2001 Transport White Paper COM(2006)314,[3] including the Commission orientation for the future EU transport policy. COM(2006)314 highlighted the importance of energy use and the environmental impact of mobility. Moreover, it identified additional instruments to be adopted at the European level in order to achieve sustainable mobility, such as a freight logistics action plan, more green and efficient transport systems, a debate on how to change mobility of people in urban areas, an action plan to boost inland waterways and an ambitious programme for green power in trucks and cars.

The Council Transport, Telecommunications and Energy of 12 October 2006 urged a revision of the environmental objectives of the 2001 White Paper of the European Commission COM(2001)370. It also supported the main goal of such an instrument, namely the focus on the environmental and social effects of mobility in the EU. Ministers called for an increase in intelligent transport systems, as well as efficiency and the sustainability of transport.

The EEA Report No 1/2007 Transport and environment: On the way to a new common transport policy confirmed that the reduction of GHG emissions from transport continued to be a priority of EU climate policy.[4] The EEA annual environment assessment of the European transport sector showed that the GHG emissions from transport had increased in the period 1990–2004 by 34 per cent, and in particular, freight transport emissions were up by 43 per cent, passenger transport by 20 per cent and aviation by 96 per cent.

The consultation document Preparation of an impact assessment on the internalisation of external costs issued by the Directorate General for Energy and Transport of the European Commission in November 2007 is at the foundations of an online consultation aimed at the preparation of a European strategy on the internalization of external costs of EU transport. The consultation document identified the inclusion of all major modes of transport – road, rail, maritime and inland waterway transport – in the EU ETS as one of the policy options to internalize the environmental costs of transport.

On 11 March 2008 the European Parliament adopted a non-legislative resolution on sustainable European transport policy calling for the transport sector to adopt a 20 per cent obligation to reduce GHG emissions by 2020 compared to 1990 levels. Emissions from road transport in key areas such as large cities and sensitive intercity corridors were considered together with the maritime transport.

Measures indicated to reduce the emissions of transport were technological improvements as well as market-based instruments.

On 8 July 2008 the European Commission released Communication Greening transport COM(2008)433, which aimed at promoting the sustainability of transport.[5] COM(2008)433 covered the following three proposals:

- Strategy to internalize the external costs of transport: focuses on making transport prices better reflect their real cost to society so that environmental damage and congestion can be reduced while boosting the efficiency of transport and ultimately the economy as a whole.
- Proposal for a Directive on road tolls for lorries: enabling Member States to reduce environmental damage and congestion through more efficient and greener road tolls for lorries.
- Rail transport and interoperability communication: setting out how to reduce the perceived noise from existing rail freight trains by 50 per cent and the measures the Commission and other stakeholders will need to take in the future to achieve this.

8.1   ROAD TRANSPORT

A typical measure of the EC to reduce CO_2 emissions from passenger cars and improve fuel economy is the regulatory self-commitment (voluntary agreement) concluded by the Community with automobile manufacturers' associations:

- Commission Communication Implementing the Community Strategy to reduce CO_2 emissions from cars: an environmental agreement with the European automobile industry COM(1998)495;[6]
- Commission Communication Implementing the Community strategy to reduce CO_2 emissions from cars: outcome of the negotiations with the Japanese and Korean automobile industries COM(1999)446;[7]
- Commission Recommendation 2000/304/EC of 13 April 2000 on the reduction of CO_2 emissions from passenger cars (Japanese Automobile Manufacturers Association – JAMA);[8]
- Commission Recommendation 2000/303/EC of 13 April 2000 on the reduction of CO_2 emissions from passenger cars (Korean Automobile Manufacturers Association – KAMA).[9]

In 1999 a similar agreement was concluded by the EC with European car manufacturers through Commission Recommendation 1999/125/EC of 5 February 1999 on the reduction of CO_2 emissions from passenger cars.[10]

All the above mentioned voluntary agreements established quantified CO_2 emissions reduction goals for the average of new passenger cars sold in the EU, i.e. 140g CO_2/km (to be achieved by 2009 by JAMA and KAMA and by 2008 by ACEA (European Automobile Manufacturers' Association).

Another strategy of the EC to reduce the contribution to global warming of the road transport sector concerns the introduction of limits and quality requirements for the fuel utilized by all different kinds of vehicles. In this context, in the late 1990s the EC introduced a provision to improve information relating to the fuel economy and CO_2 emissions of new passenger cars through the introduction of related labels attached or displayed on each new passenger car model on sale. This provision was Directive 1999/94/EC of the European Parliament and of the Council of 13 December 1999 relating to the availability of consumer information on fuel economy and CO_2 emissions in respect of the marketing of new passenger cars.[11]

Considerations on the introduction of fiscal measures to reduce CO_2 emissions from new passenger cars in the EC were initiated at the beginning of the 2000s. On this item, in January 2002 the European Commission established an expert group to study the opportunity to introduce vehicle adequate taxes, Final Report study undertaken by COWI A/S in January 2002.

Through Communication COM(2002)431 Taxation of passengers cars in the European Union – options for action at national and Community levels COM(2002)431,[12] the European Commission presented its strategy and options for future action in the field of passenger car taxation (for example, modernization and simplification of the existing vehicle taxation systems and removal of tax obstacles and distortions to free circulation of passenger cars within the internal market). On 6 November 2003 the European Parliament welcomed Communication COM(2002)431 with the resolution on the Commission communication on taxation of passenger cars in the European Union.

Commission Communication Implementing the Community Strategy to reduce CO_2 emissions from cars – Second annual report on the effectiveness of the strategy (Reporting year 2000) COM(2001)643 acknowledged the progress of the automobile manufacturers associations in the reduction of CO_2 emissions.[13]

Commission Communication Implementing the Community Strategy to reduce CO_2 emissions from cars: Fourth annual report on the effectiveness of the strategy (Reporting year 2002) COM(2004)78 indicated an average reduction of CO_2 emissions from passenger cars in the period 1995 to 2002 from 186g CO_2/km to 166g CO_2/km (10.8 per cent) and recognized that the EU was moving towards introducing additional measures to bring down CO_2 emissions from light vans (light commercial vehicles – LCVs).[14] In this respect, the European Commission referred also to the Proposal for a Directive of the European Parliament and of the Council amending Council Directives 70/156/EEC and 80/1268/EEC as regards the measurement of carbon dioxide emissions and fuel consumption of N1 vehicles COM (2001)543, 26 February 2002.[15]

According to Commission Communication Implementing the Community Strategy to reduce CO_2 emissions from cars: Fifth annual Communication on the effectiveness of the strategy, COM(2005)269 released on 22 June 2005 and based on annual figures on GHG emissions from passenger cars, European, Japanese and Korean car manufacturers still needed to double efforts to cut their emissions in order to achieve the targets established under the 1998 voluntary agreements.[16]

On 5 July 2005 the European Commission presented a proposal for a Council Directive on passenger car related taxes COM(2005)261, establishing annual circulation taxes and requiring Member States to ensure that at least 50 per cent of car taxes be based on vehicles' CO_2 emissions by 2010.[17]

On 9 November 2005 Directive 2005/55/EC on the approximation of the laws of the Member States relating to the measures to be taken against the emission of gaseous and particulate pollutants from compression-ignition engines for use in vehicles, and the emission of gaseous pollutants from positive-ignition engines fuelled with natural gas or liquefied petroleum gas for use in vehicles entered into force. Directive 2005/55/EC set 'Euro 3' and 'Euro 4' limits for lorries and paved the way for stricter 'Euro 5' regulation to be introduced in the near future.[18]

On 15 July 2005 the European Commission launched a stakeholder consultation on the new round of EU car emission standards (Euro 5 standards for new vehicles). This consultation phase was concluded by early November 2005. The Commission collected more than 50 contributions. The European car manufacturers association ACEA criticized the proposed regulation considering the limits as 'impractical' and 'unjustified', while green groups calls for even stricter limits, including a 75 per cent reduction of nitrous oxide and hydrocarbon emissions from petrol cars instead of the proposed 25 per cent cut against the current Euro 4 limits.

The Commission launched on 14 November 2005 a public consultation in preparation for a review of EU transport policies set out in the 2001 White Paper European transport policy for 2010: Time to decide.

On 21 December 2005, the European Commission presented two legislative proposals for the reduction of GHG emissions from road transport. First, the proposal for a regulation of the European Parliament and of the Council on type approval of motor vehicles with respect to emissions and on access to vehicle repair information, amending Directive 72/306/EEC and Directive../../EC, COM(2005)683,[19] namely the regulation on new Euro 5 vehicle emissions standards. The text of such a proposal was issued after the collection of feedback from stakeholders provided in the consultation period that coincided with the Euro 5 emission limits initially proposed by the European Commission in July 2005. The second legislative proposal was the Proposal for a Directive of the European Parliament and of the Council on the promotion of clean road transport vehicles COM(2005)634 establishing clean vehicle purchase quota for public authorities in compliance with a European 'enhanced environmentally friendly vehicle' (EEV) standard agreed in early 2005.[20]

The EU Environment Council of 9 March 2006 strongly supported the post-Euro 5 nitrogen oxide emission targets for diesel cars, namely the above mentioned Euro 5 regulation proposed by the European Commission.

On 24 August 2006, the Commission released Communication COM(2006)463, Implementing the Community Strategy to reduce CO_2 emissions from cars: Sixth annual Communication on the effectiveness of the strategy together with a Commission staff working document on CO_2 emissions from new cars in 2004 SEC/2006/1078 showing that EU and Japanese car makers reduced average new car CO_2 emissions by 1.2 per cent in 2004: 161g and 170g per kilometre, respectively – still being far from the target of 140g/km by 2008 and 2009.[21] European and Japanese car producers had voluntary committed to cut their CO_2 emissions annually by 3.3 per cent and 3.5 per cent, respectively. On 29 August 2006 the Commission announced in a press release that a failure to meet the above voluntary targets would force the European executive body to adopt binding legislation on car makers for the reduction of their GHG emissions.

On 5 September 2006 the European Parliament adopted a draft legislative resolution on the proposal for a Council directive on passenger car related taxes COM(2005)261 backing the document of the European Commission. The Parliament supported the introduction of more environmentally friendly national passenger car taxes and agreed on the target and timetable proposed by the Commission in 2005. Furthermore, the Parliament voted in favour of taxes based also on fuel consumption and emissions of air pollutants such as fine particles and nitrogen oxides, and not just CO_2 emissions.

On 13 September 2006 the Environment Committee of the European Parliament voted on the Euro 5 emission control proposals of the European Commission – type approval of motor vehicles with respect to emissions and on access to vehicle repair information, amending Directive 72/306/EEC – and adopted several compromise amendments to the proposal including a less stringent nitrous oxide limit for petrol cars and a more stringent limit for diesel cars. For Euro 6 standards, the Committee proposed a 'fuel neutral' 70mg/km nitrous oxide limit on all types of passenger cars. Furthermore, the Committee asked for the establishment of tougher emission limits on passenger cars starting from 2014, along with the agreed interim limits due to be adopted by 2010.

On 13 December 2006, the European Parliament adopted a legislative report including the compromise deal with the Council on Euro 5 (fine particles, hydrocarbons and nitrogen oxides) and Euro 6 (nitrous oxide) emission standards for new passenger car and van models. These limits will be launched in the EU market from 2009 and 2014, respectively.

On 16 January 2007, the Commission adopted the report on the targets contained in Article 7(2)(b) of Directive 2000/53/EC on end-of-life vehicles COM(2007)5, which confirmed the targets established by the end-of-life vehicle (ELV) directive of 2000.[22] The ELV directive set a requirement for the Member States to recycle 85 per cent and recover 95 per cent of scrap cars in Europe by 2015, which could contribute a 1 million tonnes reduction in CO_2 emissions.

On 31 January 2007 the European Commission released the Proposal for a Directive of the European Parliament and of the Council COM(2007)18 amending Directive 98/70/EC as regards the specification of petrol, diesel and gas-oil and the introduction of a mechanism to monitor and reduce greenhouse gas emissions from the use of road transport fuels and amending Council Directive 1999/32/EC, as regards the specification of fuel used by inland waterway vessels and repealing Directive 93/12/EEC.[23] The proposal required fuel companies to monitor and report the life-cycle emissions of their fuels by January 2009 and to cut CO_2 emissions by 10 per cent by 2020. A policy paper on carbon emissions from new cars announced by the European Commission was delayed because of internal disagreement in the Commission. The proposal was accompanied by the Impact assessment of the review of the fuel quality directive SEC(2007)56.

On 7 February 2007 the European Commission presented a plan for a cleaner automobile industry. Communication COM(2007)19, Results of the review of the Community Strategy to reduce CO_2 emissions from passenger cars and light-commercial vehicles proposed the establishment of binding CO_2 emission limit targets for passenger cars and LCVs of 130g/km by 2012.[24] This target represented the compromise between the environment commissioner Stavros Dimas (who had initially proposed stricter limits) and the automobile industry (which feared it would have to lay off workers as a consequence of the new EU policy). On 14 February 2007 the European Parliament adopted a resolution on climate change calling for the introduction of a binding obligation on car makers to reduce CO_2 emissions from new cars to an average 120 g/km by 2012. The Council Environment of 20 February 2007 gave strong support to binding legislation to reduce CO_2 emissions for new cars and vans as proposed by the Commission with proposal COM(2007)19.

On 3 May 2007 the Council and the European Parliament reached an agreement on the final text of the European regulation on type approval of motor vehicles with respect to emissions from light passenger and commercial vehicles (Euro 5 and Euro 6) and on access to vehicle repair and maintenance information based on Commission proposal COM(2005)683. The regulation established Euro 5 limits for fine particles, hydrocarbons and nitrogen oxides from September 2009 and Euro 6 limits for nitrous oxides from September 2014.

On 8 June 2007 EU transport and energy ministers agreed on the new EU strategy on energy and transport. This was based on four priorities: the improvement of energy efficiency in all modes of transport; the increase of the use of alternative and renewable fuels; the promotion of energy-friendly behaviours; and integrated transport systems aimed at the reduction of energy use.

Members of the European Parliament were not united on the decision over the deadline for mandatory obligations on CO_2 emission reductions for new vehicles. This emerged on 26 June 2007 within the Committee on the Environment, Public Health and Food Safety debate when Rapporteur Davies's proposal for the introduction of a mandatory target of 120g/km by 2015 (draft report on the Community Strategy to reduce CO_2 emissions from passenger cars and light-commercial vehicles) gained support from only half of MEPs. The European Commission proposed a target of 130g/km by 2012 (Communication COM(2007)19, Results of the review of the Community Strategy to reduce CO_2 emissions from passenger cars and light commercial vehicles).

On 12 July 2007 the European Parliament adopted a non-legislative resolution on keeping Europe moving urging the adoption of a strategy and an overall target for the contribution of the transport sector to the reduction of GHG emissions (−20 per cent by 2020 over 1990 levels). The document of the Parliament focused on the shift from road transport to rail and to this end it mentioned the completion of the trans-European network as well as more financing for infrastructure.

On 16 July 2007 the Commission launched a public consultation on future limits for polluting emissions from buses and trucks. The Commission was preparing a draft legislative proposal for Euro 6 standards for emissions from heavy-duty vehicles. While Euro 4 standards applied since 2006 and Euro 5 from 2009, the Commission was assessing the possibility for tougher limits for particulate matter, nitrogen oxides, hydrocarbons, carbon monoxide and ammonia from heavy-duty diesel and natural gas or liquefied petroleum gas engines.

The European Parliament's draft report of 19 July 2007 on Communication COM(2007)18 of the Commission amending directive 98/70/EC on the quality of transport fuels focused on the introduction in the Commission proposal of biofuels sustainability criteria to prevent deforestation and water shortages.

On 5 September 2007 the European Federation for Transport and Environment released the report Reducing CO_2 emissions from cars containing data on the car industry's voluntary commitment of 140g/km by 2008/2009. According to the report, CO_2 emissions from European cars fell by 0.7 per cent in 2006. On the same day, the Council agreed on a compromise text on the proposed EU fuel quality directive revision indicating a cut of 10 per cent in GHG emissions from transport fuels by 2010

On 12 September 2007 the Parliament's Environment Committee adopted a non-binding resolution urging car makers to meet the EU target of an average 120g/km by 2012 through improvements in vehicle technology. Furthermore, Rapporteur Davies's proposal included the establishment a new market mechanism for the reduction of vehicle emissions from 2011.

On 17 September 2007 the European Commission issued Communication COM(2007)541 Towards Europe-wide safer, cleaner and efficient mobility: The first intelligent car report including, inter alia, a plan to increase the environmental impact of cars.[25] The plan included the improvement of information and communication technologies, as well as the introduction of new technologies to reduce emissions from road transport.

A compromise note presented by the EU presidency (Portugal) on 2 October 2007 on the Commission proposal COM(2007)18 introducing a mechanism to monitor and reduce gas emissions from the use of road transport fuels by 10 per cent over a decade and amending Council Directive 1999/32/EC, showed that the EU Member States were still not united on the document. In particular, the major reservations, mainly expressed by the UK, Poland, Italy and Greece, concerned the doubts on the possibility to achieve the proposed targets, as well as the appropriateness of the measures.

On 18 October 2007, the European Commission released the Fourth annual report – Quality of petrol and diesel fuel used for road transport in the European Union (Reporting year 2005) COM(2007)617, on the implementation of Directive 98/70/EC on fuel quality which highlighted, inter alia, the decrease of sulphur content in road fuels sold within the EU in 2005.[26]

On 24 October 2007, following the above mentioned vote of 12 September 2007, the European Parliament adopted a resolution on the Community Strategy to reduce CO_2 emissions from passenger cars and light commercial vehicles, which referred to legally binding cuts in CO_2 emissions from new cars equivalent to 125g/km by 2015, compared with the Commission proposal of 120g/km.

On 26 November 2007 the Committee on the Environment, Public Health and Food Safety of the European Parliament adopted a set of compromise amendments to Commission proposal COM(2007)18 on the quality of road transport fuels for the new EU fuel quality directive. The amendments added minimum sustainability criteria for biofuels, such as a lower level of life-cycle emission of at least 50 per cent compared with fossil fuels. On the same topic, on 20 December 2007, the Council Environment agreed on the progress report on the proposal for the revision of the EU fuel quality directive.

On 19 December 2007 the European Commission released proposal COM(2007)856 for a Regulation of the European Parliament and of the Council setting emission performance standards for new passenger cars as part of the Community's integrated approach to reduce CO_2 emissions from light-duty vehicles.[27] The proposal was aimed at reducing CO_2 emissions from new cars to 130g/km by 2012, although the extent of the target was supposed to change according to the size and weight of vehicle. Targets had to be applied to each manufacturer and penalties set in the event that fleet emissions were above the targets.

On the same day, the European Commission adopted proposal COM(2007)817 for a Directive on the promotion of clean and energy efficient road transport vehicles setting the regulatory framework for a new clean vehicle procurement law as from 2012.[28] According to the proposal, all public authorities in the EU would be required to consider lifetime costs of fuel consumption and emissions of CO_2, nitrogen oxides, hydrocarbons and particulate matter when procuring road vehicles. Fuel consumption and pollution emissions costs would have to be considered by public authorities in the calculation of vehicle's prices for procurement decisions.

On 21 December 2007 the European Commission tabled proposal COM(2007)851 for a regulation on type-approval of motor vehicles and engines with respect to emissions from heavy duty vehicles (Euro 6) and on access to vehicle repair and maintenance information.[29] The proposal laid down the Euro 6 limits for pollutant emissions for heavy-duty vehicles, namely nitrogen oxides and particulate matter emissions from new trucks, buses and other vehicles in the EU from 2013 by amending four existing directives.

On 15 January 2008 the European Parliament adopted a resolution on the future of the EU car industry, which included the preference for a 125g/km target by 2015 for new cars and the opposition to the draft proposal of the European Commission on CARS 21 establishing the 2012 deadline for the cutting of new car CO_2 emissions.

On 29 January 2008 the Committee on Economic and Monetary Affairs of the European Parliament adopted a report on the proposal of the European Commission to harmonize tax duties of commercial road fuels. While the Commission proposed the raising of the EU minimum tax rate for diesel by 2012 in order to match with the minimum rate for unleaded petrol, the MEPs voted for the deadline of 2015 and rejected a tax increase for both fuels by 2014.

On 7 February 2008 the European Commission adopted Communication COM(2007)22 A competitive automotive regulatory framework for the 21st century – Commission's position on the CARS 21 high level group final report.[30] The CARS 21 high level group was established in 2005 with the aim of setting future automotive policy and regulation. The EU strategy towards the automotive sector set in this Communication was based on the promotion of interaction between different policy areas, on the protection of public interest in environment, health and safety and on the reduction of the regulatory burden for industry.

At the Environment Council meeting of 3 March 2008 EU ministers failed to agree on the reduction of CO_2 emissions from new cars. The main points of disagreement concerned the safeguarding of competitiveness of the automobile industry, for both large vehicles and small cars, the proposal for long-term emission reduction targets for 2020 and the level of fines.

In early April 2008 disagreement between the European Parliament and the Council emerged on long-term CO_2 emissions reduction targets for cars. On 8 April 2008 the Environment Committee of the Parliament held a first debate on the Commission's proposal to curb CO_2 emissions from passenger cars in the EU. Among others, issues discussed were long-term targets (2020 or 2024), modalities to cut emissions, penalties and flexibility. Meanwhile a progress report dated 2 April 2008 of the Slovenian presidency summarizing the discussions in the Council of Ministers showed support for a long-term target and disagreement on the starting date and modalities to reduce CO_2 emissions.

On 15 April 2008 a legal opinion prepared for the European Parliament's legal affairs following the request of German MEP Klaus-Heiner Lehne clarified that the European Commission had the right to propose legislation imposing financial penalties on car manufacturers in non-compliance with their CO_2 emissions reduction targets. The German MEP argued that penalties for car manufacturers should be under the competence of national governments rather than being adopted at the EU level. According to the legal opinion, penalties were considered to be of an administrative nature, rather than fiscal measures.

On 8 July 2008 the European Commission adopted proposal COM(2008)436 for a Directive of the European Parliament and of the Council amending Directive 1999/62/EC on the charging of heavy goods vehicles for the use of certain infrastructures.[31] The proposed directive would amend the 'Eurovignette' directive for heavy goods vehicles introducing charges for environmental costs caused by the operations of those vehicles. While the Eurovignette legislation covered only roads included in the trans-European road network, the proposal would extend the coverage to other roads and would allow national governments to charge lorries for air pollution, noise and congestion costs. The proposal was included in a package of measures aimed at the greening of freight transport including, among others, a strategy to calculate the environmental costs of transport and measures to reduce noise pollution in the rail sector.

On 1 September 2008 the Industry Committee of the European Parliament discussed the legislative proposal to reduce CO_2 emissions from new cars to 130g/km by 2012. The Committee voted for the progressive introduction of the 130g/km limit by 2015, therefore delaying the time set by the European Commission in its original proposal. Although the Industry Committee had a subsidiary role to the Environment Committee on this matter, the General Assembly considered its position in the final vote on the proposal.

On 25 September 2008 the Environment Committee of the European Parliament rejected the compromise amendments presented by Rapporteur Guido Sacconi, which would have softened the European draft legislation on the reduction of CO_2 emissions from new cars. The Environment Committee confirmed the limit on average CO_2 emissions from new cars in the EU to 120g/km by 2012. MEPs also agreed on a long-term target of 95g/km by 2020 and on the introduction of fines for every excess gram of CO_2 (excess emissions premiums) to be paid from 2012 by car manufacturers exceeding the targets. The compromise amendments reached by socialist Sacconi with his centre-right colleague Callanan would delay the introduction of the limits until 2015 and reduce significantly the amount of the payable fine.

The contents of the joint position of France and Germany reached in June 2008 on the Proposal for a Directive on CO_2 emissions from new cars were mirrored by the amendments on the proposal discussed on 1 October 2008 by a working group of Member State experts in the EU Council under the French presidency. The amendments weakened the Commission's proposal by delaying the introduction of that legislation until 2015, reducing the level of fines and postponing to 2012 the adoption of binding emissions limits for 2020.

On 22 October 2008 the European Parliament voted in favour of a legislative proposal tabled by the European Commission in January 2008 on the promotion of clean road transport vehicles. These rules would require public authorities and organizations serving a public contract to consider the costs of CO_2 emissions and air pollutants when purchasing public vehicles. The vote of the Parliament tightened the proposal of the Commission by urging the introduction of the new rules by 2010 and setting the costs of CO_2 emissions at a minimum of €30 per tonne.

On 31 October 2008 the details of the draft Council position on the legislative proposal to reduce CO_2 emissions from new cars emerged after the agreement among EU senior diplomats. The Council position supported the introduction of an emission limit for car makers (130g/km to be reached progressively: 60 per cent by 2012, 75 per cent by 2013, 85 per cent by 2014 and 100 per cent by 2015) as well as a long-term emission target for 2020 (95g/km with a review in 2013).

On 4 November 2008 Rapporteur Khadraoui presented significant amendments to the Commission proposal COM(2008)436 on the charging of heavy goods vehicles for the use of certain infrastructures. The Belgian MEP wanted to introduce climate, air and noise pollution impacts in the costs that lorries would have to pay. Furthermore, he proposed to eliminate the limit on the amount of chargeable costs introduced by the Commission.

A first reading agreement between the Parliament and the Council of Ministers on the proposal for the revision of the fuel quality directive was reached on 25 November 2008. The new directive created an obligation for fuel suppliers to reduce life-cycle GHG emissions from road fuels by 6 per cent in the period 2010–2020. This reduction will be provided by improvements in the production efficiency and the introduction of cleaner fuels such as biofuels to be produced in accordance with the same sustainability criteria agreed under the new directive on renewable energy. The results of the directive will be assessed in 2012 by the Commission, which could propose an additional 2 per cent reduction to oil companies.

On 1 December 2008 an informal agreement between the Parliament and the French presidency was reached on the CO_2 emission limits for passenger cars. The deal provided for the progressive introduction of the new limit of 130g/km of CO_2 for new passenger cars. This will apply to 65 per cent of the car fleet by 2012; 75 per cent by 2013; 80 per cent by 2014; and 100 per cent by 2015.

The European Commission will provide the formula upon which the application of the new rules will be based. Furthermore, the deal included a long-term target of average emissions of 95g CO_2/km for 2020 and the following fines (excess emissions premiums) to be paid by car manufacturers exceeding their targets in the period 2012–2018: €5 for the first gram of CO_2; €15 for second gram; €25 for the third gram; and €95 for the fourth gram of CO_2 onwards.

As of 2019 the fine will be €95 for each gram over the target. Until 2014 car manufacturers have the possibility of claiming credits expressed in g/km to be counted towards their targets by the introduction of ecological innovations such as light-emitting diode lights and solar panels and by the production of cars with CO_2 emissions lower than 50g/km in 2012. Finally, manufacturers selling less than 300,000 cars per year will exempted from the new limits and required to cut their emissions by 25 per cent from 2007 levels.

The Transport, Communications and Energy Council of 8 and 9 December 2008 failed to reach an agreement on a presidency progress report on the proposal for new Eurovignette directive (Directive amending Directive 1999/62/EC on the charging of heavy goods vehicles for the use of certain infrastructures). The failure was due to the complexity of the plan and considerations regarding its economic impact on the Member States.

On 16 December 2008 the Parliament approved the compromise text on the Regulation on type approval of motor vehicles and engines with respect to emissions from heavy duty vehicles (Euro 6) and on access to vehicle repair and maintenance information. According to regulation Euro 6, emission standards will be applied to emissions from new models of trucks, lorries and buses from 2013, notably a cap on nitrogen oxides emissions of 400mg/kWh and on particulate matter of 10mg/kWh. New vehicles of existing models are obliged to comply with the regulation by 2014.

On 16 December 2008 the European Commission presented Communication COM(2008)886 on the Action plan for the deployment of intelligent transport systems in Europe[32] and Communication COM(2008)887 on the Proposal for a Directive of the European Parliament and of the Council laying down the framework for the deployment of intelligent transport systems in the field of road transport and for interfaces with other transport modes.[33] Both the plan and the directive proposal aimed at the improvement of the use of ICTs in the transport sector in order to reduce CO_2 emissions. In particular, the plan aimed at the harmonization of intelligent transport systems in the EU and the legislative proposal at the harmonization of the EU framework for the development of those technologies.

On 17 December 2008, the European Parliament approved a legislative resolution of 17 December 2008 on the Proposal for a Regulation of the European Parliament and of the Council setting emission performance standards for new passenger cars as part of the Community's integrated approach to reduce CO_2 emissions from light-duty vehicles COM(2007)856 and a legislative resolution on the Proposal for a Directive of the European Parliament and of the Council amending Directive 98/70/EC as regards the specification of petrol, diesel and gas-oil and introducing a mechanism to monitor and reduce greenhouse gas emissions from the use of road transport fuels and amending Council Directive 1999/32/EC, as regards the specification of fuel used by inland waterway vessels and repealing Directive 93/12/EEC (COM(2007)0018).

On 23 April 2009 the European Parliament adopted a non-legislative resolution that backed the Green Paper presented by the Commission in February 2009 aimed at the integration of climate change issues into the trans-European transport network (TEN-T). According to the MEPs, all modes of transport should promote climate protection and sustainable development and future decisions on TEN-T projects should consider nature protection, pollution and environmental impact assessment.

On 12 January 2010 the European Commission released the report Monitoring the CO_2 emissions from new passenger cars in the EU: data for the year 2008 COM(2009)713, which certified the decrease of those emissions by 3.3 per cent in 2008, equivalent to 153.5 grams of CO_2/km.[34] The report identified that the gap between CO_2 emissions of passenger cars in old and new Member States decreased from 10g/km in 2005 to less than 1g/km in 2008. Furthermore, according to the Commission, the number of vehicles using alternative fuels such as LPG, natural gas, bioethanol and electricity doubled in the period 2007–2008.

8.2   AVIATION

The GHG emissions generated from the aviation sectors are rapidly increasing and even though international aviation emissions are not included in the targets of the Kyoto Protocol, the adoption of measures to reduce these emissions is one of the key priorities of the Second European Climate Change Programme (ECCP II) launched by the European Commission in 2005.

On the aviation sector and its impacts on climate change, the European Commission published on 29 July 2005 a consultancy study 'Giving wings to emissions trading', assessing the possibility of including the aviation industry in the EU ETS.[35]

The European Commission confirmed in Communication COM(2005)459 of 27 September 2005 Reducing the climate change impact of aviation that the inclusion of the aviation sector in the EU emissions trading scheme was one of the best ways for tackling the sector's climate impacts and that a specific working group focusing on the review of the ETS directive would be established.[36]

In response to the Commission' Communication, the Council Environment of 2 December 2005 provided strong support to the possibility of including the aviation sector in the European ETS. The Council conclusion called for the extension of the obligations under the EU ETS also for non-intra-EU flights, as well as for the coverage of all GHG emissions generated from air transport.

Moreover, the working group on aviation established by the Commission within the framework of the ECCP II released its final report with the purpose of identifying a set of recommendations for the integration of the aviation sector into the EU ETS. Such a document confirmed the difficulties of reaching an agreement on which flights would be covered by the scheme and the emission limits for aviation.

On 4 July 2006 the European Parliament adopted a political resolution (non-binding) calling for the establishment of a carbon trading scheme for the aviation sector separate from the EU ETS and for the adoption of a tax on kerosene or aviation fuel. According to the Parliament, such a trading system should cover flights from and to the EU and should be based on the auctioning of allowances to air companies.

On 20 December 2006, the European Commission adopted a Proposal for a Directive amending Directive 2003/87/EC so as to include aviation activities in the scheme for greenhouse gas emission allowance trading within the Community COM(2006)818.[37] The proposal called for the inclusion of both emissions from EU and foreign flights within the EU from 2011 and all flights to and from EU airports from 2012. The proposal was accompanied by a summary impact assessment study.

The Environment Council meeting of 20 February 2007 decided almost unanimously to support the inclusion of aviation in the EU ETS proposed by the Commission at the end of 2006. Open uncertainty remained on the questions of initial exemption for flights to and from non-EU countries as well as on the date of the inclusion of aviation in the EU ETS.

The Energy and Transport Council of 8 June 2007 adopted the Council Conclusions on the inclusion of aviation in the European emissions trading scheme for the position to be taken by EU Member States at the ICAO Assembly in September 2007. This resolution invited the International Civil Aviation Organization (ICAO) to adopt a more proactive engagement in the promotion of the inclusion of air transport into the EU ETS.

On 26 June 2007 at a second debate in the European Parliament on the inclusion of aviation in the EU ETS, MEPs suggested that all flights from and to the EU should be part of the EU ETS as from 2010. MEPs also proposed more stringency in the cap on GHG emissions from the air transport sector and more auctioning in the allocation process.

Support for the inclusion of aviation in the EU ETS was backed on 13 July 2007 by the rapporteur of the European Parliament's Transport Committee, whose report proposed 25 per cent more allowances for airlines in comparison with the proposal of the Commission. Less rigid rules for airlines in terms of accession to JI and CDM credits were also suggested.

The 36th Assembly of the ICAO held on 28 September 2007, discussing policies and measures to reduce air transport emissions, did not approve the EU plans to include European and foreign airlines in the EU ETS by 2011.

On 2 October 2007 the Environment Committee of the European Parliament adopted a report on the Commission's proposal COM(2006)818 introducing tougher rules for the opting in of aviation into the EU ETS. In particular, MEPs voted for tighter CO_2 emission caps (75 per cent of average levels in the period 2004–2006), the inclusion of EU and non-EU flights in the EU ETS by 2010 and more auctioning of allowances (50 per cent).

On 10 October 2007 the Committee of the Regions adopted an opinion that supported the inclusion of aviation in the EU ETS as proposed by the Commission.

On 13 November 2007, the Parliament voted its first-reading position that tightened the European Commission proposal (amendments) confirming the 2010 date for all flights, and urging a cap for airline emissions at 90 per cent of 2004–2006 levels and auctioning of 25 per cent.

Political agreement on the inclusion of the aviation sector in the EU ETS was reached by the EU Council Environment of 20 December 2007. The common position of the Council established the inclusion of flights from and to the EU into the EU ETS as of January 2012 ('sole date for all flights'). The cap was based on the average of the emission levels in the period 2004–2006. 10 per cent of EU allowances (EUAs) was proposed to be auctioned in 2013 with possible increases afterwards. Furthermore, the common position introduced a special reserve for new entrants of fast-growing aircraft operators, new monitoring and reporting plans, as well as a semi-open scheme as far as the conversion of allowances is concerned. Finally, the common position included a *de minimis* clause excluding from the scheme operators with less than 243 flights for three consecutive four-month periods.

On 23 January 2008 the European Commission presented the integrated package of legislative proposals on climate and energy, setting the road of the future EU climate and energy policy. The package included the proposal for a directive amending Directive 2003/87/EC so as to improve and extend the greenhouse gas emission allowance trading system of the Community COM(2008)16 (EU ETS Review), which set the rules for the third phase of the EU ETS running in the period 2013–2020.[38] The EU ETS proposal referred to the aviation sector, in particular to the allocation of EUAs, and the Commission made clear that the appropriate percentage of allowances to be auctioned beyond 2012 shall be defined in accordance with the criteria and the line of the general review of the directive. Furthermore, in the preamble, recital 33 proposal, COM(2008)16 stated:

(33) As regards the approach to allocation, aviation should be treated as other industries which receive transitional free allocation rather than as electricity generators. This means that 80% of allowances should be allocated for free in 2013, and thereafter the free allocation to aviation should decrease each year by equal amounts resulting in no free allocation in 2020. The Community and its Member States should continue to seek to reach an agreement on global measures to reduce greenhouse gas emissions from aviation and review the situation of this sector as part of the next review of the Community scheme.

On 18 April 2008 the Council of the EU adopted the common position on future inclusion of aviation activities in the EU ETS by amending Directive 2003/87/EC whose details are highlighted above (political agreement of December 2007). Only 3 amendments out of 60 proposed by the Parliament were adopted in the common position.

On 27 May 2008 the Environment Committee of the European Parliament voted on the draft recommendation for second reading on the Council common position. The Committee proposed 2011 as the starting date for the introduction of aviation in the EU ETS and auctioning for 25 per cent of the permits, with this percentage increasing in line with the other sectors. Still unresolved was the issue of revenues from the auctioning of allowances, with the European Parliament strongly advocating the earmarking of them.

On 26 June 2008 informal agreement between MEPs and the Slovenian presidency was reached on the inclusion of aviation in the EU ETS. The final vote was held in the European Parliament on 8 July 2008. The details of the second reading agreement deal were as it follows:

- Inclusion of all flights from 2012;
- Cap on airline emissions at 97 per cent of 2004–2006 levels in first year; and a cap of 95 per cent for the following years;
- 15 per cent auctioning of allowances;
- Revenues from auctioning: governments shall report on how they are spent;
- 15 per cent limit on the use of JI and CDM credits.

The directive including aviation activities in the EU ETS was formally adopted by the Council on 24 October 2008. All amendments voted by the Parliament in second reading were accepted.

On 19 November 2008 Directive 2008/101/EC amending Directive 2003/87/EC so as to include aviation activities in the scheme for greenhouse gas emission allowance trading within the Community was adopted by the Council and the Parliament.[39]

Following the adoption of Directive 2008/101/EC the following implementing legislation was adopted by the European institutions:

- Commission Decision 2009/339/EC on the inclusion of monitoring and reporting guidelines for emissions and tonne-kilometre data from aviation activities;[40]
- Commission Regulation No 82/2010 on the list of aircraft operators specifying the administering Member State;[41]
- Commission Decision 2009/450/EC on the detailed interpretation of the aviation activities listed in the Annex I to Directive 2003/87/EC.[42]

NOTES

1. Communication from the Commission to the Council and the European Parliament on a community strategy to reduce CO_2 emissions from passenger cars and improve fuel economy COM(1995)689, Brussels, 20 December 1995.

2. White Paper, European transport policy for 2010: Time to decide, COM(2001)370, Brussels, 12 September 2001.

3. Commission Communication Keep Europe moving – Sustainable mobility for our continent, Mid-term review of the European Commission's 2001 Transport White Paper COM(2006)314, Brussels, 22 June 2006.

4. EEA Report No 1/2007 Transport and environment: On the way to a new common transport policy, 10 October 2007.

5. Communication from the Commission to the European Parliament and the Council – Greening transport COM(2008)433, Brussels, 8 July 2008.

6. Communication from the Commission to the Council and the European Parliament, Implementing the Community Strategy to reduce CO_2 emissions from cars: an environmental agreement with the European automobile industry, COM(1998)495, 29 July 1998.

7. Communication from the Commission to the Council and the European Parliament, Implementing the Community Strategy to reduce CO_2 emissions from cars: outcome of the negotiations with the Japanese and Korean automobile industries, COM(1999)446, Brussels, 14 September 1999.

8. Commission Recommendation 2000/304/EC of 13 April 2000 on the reduction of CO_2 emissions from passenger cars (Japanese Automobile Manufacturers Association – JAMA), OJ L 100, 20 April 2000, pp57–58.

9. Commission Recommendation 2000/303/EC of 13 April 2000 on the reduction of CO_2 emissions from passenger cars (Korean Automobile Manufacturers Association – KAMA) OJ L 100, 20 April 2000, pp55–56.

10. Commission Recommendation 1999/125/EC of 5 February 1999 on the reduction of CO_2 emissions from passenger cars (notified under document number C(1999)107), OJ L 40, 13 February 1999, pp49–50.

11. Directive 1999/94/EC of the European Parliament and of the Council of 13 December 1999 relating to the availability of consumer information on fuel economy and CO_2 emissions in respect of the marketing of new passenger cars, OJ L 12, 18 January 2000, pp16–23.

12. Commission Communication Taxation of passengers cars in the European Union - options for action at national and Community levels COM(2002)431, Brussels, 6 September 2002.

13. Communication from the Commission to the Council and the European Parliament, Implementing the Community Strategy to reduce CO_2 emissions from cars – Second annual report on the effectiveness of the strategy (Reporting year 2000) COM(2001)643, 8 November 2001.

14. Communication from the Commission to the Council and the European Parliament, Implementing the Community Strategy to reduce CO_2 emissions from cars: Fourth annual report on the effectiveness of the strategy (Reporting year 2002) COM(2004)78, 23 April 2004.

15. Proposal for a Directive of the European Parliament and of the Council amending Council Directives 70/156/EEC and 80/1268/EEC as regards the measurement of carbon dioxide emissions and fuel consumption of N1 vehicles COM (2001)543, 26 February 2002, pp317–319.

16. Communication of the European Commission, Implementing the Community Strategy to reduce CO_2 emissions from cars: Fifth annual Communication on the effectiveness of the strategy, COM(2005)269, Brussels, 22 June 2005.

17. Proposal for a Council Directive on passenger car related taxes COM(2005)261, Brussels, 5 July 2005.

18. Directive 2005/55/EC of the European Parliament and of the Council of 28 September 2005 on the approximation of the laws of the Member States relating to the measures to be taken

against the emission of gaseous and particulate pollutants from compression-ignition engines for use in vehicles, and the emission of gaseous pollutants from positive-ignition engines fuelled with natural gas or liquefied petroleum gas for use in vehicles, *OJ* L 275 of 20 October 2005, pp1−163.

19. Proposal for a Regulation of the European Parliament and of the Council on type approval of motor vehicles with respect to emissions and on access to vehicle repair information, amending Directive 72/306/EEC and Directive../../EC, COM(2005)683, Brussels, 21 December 2005.

20. Proposal for a Directive of the European Parliament and of the Council on the promotion of clean road transport vehicles, COM(2005)634, Brussels, 21 December 2005.

21. Communication to the Council and the European Parliament, Implementing the Community Strategy to reduce CO_2 emissions from cars: Sixth annual communication on the effectiveness of the strategy together with a Commission staff working document on CO_2 emissions from new cars in 2004, COM(2006)463, Brussels, 24 August 2006.

22. Report on the targets contained in Article 7(2)(b) of Directive 2000/53/EC on end-of-life vehicles, COM(2007)5, Brussels, 16 January 2007.

23. Proposal for a Directive of the European Parliament and of the Council amending Directive 98/70/EC as regards the specification of petrol, diesel and gas-oil and the introduction of a mechanism to monitor and reduce greenhouse gas emissions from the use of road transport fuels and amending Council Directive 1999/32/EC, as regards the specification of fuel used by inland waterway vessels and repealing Directive 93/12/EEC, COM(2007)18, Brussels, 31 January 2007.

24. Communication of the Commission, Results of the review of the Community Strategy to reduce CO_2 emissions from passenger cars and light commercial vehicles, COM(2007)19, Brussels, 7 February 2007.

25. Communication of the Commission, Towards Europe-wide safer, cleaner and efficient mobility: The first intelligent car report, COM(2007)541, Brussels, 17 September 2007.

26. European Commission, Fourth annual report − Quality of petrol and diesel fuel used for road transport in the European Union (Reporting year 2005), COM(2007)617, Brussels, 18 October 2007.

27. European Commission, proposal for a Regulation of the European Parliament and of the Council setting emission performance standards for new passenger cars as part of the Community's integrated approach to reduce CO_2 emissions from light-duty vehicles, COM(2007)856, Brussels, 19 December 2007.

28. European Commission, Proposal for a Directive on the promotion of clean and energy efficient road transport vehicles, COM(2007)817, Brussels, 19 December 2007.

29. European Commission, proposal for a Regulation on type approval of motor vehicles and engines with respect to emissions from heavy duty vehicles (Euro 6) and on access to vehicle repair and maintenance information, COM(2007)851, Brussels, 21 December 2007.

30. European Commission Communication, A competitive automotive regulatory framework for the 21st century − Commission's position on the CARS 21 high level group final report, COM(2007)22, Brussels, 7 February 2008.

31. European Commission, Proposal for a Directive of the European Parliament and of the Council amending Directive 1999/62/EC on the charging of heavy goods vehicles for the use of certain infrastructures, COM(2008)436, Brussels, 8 July 2008.

32. European Commission, Communication on the action plan for the deployment of intelligent transport systems in Europe, COM(2008)886, Brussels, 16 December 2008.

33. European Commission, Communication on the Proposal for a Directive of the European Parliament and of the Council laying down the framework for the deployment of intelligent

transport systems in the field of road transport and for interfaces with other transport modes, COM(2008)887, Brussels, 16 December 2008.

34. European Commission, Report monitoring the CO_2 emissions from new passenger cars in the EU: data for the year 2008, COM(2009)713, Brussels, 12 January 2010.

35. European Commission, consultancy study 'Giving wings to emissions trading', 29 July 2005.

36. European Commission Communication, Reducing the climate change impact of aviation, COM(2005)459, Brussels, 27 September 2005.

37. European Commission, Proposal for a Directive amending Directive 2003/87/EC so as to include aviation activities in the scheme for greenhouse gas emission allowance trading within the Community, COM(2006)818, Brussels, 20 December 2006.

38. European Commission, Proposal for a Directive amending Directive 2003/87/EC so as to improve and extend the greenhouse gas emission allowance trading system of the Community (EU ETS Review), COM(2008)16, Brussels, 23 January 2008.

39. Directive 2008/101/EC of the European Parliament and of the Council of 19 November 2008 amending Directive 2003/87/EC so as to include aviation activities in the scheme for greenhouse gas emission allowance trading within the Community, *OJ* L 8, 13 January 2009, pp3–21.

40. Commission Decision 2009/339/EC of 16 April 2009 on the inclusion of monitoring and reporting guidelines for emissions and tonne-kilometre data from aviation activities, *OJ* L 103 of 23 April 2009, pp10–29.

41. Commission Regulation No 82/2010 of 28 January 2010 amending Regulation (EC) No 748/2009 on the list of aircraft operators which performed an aviation activity listed in Annex I to Directive 2003/87/EC on or after 1 January 2006 specifying the administering Member State for each aircraft operator on the list of aircraft operators specifying the administering Member State, *OJ* L 25 of 29 January 2010, pp12–120.

42. Commission Decision 2009/450/EC of 8 June 2009 on the detailed interpretation of the aviation activities listed in the Annex I to Directive 2003/87/EC, *OJ* L 149 of 12 June 2009, pp69–72.

CHAPTER 9

Greenhouse Gases Emissions Monitoring and Reporting

O fficial data and information on GHG emission trends in the EC and the Member States are provided by the European Environment Agency.[1] At the moment of writing, information on EU GHG emissions indicates that the EC and the Member States met the UNFCCC commitment to stabilize emissions at 1990 levels by 2000 and are well on track to achieve their quantified emission limitation and reduction commitments agreed under the Kyoto Protocol, in particular though the use of the flexible mechanisms. In this context, it should not be forgotten that a big part of the GHG emission reductions achieved in Europe between 1990 and 2000 is the indirect consequence of two events in 1990 that are not related to specific and direct climate policies at the national level. These are German state reunification between West and East and the consequent restructuring of the industrial sector in the former Democratic Republic of Germany and the shift from coal to gas in the UK due to the privatization of coal mining and the switch to natural gas.

The majority of early European legislation concerned the collection of information on the GHG emissions and reduction commitments in the Community, in compliance with the UNFCCC monitoring and reporting obligations required of all parties and laid down in Articles 4(1) and 12 of the UNFCCC. The first provision introducing reporting obligations for Member States with regard to all 'anthropogenic CO_2 and other greenhouse gas emissions not controlled by the Montreal Protocol' in the Member States, as required under the UNFCCC and the Kyoto Protocol was Council Decision 93/389/EEC of 24 June 1993 for a monitoring mechanism of Community CO_2 and other GHG emissions[2] followed by Council Decision 99/296/EC amending Council Decision 93/389/EEC for a monitoring mechanism of Community CO_2 and other GHG Emissions.[3]

In accordance with Decision 99/926/EC, EU Member States were obliged to: (1) prepare national programmes for the limitation and/or reduction of their GHG emissions (Article 2) aiming at: 'the stabilization of CO_2 emissions by 2000 at 1990 levels in the Community as a whole'; the fulfilment of the EC GHG emission reduction commitments under the UNFCCC and the Kyoto Protocol; the monitoring of the actual and projected progress of Member States towards the commitments under the UNFCCC and the Kyoto Protocol; and (2) prepare and submit to the Commission each year not later than 31 December data on 'anthropogenic CO_2 emissions and CO_2 removal by sinks for the previous calendar year' and 'a national inventory data on emissions by sources and removals by sinks of the other greenhouse gases for the previous year but one and provisional emission data (inventories) for the previous year'.

The documentation required and collected in accordance with Council Decision 99/296/EC was used by the European Commission for the preparation and submission to the UNFCCC of the Community GHG inventory report and the progress evaluation report.

In 2003, the Commission issued a proposal of a Decision of the European Parliament and of the Council for a monitoring mechanism of Community GHG emissions and the implementation of the Kyoto protocol COM(2003)51[4] replacing Council Decision 99/296/EC and addressing:

- New reporting obligations and guidelines for the implementation of the UNFCCC and the Kyoto Protocol, reflecting the political agreement and legal decisions adopted at COP6bis (Bonn) and COP7 (Marrakech);
- Additional instructions for further harmonization of emission forecasts at Member State and Community levels;
- Reporting and implementation requirements related to the ratification of the Kyoto Protocol and 'burden sharing' between the Community and its Member States under Council Decision 2002/358/EC.

The agreement on proposal COM(2003)51 was reached by the European Parliament and the Council in early 2004 with the adoption of Decision 280/2004/EC of the European Parliament and of the Council of 11 February 2004 concerning a mechanism for monitoring Community GHG emissions and for implementing the Kyoto Protocol.[5] With Decision 280/2004/EC, the obligations of the Kyoto Protocol other than the quantified emission limitation and reduction commitments established under Decision 2002/358/EC became part of Commmunity law and of the *acquis communautaire*.

On 10 February 2005 the Commission adopted Decision 2005/166/EC laying down rules implementing Decision No 280/2004/EC of the European Parliament and of the Council concerning a mechanism for monitoring Community GHG emissions and for implementing the Kyoto Protocol, which identified the parameters for projecting future emissions as well as indicators for the measurement of progress towards GHG emission commitments.[6]

Data on GHG emissions in the national inventories submitted by each Member States to the Commission every year are used by the Commission with the support of the EEA to prepare the annual report on the Community's progress towards the GHG reduction commitments.

The list of early annual reports is provided here:

- Report from the Commission under Council Decision 93/389/EEC for a monitoring mechanism of Community CO_2 and other GHG emissions – First evaluation report COM(1994)67:[7] this report presented the first evaluation of the national programmes for limiting CO_2 and other GHG emissions in the EC;
- Report from the Commission, of 14 March 1996, under Council Decision 93/389/EEC, Second evaluation report COM(1996)91:[8] this report concerned an evaluation of the national programmes of the 15 Member States relating to the mechanism for monitoring Community CO_2 and other GHG emissions. While the quality of the information used for this evaluation was better, compared with the report of 1993, the data collected in COM(1996)91 were still inadequate since there was not any Community inventory of CO_2 emissions for 1993, any Community emissions projection for 2000 and information on the implementation of the measures adopted by the Member States was insufficient. Under these premises, the European Commission in 1996 was not able to say whether the measures to reduce GHG emissions adopted by the Member States were sufficient to achieve the objectives set, notably a significant reduction in CO_2 emissions after the year 2000;
- 1st progress report Report from the Commission under Council Decision 93/389/EEC as amended by Decision 99/296/EC for a monitoring mechanism of Community GHG emissions COM(2000)749;[9]
- 2nd progress report, Report 2001 under Council Decision 93/389/EEC for a monitoring mechanism of Community GHG emissions COM(2001)708;[10]
- 3rd progress report, Report 2002 under Council Decision 93/389/EEC for a monitoring mechanism of Community GHG emissions COM(2002)702;[11]
- 4th progress report, Report 2003 under Council Decision 93/389/EEC for a monitoring mechanism of Community GHG emissions COM(2003)735;[12]
- Report of the European Commission, Catching up with the Community's Kyoto target (under Decision 280/2004/EC of the European Parliament and of the Council concerning a mechanism for monitoring Community GHG emissions and for implementing the Kyoto Protocol) COM(2004)818.[13]

In addition to the yearly report on the EC's progress towards the GHG reduction commitments established by the Kyoto Protocol, the EEA also prepares annual EC GHG inventory data and an inventory report including GHG emissions trends and projections in the Community. These two joint reports serve and complement the annual report on demonstrable progress.

The list of early annual GHG inventories and inventory reports includes:

- Annual EC GHG inventory 1990–2001 and inventory report 2003, Technical report No 95 of 4 May 2003;[14]

- GHG emissions trends and projections in Europe, Environmental Issue Report No 36/2003;[15]
- Annual EC GHG inventory 1990–2002 and inventory report 2004, Technical Report No 2/2004, 15 July 2004;[16]
- GHG emissions trends and projections in Europe, Environmental Issue Report No 5/2004, 21 December 2004;[17]
- Annual EC GHG inventory 1990–2003 and inventory report 2005, Technical Report No 4/2005, 21 June 2005;[18]
- GHG emissions trends and projections in Europe 2005, EEA Report No 8/2005, 1 December 2005;[19]
- Annual EC GHG inventory 1990–2004 and inventory report 2006, Technical Report No 6/2006, 22 June 2006;[20]
- GHG emissions trends and projections in Europe 2006, EEA Report No 9/2006;[21]
- Annual EC GHG inventory 1990–2004 and inventory report 2006, Technical Report No 10/2006, 14 December 2006;[22]
- Annual EC GHG inventory 1990–2005 and inventory report 2007, Technical Report No 7/2007, 14 June 2007;[23]
- GHG emission trends and projections in Europe 2007, EEA Report No 5/2007;[24]
- Annual EC GHG inventory 1990–2006 and inventory report 2008, Technical Report No 6/2008, 18 June 2008;[25]
- GHG emissions trends and projections in Europe 2008, EEA Report No 5/2008, 16 October 2008;[26]
- Annual EC GHG inventory 1990–2007 and inventory report 2009, Technical Report No 4/2009, 29 May 2009;[27]
- GHG emissions trends and projections in Europe 2009, EEA Report No 9/2009, 12 November 2009;[28]
- Annual EC GHG inventory 1990–2008 and inventory report 2010, Technical Report No 6/2010, 2 June 2010.[29]

In February 2004, the European Commission established, together with the EEA, the European Pollutant Emission Register (EPER), the first European register of industrial emissions into air and water that included data on CO_2 emissions and other GHG emissions at installation level.

The report European Environment Outlook 2005 released in September 2005 by the EEA confirmed that key EU environmental targets, including GHG emissions reduction and renewable energy goals, could be missed by the Member States. According to this report, which compares GHG emissions projections with the targets set in the EU 6th EAP, in 2008–2012 GHG emissions are expected to be less than 3 per cent below 1990 levels (the EC Kyoto Protocol GHG emission reduction commitment is 8 per cent below 1990 levels).

The European Environment – State and Outlook 2005 report released on 29 November 2005 by the EEA included climate change among the main challenges of European environmental protection. The document indicated an increase in the European average temperature by 0.95°C during the 20th century, 35 per cent higher than the global average increase of 0.7°C, with temperatures continuing to rise in the future.

According to the GHG emissions projections released by the European Commission on 1 December 2005, the EU was 'well on its way' to reach the Kyoto Protocol target. Communication from the Commission – Report on demonstrable progress under the Kyoto Protocol (required under Article 5(3) of Decision 280/2004/EC concerning a mechanism for monitoring Community GHG emissions and for implementing the Kyoto Protocol), COM(2005)615 indicated that the EU15 GHG emissions were 6.8 per cent below 1990 levels by 2010 and therefore on track to meet the GHG reduction obligations agreed under the Kyoto Protocol (−8 per cent). With the inclusion of the flexible mechanisms, the EU15 could reach a −9.3 per cent level of GHG emission reduction.[30]

The EC GHG inventory 1990–2003 released by the EEA on 21 June 2005 showed that, despite a slight reduction in 2002, EU GHG emissions increased in 2003: 1.5 per cent in the EU25 and 1.3 per cent in the EU15. GHG emissions rose particularly in two countries: Spain and the UK. EU15 GHG emissions in 2005 were therefore 1.7 per cent below 1990 levels compared with the Kyoto Protocol target, while EU15 CO_2 emissions continued to rise and were 3.4 per cent above the 1990 levels.[31]

The EC GHG inventory 1990–2004 released by the EEA on 7 June 2006 revealed an increase of GHG emissions in Europe in 2004 of 0.4 per cent for EU25 and 0.3 per cent for EU15. EU15 GHG emissions in 2006 were 0.9 per cent lower than in 1990.[32] The sectors responsible of the GHG emission increase reported in 2006 were road transport, steel industry and oil refining. Italy and Spain were the two Member States with the most relevant increases of GHG emissions reported in 2006.

The GHG emissions trends and projections in Europe 2006, EEA Report No 9/2006, released on 27 October 2006, showed that the Kyoto Protocol target could be met by the EC and the Member States provided that all existing and planned domestic policies and measures were implemented, as well as the Kyoto mechanisms and the inclusion of carbon sinks.[33] Despite the estimated compliance of the EC as a whole with the Kyoto Protocol obligations, seven Member States (Austria, Belgium, Denmark, Ireland, Italy, Portugal and Spain) were predicted to fail to reach their targets under the EU BSA.

The Report on the progress towards achieving the Kyoto objectives (required under Decision 280/2004/EC of the European Parliament and of the Council concerning a mechanism for monitoring Community GHG emissions and for implementing the Kyoto Protocol), COM(2006)658, released by the European Commission on 27 October 2006 confirmed that the EC Kyoto Protocol target could only be met in the event that all Member States would fully implement their current and planned policies and measures to reduce GHG emissions.[34]

On 14 December 2006, the European Commission submitted to the UNFCCC its Annual EC GHG inventory 1990–2004 and inventory report 2006 including the final estimation of the EU15 GHG emissions in 1990.[35] These data form the baseline against which the EU15 must reduce their emissions by 8 per cent by 2008–2012. Accordingly, Commission Decision of

14 December 2006 determining the respective emission levels allocated to the Community and each of its Member States under the Kyoto Protocol pursuant to Council Decision 2002/358/EC fixed at 11,393,397 tonnes of CO_2 equivalent the sum of emission levels allocated to the Community and the Member States for the first commitment period and to be converted into AAUs under the Kyoto Protocol.[36]

The summary for policy-makers of Working Group I 'The physical science basis' to the Fourth Assessment Report of the IPCC, adopted by 113 states on 2 February 2007 in Paris, provided the international community and policy-makers with the greatest warning yet to limit emissions of GHGs. The report concluded that climate change will continue for the coming centuries. A temperature rise of 1.8°C up to 4.1°C is highly probable by 2099, as well as an increase in sea level of 18 to 55cm by 2099. Finally, more scientific certainty was provided on the question of whether the increase in tropical storms such as hurricanes since 1970 is due to human activity. Two other working groups worked under the Fourth Assessment Report of the IPCC in 2007: (1) summary for policy-makers of Working Group II 'Impacts, adaptation and vulnerability' to the Fourth Assessment Report of the IPCC; and summary for policy-makers of Working Group III 'Mitigation of climate change' to the Fourth Assessment Report of the IPCC.

According to the EEA Annual EC GHG inventory 1990–2005 and inventory report 2007 released on 14 June 2007, EU15 GHG emissions dropped in 2005 by 0.8 per cent compared to the 2004 level after increases over the previous years.[37] Furthermore, EU15 and EU27 GHG emissions decreased by 2.0 per cent and 7.9 per cent respectively between 1990 and 2005.

On 31 August 2007, Commission Decision of 18 July 2007 revising the guidelines for installations to monitor and report GHG emissions pursuant to Directive 2003/87/EC was published in the *Official Journal*.[38] The new guidelines applied from January 2008 and were introduced with the aim of reducing the costs of compliance of EU operators falling under Directive 2003/87/EC.

On 27 November 2007 the EEA released GHG emission trends and projections in Europe 2007, Report No 5/2007, which shed light on the EU15 road toward the Kyoto Protocol targets.[39] According to the EEA, in 2007 the EU15 were likely to achieve the 8 per cent GHG emission reduction targets without the inclusion of flexible mechanisms and sinks, provided that existing and additional domestic policies and measures would be correctly implemented in the Member States. Three Member States were not on track with their individual GHG emission reduction targets under the Kyoto Protocol: Denmark, Italy and Spain.

In 2007 the contributions of all EU15 efforts to meet the GHG emission reduction obligations under the Kyoto Protocol in comparison with base year levels were:

- Existing domestic policies and measures: −4.0 per cent;
- Additional domestic policies and measures: −3.9 per cent;
- Kyoto Protocol flexible mechanisms: −2.5 per cent;
- Carbon sinks: −0.9 per cent.

On 27 November 2007 the European Commission released Communication COM(2007)757 on the progress towards achieving the Kyoto objectives required under Decision 280/2004/EC.[40] In accordance with the projections in COM(2007)757, the European Commission confirmed that additional policies and measures were needed for the EC to meet its targets under the Kyoto Protocol.

For the second year in a row, EU GHG emissions continued to fall in 2006. This emerged from the Annual EC GHG inventory 1990–2006 and inventory report 2008, Technical Report No 6/2008, published by the EEA on 18 June 2008 in accordance with the MRV obligations created by the UNFCCC and the Kyoto Protocol for Annex I Parties.[41] The main outcomes of the EEA report were:

- GHG emissions EU15 2005–2006: −0.8 per cent;
- GHG emissions EU15 1990–2006: −2.7 per cent;
- GHG emissions EU27 2005–2006: −0.3 per cent;
- GHG emissions EU27 1990–2006: −7.7 per cent.

Sectors that contributed more to the cut of GHG emissions in 2006 were households and offices, which registered lower consumption of gas and oil in 2006. France, Italy and the UK reported a higher level of GHG emissions due to a cold winter and higher gas prices. CO_2 emissions from electricity, heat production and transport increased in 2006.

On 16 October 2008 the EEA released the GHG emissions trends and projections in Europe 2008, EEA Report No 5/2008, which confirmed the reduction trend of the EU Member States in terms of GHG emissions.[42] In particular, the findings of report No 5/2008 can be summarized as:

- GHG EU27 2005–2006: −0.3 per cent;
- GHG EU15 1990–2006: −2.7 per cent;
- GHG EU27 1990–2006: −7.7 per cent.

According to Report No 5/2008 the reduction obligations of the EC and the Member States under the Kyoto Protocol could be overachieved (−11.3 per cent) by 2012 only in the event that: all existing and additional measures were implemented; flexible mechanisms and carbon sinks were applied; there was overachievement of individual targets by some Member States.

On 29 May 2009, the EEA published Annual EC GHG inventory 1990–2007 and inventory report 2009, Technical Report No 4/2009, as required by Council Decision 280/2004/EC.[43] According to the report, EU27 GHG emissions excluding LULUCF decreased by 9.3 per cent between 1990 and 2007 and by 1.2 per cent between 2006 and 2007 (see Figure 9.1).

In 2007 EU15 GHG emissions without LULUCF were 4.3 per cent below the 1990 level and emissions decreased by 1.6 per cent between 2006 and 2007. Once again, the data collected by the EEA confirmed that a combination of existing and planned domestic policies and measures, carbon sinks and Kyoto

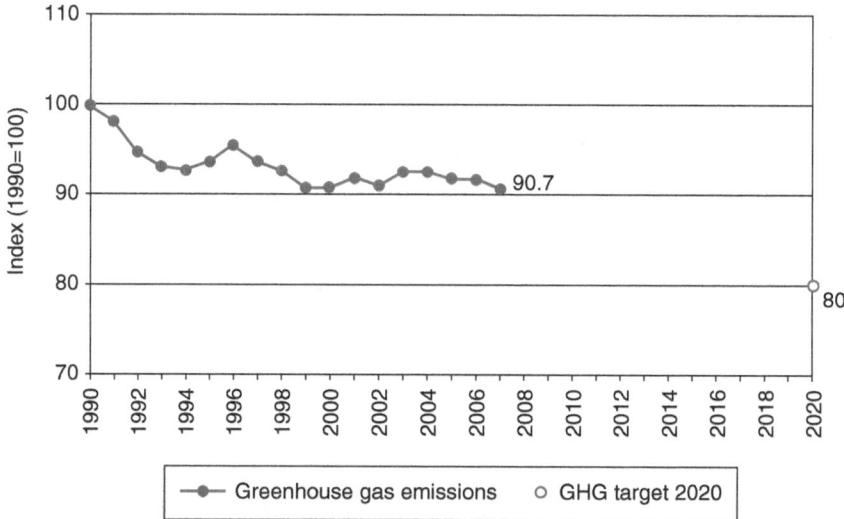

Figure 9.1 *EU27 GHG emissions 1990–2007*

Source: EEA (2009a) Technical Report No 4/2009

mechanisms would be needed to achieve the 8 per cent target agreed in Kyoto (see Figure 9.2).

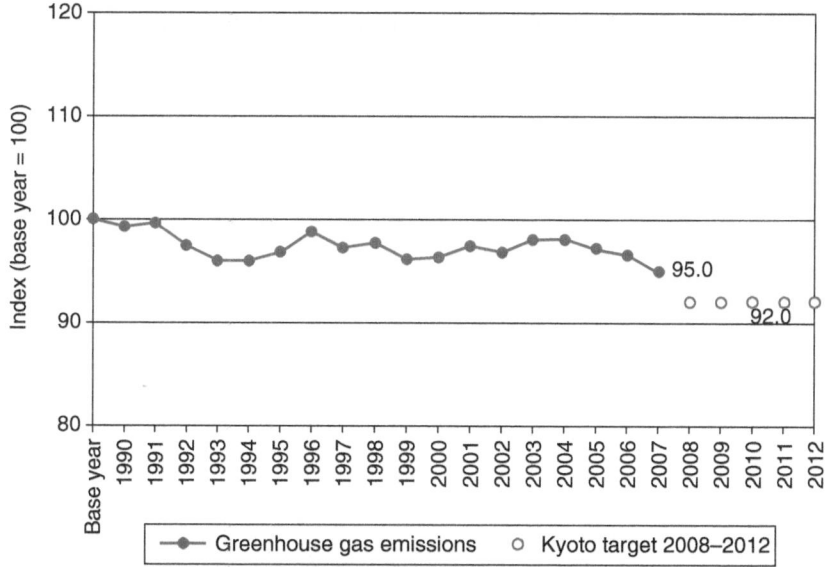

Figure 9.2 *EU15 GHG emissions 1990–2007 (excluding LULUCF) compared with the target for 2008–2012*

Source: EEA (2009a) Technical Report No 4/2009

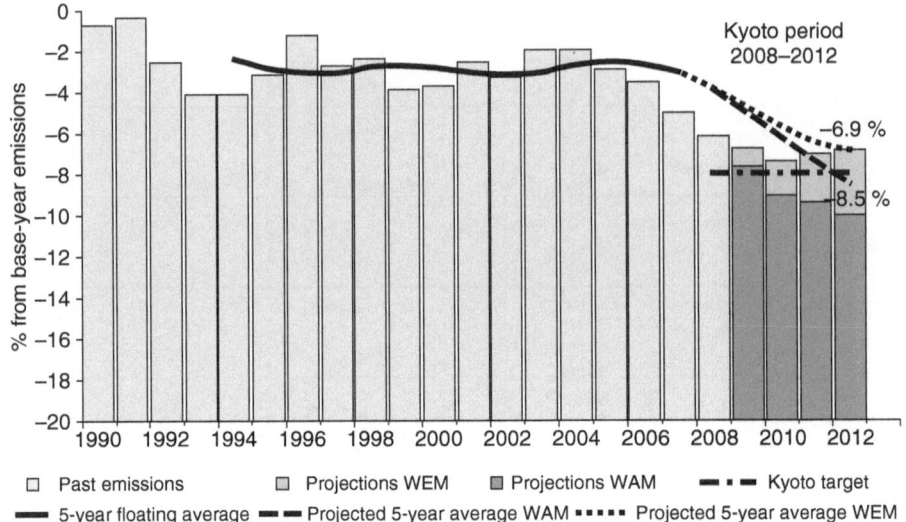

Figure 9.3 *Projected emission scenarios in the EU15*
Source: EEA (2009b) Report No 9/2009

On 12 November 2009 the EEA released GHG emissions trends and projections in Europe 2009, EEA Report No 9/2009, which confirmed the trend in GHG emission levels in the recent past, namely the overall decrease although emissions from transport and fluorinated gases were still growing.[44] In accordance with EEA Report No 9/2009, in 2008 'emissions in the EU decreased to reach their lowest level since 1990'. Five EU15 Member States (France, Germany, Greece, Sweden and the UK) had already reached average GHG emission levels below their targets under the Kyoto Protocol. The report indicated that the EU15 could reduce its GHG emissions levels to 8.5 per cent below the Kyoto base year (see Figure 9.3).

On 12 November 2009 the European Commission released the report Progress towards achieving the Kyoto objectives COM(2009)630,[45] which indicated that in 2007 EU15 GHG emissions decreased by 1.6 per cent compared to 2006 while EU15 GDP increased by 2.7 per cent in the same period. The European Commission reiterated that the EU15 would reach its Kyoto Protocol target and showed that GHG emission in the EU15 decreased between 1990 and 2005 by 7.6 per cent (see Figure 9.4).

An Assessment of GHG emission projections in the EU until 2010 published in March 2009 by the EEA indicated that by 2010 the EU15 would meet its Kyoto Protocol target by a discrete margin provided that existing and additional measures were implemented and carbon sinks and flexible mechanisms included.[46] The EU15 could therefore reach a decrease of GHG emissions of 11.3 per cent by 2010 compared to its 1990 level.

In accordance with the Annual EU GHG inventory 1990–2008 and inventory report 2010, Technical Report No 6/2010,[47] in 2008 EU27 GHG emissions

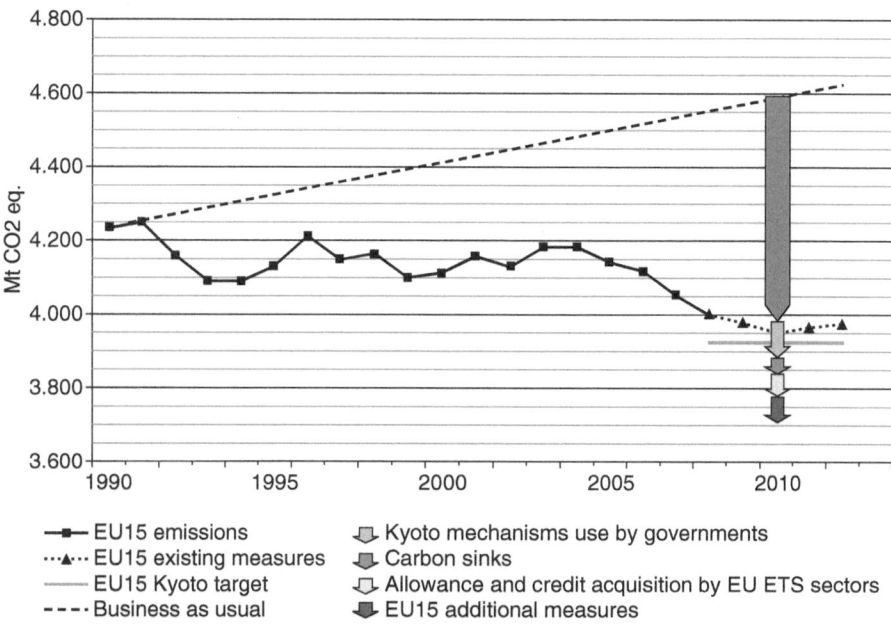

Figure 9.4 *Projected emission scenarios in the EU15*

Note: WEM = with existing measures; WAM = with additional measures.

Source: European Commission (2009), COM(2009)630

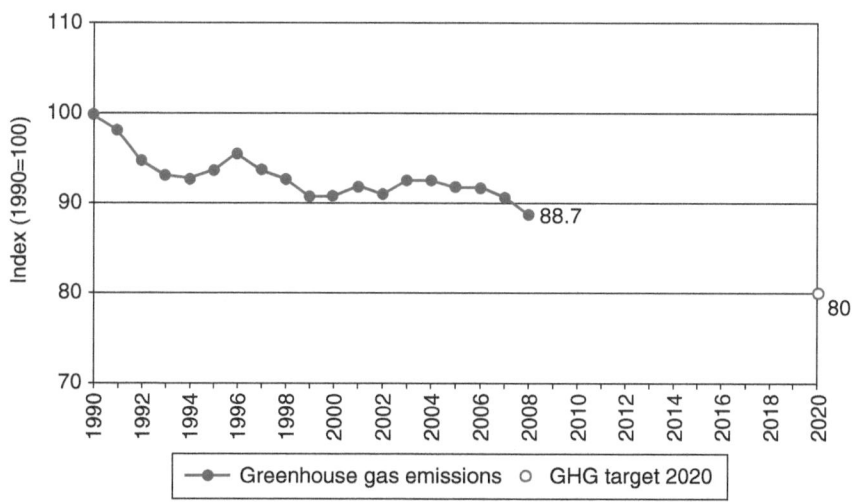

Figure 9.5 *EU27 GHG emissions 1990–2008 (excluding LULUCF)*

Source: EEA (2010)

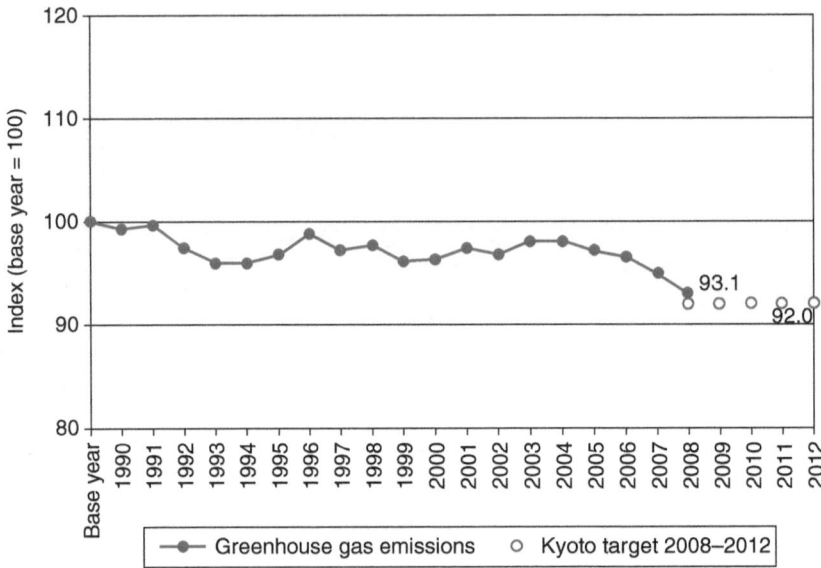

Figure 9.6 *EU15 GHG emissions, base year 2008 compared with target for 2008–2012 (excluding LULUCF)*

Source: EEA (2010)

without LULUCF decreased by 11.3 per cent between 1990 and 2008 and by 2.0 per cent between 2007 and 2008, while EU15 GHG emissions without LULUCF were 6.5 per cent below 1990 and decreased by 1.9 per cent between 2007 and 2008. Between 1990 and 2008, EU15 emissions decreased by 6.5 per cent, while in the EU27 they decreased by 11.3 per cent. In the EU27, the most relevant decrease of GHG emissions occurred in public electricity and heat production, energy use in manufacturing industries and households, and agriculture (see Figures 9.5 and 9.6)

In accordance with the information on GHG emissions in the EU collected by the registries of the Member States, verified emissions of GHGs of all the EU ETS installations fell 11.6 per cent in 2009 compared with 2008. Such a drop in GHG emissions for EU ETS installations was mainly due to the economic recession and to the low level of gas prices in 2009. Only 2 per cent of all installations did not surrender the expected quantity of EU allowances by 1 May 2010. Out of the total quantity of EU allowances surrendered in 2009, 4.1 per cent were CERs, while 0.17 were ERUs. Almost two-thirds of the 11 per cent reduction took place in Germany, the UK, Italy and Spain.

The GHG emissions scenario of the EU in respect of the Kyoto Protocol's first commitment period are provided in Table 9.1 on the basis of data presented by the European Commission and the EEA since 1996.

Table 9.1 *GHG emissions scenario of the EU in respect of the Kyoto Protocol's first commitment period*

DATA[48 49]	Base year (million tonnes CO_2 equivalent)[50]	EU15 GHG emissions			
		1990–XX (%) (CO_2 equivalent)	1990–2010 Business-as-usual (%)	1990–2010 Additional policies and measures (%)	1990–2010 flexible mechanisms (%)
European Commission COM(1994)67[51]	—	No guarantees that commitments referred to in Article 2 of Decision 93/389/EEC (stabilization of CO_2 emissions by 2000 at 1990 levels in the Community as a whole) will be achieved	—	—	—
European Commission COM(1996)91[52]	CO_2 EU15	1990–1993	It cannot be excluded that CO_2 emissions of the EU15 will increase within a range between 0–5 per cent by 2000 over 1990 levels	—	—
	3329.7	–2.2 energy CO_2 EU15			
EEA 41/2000[53]	4149.5	1990–1998	—	—	—
EEA 6/2000[54]	4149.5	–2.5 EU	—	—	—
European Commission COM(2000)749[55]	4194.5	1990–1998 –2.5 EU15	–1.4	–7.0	—

(continued)

Table 9.1 – continued

DATA[48 49]	Base year (million tonnes CO_2 equivalent)[50]	1990–XX (%) (CO_2 equivalent)	EU15 GHG emissions 1990–2010 Business-as-usual (%)	1990–2010 Additional policies and measures (%)	1990–2010 flexible mechanisms (%)
EEA 60/2001[56]	4199	—	—	—	—
EEA 10/2001[57]	4198.7	1990–1999 -4 EU15	DTT* -0.4	—	—
European Commission COM(2001)708[58]	4198.6	1990–1999 -4 EU15	Stabilization at 1990 levels (at best)	-5.0	—
EEA 7/2002[59]	4207.6	1990–2000 -3.5 EU15	2000 DTT 0.5	—	
EEA 75/2002[60]	4207.6	1990–2000 -3.5 EU15	—	—	
EEA 1/2002[61]	—		At best stabilization GHG emissions at 1990 level for the EU15	—	
EEA 33/2002[62]	—	1990–2000 -3.5 EU15	-4.7 EU15 DTT +0.5	—	
European Commission COM(2002)702[63]	4207.6	1990–2000 -3.5 EU15	-4.7	-12.0	
EEA 95/2003[64]	4192	1990–2001 -2.3 EU15	DTT +2.1	—	—

EEA 77/2003[65]	—	—	-4.7	-6.0	—
European Commission COM(2003)735[66]	4204	1990–2001 -2.3 EU15	-0.5	-7.2	-0.5
EEA 36/2003[67]	4204	1990–2001 -2.3 EU15	-0.2 DTT 2.1	-7.2	—
EEA 4/2004[68]	4204	1990–2001 -2.3 EU15	-0.5 DTT +2.1	-7.2	—
EEA 2/2004[69]	4245.2	1990–2002 -2.9 EU15	DTT +1.9	—	—
European Commission COM(2004)818[70]	4245.2	1990–2002 -2.9 EU15 / -2.9 EU15 / -9.0 EU25	DTT +1.9	—	-8.0
EEA 5/2004[71]	—	1990–2002 -2.9 EU15 / -9.0 EU23	1990–2002 / -1.0 EU15 / -9.0 EU23	-7.7	-8.8
EEA 4/2005[72]	4252.5	1990–2003 -1.7 EU15 / -5.5 EU25	-4.5 EU15 / DTT +3.5	—	—
EEA 8/2005[73]	4252.4	1990–2003 -1.7 EU15 / -8.0 EU25	-1.6 EU15 DTT +3.5 / -5.0 EU25 DTT -2.9	-6.8 EU15 / -9.3 EU25	-9.3 EU15 / DTT +1.9 / DTT -4.2 / EU25

(continued)

Table 9.1 – *continued*

DATA[48,49]	Base year (million tonnes CO₂ equivalent)[50]	EU15 GHG emissions			
		1990–XX (%) (CO₂ equivalent)	1990–2010 Business-as-usual (%)	1990–2010 Additional policies and measures (%)	1990–2010 flexible mechanisms (%)
European Commission COM(2005)655[74]	4179.6	1990–2003 −1.7 EU15 −8.0 EU25	−1.6 EU15 DTT +1.9 −5.0 EU25	−6.8 EU15 −9.0 EU25	−9.3 EU15
EEA 6/2006[75]	4265.7	1990–2004 −0.6 EU15 −4.8 EU25	−0.9 EU15 DTT +4.7	—	—
European Commission COM(2006)658[76]	4266.4	1990–2004 −0.9 EU15 −7.3 EU25	−0.6 EU15 DTT +4.7 −4.6 EU25	−4.6 EU15 −8.1 EU25	−7.2 EU15 (−8.0 sinks) −10.8 EU25
EEA 9/2006[77]	4266.4	1990–2004 −0.9 EU15 −7.3 EU25	−0.6 EU15 DTT +4.7 −2.1 EU23	−4.6 EU15 −5.6 EU23	−7.2 EU15 (−8.0 sinks) DTT +2.3
EEA 6/2006[78]	4280.4	1990–2004 −1.1 EU15 (−0.8 to 1990) −4.9 EU25	DTT +4.5	—	—

EEA 10/2006[79]	4263 (1990)	1990–2004 −0.8 EU15 −4.9 EU25	—	—	—
EEA 7/2007[80]	4278.8	1990–2005 −1.5 EU15 −7.9 EU27	—	—	—
EEA 5/2007[81]	—	1990–2005 EU15 EU27	−4.0 EU15	−7.9 EU15	−10.4 EU15
European Commission COM(2007)757[82]	—	1990–2005 −2.0 EU15 −11.0 EU27	−4.0 EU15 −10.7 EU27	−7.4 EU15 −13.2 EU27 (flexible mechanisms and sinks)	−11.4 EU15 −16.7 EU27 (additional PAMs)
EEA 6/2008[83]	4265.5	1990–2006 −2.7 EU15 −7.7 EU27	—	—	—
EEA 5/2008[84]	4266.0	1990–2006 −2.7 EU15 −7.7 EU27	−3.6 EU15	−6.9 EU15	−9.9 EU15
EEA 4/2009[85]	4265.5	1990–2007 −4.3 EU15 −9.3 EU27	—	—	—

DTT – distance to the target
Source: Massai (2010a)

NOTES

1. Agency of the EU under regulation EC No 401/2009 of the European Parliament and of the Council of 23 April 2009 establishing the European Environment Agency and the European Environment Information and Observation Network.

2. Council Decision 93/389/EEC of 24 June 1993 for a monitoring mechanism of Community CO_2 and other GHG emissions, OJ L 167, 9 July 1993, pp31−33.

3. Council Decision 99/296/EC amending Council Decision 93/389/EEC for a monitoring mechanism of Community CO_2 and other GHG Emissions, OJ L 117 of 5 May 1999, p35.

4. Proposal of a Decision of the European Parliament and of the Council for a monitoring mechanism of Community GHG emissions and the implementation of the Kyoto protocol COM(2003)51, Brussels, 5 February 2003.

5. Decision 280/2004/EC of the European Parliament and of the Council of 11 February 2004 concerning a mechanism for monitoring Community GHG emissions and for implementing the Kyoto Protocol, OJ L49, 19 February 2004, pp1−8.

6. Decision 2005/166/EC laying down rules implementing Decision No 280/2004/EC of the European Parliament and of the Council concerning a mechanism for monitoring Community GHG emissions and for implementing the Kyoto Protocol, Brussels, 10 February 2005.

7. Report from the Commission under Council Decision 93/389/EEC for a monitoring mechanism of Community CO_2 and other GHG emissions − First evaluation report, COM(1994)67, Brussels, 10 March 1994.

8. Report from the Commission of 14 March 1996, under Council Decision 93/389/EEC, Second evaluation report, COM(1996)91.

9. Report from the Commission under Council Decision 93/389/EEC as amended by Decision 99/296/EC for a monitoring mechanism of Community GHG emissions, COM(2000)749, Brussels, 22 November 2000.

10. Report 2001 under Council Decision 93/389/EEC for a monitoring mechanism of Community GHG emissions, COM(2001)708, Brussels, 30 November 2001.

11. Report 2002 under Council Decision 93/389/EEC for a monitoring mechanism of Community GHG emissions, COM(2002)702, Brussels, 9 December 2002.

12. Report 2003 under Council Decision 93/389/EEC for a monitoring mechanism of Community GHG emissions, COM(2003)735, Brussels, 28 November 2003.

13. Catching up with the Community's Kyoto target (under Decision 280/2004/EC of the European Parliament and of the Council concerning a mechanism for monitoring Community GHG emissions and for implementing the Kyoto Protocol), COM(2004)818, Brussels, 20 December 2004.

14. Annual EC GHG inventory 1990−2001 and inventory report 2003, Technical report No 95 of 8 July 2003, EEA.

15. GHG emissions trends and projections in Europe, Environmental Issue Report No 36/2003, EEA.

16. Annual EC GHG inventory 1990−2002 and inventory report 2004, Technical Report No 2/2004, 15 July 2004, EEA.

17. GHG emissions trends and projections in Europe, Environmental Issue Report No 5/2004, 21 December 2004, EEA.

18. Annual EC GHG inventory 1990−2003 and inventory report 2005, Technical Report No 4/2005, 21 June 2005, EEA.

19. GHG emissions trends and projections in Europe 2005, Report No 8/2005, 1 December 2005, EEA.

20. Annual EC GHG inventory 1990–2004 and inventory report 2006, Technical Report No 6/2006, 22 June 2006, EEA.

21. GHG emissions trends and projections in Europe 2006, EEA Report No 9/2006.

22. Annual EC GHG inventory 1990–2004 and inventory report 2006, Technical Report No 10/2006, 14 December 2006, EEA.

23. Annual EC GHG inventory 1990–2005 and inventory report 2007, Technical Report No 7/2007, 14 June 2007, EEA.

24. GHG emission trends and projections in Europe 2007, EEA Report No 5/2007.

25. Annual EC GHG inventory 1990–2006 and inventory report 2008, Technical Report No 6/2008, 18 June 2008, EEA.

26. GHG emissions trends and projections in Europe 2008, EEA Report No 5/2008, 16 October 2008.

27. Annual EC GHG inventory 1990–2007 and inventory report 2009, Technical Report No 4/2009, 29 May 2009, EEA.

28. GHG emissions trends and projections in Europe 2009, EEA Report No 9/2009, 12 November 2009.

29. Annual EC GHG inventory 1990–2008 and inventory report 2010, Technical Report No 6/2010, 2 June 2010, EEA.

30. Communication from the Commission of 15 December 2005 – Report on demonstrable progress under the Kyoto Protocol (required under Article 5(3) of Decision 280/2004/EC concerning a mechanism for monitoring Community GHG emissions and for implementing the Kyoto Protocol), COM(2005)615, Brussels,

31. Annual EC GHG inventory 1990–2003 and inventory report 2005, Technical Report No 4/2005, EEA, 21 June 2005.

32. Annual EC GHG inventory 1990–2004 and inventory report 2006, Technical Report No 6/2006, EEA, 7 June 2006.

33. GHG emissions trends and projections in Europe 2006, EEA Report No 9/2006, 27 October 2006.

34. Report on the progress towards achieving the Kyoto objectives (required under Decision 280/2004/EC of the European Parliament and of the Council concerning a mechanism for monitoring Community GHG emissions and for implementing the Kyoto Protocol), COM(2006)658, 27 October 2006, Brussels.

35. Annual EC GHG inventory 1990–2004 and inventory report 2006, 12 December 2006.

36. Commission Decision of 14 December 2006 determining the respective emission levels allocated to the Community and each of its Member States under the Kyoto Protocol pursuant to Council Decision 2002/358/EC, OJ L 358 of 16 December 2006, pp87–89.

37. Annual EC GHG inventory 1990–2005 and inventory report 2007, 27 May 2007, EEA.

38. Commission Decision 2007/589/EC of 18 July 2007 revising the guidelines for installations to monitor and report GHG emissions pursuant to Directive 2003/87/EC, OJ L 229 of 31 August 2007, pp1–85.

39. GHG emission trends and projections in Europe 2007, Report No 5/2007, 27 November 2007, EEA.

40. Communication of the European Commission on the progress towards achieving the Kyoto objectives, COM(2007)757, Brussels, 27 November 2007.

41. Annual EC GHG inventory 1990–2006 and inventory report 2008, Technical Report No 6/2008, EEA.

42. GHG emissions trends and projections in Europe 2008, EEA Report No 5/2008, 16 October 2008.

43. Annual EC GHG inventory 1990–2007 and inventory report 2009, Technical Report No 4/2009, 29 May 2009, EEA.

44. GHG emission trends and projections in Europe 2009, Tracking progress towards Kyoto targets, EEA Report No 9/2009, EEA, 12 November 2009.

45. Report from the Commission to the European Parliament and the Council on the Progress towards achieving the Kyoto objectives, required under Article 5 of Decision 280/2004/EC of the European Parliament and of the Council concerning a mechanism for monitoring Community GHG emissions and for implementing the Kyoto Protocol, COM(2009)630, Brussels, 12 November 2009.

46. CSI 011, GHG emission projections assessment published March 2009 by the Agency of the EU under regulation EC No 401/2009 of the European Parliament and of the Council of 23 April 2009 establishing the European Environment Agency and the European Environment Information and Observation Network.

47. Annual EU GHG inventory 1990–2008 and inventory report 2010, Technical Report No 6/2010, EEA, 2 June 2010.

48. The difference in data referring to the base year emissions reported in this and forthcoming tables relates to the fact that before 2006 final data on EU Member States' GHG emissions were not available in the reports on the EU's assigned amount (required under Article 3(7) and (8) of the Kyoto Protocol) under the UNFCCC, as indicated in Commission Decision 2006/944/EC of 14 December 2006 determining the respective emission levels allocated to the Community and each of its Member States under the Kyoto Protocol pursuant to Council Decision 2002/358/EC, OJ L 358, 16 December 1996, pp87–89.

49. Unless otherwise specified, the data in this table do not take into consideration offsets of GHG emissions from land use, land use change and forestry activities.

50. The base year emissions of the EC are the sum of the respective base year emissions of the 15 Member States that agreed to jointly fulfil their respective commitments under Article 3 of the KP. This rule also refers to the base year for hydrofluorocarbons (HFCs), perfluorocarbons (PFCs) and sulphur hexafluoride (SF_6). For these gases, parties included in Annex I of the Kyoto Protocol can select either 1995, in accordance with Article 3(8) of the Kyoto Protocol, or the base year for other Annex A GHGs (mostly 1990). Austria, France and Italy selected 1990 as the base year for these gases, all other Member States chose 1995. The base year for all other GHG emissions under the Protocol is 1990 for the EU15. The base year for HFCs, PFCs and SF_6 for the new Member States with commitments under Kyoto is 1995, except for Slovakia, which chose 1990 as base year for fluorinated gases. The base year for all other GHG emissions under Kyoto is 1990, except for Poland (1988), Slovenia (1986) and Hungary (1985–1987).

51. Report from the Commission of the European Communities under Council Decision 93/389/EEC – First Evaluation of existing national programmes under the Monitoring Mechanism of Community CO_2 and Other GHG emissions, COM(1994)67, Brussels, 10 March 1994.

52. Report from the Commission under Council Decision 93/389/EEC – Second Evaluation of National Programmes under the Monitoring Mechanism of Community CO_2 and Other GHG emissions: Progress towards the Community CO_2 Stabilisation Target, COM(1996)91, Brussels, 14 March 1996.

53. EEA Annual EC GHG Inventory 1990–1998, Technical Report No 41, May 2000.

54. EEA EC and Member States GHG Emission Trends 1990–1998, Topic Report No 6/2000, July 2000.

55. Commission of the European Communities Report under Council Decision 1999/296/EC for a monitoring mechanism of Community GHG emissions, COM(2000)749, Brussels, 22 November 2000.

56. EEA Annual EC GHG Inventory 1990–1999, Technical Report No 60, 22 April 2001.

57. EC and Member States GHG emission trends 1990–1999, Topic Report 10/2001, EEA, 23 October 2001.

58. Report 2001 under Council Decision 93/389/EEC for a monitoring mechanism of Community GHG emissions, COM(2001)708, Brussels, 30 November 2001.

59. EEA GHG Emission Trends in Europe, 1990–2000, Topic Report No 7/2002, 9 February 2002.

60. Annual EC GHG Inventory 1990–2000 and Inventory Report 2002, Technical Report 75/2002, EEA, 28 April 2002.

61. EEA Analysis and Comparison of National and EU-Wide Projections of GHG Emissions, Topic Report 1/2002, 22 August 2002.

62. EEA GHG Emission Trends and Projections in Europe, 1990–2000, Environmental Issue Report 33/2002, 4 December 2002.

63. Report 2002 under Council Decision 93/389/EEC for a monitoring mechanism of Community GHG emissions, COM(2002)702, Brussels, 9 December 2002.

64. Annual EC GHG Inventory 1990–2001 and Inventory Report 2003, Technical Report 95/2003, EEA, 4 May 2003.

65. GHG Emission Projections for Europe, Technical Report 77/2003, EEA, 30 June 2003.

66. Report from the Commission under Council Decision 93/389/EEC as amended by Decision 99/296/EC for a monitoring mechanism of Community GHG emissions, COM(2003)735, Brussels, 28 November 2003.

67. GHG Emission Trends and Projections in Europe 2003, Technical Report 36/2003, EEA, 4 December 2003.

68. Analysis of GHG emissions trends and projections in Europe 2003, Technical Report No 4/2004, EEA, 13 July 2004, EEA.

69. Annual EC GHG Inventory 1990–2002 and Inventory Report 2004, Technical Report No 2/2004, EEA, 25 October 2004, EEA.

70. Commission Report of 20 December 2004: Catching up with the Community's Kyoto target (under Decision 280/2004/EC of the European Parliament and of the Council concerning a mechanism for monitoring Community GHG emissions and for implementing the Kyoto Protocol), COM(2004)818, Brussels, 20 December 2004.

71. EEA GHG Emission Trends and Projections in Europe 2004, Report No 5/2004, 21 December 2004.

72. Technical Report No 4/2005, Annual EC GHG Inventory 1990–2003 and Inventory Report 2005, EEA.

73. GHG Emission Trends and Projections in Europe 2005, Technical Report No 8/2005, EEA, 1 December 2005.

74. Report from the European Commission: Progress towards Achieving the Community's Kyoto Target, COM(2005)655, Brussels, 15 December 2005.

75. Annual EC GHG Inventory 1990–2004 and Inventory Report 2006, Technical Report No 6/2006, EEA, 22 June 2006.

76. Report from the Commission: Progress towards Achieving the Kyoto Objectives, COM(2006)658, Brussels, 27 October 2006.

77. GHG Emission Trends and Projections in Europe 2006, Technical Report No 9/2006, 27 October 2006, EEA.

78. Annual EC GHG Inventory Report 1990–2004 and Inventory Report 2006, Technical Report No 6/2006, EEA, version of 14 December 2006.

79. The EC's Initial Report under the Kyoto Protocol, Technical Report No 10/2006, 2 February 2007.

80. Annual EC GHG Inventory 1990–2005 and Inventory Report 2007, EEA Technical Report No 7/2007, 14 June 2007.

81. GHG Emission Trends and Projections in Europe 2007, EEA Report No 5/2007, 27 November 2007.

82. Communication from the Commission: Progress towards Achieving the Kyoto Objectives, COM(2007)757, Brussels, 27 November 2007.

83. Annual EC GHG Inventory 1990–2006 and Inventory Report 2008, EEA Technical Report No 6/2008, 18 June 2008.

84. GHG Emission Trends and Projections in Europe 2008, EEA Report No 5/2008, 16 October 2008.

85. European Environment Agency, Annual EC GHG inventory 1990–2007 and inventory report 2009, Technical report No 4/2009, 29 May 2009.

CHAPTER 10

European Climate Change Programme

S ince 1991 the EU has been developing a strategy to reduce CO_2 emissions and improve the level of energy efficiency and renewable energy. The ECCP launched by the European Commission in 2000 through Communication from the Commission to the Council and the European Parliament on EU policies and measures to reduce greenhouse gas emissions: Towards a European Climate Change Programme (ECCP), COM(2000)88, was established with the aim of coordinating the different areas of European climate and clean energy policy.[1]

The ECCP is one of the best examples of the application of the principle of integration of environmental concerns into other policies. ECCP activities were aimed at the identification of adequate policies and measures to reduce GHG emissions throughout the Community, which were discussed and designed inside the thematic working groups composed of representatives of the European Commission, interested directorate generals, representatives of existing and new Member States, private sector, NGOs and other stakeholders.

10.1   ECCP I (2000–2001)

The first phase of ECCP I (2000–2001) focused on cost-effective policies and measures to be introduced in the energy, transport and industry sectors. The six working groups established within the ECCP were: flexible mechanisms, energy supply, energy consumption, transport, industry and research. A range of 40 EU-wide common and coordinated policies and measures (CCPMs) were identified in the ECCP Report released by the European Commission in June 2001.

Some of the provisions indicated in the first ECCP report were integrated in the legislation package (2001–2003) included in Communication from the

Commission on the implementation of the first phase of the European Climate Change Programme COM(2001)580, indicating four different areas to implement the first phase of the ECCP:[2]

- Cross-cutting issues: correct and effective implementation of the IPPC directive, Proposal for a Directive on linking project-based mechanisms including JI and CDM to the EU ETS, proposal for a review of the EC GHG monitoring mechanism;
- Energy sector: proposal for a framework directive introducing minimum efficiency for end-use equipment, Proposal for a Directive on energy demand management, Proposal for a Directive on the promotion of CHP, additional non-legislative proposals;
- Transport sector: measures in line with the White Paper on European transport policy for 2010: time to decide COM(2001)370 on the shift of the balance between ways of transport,[3] proposal for improvement of infrastructure use and charging, proposal for the promotion of the use of biofuels;
- Industry sector: proposal for regulation on fluorinated gases.

10.2   SECOND PHASE OF ECCP I (2001–2003)

In the second phase of ECCP I (2001–2003) the following 11 working groups were established: linking JI and CDM with the EU ETS, agriculture, forest-related sinks, sinks in agricultural soils, fluorinated gases, energy supply, energy demand, transport, industry, waste and research.

In April 2003 the European Commission released its Second ECCP Progress Report: Can we meet our Kyoto targets?, which provided an overview of the activities within the working groups and the overview of the follow-up work in terms of implementation of measures that were identified in the first phase of the ECCP.[4]

Policies and measures identified in the second phase of ECCP I were:

- On flexible mechanisms: Directive 2003/87/EC establishing a EU Emissions Trading system and Directive 2004/101/EC on linking project-based mechanisms including JI and CDM to the EU ETS.
- On energy supply (150 million tonnes CO_2 equivalent reduction potential) the following measures were identified:
  - Directive 2001/77/EC of the European Parliament and of the Council of 27 September 2001 on the promotion of electricity produced from renewable energy sources in the internal electricity market;[5]
  - Communication from the Commission on alternative fuels for road transportation and on a set of measures to promote the use of biofuels COM(2001)547;[6]
  - Directive 2003/30/EC of the European Parliament and of the Council of 8 May 2003 on the promotion of the use of biofuels or other renewable fuels for transport.[7]

- On CHP:
  - Communication from the Commission on a Community strategy to promote CHP and to dismantle barriers to its development COM(1997)514;[8]
  - Proposal of the European Commission of a Directive of the European Parliament and of the Council on the promotion of cogeneration based on a useful heat demand in the internal energy market COM(2002)415;[9]
  - Common Position (EC) No 52/2003 of 8 September 2003 adopted by the Council with a view to adopting a directive of the European Parliament and of the Council on the cogeneration based on a useful heat demand in the internal energy market and amending Directive 92/42/EEC;[10]
  - Directive 2004/8/EC of the European Parliament and of the Council of 11 February 2004 on the promotion of cogeneration based on a useful heat demand in the internal energy market and amending Directive 92/42/EEC.[11]
- On the internal market in electricity and gas:
  - Directive 2003/54/EC of the European Parliament and of the Council of 26 June 2003 concerning common rules for the internal market in electricity and repealing Directive 96/92/EC;[12]
  - Directive 2003/55/EC of the European Parliament and of the Council of 26 June 2003 concerning common rules for the internal market in natural gas and repealing Directive 98/30/EC.[13]
- On renewable energy:
  - The second phase of the ECCP focused on the promotion of energy from renewable sources in heating applications ('RES-H'). The Commission assessed the potential for increased uptake and the ways in which both existing (directive on energy performance of buildings or the proposal for a directive on CHP) and new measures could contribute to the promotion of RES-H;
  - Common Position (EC) No 5/2003 of 3 February 2003 adopted by the Council with a view to adopting a Directive of the European Parliament and of the Council concerning common rules for the internal market in electricity and repealing Directive 96/92/EC;[14]
  - Common Position (EC) No 6/2003 of 3 February 2003 adopted by the Council with a view to adopting a Directive of the European Parliament and of the Council concerning common rules for the internal market in natural gas and repealing Directive 98/30/EC.[15]
- On energy demand:
  - Directive 2002/91/EC of the European Parliament and of the Council of 16 December 2002 on the energy performance of buildings.[16]
- On labelling equipment:
  - Commission Directive 2002/40/EC of 8 May 2002 implementing Council Directive 92/75/EEC with regard to energy labelling of household electric ovens;[17]

- Corrigendum to Commission Directive 2002/40/EC of 8 May 2002 implementing Council Directive 92/75/EEC with regard to energy labelling of household electric ovens;[18]
- Commission Directive 2002/31/EC of 22 March 2002 implementing Council Directive 92/75/EEC with regard to energy labelling of household air-conditioners;[19]
- Corrigenda to Commission Directive 2002/31/EC of 22 March 2002 implementing Council Directive 92/75/EEC with regard to energy labelling of household air-conditioners;[20]
- Regulation (EC) No 2422/2001 of the European Parliament and of the Council of 6 November 2001 on a Community energy efficiency labelling programme for office equipment;[21]
- Proposal for a Directive of the European Parliament and of the Council on establishing a framework for the setting of Eco-design requirements for Energy-Using Products and amending Council Directive 92/42/EEC COM(2003)453;[22]
- Agreement in the EU Council of Ministers of 11 June 2004, Luxembourg on proposal COM(2003)453 establishing environmental standards for energy using products such as electrical appliances.
- On the transport sector: no major developments apart from the above mentioned strategy on the reduction of CO_2 from passenger cars.
- On fluorinated gases: see Chapter 12.

With regard to the working groups established in the second phase of the ECCP, namely Sinks – Sub-Group on Agricultural Soils and Forest-Related Sinks, progress has been reported on the mitigation potential of improved use and management of agricultural soils and on the potential for carbon sequestration in EU forests. The second ECCP progress report confirmed that additional and more effective policies and measures were required in order to achieve the EC GHG emissions reduction obligations under the Kyoto Protocol.

10.3   ECCP II

ECCP II was launched by the European Commission on 24 October 2005 with its view on further development of EU climate policy. The ECCP II referred to Communication COM(2005)35 'Winning the battle against climate change'[23] and it was supposed to deliver the first recommendations on the adoption of specific European legislation to combat climate change by mid-2006.

The ECCP II mainly focused was on:

- ECCP I review;
- Geological CCS;
- Adaptation to climate change;
- Aviation;

- Integrated approach to reduce CO_2 emissions from light duty vehicles;
- Energy efficiency;
- Renewable energy;
- Technology policy.

NOTES

1. Communication from the Commission to the Council and the European Parliament on EU policies and measures to reduce greenhouse gas emissions: Towards a European Climate Change Programme (ECCP), COM(2000)88, Brussels, 8 May 2000.

2. Communication from the Commission to the Council and the European Parliament on the implementation of the first phase of the European Climate Change Programme, COM(2001)580, Brussels, 23 October 2001.

3. White Paper on European transport policy for 2010: Time to decide, COM(2001)370, Brussels, 12 September 2001.

4. European Commission, Second ECCP Progress Report: Can we meet our Kyoto targets?, Brussels, April 2003.

5. Directive 2001/77/EC of the European Parliament and of the Council of 27 September 2001 on the promotion of electricity produced from renewable energy sources in the internal electricity market, OJ L 283 of 27 October 2001, pp33–40.

6. Communication from the Commission to the European Parliament, the Council, the Economic and Social Committee and the Committee of the Regions on alternative fuels for road transportation and on a set of measures to promote the use of biofuels, COM(2001)547, Brussels, 7 November 2001.

7. Directive 2003/30/EC of the European Parliament and of the Council of 8 May 2003 on the promotion of the use of biofuels or other renewable fuels for transport, OJ L 123 of 17 May 2003, pp42–46.

8. Communication from the Commission on a Community strategy to promote CHP and to dismantle barriers to its development, COM(1997)514, Brussels, 15 October 1997.

9. Proposal of the European Commission of a Directive of the European Parliament and of the Council on the promotion of cogeneration based on a useful heat demand in the internal energy market, COM(2002)415, Brussels, 22 July 2002.

10. Common Position (EC) No 52/2003 of 8 September 2003 adopted by the Council, acting in accordance with the procedure referred to in Article 251 of the Treaty establishing the EC, with a view to adopting a directive of the European Parliament and of the Council on the cogeneration based on a useful heat demand in the internal energy market and amending Directive 92/42/EEC, OJ C 258 E of 28 October 2003, pp1–17.

11. Directive 2004/8/EC of the European Parliament and of the Council of 11 February 2004 on the promotion of cogeneration based on a useful heat demand in the internal energy market and amending Directive 92/42/EEC, OJ L 52 of 21 February 2004, pp50–60.

12. Directive 2003/54/EC of the European Parliament and of the Council of 26 June 2003 concerning common rules for the internal market in electricity and repealing Directive 96/92/EC, OJ L 176 of 15 July 2003, pp37–56.

13. Directive 2003/55/EC of the European Parliament and of the Council of 26 June 2003 concerning common rules for the internal market in natural gas and repealing Directive 98/30/EC, OJ L 176 of 15 July 2003, pp57–78.

14. Common Position (EC) No 5/2003 of 3 February 2003 adopted by the Council, acting in accordance with the procedure referred to in Article 251 of the Treaty establishing the EC, with a view to adopting a Directive of the European Parliament and of the Council concerning common rules for the internal market in electricity and repealing Directive 96/92/EC, *OJ* C 050 E of 4 March 2003, pp15–35.

15. Common Position (EC) No 6/2003 of 3 February 2003 adopted by the Council, acting in accordance with the procedure referred to in Article 251 of the Treaty establishing the EC, with a view to adopting a Directive of the European Parliament and of the Council concerning common rules for the internal market in natural gas and repealing Directive 98/30/EC, *OJ* C 050 E of 4 March 2003, pp36–58.

16. Directive 2002/91/EC of the European Parliament and of the Council of 16 December 2002 on the energy performance of buildings, *OJ* L 1 of 4 January 2003, pp65–71.

17. Commission Directive 2002/40/EC of 8 May 2002 implementing Council Directive 92/75/EEC with regard to energy labelling of household electric ovens, *OJ* L 128 of 15 May 2002, pp45–56.

18. Corrigenda to Commission Directive 2002/40/EC of 8 May 2002 implementing Council Directive 92/75/EEC with regard to energy labelling of household electric ovens, *OJ* L 33 of 8 February 2003, p43.

19. Commission Directive 2002/31/EC of 22 March 2002 implementing Council Directive 92/75/EEC with regard to energy labelling of household air-conditioners, *OJ* L 86 of 3 April 2002, pp26–41.

20. Corrigenda to Commission Directive 2002/31/EC of 22 March 2002 implementing Council Directive 92/75/EEC with regard to energy labelling of household air-conditioners, *OJ* L 34 of 11 February 2003, p30.

21. Regulation (EC) No 2422/2001 of the European Parliament and of the Council of 6 November 2001 on a Community energy efficiency labelling programme for office equipment, *OJ* L 332 of 15 December 2001, pp1–6.

22. Proposal for a Directive of the European Parliament and of the Council on establishing a framework for the setting of Eco-design requirements for Energy-Using Products and amending Council Directive 92/42/EEC, COM(2003)453, Brussels, 1 August 2003.

23. Communication of the European Commission 'Winning the battle against climate change', COM(2005)35, Brussels, 9 February 2005.

CHAPTER 11

Carbon Capture and Storage

*I*n July 2005 the EC Directorate General for Energy and Transport announced the establishment by Norway, the UK and Denmark of an informal forum to exchange information and discuss issues in relation to the use of CO_2 for enhanced oil recovery and storage in the North Sea.

On 2 December 2005 a working group on carbon capture development in the EU was launched in Brussels by the Directorate General on Research. The Zero Emission Technology Platform established in 2005 was a stakeholder platform for CCS of CO_2 from fossil fuel power generation with the objective of coordinating research and activities in the sector of CCS at the European level. The Zero Emission Technology Platform received a sum of €70 million of EU research funding and was composed of representatives of the industrial sector, EU and government officials, academics and NGOs. It met on 12 September 2006 for the adoption of two CCS action plans aimed at the promotion of the specific technology through policy framework recommendations to accelerate the market developments in the sector, and research recommendations to foster research on CCS.

On 16 February 2007 the European Commission opened a public consultation aimed at the development of plans for the adoption of an EU legal framework for CCS. The results of this consultation were collected by the European Commission and published on 26 September 2007 in the Report of the Public Consultation on the European Strategic Energy Technology Plan (SET-Plan). According to the report, respondents to the public consultation appeared doubtful on the real impact of CCS in the reduction of GHG emissions in Europe of 20 to 30 per cent by 2020. The majority of voters considered initiatives such as technologies to improve energy efficiency in buildings and transport more suitable to reach these targets.

The Proposal for a Directive on the geological storage of carbon dioxide and amending Council Directives 85/337/EEC, 96/61/EC, Directives 2000/60/EC,

2001/80/EC, 2004/35/EC, 2006/12/EC and Regulation (EC) No 1013/2006, COM(2008)18[1] as well as Communication COM(2008)13 on supporting early demonstration of sustainable power generation from fossil fuels (CO_2 capture and storage)[2] were adopted on 23 January 2008 by the European Commission aiming at the establishment of a regulatory framework for the inclusion of the capture and storage of CO_2 in the EU ETS. The proposals included in the 2008 integrated climate and energy package highlighted the details of exploration and storage permits within the EU ETS with the aim of boosting investment in this sector. Installations opting for the capture of CO_2 would receive an equivalent sum of EU allowances to be traded in the European carbon market. The main goal of the proposals was fostering the capture, transport and storage of CO_2 emissions in geological formations in Europe provided that the management of environmental risk and the barriers in existing legislation were adequately regulated. The CCS activities should be included in the 3rd period of the EU ETS (2013–2020) on a voluntary basis, and demonstration projects on CCS are funded through an ad hoc reserve of €300 million EU allowances. The system of permits should be managed by the European Commission and new power plants with a capacity bigger than 300 megawatts will be required to reserve specific space for the establishment of CCS systems. On the issue of responsibility, once the storage site is closed it should pass to the Member State where the plant is located.

Proposal COM(2008)18 was accompanied by the following documents: Impact Assessment with Summary Impact Assessment; Annex XII to the Impact Assessment: Indicative transport and storage networks for CO_2 at Member States level; and Annex XIII to the Impact Assessment: Indicative transport and storage networks for CO_2 across the EU.

On 8 September 2008 the Environment Committee of the European Parliament debated on a draft non-legislative resolution responding to proposal COM(2008)18 of the Commission. The resolution claimed that CCS technology should be financed by the power sector rather than by public subsidies that should be used only to finance research on safety, environmental and monitoring requirements.

On 18 September 2008 the European Commission presented a services paper on energy-intensive industries exposed to significant risk of carbon leakage, which contained a methodology on a three-stage assessment of the exposure of industrial sectors to 'carbon leakage'. According to the Commission, the European aluminium, steel and cement industries were likely to qualify for the free allocation of carbon allowances in compensation for the lost of international competitiveness in the third phase of the EU ETS.

On 6 October 2008 the Environment Committee voted on the compromise amendments of Rapporteur Davies on the draft directive on CCS. MEPs set a cap of 500g of CO_2 emissions per kilowatt hour for large new power plants as from 2015, which in practice would oblige those plants to use the option of CCS. MEPs' intention was to speed up the process for the introduction of this technology. Furthermore, MEPs agreed on a funding mechanism to support the CCS technology financed by the new entrants' reserve under the EU ETS.

On 29 October 2008 the Council agreed on the compromise text of the amended text of the directive proposal on CCS. The Council did not agree with two major proposals adopted by the Environment Committee of the Parliament, namely the introduction of a binding cap on CO_2 emissions from large power plants and the set-aside in the new entrants' reserve under the EU ETS to finance 12 CCS demonstration projects by 2015.

On 12 November 2008 the Commission presented an informal policy options paper that contained a request for the reduction of the amount of carbon allowances to be earmarked for CCS demonstration projects in the EU ETS. According to the Commission this reduction was essential in order to provide funding for CCS from a combination of different sources. In response to the Commission's paper, MEPs adopted on 18 November 2008 a non-legislative resolution calling for a 'direct financial commitment' of the EU to finance CCS projects and considering the incentives proposed by the Commission as 'insufficient'.

On 17 December 2008 the European Parliament approved with a big majority the integrated climate and energy package including, amongst others, a legislative resolution on the Proposal for a Directive on the geological storage of carbon dioxide COM(2008)18.[3]

On 25 June 2009 Directive 2009/31/EC of the European Parliament and of the Council on the geological storage of carbon dioxide entered into force and established a legal framework for the geological storage of CO_2 with the aim of contributing to the fight against climate change.[4] In particular, Directive 2009/31/EC introduced a procedure designed to ensure the integrity of a project for the storage of CO_2, in terms of avoiding any risk of leakage or damage to human health or the environment by issuing storage permits for each site and by guaranteeing the security of the transport network or the storage site. Furthermore, Directive 2009/31/EC established clear rules for the monitoring of the sites, for adjustment measures to be introduced in the event of leakage and for closure and post-closure obligations. Responsibility for the sites is transferred from the Member States to the operators and liability for local damage is regulated by Directive 2004/35/EC on environmental liability. As to the relationship between CCS and other existing EU legislation, barriers in existing waste and water legislation were removed and the large combustion plants directive amended with the introduction of an assessment of capture readiness for large plants. In particular, Directive 2009/29/EC on the EU ETS review was amended with the inclusion of CCS in Annex I, with the reference to emissions captured, transported and stored not emitted.

A set of guidance documents was prepared by the European Commission in order to assist stakeholders and Member States with the implementation of the CCS directive. The documents provided a unified overall methodological approach to be adjusted on a case-by-case basis:

- Guidance document 1: CO_2 storage life-cycle approach to risk management;
- Guidance document 2: Specific approaches to key stages of the CO_2 storage life-cycle:
    - Selection of the storage site;
    - Composition of the CO_2 stream;
    - Monitoring of the storage site;
    - Corrective measures;

- Guidance Document 3: Transfer of responsibility (Article 18);
- Guidance Document 4: Financial security (Article 19) and financial transfer (Article 20).

On 3 February 2010 a committee of Member States agreed on the European Commission draft decision laying down criteria and measures for the financing of commercial demonstration projects that aim at the environmentally safe capture and geological storage of CO_2 as well as demonstration projects of innovative renewable energy technologies under the scheme for GHG emission allowance trading within the Community established by Directive 2003/87/EC (EU ETS). CCS and renewable energy demonstration projects shall receive the carbon price equivalent of 300 million EU allowances (€4.5 at current prices in the market) as set aside by the EU ETS new entrants reserve in the period 2013–2020. The draft decision established that the European Commission will decide on the eligibility of projects and not the Member States, as proposed by some of them. According to the estimation of the Commission, eight CCS projects covering two-thirds of the set aside should receive funding by the end of 2011, with at least one renewable project in each Member State and with a maximum of three allowed. The European Parliament will have three months' time to scrutiny the draft decision in accordance with the comitology procedure.

NOTES

1. European Commission, Proposal for a Directive on the geological storage of carbon dioxide and amending Council Directives 85/337/EEC, 96/61/EC, Directives 2000/60/EC, 2001/80/EC, 2004/35/EC, 2006/12/EC and Regulation (EC) No 1013/2006, COM(2008)18, Brussels, 23 January 2008.

2. Communication of the European Commission COM(2008)13 on supporting early demonstration of sustainable power generation from fossil fuels (CO_2 capture and storage), Brussels, 23 January 2008.

3. Legislative resolution of the European Parliament on the Proposal for a Directive of the European Parliament and of the Council on the geological storage of carbon dioxide and amending Council Directives 85/337/EEC, 96/61/EC, Directives 2000/60/EC, 2001/80/EC, 2004/35/EC, 2006/12/EC and Regulation (EC) No 1013/2006, COM(2008)18, Brussels, 23 January 2008.

4. Directive 2009/31/EC of the European Parliament and of the Council of 23 April 2009 on the geological storage of carbon dioxide and amending Council Directive 85/337/EEC, European Parliament and Council Directives 2000/60/EC, 2001/80/EC, 2004/35/EC, 2006/12/EC, 2008/1/EC and Regulation (EC) No 1013/2006, OJ L 140 of 5 June 2009, pp114–135.

CHAPTER 12

Fluorinated and Other Gases

The most relevant GHG in the atmosphere is CO_2. However the Kyoto Protocol also regulates three fluorinated gases that have a high global warming potential (GWP): hydrofluorocarbons (HFCs), perfluorocarbons (PFCs) and sulphur hexafluoride (SF_6). Since the early 1990s, when fluorinated gas utilization was linked with the replacement of ozone-depleting substances, the gases have been mainly used as refrigerants, fire extinguishing agents, dielectric media, solvents, foam blowing agents and in other applications.

The reduction of fluorinated gas use is another priority of European climate policy and the first provision in this respect was the European Commission proposal for a regulation of the European Parliament and of the Council on certain fluorinated gases, COM(2003)492 of 11 August 2003.[1]

On 31 March 2004 the European Parliament rejected the limits on fluorinated gases proposed by the European Commission, leaving the draft EU directive fundamentally based on the containment of HFCs and other fluorinated gases rather than on their substitution. The amendments adopted by the Parliament introduced, amongst other measures, quotas for fluorinated gases in the air conditioning units of new cars. The discussion was based on the possibility of creating two separate legal instruments (instead of the single fluorinated gas directive initially proposed by the Commission): one internal market-based directive aiming at the phase-out of HFCs in vehicle air conditioning and a second more environment-friendly provision on fluorinated gases used in stationary applications. The compromise deal presented by the Dutch EU presidency a few months later contained the proposals for the phase-out of these gases in the period between 2011 and 2017. The final political agreement was reached by EU environment ministers in Luxembourg on 14 October 2004: an 'internal market' directive establishing a six-year phase-out period for HFCs from mobile air conditioning systems starting in 2011 and a regulation with either an internal market or environmental legal basis on fluorinated gases in other applications, such as stationary sources.

On 20 June 2005 the Council of the EU adopted a common position on the directive proposal and on the regulation proposal, which was accompanied by the following country statements: Austria and Denmark recalled a single environmental legal base for the regulation on stationary fluorinated gas uses, and Belgium, Portugal and Sweden stressed that the regulation should ban the gases 'where cost-effective alternatives exist'.

The report of Rapporteur Avril Doyle to the Environment, Public Health and Food Safety Committee of the European Parliament released in September 2005 included a proposal for the protection of the environment as a single legal base for an EU regulation on limiting emissions and uses of fluorinated GHGs. This change would have made it easier for EU countries to ban certain fluorinated gas uses unilaterally, even if this could fragment the single market.

On 11 October 2005 the European Parliament's Environment Committee voted on a draft regulation ban of all controlled fluorinated gases in composite foams from January 2009 and in stationary air conditioning units a year later. The committee called for much stronger EU controls on fluorinated GHGs in stationary applications and reiterated the necessity for stricter national bans to be maintained by the Member States. According to such a regulation, HFCs would have been banned in household refrigerators in four years from its entry into force, by 2010 in commercial and industrial refrigeration, and by 2006 in aerosols.

In a second reading vote on 26 October 2005, the Parliament failed to agree on a draft regulation imposing stricter controls on fluorinated gases and related emissions. The Parliament legislative resolution on the Council common position for adopting a directive relating to emissions from air conditioning systems in motor vehicles[2] adjusted the Council requests, placing the regulation on fluorinated gases within the framework of an internal market legal basis, instead of adopting the environmental legal base proposed by a part of the same Parliament. The Parliament also rejected an amendment calling for fluorinated gases to 'only be used where other safe, technically feasible and environmentally acceptable alternatives do not exist'. Finally the Parliament rejected a series of proposals to extend the regulation's limited number of specific use bans and it adopted 20 amendments including:

- Labelling requirements for fluorinated gas containing equipment;
- A requirement that Member States facilitate cross-border transport of recovered fluorinated gases for destruction or reclamation;
- The option for the European Commission to propose by 31 December 2007 new containment requirements for the gases in non-mobile air conditioning and mobile refrigeration;
- A directive on fluorinated gases in automotive air conditioning.

On 31 January 2006 the first conciliation meeting between Council and Parliament (provided for in Article 251(4) of the EC Treaty) agreed on two laws.

Under the deal, which did not differ very much from the common position of 20 June 2005, any Member State was allowed to keep its stricter national existing gas limits adopted by December 2005 until 2012 (Austria and Denmark). All the other Member States, with existing less stringent limits, were required to comply with the EU standards. By 2012 a review of the regulation was expected. The Conciliation Committee adopted a joint text on 17 March 2006 on the Regulation of the European Parliament and of the Council on certain fluorinated GHGs. On 6 April 2006 the European Parliament adopted the result of the Conciliation Committee, but only after the withdrawal by the Commission of the provision concerning the right of the Member States to maintain until 2012 more stringent national measures.

The process for the adoption of legislation curbing fluorinated gases was finally concluded on 4 July 2006, when the two following important provisions aimed at the reduction of the three gases covered by the Kyoto Protocol entered into force: (1) Directive 2006/40/EC of the European Parliament and of the Council of of 17 May 2006 relating to emissions from air conditioning systems in motor vehicles and amending Council Directive 70/156/EEC: HFCs having a GWP higher than a level of 150 are banned from use in air conditioning systems in cars (Member States are due to implement the directive from January 2008);[3] and (2) Regulation (EC) No 842/2006 of the European Parliament and of the Council of 17 May 2006 on certain fluorinated GHGs: this provision introduced control, handling and recovery obligations, labelling and reporting rules, as well as limitations on some applications of HFCs, PFCs and SF_6. The majority of the prohibitions included in the regulation took effect from 4 July 2007.[4]

On 17 December 2007 the European Commission adopted legislation setting minimum standards on data reporting (Commission Regulation 1493/2007),[5] labelling (Commission Regulation 1494/2007)[6] and leakage control in refrigeration, air conditioning and heat pumps (Commission Regulation 1516/2007)[7] as required under Regulation (EC) No 842/2006 on certain fluorinated GHGs.

On 2 April 2008 the European Commission adopted seven regulations establishing minimum requirements and conditions for mutual recognition for the certification of companies and personnel handling fluorinated gases in the Member States. The regulations include installation, maintenance and servicing of refrigeration, air conditioning, heat pump and fire protection systems, and recovery of fluorinated gases from high voltage switchgear. These were:

- Commission Regulation (EC) No 303/2008 of 2 April 2008 establishing, pursuant to Regulation (EC) No 842/2006 of the European Parliament and of the Council, minimum requirements and the conditions for mutual recognition for the certification of companies and personnel as regards stationary refrigeration, air conditioning and heat pump equipment containing certain fluorinated GHGs;[8]
- Commission Regulation (EC) No 304/2008 of 2 April 2008 establishing, pursuant to Regulation (EC) No 842/2006 of the European Parliament and of the Council, minimum requirements and the conditions for mutual recognition

for the certification of companies and personnel as regards stationary fire protection systems and fire extinguishers containing certain fluorinated GHGs;[9]

- Commission Regulation (EC) No 305/2008 of 2 April 2008 establishing, pursuant to Regulation (EC) No 842/2006 of the European Parliament and of the Council, minimum requirements and the conditions for mutual recognition for the certification of personnel recovering certain fluorinated GHGs from high voltage switchgear;[10]
- Commission Regulation (EC) No 306/2008 of 2 April 2008 establishing, pursuant to Regulation (EC) No 842/2006 of the European Parliament and of the Council, minimum requirements and the conditions for mutual recognition for the certification of personnel recovering certain fluorinated GHG-based solvents from equipment;[11]
- Commission Regulation (EC) No 307/2008 of 2 April 2008 establishing, pursuant to Regulation (EC) No 842/2006 of the European Parliament and of the Council, minimum requirements for training programmes and the conditions for mutual recognition of training attestations for personnel as regards air conditioning systems in certain motor vehicles containing certain fluorinated GHGs;[12]
- Commission Regulation (EC) No 308/2008 of 2 April 2008 establishing, pursuant to Regulation (EC) No 842/2006 of the European Parliament and of the Council, the format for notification of the training and certification programmes of the Member States;[13]
- Commission Regulation (EC) No 309/2008 of 2 April 2008 entering a name in the register of protected designations of origin and protected geographical indications Isle of Man Manx Loaghtan Lamb (PDO).[14]

NOTES

1. European Commission proposal for a regulation of the European Parliament and of the Council on certain fluorinated gases, COM(2003)492, Brussels, 11 August 2003.

2. Legislative resolution of the European Parliament on the Council common position for adopting a directive of the European Parliament and of the Council relating to emissions from air conditioning systems in motor vehicles and amending Council Directive 70/156/EEC (16182/4/2004 – C6-0222/2005 – 2003/0189B(COD)), 26 October 2005.

3. Directive 2006/40/EC of the European Parliament and of the Council of 17 May 2006 relating to emissions from air-conditioning systems in motor vehicles and amending Council Directive 70/156/EEC, OJ L 161 of 14 June 2006, pp12–18.

4. Regulation (EC) No 842/2006 of the European Parliament and of the Council of 17 May 2006 on certain fluorinated GHGs, OJ L 161 of 14 June 2006, pp1–11.

5. Commission Regulation (EC) No 1493/2007 of 17 December 2007 establishing, pursuant to Regulation (EC) No 842/2006 of the European Parliament and of the Council, the format for the report to be submitted by producers, importers and exporters of certain fluorinated GHGs, OJ L 332 of 18 December 2007, pp7–24.

6. Commission Regulation (EC) No 1494/2007 of 17 December 2007 establishing, pursuant to Regulation (EC) No 842/2006 of the European Parliament and of the Council, the

form of labels and additional labelling requirements as regards products and equipment containing certain fluorinated GHGs, *OJ* L 332 of 18 December 2007, pp25–26.

7. Commission Regulation (EC) No 1516/2007 of 19 December 2007 establishing, pursuant to Regulation (EC) No 842/2006 of the European Parliament and of the Council, standard leakage checking requirements for stationary refrigeration, air conditioning and heat pump equipment containing certain fluorinated GHGs, *OJ* L 335 of 18 December 2007, pp10–12.

8. Commission Regulation (EC) No 303/2008 of 2 April 2008 establishing, pursuant to Regulation (EC) No 842/2006 of the European Parliament and of the Council, minimum requirements and the conditions for mutual recognition for the certification of companies and personnel as regards stationary refrigeration, air conditioning and heat pump equipment containing certain fluorinated GHGs, *OJ* L 92 of 3 April 2008, pp3–11.

9. Commission Regulation (EC) No 304/2008 of 2 April 2008 establishing, pursuant to Regulation (EC) No 842/2006 of the European Parliament and of the Council, minimum requirements and the conditions for mutual recognition for the certification of companies and personnel as regards stationary fire protection systems and fire extinguishers containing certain fluorinated GHGs, *OJ* L 92 of 3 April 2008, pp12–16.

10. Commission Regulation (EC) No 305/2008 of 2 April 2008 establishing, pursuant to Regulation (EC) No 842/2006 of the European Parliament and of the Council, minimum requirements and the conditions for mutual recognition for the certification of personnel recovering certain fluorinated GHGs from high-voltage switchgear, *OJ* L 92 of 3 April 2008, pp17–20.

11. Commission Regulation (EC) No 306/2008 of 2 April 2008 establishing, pursuant to Regulation (EC) No 842/2006 of the European Parliament and of the Council, minimum requirements and the conditions for mutual recognition for the certification of personnel recovering certain fluorinated GHG-based solvents from equipment, *OJ* L 92 of 3 April 2008, pp21–24.

12. Commission Regulation (EC) No 307/2008 of 2 April 2008 establishing, pursuant to Regulation (EC) No 842/2006 of the European Parliament and of the Council, minimum requirements for training programmes and the conditions for mutual recognition of training attestations for personnel as regards air-conditioning systems in certain motor vehicles containing certain fluorinated GHGs, *OJ* L 92 of 3 April 2008, pp25–27.

13. Commission Regulation (EC) No 308/2008 of 2 April 2008 establishing, pursuant to Regulation (EC) No 842/2006 of the European Parliament and of the Council, the format for notification of the training and certification programmes of the Member States, *OJ* L 92 of 3 April 2008, pp28–34.

14. Commission Regulation (EC) No 309/2008 of 2 April 2008 entering a name in the register of protected designations of origin and protected geographical indications Isle of Man Manx Loaghtan Lamb (PDO), *OJ* L 92 of 3 April 2008, pp35–36.

CHAPTER 13

Agriculture and Forestry

*T*he EU Forestry Strategy initiated in 1998 aimed at the establishment of policies and measures to achieve sustainable forest management (SFM), based on the coordination of the forest policies of the Member States and Community policies and initiatives relevant to forests and forestry.

The main bases of the EU Forestry Strategy were:

- Council Resolution of 15 December 1998 on a Forestry Strategy for the European Union;[1]
- Communication from the Commission to the Council and the European Parliament reporting on the implementation of the EU Forestry Strategy COM(2005)84;[2]
- Commission Staff Working Document, Annex to the Communication on the implementation of the EU Forestry Strategy SEC(2005)333.[3]

On 1 March 2010 the European Commission issued the Green Paper on Forest Protection and Information in the EU: Preparing forests for climate change COM(2010)66 designed to launch the public consultation on options for an EU approach to forest protection and information.[4] The Green Paper is part of the EU Forest Action Plan announced by the Commission in the White Paper 'Adapting to Climate Change: Towards a European Framework for action' COM(2009)147. The Green Paper recognized the primary role of the Member States in forest protection and identified information sharing as the main area of action at the EU level. In particular, the Green Paper focused on the need for harmonization of national forest registries in order to meet the reporting obligations of the UNFCCC. Furthermore, it explored the possibility of expanding the coverage of national inventories to wood production aspects such as forest management criteria and socio-economic indicators.

The Environment Council of 16 March 2010 adopted a resolution confirming that the 'fast track' funding for climate change mitigation and adaptation in developing countries in the period 2010–2012, established by the Copenhagen Accord, will be mainly allocated to forest protection and the transfer of technology. The EU environment ministers suggested that the parties in Cancun should agree on targets such as halving gross tropical deforestation by 2020 and stopping global forest cover loss by 2030. Furthermore, they stressed the importance of financing actions on REDD and welcomed the REDD Plus Interim Partnership.

On 11 June 2010 the Environment Council adopted Council conclusions on Climate change: Follow-up to the Copenhagen Conference[5] that included conclusions on preparing forests for climate change, calling for the 'strenghtening of coopeation at EU and international level on sustainable forest management and highlights the need to further mainstream this issue in relevant policies'.

NOTES

1. Council Resolution of 15 December 1998 on a Forestry Strategy for the European Union (1999/C 56/01), *OJ* C 56 of 26 February 1999, pp1–5.

2. Communication from the Commission to the Council and the European Parliament reporting on the implementation of the EU Forestry Strategy, COM(2005)84, Brussels, 10 March 2005.

3. Commission Staff Working Document, Annex to the Communication on the implementation of the EU Forestry Strategy SEC(2005)333, Brussels, 10 March 2005.

4. European Commission, Green Paper on Forest Protection and Information in the EU: Preparing forests for climate change, COM(2010)66, Brussels, 1 March 2010.

5. 3021st Council meeting, Council Environment Conclusions, Brussels, 11 June 2010.

CHAPTER 14

European Allowance Trading Directive 2003/87/EC and its Amendments

*T*he EU ETS is the first worldwide regional GHG emissions trading system, a
market-based instrument developed under the premises of Article 17 of the
Kyoto Protocol establishing the IET designed to contribute to the reduction of
GHG emissions by the EU and the Member States. The EU ETS is one of the
centrepieces of EU climate and clean energy policy and it covers a large part
(roughly 25 per cent) of GHGs emitted in Europe.

A first indirect reference to the establishment of a European-wide ETS can be
found in the Fifth EC Environmental Action Programme in which the European
Commission explicitly suggested the introduction of market-based instruments as
new tools to face environmental problems.

Communication of the European Commission Climate Change: Towards
an EU Post-Kyoto Strategy COM(1998)353 of 3 June 1998[1] recognized the
possibility to set up a European ETS by 2005 as an initial step to introduce flexible
mechanisms in European legislation.

Linked to this communication, the first important and direct input to the
establishment of a European carbon market was the Green Paper on GHG emissions
trading within the European Union, COM(2000)87 of 8 March 2000.[2] Green Paper
COM(2000)87 was intended to illustrate the key points and functioning of the ETS
and to begin the debate within the European institutions and stakeholders on the
possibility of introducing it in the EU. The Green Paper initially considered only CO_2
emissions and a group of installations already included under the IPPC directive.

Within the framework of the European Climate Change Programme, Working
Group I on flexible mechanisms was set up in order to explore first ideas on the EU
ETS among EU institutions, practitioners, stakeholders and any interested parties.
This group met several times between July 2000 and May 2001. On 25 October
2001 Working Group I on flexible mechanisms released its interim report.[3]

As a result of the activities of ECCP Working Group I, the European Commission released in October 2001 Proposal for a framework directive for GHG emissions trading within the EC COM(2001)581,[4] issued shortly after the US public announcement to withdraw from the Kyoto Protocol and just before the beginning of the Seventh Conference of the Parties to the UNFCCC in Marrakech (COP7). Draft proposal COM(2001)581 confirmed the strong commitment of the EC towards the fight against global warming and the intention of the EU and the Member States to take the lead on the matter at the international level.

On 19 February 2002 the Commission released a non-paper on synergies between the EC emissions trading proposal (COM(2001)581) and the IPPC Directive, as well as a list of replies to frequently asked questions on the emissions trading proposal (FAQ).

Following the opinion of the European Parliament of 10 October 2002 (first reading), the Commission presented in November 2002 an amended Proposal for a Directive of the European Parliament and of the Council establishing a scheme for GHG emission allowance trading within the Community and amending Council Directive 96/61/EC COM(2002)680.[5]

On 18 March 2003 the Council adopted its Common Position on the adoption of a directive establishing a scheme for GHG emission allowance trading within the Community and amending Council Directive 96/61/EC proposing several amendments to the following parts of the Parliament's document: allocation method, opting out, opting in, pooling system, linking to JI/CDM and penalties.

On 25 March 2003 the Commission issued Communication to the European Parliament concerning the Council's Common Position on the adoption of a directive establishing a scheme for GHG emission allowance trading within the Community and amending Council Directive 96/61/EC – SEC(2003)364 of 18 March 2003.[6]

The agreement between the Council and the Parliament opened the way for the final text of the European Allowance Trading Directive (EATD) based on the European Parliament's amendments to the Council's common position adopted at the sitting of 2 July 2003 (second reading), followed by the Opinion of 18 July 2003 of the Commission on the European Parliament's amendments to the Council's common position regarding the Proposal for a Directive of the European Parliament and of the Council COM (2003)463 and accepted by the Council at its meeting of 22 July 2003.

Directive 2003/87/EC of the European Parliament and of the Council of 13 October 2003 establishing a scheme for GHG emission allowance trading (EATD) within the Community and amending Council Directive 96/61/EC on integrated pollution and prevention control was adopted in accordance with the co-decision procedure under Article 251 of the TEC, and it was the result of a long series of negotiations and consultations among the EU institutions and several stakeholders at the European level.[7]

The EATD set up a domestic, mandatory and entity-based cap-and-trade system, compatible with the IET created by Article 17 of the Kyoto Protocol. The system is based on the distribution of EUAs equal to the emissions of 1 tonne of

CO_2 equivalent to the installations falling under the categories indicated in Annex I to the EATD. Articles 3a and 3e of Directive 2003/87/EC define the terms 'allowance' and 'installation': the former is a right to 'emit one tonne of carbon dioxide equivalent during a specified period, which shall be valid only for the purposes of meeting the requirements of this Directive and shall be transferable in accordance with the provisions of this Directive' (3a), while the latter is 'a stationary technical unit where one or more activities listed in Annex I are carried out and any other directly associated activities which have a technical connection with the activities carried out on that site and which could have an effect on emissions and pollution' (3e). Installations with a CO_2 emission level below their assigned amounts may sell emission allowances to other installations that are in danger of exceeding their quotas. Operators that do not surrender sufficient allowances in a defined period are subject to financial penalties. Directive 2003/87/EC established two different periods of trading: 2005–2007 as a first learning-by-doing phase, and 2008–2012 in conjunction with the Kyoto Protocol's first commitment period. In accordance with Article 9 of the EATD, Member States were required to prepare and submit NAPs to the Commission by 31 March 2004 (NAP I) and by 30 June 2006 (NAP II). NAPs had to be designed in accordance with the criteria outlined in Annex III of EATD and had to be approved by the Commission. The NAPs contained the details of the Member States' caps to the GHG emissions of the EU ETS.

Second and final written warnings for court action (following first written warnings sent to all EU15 countries at the beginning of 2004) for failing to meet the deadline of 31 December 2003 for the transposition of the EATD into national law were sent by the Commission to all EU15 countries except Austria, Germany, France and Sweden.

A questionnaire included in the Commission Decision of 4 May 2005 aimed at the monitoring of the implementation of the EATD in the Member States was issued by the Commission in May 2005. National authorities were required to provide annual reports – the first due by 30 June 2005 – highlighting the procedure for issuing permits under the EU ETS as well as verification measures.

Directive 2003/87/EC also contained MRV requirements based on the principles set out in Annex IV and V and in the Commission Decision of 29 January 2004 establishing guidelines for the monitoring and reporting of GHG emissions pursuant to Directive 2003/87/EC of the European Parliament and of the Council C(2004)130.[8] Member States are therefore required to ensure that installations' GHG emissions are monitored and that each operator reports every year to the national competent authority. Verification of the reported emissions is ensured by independent entities

14.1 NATIONAL ALLOCATION PLANS

The efficacy of the European allowance trading system in terms of achieving real GHG emission reductions depends in large part on the ambition of the NAP to be elaborated by the Member States and notified to the Commission in accordance with Article 9 of the EATD. First phase NAPs (2005–2007) had to be submitted by the Member States by 31 March 2004, and the second phase NAPs (2008–2012) by 30 June 2006. The NAPs defined the quantity of allowances initially allocated to the operators of installations covered by the directive and had to be based on 'objective and transparent criteria' (Article 9 Directive 2003/87/EC).

Only Germany, Austria, Ireland, Finland and Denmark complied with the first deadline of 31 May 2004, while the Netherlands, Luxembourg, Sweden and the UK submitted their NAPs with a slight delay. Portugal, Belgium, Italy, France, Spain and Greece did not submit the final version of their NAPs on time. All new Member States (deadline extended to 31 May 2004) with the exception of the Czech Republic and Hungary submitted their NAP on time.

The Commission, which had the power to reject those NAPs in non-compliance with the criteria listed in Annex III of the directive, developed the following guidelines with the aim of assisting the Member States in the correct implementation of the Annex III criteria:

- Communication from the Commission on guidance to assist Member States in the implementation of the criteria listed in Annex III to Directive 2003/87/EC establishing a scheme for GHG emission allowance trading within the Community and amending Council Directive 96/61/EC, and on the circumstances under which force majeure is demonstrated, COM(2003)830, 7 January 2004;[9]
- Non-Paper on National Allocation Plans for the purposes of the EU emissions allowance trading scheme, Brussels 1 April 2003;
- Questions & Answers (FAQs) on Emission Trading and National Allocation Plans, Brussels, updated version, 20 June 2005;
- Questions & Answers (FAQs) on Emission Trading and National Allocation Plans for 2008–2012, Brussels, updated version, 29 November 2006.

On 20 June 2005 with the approval of the NAP of Greece by the Commission the process of approval of the NAPs for the first phase by the European Commission was completed.

The chronological order of the approval procedure concerning the first set of NAPs (2005–2007) was as follows:

- First wave (Commission Communication COM(2004)500 of 7 July 2004): the Netherlands, Sweden, Denmark, Ireland, Slovenia, Germany, Austria and the UK (with conditions);
- Second wave (Commission Communication COM(2004)681 of 20 October 2004): Belgium, Estonia, Latvia, Luxembourg, Portugal and Slovak Republic (approved), as well as Finland and France (with conditions);
- The remaining NAPs were notified by the following Member States in late 2004/early 2005: Spain; Poland (draft 20 July 2004); Lithuania; Hungary; Malta (18 October 2004); Cyprus; Czech Republic (12 April 2005); Italy (25 May 2005 with conditions); Greece (20 June 2005);

- Poland approved its NAP on 27 December 2005 providing the list of installations (876) included in the scheme.

In the first learning-by-doing phase (2005–2007) the ETS covered 11,428 installations responsible for 6.57 billion tonnes of CO_2 emissions. The Commission finally rejected 13 NAPs by requesting ex-post adjustments equivalent to a total cut of 4 per cent (290 million tonnes CO_2 equivalent) from the initial proposal.

Table 14.1 indicates the information on EUAs per Member State for the period 2005–2007 on the basis of NAPs approved by the European Commission.

On 28 November 2005 the European Commission published the first results of a survey carried out from June to September 2005 on the expectations and comments from companies participating in the ETS. The survey was aimed at the collection of views from stakeholders with the aim of improving the EU ETS in the next phase.

Table 14.1 *Members States' EUAs based on NAPs approved by the European Commission, 2005–2007*

Member State	CO_2 allowances in million tonnes	Share in EU allowances (%)	Installations covered	Registry functional	Kyoto target (%)
Austria	99.0	1.5	205	Yes	−13
Belgium	188.8	2.9	363	No	−7.5
Czech Republic	292.8	4.4	435	No	−8
Cyprus	16.98	0.3	13	No	−
Denmark	100.5	1.5	378	Yes	−21
Estonia	56.85	0.9	43	No	−8
Finland	136.5	2.1	535	Yes	0
France	469.5	7.1	1172	Yes	0
Germany	1497.0	22.8	1849	Yes	−21
Greece	223.2	3.4	141	No	+25
Hungary	93.8	1.4	261	No	−6
Ireland	67.0	1.0	143	No	+13
Italy	697.5	10.6	1240	No	−6.5
Latvia	13.7	0.2	95	No	−8
Lithuania	36.8	0.6	93	No	−8
Luxembourg	10.07	0.2	19	No	−28
Malta	8.83	0.1	2	No	−
Netherlands	285.9	4.3	333	Yes	−6
Poland	717.3	10.9	1166	No	−6
Portugal	114.5	1.7	239	No	+27
Slovak Republic	91.5	1.4	209	No	−8
Slovenia	26.3	0.4	98	No	−8
Spain	523.3	8.0	819	Yes	+15
Sweden	68.7	1.1	499	Yes	+4
United Kingdom	736.0	11.2	1078	Yes	−12.5
Total	**6572.4**	**100.0**	**11,428**		

Source: European Commission

On 7 April 2006 the EEA released Technical Report No 2/2006, Application of the emissions trading directive by EU Member State, prepared on the basis of information reported by the Member States on the application of the EU ETS in the first year (in accordance with Article 21 of the EATD). The report contained information on the allocation process, on competent authorities and legislation adopted and it called for improvements in the quality of implementation in the Member States as well as in the establishment of more harmonized rules.

The EU central emission registry, the Community Independent Transaction Log (CITL), collects all information regarding the issuance, transfer, cancellation, retirement and banking of EUAs that occur in the national registries.

The European Commission released on 15 May 2006 externally verified data on the CO_2 emissions produced in 2005 by the installations covered under the EU ETS. The emission figures (available for 21 Member States) covered 9000 installations that emitted in the first year of the EU ETS 2.5 per cent less CO_2 than received by the government through the NAPs. Only six Member States (Ireland, the UK, Italy, Spain, Sweden and Austria) reported excess emissions in comparison with the allowances received at the beginning of the year.

According to the CITL of 2 April 2007, data for verified emissions of CO_2 in 2006 (accounting for almost 93 per cent of emissions reported in 2005) increased by 1–1.5 per cent relative to 2005 levels. However, CO_2 emissions of EU ETS installations were still below the total cap set by the NAPs for 2006.

Data released by the European Commission on 23 May 2008 on 2007 verified emissions from EU ETS businesses showed an increase of industrial CO_2 emissions by 0.68 per cent compared to 2006. These data confirmed the necessity for tighter caps in NAP II. Among the biggest emitters in the EU ETS were German firms, followed by UK, Italian and Polish enterprises.

14.2 SECOND PHASE NATIONAL ACTION PLANS

On 9 January 2006 the European Commission released Communication Further guidance on allocation plans for the 2008 to 2012 trading period of the EU ETS COM(2005)703 providing guidance to the Member States on the preparation of second round NAPs, which run from 2008 to 2012 in conjunction with the Kyoto Protocol first commitment period.[10] The Commission was responsible for the approval or rejection of the NAPs within three months of their submission. A final decision on NAPs II had to be taken before 31 December 2006. On 9 January 2006, the Commission also published the Questions & answers on national allocation plans for 2008–2012 in order to facilitate the preparation of the second phase NAPs.

Nearly all the 25 EU Member States did not meet the 30 June 2006 deadline for the submission of the second phase NAPs (only Estonia was on time). 'Pre-infringement letters' were sent by the Commission to 14 Member States, namely Austria, Belgium, Cyprus, the Czech Republic, Denmark, Finland, Hungary, Latvia, Malta, the Netherlands, Portugal, Slovenia, Slovakia and Sweden.

The timing of the notification of the second phase NAPs by the Member States to the European Commission was as follows:

- Estonia, June 2006
- Poland, 23 June 2006
- Germany, 28 June 2006
- Ireland, 12 July 2006
- Lithuania,
- Luxembourg, 18 July 2006
- Slovakia, July 2006
- Latvia, 9 August 2006
- UK, 21 August 2006
- Sweden, 31 August 2006
- Greece, 5 September 2006
- Netherlands, 26 September 2006
- Belgium, September 2006
- Malta, September 2006
- Finland, 29 September 2006
- Portugal,
- Slovenia, October 2006
- Spain, 25 November 2006
- Czech Republic, 30 November 2006
- Romania, 12 December 2006
- Italy, 18 December 2006
- France, 20 April 2007
- Austria, 10 January 2007
- Hungary, 17 January 2007
- Bulgaria, 30 January 2007
- Cyprus, 26 February 2007
- Denmark, 6 March 2007

On 5 September 2006 the European Parliament debated the delays in the submission of the second phase NAPs by the Member States and several MEPs expressed concerns on the overallocation in the first phase. On 12 October 2006 the European Commission announced infringement proceedings against Austria, the Czech Republic, Denmark, Hungary, Italy, Portugal, Slovenia and Spain for the failure to submit their NAP II on time. The Commission sent the first written warnings to these countries asking for an explanation and establishing a two-month deadline before the start of legal action.

On 29 November 2006 the Commission released Communication COM(2006)725 on the assessment of national allocation plans for the allocation of GHG emission allowances in the second period of the EU Emissions Trading Scheme accompanying Commission Decisions of 29 November 2006 on the national allocation plans of Germany, Greece, Ireland, Latvia, Lithuania, Luxembourg, Malta, Slovakia, Sweden and the United Kingdom in accordance with Directive 2003/87/EC.[11] With this communication the Commission urged the ten Member States to cut their allowances by 7 per cent below the emissions proposed in the NAPs, corresponding to a reduction of 7 per cent of 2005 levels of emissions.

The decisions adopted by the European Commission on NAP II were as follows:

- 5 February 2007, Slovenia
- 16 February 2007, Belgium and the Netherlands
- 26 February 2007, Spain
- 26 March 2007, France, Poland and Czech Republic
- 2 April 2007, Austria
- 16 April 2007, Hungary
- 4 May 2007, Estonia
- 4 June 2007, Finland
- 18 July 2007, Cyprus
- 31 August 2007, Denmark
- 18 October 2007, Portugal
- 26 October 2007, Romania and Bulgaria
- 26 October 2007, amendment Germany
- 7 December 2007, amendment Slovakia
- 12 February 2008, amendment Poland

The revised NAP II for Poland included the increase of EUAs for electricity and heat producers of 12 per cent and 20 per cent respectively. Allowances were redistributed away from steel, chemicals and paper sectors. On 3 July 2008 Poland adopted the final NAP for the second phase of the EU ETS after six months from the beginning of the second trading period. The new NAP allocated 208.5 million tonnes of CO_2 emissions per year to national installations, with a cut of 11 per cent for the emissions forecast by the power sector.

Table 14.2 contains the summary information on the 27 NAP II assessed by the Commission as of 7 December 2007.

According to the data of the European Commission of 7 June 2007, CO_2 emissions from EU ETS installations rose in 2006 although remaining below the allowed levels. Only 220 installations (0.2 per cent of total emissions) failed to surrender the due allowances by 1 May 2007.

On 13 July 2007 the Commission decided on the amendments to NAP II proposed by Ireland, Latvia, Lithuania, Luxembourg and Sweden before the deadline for the submission of amendments (31 December 2006). For the first time in the brief history of the EU ETS, the Commission agreed to revise upward the NAP of the Member States. Ireland and Latvia were allowed to increase their amount of EUAs by 1.18 and 0.14 million tonnes per year, respectively.

The following Member States decided to challenge the European Commission's decisions on the NAP II in front of the ECJ, mostly launching legal proceedings against the cuts in EUAs proposed by the European Commission (see Chapter 18): Slovakia; Poland; the Czech Republic; Hungary; Estonia; Latvia; Lithuania; and Malta.

Table 14.2 *27 NAP II assessed by the European Commission in December 2007 (units of measurement, million tonnes of carbon dioxide equivalent, MtCO$_2$eq.)*

Member State	1st period cap (2005–2007)	2005 verified emissions	Proposed cap 2008–2012	Cap allowed 2008–2012 (in relation to proposed, %))	Emissions from additional installations in 2008–2012[1]	JI/CDM limit 2008–2012 (%)[2]
Austria	33.0	33.4	32.8	30.7 (93.6)	0.35	10
Belgium	62.1	55.58[3]	63.3	58.5 (92.4)	5.0	8.4
Bulgaria	42.3	40.6[4]	67.6	42.3 (62.6)	n.a	12.55
Cyprus	5.7	5.1	7.12	5.48 (77)	n.a.	10
Czech Rep.	97.6	82.5	101.9	86.8 (85.2)	n.a.	10
Denmark	33.5	26.5	24.5	24.5 (100)	0	17.01
Estonia	19	12.62	24.38	12.72 (52.2)	0.31	0
Finland	45.5	33.1	39.6	37.6 (94.8)	0.4	10
France	156.5	131.3	132.8	132.8 (100)	5.1	13.5
Germany	499	474	482	453.1 (94)	11.0	20[5]
Greece	74.4	71.3	75.5	69.1 (91.5)	n.a.	9
Hungary	31.3	26.0	30.7	26.9 (87.6)	1.43	10
Ireland	22.3	22.4	22.6	22.3 (98.6)	n.a.	10
Italy	223.1	225.5	209	195.8 (93.7)	n.k. [6]	14.99
Latria	4.6	2.9	7.7	3.43 (44.5)	n.a.	10
Lithuania	12.3	6.6	16.6	8.8 (53)	0.05	20
Luxembourg	3.4	2.6	3.95	2.5 (63)	n.a.	10
Malta	2.9	1.98	2.96	2.1 (71)	n.a.	Tbd
Netherlands	95.3	80.35	90.4	85.8 (94.9)	4.0	10
Poland	239.1	203.1	284.6	208.5 (73.3)	6.3	10
Portugal	38.9	36.4	35.9	34.8 (96.9)	0.77	10
Romania	74.8	70.8[7]	95.7	75.9 (79.3)	n.a	10
Slovakia	30.5	25.2	41.3	32.6 (78.9)	1.78	7
Slovenia	8.8	8.7	8.3	8.3 (100)	n.a.	15.76
Spain	174.4	182.9	152.7	152.3 (99.7)	6.7[8]	ca. 20
Sweden	22.9	19.3	25.2	22.8 (90.5)	2.0	10
UK	245.3	242.4[9]	246.2	246.2 (100)	9.5	8
SUM	**2298.5**	**2122.16[10]**	**2325.34**	**2082.68 (89.56)**	**54.69**	–

Source: European Commission

The EATD and its implementing measures were incorporated into the EEA Agreement by Joint Committee Decision 146/2007 of 26 October 2007, which entered into force on 29 December 2007. The EFTA (European Free Trade Area) Surveillance Authority provides for the assessment of the NAPs notified by the EEA states.

In accordance with Article 9 of the EATD, the EFTA Surveillance Authority is required to decide on the national allocation plans presented by the EEA countries. The following NAPs have been assessed by the EFTA so far: Liechtenstein, approved 19 December 2007, and Norway, 27 February 2009.

As of 2 April 2008 the verified GHG emissions of companies under the EU ETS are visible in the CITL. Installations emitted in 2007 1 per cent more CO$_2$

than in 2006. Reporting installations numbered 9962, with 1914 billion tonnes of CO_2 emitted.

On 11 December 2009 the European Commission rejected for the second time the NAPs submitted by Estonia and Poland in the second phase of the EU ETS. The main reasons for the rejection of the two NAPs were the excessive amount of EUAs distributed by the two Member States to their operators and the violation of several criteria indicated by Directive 2003/87/EC. On 23 September 2009 the Court of First Instance (CFI) annulled the first decision of the Commission to reject the two plans because of their incompatibility with the EU ETS directive. According to the CFI the Commission had infringed a number of obligations and exceeded its powers, especially in establishing more stringent caps for the two countries. On 3 December 2009 the Commission lodged an appeal against that decision, which is still pending. Estonia and Poland are now required to submit new NAPs that will be reviewed by the Commission.

In accordance with Article 21 of the EATD, Member States 'shall submit to the Commission a report on the application of this directive'. These reports are used to monitor and enhance the implementation of the EU ETS by the Member States, in particular on the allocation of allowances, the operation of registries, the application of MRV requirements and issues on compliance.

The main legal challenges encountered by the EU institutions and the Member States in the first two years of implementation of the EU ETS were:[12]

- The principle of subsidiarity: the implementation and administration of the EU ETS takes place at the Member State level and is based on different types of decentralization, namely on the involvement of regional and local authorities. Delocalization sometimes contributed to a certain degree of uncertainty among the EU institutions and the Member States, in particular in relation to the identification of the competent authorities to be addressed for the implementation of the EU ETS;
- The definition of combustion installations: several Member States encountered problems in the identification of combustion installations according to the definition provided by Annex I of Directive 2003/87/EC;
- The legal nature of EUAs: there was a lack of uniformity in the legal treatment of allowances in the Member States with a few states considering allowances as financial instruments, others as normal commodities.

Finally, it is important to emphasize the role of the European Commission in ensuring the correct implementation of the EATD. In this respect, the European Commission is lacking an adequate structure and resources, and the need for a more decentralized administration is evident in relation to the experience of the first and second phases of the EU ETS. Furthermore, the infringement procedure designed under EC law appears too slow (an average of two to four years is required for the formulation of a judgment by the ECJ) and too centralized to ensure the effective enforcement of the EU ETS.[13]

14.3   EU ETS REVIEW

Article 30 of Directive 2003/87/EC introduced two options for the European Commission to amend the EU ETS. The first option, namely the possibility for the European Commission to present a proposal to the European Parliament and the Council by 31 December 2004 to amend Annex I of the EATD as to the inclusion of other activities and emissions of other GHGs in accordance with Article 30(1) was not followed. Instead, the Commission opted to refer to Article 30(2) and submit a report to the European Parliament and the Council by 30 June 2006 on the application of Directive 2003/87/EC with the view to introduce some modifications. The report was presented by the European Commission on 13 November 2006 on the basis of the experience of the implementation of the EU ETS and in relation to the development in the international community on climate policy and the progress in the monitoring of GHG emissions. The report was included in Communication COM(2006)676 Building a global carbon market – Report pursuant to Article 30 of Directive 2003/87/EC (EATD).[14] This document set the agenda for the revision of the EATD and it referred to a legislative proposal to be presented by the Commission by June 2007 as discussed and defined in the consultative process established under the ECCP II. COM(2006)676 focused on four aspects of the EU ETS:

- Scope: providing clarity as to the definition of combustion installations, in particular small installations and expanding the system to other sectors and gases;
- Further harmonization within the EU and increased predictability of the market: studying the possibility of introducing a single EU-wide cap and further developing the concepts of auctioning and benchmarking;
- Robust compliance and enforcement systems: further harmonization requirements in particular on monitoring and reporting of emissions and third party verification;
- Linking with third countries and consequential arrangements.

The consultancy report Greenhouse Gas Emissions for Shipping and Implementation Guidance for the Marine Fuel Sulphur Directive commissioned by the European Commission and released in January 2007 suggested the inclusion of shipping into the EU ETS along with the aviation sector. The report identified the EU ETS as the adequate solution to reduce the impact of GHG emissions from shipping. The inclusion of shipping into the EU ETS would require the development of new allocation methods as well as the establishment of CO_2 emissions caps for all ships transiting European ports.

On 12 July 2007 the European Parliament adopted a resolution on future maritime policy for the EU welcoming the Commission Green Paper 'Towards a future maritime policy for the Union: A European vision for the oceans and seas' COM(2006)275, which aimed at the integration of fishery, marine environment and tourism policies.[15] In the resolution, the Parliament invited the Commission to propose legislation to significantly reduce CO_2, sulphur dioxide and nitrogen oxides emissions from ships and urged the EU to take the lead at the international level by proposing the inclusion of the maritime sector in the UNFCCC. Furthermore, the Parliament identified emissions trading and 'renewable energies such as wind and solar power' as green measures to be 'introduced for shipping'.

On 26 October 2007 the European Commission announced the conclusion of an agreement with Norway, Liechtenstein and Iceland on the enlargement of the EU ETS. These countries were therefore required to link their emissions trading systems to the EU ETS through the inclusion of Directive 2003/87/EC in the European Economic Agreement. The legal basis of this decision was based on Articles 25 and 30 of Directive 2003/87/EC and was subject to the approval procedures foreseen in Norway, Iceland and Liechtenstein. As far as the assessment of the NAP is concerned, the decision will be taken by the European Commission together with the EFTA Surveillance Authority.

On 23 January 2008 the European Commission presented the integrated package of legislative proposals on climate and energy setting the long-term strategy of the future EU climate and clean energy policy. The package translated into legal text the political commitments agreed by the European Council of March 2007 where the EU heads of state and government decided for a 20 per cent reduction of GHG emissions below 1990 levels by 2020 with the objective of scaling up to 30 per cent in the event that the international community reached a satisfactory agreement in Copenhagen at the end of the process launched by the BAP with the aim of indentifying the international climate change regime in the post-2012 phase. The package of legislative proposals included the Proposal for a directive of the European Parliament and of the Council amending Directive 2003/87/EC so as to improve and extend the EU GHG emission allowance trading system, COM(2008)16.[16] Proposal COM(2008)16 was aimed at the improvement of the EU ETS on the basis of the experience in the two phases within the period 2005–2012 and the establishment of a EU-wide cap for emissions from EU ETS installations of 21 per cent below 2005 levels by 2020. The proposed amendments included:

- Extension of the trading period to eight years (2013–2020);
- EU-wide cap (21 per cent) instead of 27 national caps;
- Extension of the directive's coverage to other sectors (chemicals, non-ferrous metals and aluminium) and gases (nitrous oxide and PFCs);
- Opting-out option for small emitters (less than 10,000 tonnes of CO_2 equivalent/year);
- Harmonized rules for the allocation of EUAs;
- Auctioning as the main allocation method (60 per cent by 2013);
- At least one fifth of revenue from auctioning to be earmarked for measures against climate change;
- More harmonized rules on monitoring and reporting;
- JI and CDM: banking of CERs (certified emission reductions) and EURs (emission reduction units) from the second to the third phase with no additional budget provided unless international agreement is reached;
- Review on industrial sectors' potential carbon leakage by mid-2010.

The proposal of the European Commission to earmark a portion (at least 20 per cent) of the revenues from the auctioning of EUAs for climate-friendly initiatives was rejected on 12 February 2008 by the EU Ministers of Economic and Financial Affairs in the Council conclusions on efficiency of economic instruments to reach energy and climate change targets.

On 27 February 2008 the European Commission adopted a plan of 19 points focusing on the protection of competitiveness of the EU ETS sectors, in particular forest-based industries. The paper considered that in case no international agreement was reached, forest-based and metals industries could receive EUAs free of charge in order to combat competition with similar sectors outside the EU. Emphasis was also put on the recovery levels of paper and wood products, on the carbon storage potential of harvested wood products and finally on the need for more research in the field of biofuels and chemicals from wood.

A set of compromise amendments to the proposal on the EU ETS review adopted by the Industry Committee of the European Parliament on 11 September 2008 opposed the automatic scaling up of the EU GHG emissions reduction obligations from 20 to 30 per cent by 2020. According to the MEPs, the decision on the increase of the EU target should be subject to a formal adoption by the European institutions. The vote was in line with the objective and intention of the Industry Committee to protect the energy-intensive sectors that could suffer from international competition. To this aim, the Committee also proposed to allocate carbon allowances to installations under the EU ETS free of charge in 2013.

On 6 October 2008 the Environment Committee of the Parliament adopted all but one (amendment 16) the compromise amendments (1–25, 26–35, 39–131, 132–216, 217–292, 293–362, 363–471, 472–580, 581–674, 675–782, 783–823, 824–827) proposed by Rapporteur Avril Doyle on the review of the EU ETS. The amended proposal was supposed to be the basis of a deal between the Parliament and the Council. Among the biggest changes to the proposal of the Commission were:

- The introduction of a funding mechanism for CCS derived from the new entrants reserve to co-finance CCS demonstration projects;
- The earmarking of revenues from carbon auctioning on climate change measures (50 per cent in developing countries);
- The strengthening of quality criteria to accept JI and CDM credits (representing at least 40 per cent of the reduction effort);
- An increase of threshold for exempting small- and medium-sized firms from the EU ETS (from 10,000 to 25,000 tonnes of annual emissions).

MEPs backed strongly the full auctioning of EUAs to the power sector from 2013 and the identification of industries suffering from carbon leakage only after the conclusion of the international climate treaty.

On 17 October 2008 the European Commission tabled a policy paper on reducing deforestation where it proposed to establish a forest carbon mechanism at the global level to be included in the international climate deal and to use 5 per cent of revenues from auctioning in the EU ETS to finance actions on avoided deforestation. Furthermore, the Commission confirmed that European companies will not be allowed to use carbon credits for avoided deforestation in the EU ETS before 2020 at the earliest.

On 19 November 2008 the French presidency tabled a compromise proposal aimed at resolving the deadlock in the negotiations among the Member States for the adoption of the proposal on the review of the EU ETS. According to the French proposal, installations from countries producing more than 60 per cent of their electricity from coal (Poland, the Czech Republic, Estonia and Greece) and countries poorly connected to the grid (the Baltic states) would receive 50 per cent of their EUAs for free in 2013, this decreasing to 0 by 2016.

On 21 November 2008 the European Commission released a non-paper on the determination of energy-intensive industries exposed to a significant risk of carbon leakage. This non-paper was the result of the first part assessment of the risk of carbon leakage for the industrial sector following the review of the EU ETS. The study looked at the risk of manufacturing capacity leaving the EU from 2013 due to production cost increases and consequently the increase of product prices. The result of the assessment was an exposure to international trade of less than 50 per cent for most sectors and a rise in costs of less than 10 per cent.

On 27 November 2008 the European Parliament and the French presidency of the Council of the EU reached an agreement on the cap for EU installations to use JI and CDM reduction credits. The deal, which is part of the discussion on the EU ETS review, set the limit at 50 per cent of each installation's emission reduction effort. In the event of an international agreement concluded in Copenhagen in December 2009 the cap was supposed to be reviewed.

On 28 November 2008 the French presidency of the Council of the EU tabled a proposal on the free allocation of EUAs by 2013 to industrial sectors suffering from carbon leakage. In particular, the plan introduced a system of thresholds based on (1) the increase of carbon-related costs by 5 per cent of gross value added, and (2) a trade intensity with non-EU countries above 10 per cent. The proposal would create four risk categories of carbon leakage corresponding to a different quantity of free allowances to be received.

The details of the final compromise regarding the energy and climate change package, as agreed by the European Council at its meeting on 11 and 12 December 2008, considered the major aspects of the EU ETS review. On 17 December 2008, the European Parliament adopted a legislative resolution on the Proposal for a Directive of the European Parliament and of the Council amending Directive 2003/87/EC so as to improve and extend the GHG emission allowance trading system of the Community (COM(2008)0016 – C6-0043/2008 – 2008/0013(COD)) endorsing the final compromise of 12 December 2008.

On 25 June 2009 the consolidated version of directive 2003/87/EC establishing the EU ETS was published and included all the amendments introduced by the following provisions:[17]

- Directive 2004/101/EC of the European Parliament and of the Council of 27 October 2004;
- Directive 2008/101/EC of the European Parliament and of the Council of 19 November 2008;

- Regulation (EC) No 219/2009 of the European Parliament and of the Council of 11 March 2009;
- Directive 2009/29/EC of the European Parliament and of the Council of 23 April 2009.

On 23 April 2009 Directive 2009/29/EC of the European Parliament and Council amending Directive 2003/87/EC so as to improve and extend the GHG emission allowance trading scheme of the Community (EU ETS review directive) was adopted with the aim of enhancing harmonization, avoiding distortion in the internal market and fostering the linking of different existing emission trading systems.[18]

The importance of the EU ETS review in the framework of the EU climate and clean energy policy is emphasized by the fact that the effort of the EU to mitigate GHG emissions by 20 per cent below the 1990 levels by 2020 is divided into two parts. The contribution to the 20 per cent reduction of GHG emission is shared between the sectors covered by the EU ETS and those sectors that do not belong to the categories of activities under Annex I (non-ETS sectors) and therefore are not obliged to obtain an emission permit to operate. Figure 14.1 indicates the measure of such a distribution, −21 per cent and −10 per cent, respectively, compared to 2005 levels.[19]

The details of the EU ETS review directive are summarized below:

- 21 per cent reduction of GHG emissions in the period 2013–2020 compared with 2005 from industrial sectors capped under the EU ETS review (Recitals 5 and 14);
- Allowances to be distributed by the EC and abolition of NAPs to be prepared by the Member States (Article 9);

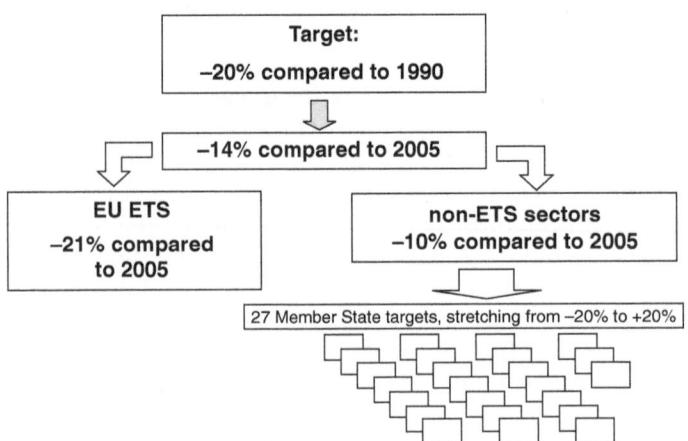

Figure 14.1 *Distribution of the effort among sectors to achieve the 20 per cent GHG emission reduction target of the EU*

Source: European Commission

- Allowances distributed through auctioning starting from 20 per cent in 2013 and increasing to 70 per cent by 2020 (Recital 11);
- Full auctioning of allowances as of 2013 for the power sector (Recital 19);
- Free distribution for power stations in a number of new Member States (70 per cent free of charge in 2013 declining to zero in 2020) (Article 10c(2));
- Industrial sectors exposed to the risk of carbon leakage eligible to receive EUAs for free from 2013 until an international climate agreement is reached (Article 10a);
- Establishment of eligibility criteria to qualify industrial sectors or subsectors exposed to the risk of carbon leakage (Recitals 24 and 45 and Article 10a(8) (14 to 17));
- Revenue from 8 per cent of auctioned EUAs distributed to Member States in accordance with the emissions of the EU ETS sectors in 2005 or the average of 2005–2007 (Article 10(2)a);
- Revenue from 10 per cent of auctioned EUAs distributed to the EU Member States with less GDP per capita for 'solidarity and growth', with an additional 2 per cent distributed to Poland, Hungary, Slovakia, the Czech Republic, Estonia, Latvia, Lithuania, Romania and Bulgaria;
- Revenue resulting from 50 per cent of auctioned EUAs to finance adaptation and mitigation actions in the EU and developing countries (Article 10(2)b);
- Installations allowed to use international reduction credits from JI and CDM up to 50 per cent of their targets (Article 10(2)c);
- Up to 300 million EUAs set aside to finance, amongst others, CCS until the end of 2015 (Article 10a(8)).

The NAPs to be submitted to the Commission in accordance with Article 9 of Directive 2003/87/EC (2005–2007 and 2008–2012) were substituted by the Community-wide quantity of allowances to be decided in Brussels for 2013 onwards and published by the European Commission by 30 June 2010 (Article 9(2) of amended Directive 2003/87/EC). In accordance with Article 9a of the EU ETS review directive some adjustments of the Community-wide cap were taken into consideration by the European Commission and published by 30 September 2010, for instance in respect of new sectors and gases not covered by Directive 2003/87/EC.

Contrary to what happened in the EU ETS phase I and II, 'auctioning should be the basic principle for allocation' in the new EU ETS after 2013 in order to ensure 'the highest possible degree of economic efficiency' among Member States and sectors covered by amended Directive 2003/87/EC.[20] In order to ensure economic efficiency and to avoid the distortion of competition within the Community, and with the impossibility of treating economic sectors in a different manner, the total amount of EUAs to be auctioned is divided into percentages distributed to the Member States according to the following:[21]

- 88 per cent of EUAs is distributed on the basis of the Member States' share of verified emissions under the EU ETS in 2005 or the average of the period 2005–2007 (Annex IIa revised Directive 2003/87/EC);

- 10 per cent of EUAs is distributed to certain Member States according to the principles of solidarity and growth in the Community;
- 2 per cent of EUAs is distributed to the Member States with a level of GHG emissions in 2005 20 per cent below the base year level (Annex IIb revised Directive 2003/87/EC).

The allocation of EUAs through auctioning should be achieved by gradual steps, namely with a progressive phasing out of free allocation ensuring that at least half of allowances is auctioned in 2013 and that full auctioning is achieved by 2020.[22] Free allocation should respond to a distribution based on Community-wide ex-ante product-specific benchmarks applicable to individual sectors and subsectors in order to reduce the risk of distortion of competition among sectors.[23] Contrary to Directive 2003/87/EC, Directive 2009/29/EC did not foresee any type of ex-post adjustment with the exception of the closure of installations. Community-wide rules for free allocation of EUAs through benchmarks will be defined by the European Commission in the comitology setting. The final benchmarks to be determined will take into consideration various factors, namely 'the average performance of the 10% most efficient installations in a sector or subsector',[24] 'the most efficient techniques, substitutes, alternative production processes, high efficiency cogeneration, efficient recovery of waste gases, use of biomass and capture and storage of CO_2, where such facilities are available'.[25]

No free allocation was foreseen for installations producing electricity, in particular electricity generators, and for installations capturing CO_2, pipelines for the transport of CO_2 or CO_2 storage sites, with the exception of installations in district heating and highly efficient cogeneration, as well as industries exposed to the risk of carbon leakage.[26]

Directive 2009/29/EC defined carbon leakage under Recital 24 of the preamble, which referred to 'an increase in greenhouse gas emissions in third countries where industry would not be subject to comparable carbon constraints'. Such a phenomenon could create an economic disadvantage for certain energy-intensive sectors or subsectors covered by the EATD and its review that could suffer from the international competition of similar sectors operating without the obligation to meet the same level of requirements to limit GHG emissions.

On 14 July 2009 the list of industries eligible to qualify for free carbon allowances under the EU ETS from 2013 was expanded to six new sectors, among which were fertilizer and nitrogen compound manufacturers. For sectors where sufficient data were lacking, a qualitative analysis (I) and (II) was carried out that considered, inter alia, technological improvement potential and the potential impact of carbon costs on profit margins.

On 21 September 2009 the Member States agreed on the draft list of 164 industry sectors at risk of carbon leakage under the scope of Directive 2003/87/EC. The draft list adopted within the Commission's Climate Change Committee was supposed to be reviewed and approved by the European Commission after scrutiny of the Council and the Parliament by the end of 2009. The result was the Commission Decision 2010/2/EU of 24 December 2009 determining, pursuant to Directive 2003/87/EC of the European Parliament and of the Council, a list of sectors and subsectors which are deemed to be exposed to a significant risk of carbon leakage and which are eligible for receiving free allowances by meeting the quantitative criteria specified in paragraphs 14 to 17 of Article 10a of Directive 2009/29/EC.[27]

The coverage of sectors and gases of Directive 2003/87/EC was extended by Directive 2009/29/EC by an increase of the scope of the directive of 6–7 per cent in terms of the GHG emissions covered by Directive 2003/87/EC. In particular, Directive 2009/29/EC included other relevant sectors such as the production of primary and secondary aluminium, basic chemical production (nitric acid, adipic acid, glyoxal and glyoxylic acid, ammonia, hydrogen, soda ash, etc.) capture, transport and geological storage of GHGs. Finally, Directive 2009/29/EC applies from 1 January 2012 to the aviation sector, notably to all flights departing or arriving in an airport located in the EU territory.

Annex I also expands the list of gases to which the EU ETS review directive applies, notably to PFCs generated from the production of primary aluminium and nitrous oxide generated by the production of nitric acid, adipic acid and glyoxal and glyoxylic acid. In this respect, the EU ETS is no longer limited to CO_2 emissions and is more in line with the Kyoto Protocol, which regulates six GHGs listed in Annex A.

As to the rule for the closure of installations, Article 10a(19) of Directive 2009/29/EC established that 'no free allocation' is distributed to installations that have ceased operation unless production is resumed within a 'specified and reasonable time'.

In accordance with Article 10a(7), 5 per cent of the Community-wide quantity of EUAs will be put into a reserve for new installations or airlines ('new entrants') entering the EU ETS after 2013. Allowances will be distributed to installations according to the same procedure used for existing installations. Allowances included in the reserve that are neither allocated to new entrants nor used will be distributed to the Member States for auctioning. Harmonized rules for the application of the definition of 'new entrants' will be adopted by the European Commission through comitology by 31 December 2011.[28] Part of the allowances in the Community-wide reserve, namely 300 million EUAs, will be used to support up to 12 demonstration projects on CCS or innovative renewable energy technologies on the basis of a fair geographical distribution.[29]

Finally, Directive 2009/29/EC introduces a broad interpretation of 'combustion' to be added to the definitions of Article 3 of Directive 2003/87/EC. According to Article 3t of Directive 2009/29/EC combustion means 'any oxidation of fuels, regardless of the way in which the heat, electrical or mechanical energy produced by this process is used, and any other directly associated activities, including waste gas scrubbing'.

Directive 2009/29/EC introduced the procedure for the replacement of the existing MRV and accreditation rules under the EU ETS. Article 14 of Directive 2003/87/EC has been replaced by a revised Article 14 introduced by Directive 2009/29/EC, which established that the European Commission will adopt a regulation through the comitology procedure aimed at replacing the current guidelines for the monitoring and reporting of emissions by 31 December 2011. The new monitoring and reporting regulation will consider the updated scientific evidence available, in particular from the IPCC. Furthermore, the new regulation may establish specific requirements for operators of energy-intensive industries producing goods subject to international competition and finally it will apply to all aircraft operators covered by the EU ETS.

Directive 2009/29/EC introduced a Community-wide harmonized penalty of €100 per allowance for non-compliance by installations. More exactly, as of 1 January 2013 penalties applicable to infringements of the provisions adopted under Directive 2009/29/EC will be increased 'in accordance with the European index of consumer prices'. The obligation to surrender allowances on top of the applicable penalties remains.[30] Finally, Directive 2009/29/EC established a single Community registry distinct from the registry of the Member States. The Community registry will serve to account, hold, transfer and cancel allowances issued from 1 January 2013.[31]

Directive 2009/29/EC provided guidance also on the use of revenue generated from the auctioning of EUAs. According to Article 10(3) of revised Directive 2003/87/EC, Member States will decide on the use of such proceeds for projects covering the following areas:

- Reduction of GHG emissions and adaptation to climate change;
- Development of renewable energies and increase of energy efficiency;
- Reduction of deforestation and increase of afforestation and reforestation;
- Forestry sequestration;
- Capture and geological storage of CO_2;
- Low-emission transport forms;
- R&D in energy efficiency and clean technologies.

On 6 April 2010 the European Commission released a draft proposal for a regulation on auctioning to allow any Member State to opt out of a single EU platform for the auctioning of EUAs under the EU ETS and create its own platform following notification to the Commission.[32] In this way, governments will be free to decide the auctioning rules and to determine the quantity and the form of the EU allowances to be auctioned.

On 14 July Member States in the Climate Change Committee voted in support of the draft Commission Regulation (EU) on the timing, administration and other aspects of auctioning of GHG emission allowances pursuant to Directive 2003/87/EC of the European Parliament and the Council establishing a scheme for GHG emission allowances trading within the Community.

14.4 REGISTRIES

In relation with the EC system for the MRV of GHG emissions and in order to comply with international obligations (Article 7(4) of the Kyoto Protocol) and European law (Article 19 of Directive 2003/87/EC and Article 6 of Monitoring

Decision), the Commission established a standardized system of national registries designed to keep track of assigned, held, transferred and cancelled AAUs, ERUs, CERs and RMUs. The Member States are required to set up national registries in order to let national installations trade and track all the different carbon credits. The CITL shows the latest news on operational registries put in place by the Member States.

Commission Regulation 2216/2004 of 21 December 2004 for a standardised and secured system of registries pursuant to Directive 2003/87/EC of the European Parliament and of the Council and Decision 280/2004/EC of the European Parliament and of the Council was the legislative response to the above mentioned legislation.[33]

On 31 July 2007 the Commission adopted Regulation 916/2007 amending Regulation (EC) No 2216/2004 for a standardised and secured system of registries pursuant to Directive 2003/87/EC and Decision 280/2004/EC, introducing more technical details related to the connection of the CITL with the International Transaction Log (ITL).[34]

Commission Regulation (EC) No 994/2008 of 8 October 2008 for a standardised and secured system of registries pursuant to Directive 2003/87/EC of the European Parliament and of the Council and Decision No 280/2004/EC of the European Parliament and of the Council included a comprehensive revision of the system of EU registries in the EU ETS from 1 January 2012 onwards.[35] This regulation contained 'general as well as operational and maintenance require-ments', which will ensure the independence of the EU ETS and therefore ease the inclusion of the aviation sector in the EU ETS from 2012 and the linking of the EU ETS with other emissions trading systems.

On 17 February 2010 representatives of Member States in the Climate Change Committee approved the proposal of the European Commission for a Commission Regulation (EU) for a standardised and secured system of registries pursuant to Directive 2003/87/EC of the European Parliament and of the Council and Decision 280/2004/EC of the European Parliament and of the Council, revising the registry system of the EU ETS. The revision would include the aviation sector allowances in the system and enhance the data security of the registry. On 16 April 2010 the Climate Change Committee gave a positive opinion on amendments to the proposed Commission Regulation for a standardised and secured system of registries approved on 17 February 2010. The consolidated text was submitted to the European Parliament and Council for scrutiny before adoption by the Commission.

NOTES

1. Communication of the European Commission Climate Change: Towards an EU Post-Kyoto Strategy, COM(1998)353, Brussels, 3 June 1998.

2. European Commission, Green Paper on GHG emissions trading within the European Union, COM(2000)87, Brussels, 8 March 2000.

3. Interim report: ECCP Working Group I 'Flexible Mechanisms', Brussels, 25 October 2000.

4. European Commission, Proposal for a framework Directive for GHG emissions trading within the EC, COM (2001)581, Brussels, 23 January 2001.

5. European Commission, amended Proposal for a Directive of the European Parliament and of the Council establishing a scheme for GHG emission allowance trading within the Community and amending Council Directive 96/61/EC COM(2002)680, Brussels, 27 November 2002.

6. Commission Communication to the European Parliament concerning the Council's Common Position on the adoption of a directive establishing a scheme for GHG emission allowance trading within the Community and amending Council Directive 96/61/EC – SEC(2003)364 of 18 March 2003, Brussels, 25 March 2003.

7. Directive 2003/87/EC of the European Parliament and of the Council of 13 October 2003 establishing a scheme for GHG emission allowance trading within the Community and amending Council Directive 96/61/EC, OJ of the EU of 25 October 2003, L 275, pp32–46.

8. Commission Decision of 29 January 2004 establishing guidelines for the monitoring and reporting of GHG emissions pursuant to Directive 2003/87/EC of the European Parliament and of the Council C(2004)130, OJ of the EU of 26 February 2004, L 59, pp1–74.

9. Communication from the Commission on guidance to assist Member States in the implementation of the criteria listed in Annex III to Directive 2003/87/EC establishing a scheme for GHG emission allowance trading within the Community and amending Council Directive 96/61/EC, and on the circumstances under which force majeure is demonstrated, COM(2003)830, Brussels, 7 January 2004.

10. Communication Further guidance on allocation plans for the 2008 to 2012 trading period of the EU Emission Trading Scheme, COM(2005)703, Brussesl, 9 January 2006.

11. Communication to the Council and to the European Parliament on the assessment of national allocation plans for the allocation of GHG emission allowances in the second period of the EU Emissions Trading Scheme accompanying Commission Decisions of 29 November 2006 on the national allocation plans of Germany, Greece, Ireland, Latvia, Lithuania, Luxembourg, Malta, Slovakia, Sweden and the United Kingdom in accordance with Directive 2003/87/EC, COM(2006)725, Brussels, 11 November 2006.

12. EEA Technical report No 4/2007, Application of the Emissions Trading Directive by EU Member States – Reporting year 2006, European Environment Agency and EEA Technical report No 13/2008 and Application of the Emissions Trading Directive by EU Member States – Reporting year 2006, European Environment Agency.

13. Eritja (2006).

14. Communication from the Commission to the Council, the European Parliament, the European Economic and Social Committee and the Committee of the Regions, Building a global carbon market – Report pursuant to Article 30 of Directive 2003/87/EC, COM(2006)676, Brussels, 13 November 2006.

15. Commission Green Paper 'Towards a future maritime policy for the Union: a European vision for the oceans and seas', COM(2006)275, Brussels, 7 June 2006.

16. Proposal for a Directive of the European Parliament and of the Council amending Directive 2003/87/EC so as to improve and extend the EU GHG emission allowance trading system, COM(2008)16, Brussels, 23 January 2008.

17. Directive 2003/87/EC of the European Parliament and of the Council of 13 October 2003 establishing a scheme for GHG emission allowance trading within the Community and amending Council Directive 96/61/EC (Text with EEA relevance) (OJ L 275, 25 October 2003, p32), amended by Directive 2004/101/EC of the European Parliament and of the

Council of 27 October 2004, Directive 2008/101/EC of the European Parliament and of the Council of 19 November 2008, Regulation (EC) No 219/2009 of the European Parliament and of the Council of 11 March 2009 and Directive 2009/29/EC of the European Parliament and of the Council of 23 April 2009.

18. Directive 2009/29/EC of the European Parliament and of the Council amending Directive 2003/87/EC so as to improve and extend the GHG emission allowance trading scheme of the Community (revised Directive 2003/87/EC) (revised or amended Directive 2003/87/EC), *OJ* of the EU of 5 June 2009, L 140, pp63−87. Several parts of this chapter are based on Massai (2010b).

19. The −20 per cent target below 1990 levels is equal to −14 per cent compared to 2005 levels.

20. Recital 15, preamble of Directive 2009/29/EC.

21. Recital 17 and Article 10(2) of Directive 2009/29/EC.

22. Article 10b(2) of Directive 2009/29/EC.

23. Recital 23 and Article 10a(1) of Directive 2009/29/EC

24. Article 10a(2) of Directive 2009/29/EC.

25. Article 10a(1) of Directive 2009/29/EC.

26. Article 10a(1) and (3) of Directive 2009/29/EC.

27. Commission Decision 2010/2/EU of 24 December 2009 determining, pursuant to Directive 2003/87/EC of the European Parliament and of the Council, a list of sectors and subsectors which are deemed to be exposed to a significant risk of carbon leakage, *OJ* L1 of 5 January 2010, pp10−18.

28. Article 10a(7) of revised Directive 2003/87/EC.

29. Article 10a(8) of revised Directive 2003/87/EC.

30. Article 16(4) revised Directive 2003/87/EC.

31. Article 19(1) revised Directive 2003/87/EC.

32. Auctioning Regulation − Commission proposal, 6 April 2010.

33. Commission Regulation 2216/2004 of 21 December 2004 for a standardised and secured system of registries pursuant to Directive 2003/87/EC of the European Parliament and of the Council and Decision 280/2004/EC of the European Parliament and of the Council, *OJ* L 386 of 29 December 2004, pp1−77.

34. Commission Regulation 916/2007 of 31 July 2007 amending Regulation (EC) No 2216/2004 for a standardised and secured system of registries pursuant to Directive 2003/87/EC and Decision 280/2004/EC, *OJ* L 200 of 1 August 2007, pp1−5.

35. Commission Regulation (EC) No 994/2008 of 8 October 2008 for a standardised and secured system of registries pursuant to Directive 2003/87/EC of the European Parliament and of the Council and Decision No 280/2004/EC of the European Parliament and of the Council, *OJ* L 271 of 11 October 2008, pp1−3.

CHAPTER 15

Linking Directive

S ince the preparation of the Proposal for a framework directive for GHG emissions trading within the European Community COM(2001)581, the Commission has considered the necessity to adopt a specific document linking JI and CDM credits to the EATD. Such a document, assuming the form of an amendment of Directive 2003/87/CE, was aimed at the inclusion of carbon credits from JI and CDM projects based on the promotion of transfer of technology from developed countries to other developed countries (JI) and/or developing countries (CDM).

A Technical Report on the impacts of linking JI and CDM credits to the European Emission Allowance Trading Scheme commissioned by the Directorate General of the Environment was released in May 2003.

On 23 July 2003 following the final agreement on the EATD Directive, the Commission issued the Proposal for a Directive amending the Directive establishing a scheme for GHG emission allowance trading within the Community, in respect of the Kyoto Protocol's project mechanisms COM(2003)403[1] and a Commission staff working paper on the Extended Impact Assessment on the Commission Proposal for a Directive amending the Directive establishing a scheme for GHG emission allowance trading within the Community, in respect of the Kyoto Protocol's project mechanisms SEC(2003)785, Brussels, 23 July 2003. On the possibility of using ERUs generated by JI and CERs by CDM projects in the EU ETS, Directive 2003/87/EC acknowledged under Article 30(3) that these credits 'will be recognized for their use in this scheme subject to provisions adopted by the European Parliament and the Council on a proposal from the Commission, which should apply in parallel with the Community scheme in 2005'.

The Presidency Conclusions of the Brussels European Council of 16 and 17 October 2003 (page 10, point 24) called for an early approval of the above mentioned Commission proposal in order to promote the diffusion of clean technologies as well as to safeguard competitiveness of the European industry.

At the Environment Council of 27 October 2003 (2536th Council meeting – Luxembourg) Ministers of the Environment of the EU did not find an agreement on the proposal COM(2003)403 and the decision was postponed. Some of the Member States pushed for the possibility to convert JI/CDM credits into European allowances under the EATD already in the first phase (2005–2007) and not by 1 January 2008, as indicated in the Commission proposal.

The European Parliament's Committee on the Environment, Public Health and Consumer Policy adopted on 16 March 2004 its report on the Directive to link the EATD to Kyoto Protocol project mechanisms containing several amendments to the original proposal:

- Linking CDM credits (CERs) from 1 January 2005 and JI credits (ERUs) from 1 January 2008;
- Linking not dependent on the entry into force of the Kyoto Protocol;
- Limit on the size of hydropower plants eligible for the linking lowered from 20MW to 10MW;
- Cap on the amount of CDM credits (CERs) and JI credits (ERUs) likely to be converted into EU allowances;
- Use of the Kyoto mechanisms restricted to 50 per cent of domestic measures to reduce GHG emissions.

The Council, with the approval of the Parliament, agreed on 7 April 2004 on the text of the linking directive, whose compromise deal was formally adopted by the Parliament on 20 April 2004 at its first reading in Strasbourg. Directive 2004/101/ EC of 27 October 2004 amending Directive 2003/87/EC establishing a scheme for GHG emission allowance trading within the Community, in respect of the Kyoto Protocol's project mechanisms (Linking) of 27 October 2004 amended Directive 2003/87/EC and introduced the direct linking with the EU ETS of ERUs and CERs generated respectively through JI and CDM.[2] Directive 2004/101/EC was adopted aiming at safeguarding the environmental integrity and effectiveness of the EU ETS and allowed for:

- Use of CERs (CDM) in the EATD from 2005 and ERUs (JI) from 2008;
- No formal limitation of the quantity of credits to be included in the EATD although governments are required to consider the issue of supplementarity and give precedence to domestic actions;
- Specific provisions on supplementarity to be included in NAPs for the second phase (2008 to 2012);
- Exclusion of credits from nuclear projects and from sinks in the first phase;
- Hydropower projects to be implemented in compliance with the international rules on dams.

Furthermore, Directive 2004/101/EC tackled a range of technical issues such as the conversion of ERUs and CERs into newly issued EUAs, the different regulatory framework of the EATD and project-based mechanisms and the different nature and type of units. The use of CERs and ERUs from project activities is allowed through the 'issue and immediate surrender of one allowance by the Member State in exchange for one CERs or ERUs held by the operator in the national registry of its Member State'. The quantity of CERs and ERUs to be used must not go beyond 'a percentage of the allocation of allowances to each installation, to be specified by each Member State in its national allocation plan for that period'. The deadline for the transposition of the linking directive into national legislation was 13 November 2005. Only Belgium, Denmark, Spain and Poland complied with that date.

On 13 November 2006, the Commission agreed on Decision C(2006)5362 on avoiding double counting of GHG emission reductions under the Community emissions trading scheme for project activities under the Kyoto Protocol pursuant to Directive 2003/87/EC, obliging the Member States to cancel a number of EUAs equal to the existing and planned JI and CDM credits from the NAP.

On the issue of the use of GHG emission reduction units generated by JI and CDM (respectively ERUs and CERs), Directive 2009/29/EC confirmed that operators can use those credits to cover part of their emissions obligations under the EU ETS. Article 11a(8) of Directive 2009/29/EC set a limit of 50 per cent of the EU-wide reductions in the period 2008–2020 for the use of credits allowed for compliance by installations. For installations covered by old Directive 2003/87/EC that cap is equal to a minimum of 11 per cent of their allocation during the period 2008–2012. For new sectors and new entrants the limit is equal to a minimum of 4.5 per cent of their verified emissions of the period 2013–2020, while for the aviation sector the cap is a minimum of 1.5 per cent. The detailed percentages will be identified by the European Commission through the comitology procedure. In this respect two different scenarios that will significantly alter the rules governing the use of JI and CDM credits to comply with obligations under the EU ETS were identified by Directive 2009/29/EC. The first scenario was linked with the failure of the Copenhagen Summit of December 2009 to establish an international agreement for the period post-2012 by 31 December 2009, and implied significant limitations on the use of Kyoto project reduction units for compliance purposes within the EU ETS. The second scenario referred to the scaling up of the EU commitment for the reduction of GHG emissions to 30 per cent below 1990 levels by 2020 and it implied additional CERs and ERUs (i.e. beyond what was previously permitted in the second phase) to be allowed during the third phase.[3]

The EU ETS review directive indicated some additional qualitative requirements imposed on project offsets. It reinforced existing limitations, such as the limitation of large hydropower plants, by adding a requirement that only credits from project classes approved by all Member States can be imported into the system. The proposal also made explicit reference to high-quality CERs and ERUs in relation to the credits issued after 2013 for projects established during the second phase of the ETS.[4]

As for the linking with other ETSs, Directive 2009/29/EC provided for the linking with any compatible scheme outside the EU. In particular Article 25 of

Directive 2003/87/EC was amended providing for the 'recognition of allowances between the Community scheme and compatible mandatory greenhouse gas emissions trading systems with absolute emissions caps established in any other country or in sub-federal or regional entities' and for the possibility to link the EU ETS with third countries or sub-federal or regional entities 'to provide for administrative and technical coordination in relation to allowances in the Community scheme or other mandatory greenhouse gas emissions trading systems with absolute emissions caps', (paragraphs 1a and 1b, respectively).

On 19 March 2010, the European Commission decided to block the surrendering of CERs and ERUs to prevent their recycling for compliance with the EU ETS. The decision of the Commission was taken after BlueNext and Nordpool, two of largest electronic platforms for the spot market of carbon credits, had suspended the trading of those credits earlier in March. The interruption of the surrendering of CERs and ERUs by companies falling under the EU ETS will apply until the entry into force of the new regulation on registries in August 2010. The problem of recycled CERs and ERUs concerns the possibility that some Member States may sell used credits (already surrendered) to companies or intermediate market players into the EU ETS simply because they are worth more than AAUs.

NOTES

1. Proposal of the European Commission for a Directive amending the Directive establishing a scheme for GHG emission allowance trading within the Community, in respect of the Kyoto Protocol's project mechanisms COM(2003)403, Brussels, 23 July 2003.

2. Directive 2004/101/EC of the European Parliament and of the Council of 27 October 2004 amending Directive 2003/87/EC establishing a scheme for GHG emission allowance trading within the Community, in respect of the Kyoto Protocol's project mechanisms, *OJ* of the EU L 338, Brussels, 13 November 2004, pp18–23.

3. Article 11a(5) of Directive 2009/29/EC.

4. Recitals 30 and 32 of Directive 2009/29/EC.

CHAPTER 16

EU and post-2012

*T*he Brussels European Council of 25–26 March 2004 was one of the first meetings where EU leaders addressed formally the issue of future actions (post-2012) on climate change to be taken either at the international or at the European level. The European Council decided to consider 'medium and longer term emission reduction strategies, including targets' at the European Council at its meeting in spring 2005.

By September 2004 the Commission had invited all stakeholders to provide opinions and comments on the Community future action and targets on climate change. Although no specific indications were given as to policies and measures to be pursued at the Community level, the Commission mentioned in a press release on 14 September 2004 the possibility of introducing GHG emission reduction targets of approximately '30 per cent by 2025 and 65 per cent by 2050'. The discussion on post-2012 GHG emissions targets was based on:

- The web-based forum 'Future Action on Climate Change';
- A stakeholder conference held on 22 November 2004 that underlined the importance of setting up a new climate change regime involving directly and indirectly all actors for the post-2012 period;
- A background paper on 'Future action on climate change: A stakeholder consultation on the EU's contribution to shaping the future global climate change regime'.

The agenda of the European Commission for COP10, to be held in Buenos Aires from 6 to 17 December 2004, focusing on post-2012 targets, developing countries and emissions trading, was backed by the European Parliament with the adoption of a resolution on climate change on 17 November 2004.

On 9 February 2005 the Commission issued the Communication COM(2005)35 Winning the Battle Against Global Climate Change[1] and a Staff Working Paper on the EU policy after 2012, which outlined the key elements for the European post-2012 strategy, namely the necessity of extending the coverage of countries and sectors committed to GHG emissions reductions, the enhancement of green technologies as well as the use of flexible mechanisms and actions on adaptation.

On 27 April 2005 the European Parliament's Environment Committee adopted a resolution on Communication COM(2005)35 Winning the Battle Against Global Climate Change (2005/2049(INI)) focusing on international negotiations on the post-2012 phase, indicating the possibility of establishing trade sanctions for states who have not ratified the Kyoto Protocol, and promoting green technologies on energy and transport.

At the Environment Council of 10 March 2005 in Brussels, the EU25 agreed on a resolution on climate strategy after 2012, urging the industrialized world to consider cuts in their GHG emissions 'in the order of 15–30% by 2020 and 60–80% by 2050'. On 22 and 23 March 2005 at the Brussels European Council, the EU heads of state and government provided support in the Presidency Conclusions to the Environment Council calling for a 15–30 per cent reduction in CO_2 emissions from developed countries by 2020. The target for 2050, namely the 60–80 per cent cut in GHG emissions, was dropped from the conclusions of the presidency.

The European Parliament resolution of 12 May 2005 called for the introduction of tariffs on energy-intensive goods imported from countries not parties to the Kyoto Protocol and for the European Commission as well as EU Member States to take the lead on the future climate strategy and set GHG emissions reduction targets of 15–30 per cent by 2020.

According to EEA Report No 1/2005, the EU could feasibly assume GHG emissions reduction targets of 40 per cent (over 1990 levels) by 2030 with a large part of this reached through IET.

Voting on the Commission proposals for post-2012 climate strategy COM(2005)35, the Parliament agreed on 11 October 2005 on a draft resolution urging stronger EU climate policies. According to the resolution, all developed countries should aim at a cut of GHG emissions of 15–30 per cent by 2020 and 60–80 per cent by 2050. Other targets and measures on post-2012 were envisaged by the resolution, especially in terms of energy efficiency (cut in energy intensity by 2.5–3 per cent per year) as well as in terms of reductions of new passenger car CO_2 emissions.

The Environment Council of 17 October 2005 agreed on the EU position for COP/MOP1 to be held from 28 November to 9 December 2005 in Montreal, Canada. The EU position paper for COP/MOP1 did not contain the reference to new binding targets for industrialized countries (30 per cent by 2020) as in part expected (such a target was agreed by the Council in March 2005), aiming at facilitating international agreements on the future of the Kyoto Protocol.

On 16 November 2005 the Parliament agreed on a resolution identifying a series of priorities for EU negotiators at COP/MOP1 in Montreal. The MEPs agreed to urge the international community to find an agreement on a global climate change framework for the period after 2012 by 2008. Neither a formal mandate, nor a timetable for the future obligations was agreed by the Parliament.

On 11 December 2005 European Commissioner Stavros Dimas welcomed the positive outcome of COP11 and COP/MOP1 in Montreal from 28 November to 9 December 2005.

The results of the Montreal summit (COP11 and COP/MOP1) were discussed in the European Parliament on 18 January 2006 and the MEPs adopted a resolution urging the EU to start 'intensive dialogue' on future actions for the reduction of global warming, focusing on the international talks over the post-2012 phase starting in May 2006.

The EU Environment Council of 9 March 2006 did not reach unanimity on the resolution of the EU position on long-term targets to reduce GHG emissions. The resolution required unanimity in order to be adopted and represent the EU position towards the post-2012 negotiations starting in May 2006, but due to the opposition of Italy, it did not include any specific reference to the 60–80 per cent emission cuts in GHG emissions in developed countries by 2050.

On 22 March 2006 Austria, on behalf of the EC and the Member States, submitted to the UNFCCC Secretariat its Views regarding Article 3, paragraph 9, of the Kyoto Protocol, to be compiled and made available to the open-ended ad hoc working group, established under decision 1/CMP.1, paragraph 2, prior to its first meeting. The EU submission was discussed in the first meeting of the AWG-KP, which met in Bonn between 17 and 25 May 2006. The EU submission invited all industrialized countries to significantly cut GHGs in the post-2012 period and insisted on the establishment of new legally binding GHG reduction targets for the subsequent commitment periods of the Kyoto Protocol.

In May 2006 the European Commission launched a climate change awareness campaign to raise public attention on the issue of global warming. In September 2006 the campaign was expanded to teachers and pupils in schools, calling for environmentally friendly changes in daily behaviour.

European and Asian heads of government convened at the sixth Asia-Europe meeting (Asem6) in Finland (EU presidency) on 10 September 2006 and confirmed the joint commitment of Asia and Europe to an effective implementation of the Kyoto Protocol. The ASEM6 declaration on climate change did not mention any binding obligation but reaffirmed the importance of the principle of 'common but differentiated responsibilities' and the need for a 'successful outcome of our discussions on further commitments'.

The High Level Group on Competitiveness, Energy and the Environment established by the European Commission and the Member States in February 2006 released on 30 October 2006 its second report, urging the EU to set a long-term reduction target for GHG and CO_2 emissions independently from the outcome of the international talks.

On 17 November 2006 the European Union welcomed the outcome of COP12 and COP/MOP2 held in Nairobi from 7 to 16 November 2006. The parties of the UNFCCC and the Kyoto Protocol agreed on a long list of decisions and other actions taken by COP 12 and COP/MOP 2, mainly focusing on support to developing countries, on adaptation, on assistance to African states and on a work plan for the revision of the Kyoto Protocol to be continued in 2008.

Both the EU Council Environment of 18 December 2006 and the EU Presidency Conclusions of 14/15 December 2006 welcomed the results of the Nairobi climate summit. At the Environment Council, unanimity was shown in its support for long-term commitments at the global level for the reduction of GHG emissions (30 per cent by 2020).

The main goal of the Energy Policy for Europe of 10 January 2007 was the reduction of GHG emissions by 20 per cent by 2020 in order to ensure a limit to global warming of maximum of 2°C above pre-industrial temperatures. Together with the Energy Policy for Europe, the Commission released Communication Limiting Global Climate Change to 2° Celsius − The way ahead for 2020 and beyond COM(2007)2, including the proposed options for the limitation of global warming.[2] Communication COM(2007)2 proposed a set of actions to be implemented by both developed and developing countries in order to limit global warming to a maximum of 2°C above pre-industrial temperatures. This Communication was accompanied by the Impact Assessment Summary SEC(2007)7 of the European Commission.[3]

On 14 February 2007 the European Parliament adopted a resolution on climate change in response to Communication COM(2007)2. The text adopted by the Parliament focused on two major objectives: limitation of the average global temperature increase to 2°C above pre-industrialization levels and adoption of a GHG emission reduction target for all industrialized countries of 30 per cent in comparison with 1990 levels by 2020. Furthermore, the Parliament urged the inclusion of aviation and maritime transport in the UNFCCC negotiations in the post-2012 phase.

The Environment Council of 20 February 2007 backed the position of the Parliament and the Commission on the post-2012 targets, by reaching an agreement on a 20 per cent reduction of GHG emissions by 2020 compared to 1990 levels with the intention of moving to 30 per cent in the event that other industrialized countries agree to similar commitments.

The fight against global warming was one of the top priorities of the Commission plan of action for 2008. This was included in the Commission's long-term strategy, namely Communication COM(2007)65 Annual Policy Strategy for 2008, released on 21 February 2007.[4]

The European Council of 8−9 March 2007 in Brussels, Presidency Conclusions agreed on the following targets for the EU climate policy after 2012:

• Binding target for the Member States to reduce GHG emissions by at least 20 per cent by 2020 in respect of 1990 levels;
• Binding target for all industrialized countries to reduce GHG emissions by 30 per cent by 2020 if other developed countries agree;
• Overall objective for all developed countries to cut their GHG emissions by 60−80 per cent by 2050.

On 25 April 2007 a Temporary Committee on Climate Change composed of 60 members was created by the European Parliament. The committee was given a one year mandate to define the Parliament's position for the post-2012 international negotiations.

On 20 September 2007 the EU presidency (Portugal) released the EU Position Paper on Climate Change and Related Events, highlighting the EU approach to the planned multilateral talks on the future of the international climate regime. The paper confirmed the commitment of the EU to limit the global temperature rise to 2°C and to aim at the reduction of GHGs of 50 per cent by 2050.

At the Joint Parliamentary Meeting on Climate Change: Rising to the Challenge held on 1 October 2007 in the European Parliament, the EU environment commissioner highlighted the EU position on the Road to Bali and beyond.

At the Informal European Council held on 18–19 October 2007 in Lisbon, EU leaders tackled the issue of climate change and COP13 in Bali and stressed the EU role in the international community as the leading 'competitive, energy-secure, low-carbon economy' worldwide.

On 23 October 2007 the Temporary Committee on Climate Change of the European Parliament adopted the Bali resolution urging COP13 in Bali to consider GHG emissions cuts of 50 per cent by 2050 for industrialized countries, as well as a global cap-and-trade mechanism.

The Environment Council of 30 October 2007 adopted the Council Conclusions on climate change highlighting the EU common position for the COP 13 and COP/MOP3 to be held in Bali in the period 3–14 December 2007. The Council Conclusions were based on the statement presented by the EU presidency on 21 September 2007. The EU environment ministers reaffirmed the urgency 'to limit global warming to no more than 2°C above pre-industrial levels' and recognized the main findings of the Working Groups to the Fourth Assessment Report of the IPCC. On the post-2012 policy approach, the Council confirmed the commitment to reaffirm the UNFCCC and Kyoto Protocol architecture, aiming at the enhancement of the implementation and the broadening of the Kyoto Protocol. As to the concrete steps to be taken by industrialized countries to fight against climate change, the EU confirmed the unilateral commitment to reduce GHG emissions by 20 per cent by 2020, including a deeper absolute reduction target of 30 per cent provided that other developed countries agreed on the same level of ambition. Finally, the Council Conclusions recalled the principle of common but differentiated responsibilities, the importance of flexible mechanisms, the strong focus on adaptation and avoidance of deforestation, as well as concerns on emissions from international aviation and maritime transport.

On 15 November 2007 the European Parliament adopted a resolution setting the main EU objectives for the discussion over the post-2012 solution in Bali at COP13. According to the MEPs the EU should aim at an agreement based on binding emission reduction targets for all developed countries, the establishment of a global IET system and the financial support to adaptation.

On 22 November 2007 the European Commission released Communication COM(2007)723 on the European Strategic Energy Technology Plan (SET-PLAN), aimed at the boosting of the development of low-carbon technologies in the EU in order to reach the future GHG emission reduction targets.[5] The SET-PLAN highlighted the importance of a coordinated approach to improve European technological development, in particular in areas such as wind, solar energy, bioenergy, carbon capture, electricity grids and nuclear fission.

On 29 November 2007 the European Parliament adopted a non-legislative resolution on the relation between trade and environment, which suggested the establishment of a European-wide labelling scheme indicating the environmental impact of products in respect of climate change, as well as the inclusion of specific climate-friendly requirements in the international trade agreement signed with third countries.

In the conclusions of the High Level Group on Competitiveness, Energy and Environment published on 30 November 2007 after two years of work, the European Commission announced its support for sector targets through voluntary international agreements to curb global warming as part of the solution for the post-2012 negotiations.

The EU presidency revealed on 3 December 2007 a Vision Paper for EU Strategic Energy Technology Plan indicating the pathway to achieve the EU objective to reduce GHG emission by 60–80 per cent by 2050. The plan called for a 40 per cent reduction in CO_2 emissions in the transport sector and stronger energy efficiency targets, especially in the housing and power sectors.

On 15 December 2007 the European Commission and the European Parliament welcomed the Bali agreement. The EU expressed its satisfaction that Bali launched the official start of formal negotiations on a global climate regime for post-2012.

On 1 February 2008 the European Parliament adopted a resolution on the outcome of the Bali meeting (COP13). MEPs welcomed the BAP but expressed concern over the failure of the international community to adopt concrete steps for the adoption of GHG emission reductions targets.

On 17 September 2008 the Temporary Committee on Climate Change of the European Parliament discussed the report '2050: The future begins today – Recommendations for the EU's future integrated policy on climate change'. This draft non-legislative resolution recognized the importance of the 2008 integrated climate and energy package but urged the EU to engage in an even stronger commitment to the fight against global warming. The report proposed the introduction of legally binding targets for the improvement of energy efficiency (20 per cent by 2020), a ban on standby in all new electronic appliances, as well as mandatory switch off and energy saving modes for industrial equipment.

On 6 October 2008 the Environment Committee voted on the compromise amendments of Rapporteur Hassi on the Proposal for a Directive on efforts sharing COM(2008)17. MEPs backed the GHG reduction targets proposed by the Commission in January 2008, confirmed the automatic increase of the EU reduction target from 20 to 30 per cent in the event of the adoption of an international agreement, proposed sanctions for non-compliance and tightened the cap on the use of JI and CDM credits (8 per cent of 2005 emissions in the period 2013–2020).

In a draft resolution tabled on 10 October 2008 by the EU presidency (France) the EU leaders rejected the proposal included in the integrated climate and energy package (2008) to automatically scale up from a 20 per cent to a 30 per cent CO_2 reduction target in the event an international agreement on post-2012 was reached. According to the resolution, such a decision shall be agreed in a new legislative provision adopted by EU in accordance with the co-decision procedure. The resolution also contained special treatment and concessions (free allocation) for countries and industrial sectors most exposed to carbon leakage.

The Council Environment of 20 October 2008 in Luxembourg agreed on the EU position for negotiations under COP14 and COP/MOP4 in Poznan in December 2008. The EU position included a request for the inclusion in the next climate agreement to be finalized in Copenhagen in 2009 of the obligation for developing countries to reduce their GHG emissions by 15 to 30 per cent below business-as-usual levels by 2020. The Council conclusions also included a reference to the global convergence of per capita carbon emissions to be considered in the climate agreement.

On 2 December 2008 the final outcome of the Temporary Committee on Climate Change of the European Parliament was debated and the committee adopted a non-legislative resolution including a set of recommendations on the future goals of EU climate policy. The resolution urged the EU to keep the global average temperature increase below 2°C and called for a target of a 25–40 per cent GHG emissions reduction by 2020. Furthermore, the resolution called on the European Commission to establish interim targets to measure the energy efficiency obligations of 20 per cent by 2020, to introduce minimum efficiency criteria in public procurement policy for buildings and services and to consider the introduction of binding GHG emissions reduction targets for the agricultural sector.

On 28 January 2009 the European Commission released Communication COM(2009)39 Towards a comprehensive climate change agreement in Copenhagen, highlighting the position of the EC on the international climate treaty to be adopted in Copenhagen by the end of 2009.[6] The paper focused on the extent to which developed countries should assist developing countries in financing mitigation and adaptation activities. The draft version of the paper included the suggestion to oblige developed countries to buy carbon allowances at a fixed price of €1 per tonne in 2013, increasing to €3 in 2020 and generating more than €160 billion. The paper released by the Commission did not refer to fixed prices but rather indicated an 'annual financial commitment' for developed countries on the basis of a commonly agreed formula. Communication COM(2009)39 also contained a request for a global climate financing mechanism for developing countries, an indication of the expected actions by developing and developed countries and urged the inclusion of aviation and shipping in the Copenhagen agreement.

The Environment Council held on 21 October 2009 in Luxembourg adopted the Council Conclusions on EU position for the Copenhagen Climate Conference (7–18 December 2009)[7] that focused on the following main pillars:

- Mitigation;
- Adaptation;

- REDD, promotion of conservation, sustainable management of forests and enhancement of forest carbon stocks (REDD Plus), accounting of LULUCF, AAUs and commitment periods;
- Low-carbon development strategies/growth plan and NAMAs;
- Carbon market;
- Climate finance, governance and delivery;
- Technology;
- Legal issues.

On 30 October 2009 the European Council officially took note of the European Commission Communication Stepping up international climate finance: A European blueprint for the Copenhagen deal COM(2009)475/3, unveiled on 10 September 2009.[8] The document of the European Commission recognized the need for €100 billion annually for developing countries by 2020 in order to assist them in mitigation and adaptation to climate change. The Presidency Conclusions agreed with the Commission on the fact that €22−50 billion should be provided by the most economically advanced nations to establish an international climate fund. Carbon markets should provide €38 billion while €12−40 billion should come from domestic private and public sources. On the contrary, the Council diverged from the Commission paper as to the contribution of the EU. On this point the Commission had identified a contribution of the EU to developing countries equal to €2−15 billion by 2020. The Council agreed that €5−7 billion upfront may be needed by developing countries in the period 2010−2013 but left open any other decision on further contributions, this being dependent on the outcomes of the climate change summit of Copenhagen. According to the European leaders, Member States should contribute to that upfront funding on a voluntary basis. Furthermore, the Council did not agree on the burden sharing mechanism that should define the single contributions of the Member States. Such a mechanism will be discussed and identified in the future with the aim of taking into consideration the different economic situations of the EU27.

On 2 December 2009 the European Commission highlighted the importance that the Copenhagen conference reached a global, ambitious and comprehensive agreement to avert dangerous climate change. On 10 and 11 December 2009 the European Council adopted its conclusions that reiterated the EU position agreed in the Council of October 2009 ('The agreement should lead to finalizing a legally binding instrument, preferably within six months after the Copenhagen Conference, for the period starting on 1 January 2013') and referred to climate change financing, namely to 'the need for a significant increase in public and private financial flows to 2020 and reiterates its conclusions of October 2009 [as well as] reconfirms its commitment to provide its fair share of international public support'.

In Copenhagen, the following statements of the EU officials were provided publicly:

- Statement by Sweden on behalf of the European Union and its Member States: Speech by the Minister for the Environment Andreas Carlgren, Joint high level segment of COP and COP/MOP, 16 December 2009;

- Statement of President Barroso at the UN High Level Segment, COP15, Copenhagen, 16 December 2009;
- Statement of President Barroso on the Copenhagen Climate Accord, COP15, Copenhagen, 19 December.

The main official reactions at the EU level on the results of the Copenhagen Summit were:

- European Commission President Barroso's statement in Copenhagen (19 December 2009);
- EU Presidency (Sweden) conclusions (22 December 2009);
- European Commission press release (22 December 2009).

On 28 January 2010 Spain and the European Commission on behalf of the EU and its Member States submitted a joint letter to the UNFCCC Secretariat as an expression of willingness to be associated with the Copenhagen Accord and notification of the quantified economy-wide emissions reduction targets for 2020. The joint letter was the result of a compromise agreement reached on 21 January 2010 by representatives of the Member States in Brussels. The EU confirmed the GHG emission reductions target of 20 per cent by 2020 with the offer to reduce GHG emissions of 30 per cent by 2020 provided that other developed countries commit to comparable emissions reduction efforts and developing countries contribute as well in accordance with their responsibilities and capabilities. Member States were divided on the insertion of numerical targets in a joint letter, with the UK, Germany, France and the Netherlands supporting the inclusion of only the 30 per cent conditional target and Italy and Poland sustaining the 20 per cent target. Furthermore, the EU reiterated its commitment to negotiate a legally binding agreement for the period starting 1 January 2013.

Table 16.1 sets out the EU and Member States commitment in the format of Appendix 1 of the Copenhagen Accord.

Table 16.1 *EU and Member States commitment under Appendix 1 of the Copenhagen Accord*

| Annex I parties | Quantified economy-wide emissions targets for 2020 | |
	Emissions reduction in 2020	Base year
EU and its Member States (Belgium, Bulgaria, Czech Republic, Denmark, Germany, Estonia, Ireland, Greece, Spain, France, Italy, Cyprus, Latvia, Lithuania, Luxembourg, Hungary, Malta, Netherlands, Austria, Poland, Portugal, Romania, Slovenia, Slovakia, Finland, Sweden, UK) acting in common	20 per cent/30 per cent As part of a global and comprehensive agreement for the period beyond 2012, the EU reiterates its conditional offer to move to a 30 per cent reduction by 2020 compared to 1990 levels, provided that other developed countries commit themselves to comparable emission reductions and that developing countries contribute adequately according to their responsibilities and respective capabilities.	1990

On 9 March 2010 the European Commission released the policy paper International climate policy post-Copenhagen: Acting now to reinvigorate global action on climate change COM(2010)86 designed with the aim of maintaining the focus of the international community on climate change.[9] The Commission paper recognized the importance of the Copenhagen Accord as well as the concrete action and commitments of the EU on the reduction of GHG emissions: 20 per cent by 2020 with the possibility of increasing the target to 30 per cent provided that certain conditions are satisfied. Furthermore, in the document the European Commission reiterated that the main objective of the EU is the adoption of a legally binding agreement at the latest by 2011 in South Africa. In the view of the European executive, COP16 in Cancun should aim at integrating the Copenhagen Accord into the UNFCCC process and delivering specific decisions on key issues such as technology transfer, climate adaptation and forestry. Finally, the paper suggested the establishment of a sectoral carbon market mechanism and referred to the commitment of disbursing fast track climate funding to developing countries.

On 15 March 2010 the Environment Council adopted Council conclusions on Climate change: Follow-up to the Copenhagen Conference, and recalled the urgency of an immediate implementation of the Copenhagen Accord.[10] The Environment Council requested the Commission prepare a study of the comparability of GHG emission reductions offered by third countries as well as an impact assessment of the EU's conditional move to a 30 per cent emissions cut.

The European Council conclusions of 26 March 2010 considered the post-2012 climate change talks and referred to a stepwise approach to be adopted by the parties in the international negotiations that should build on the Copenhagen Accord and aim at the conclusion of a global and comprehensive legal agreement to reach the objective of staying below the 2°C increase in global temperatures compared to pre-industrial levels. According to the EU leaders, COP16 in Cancun should provide specific decisions 'anchoring the Copenhagen Accord to the UN negotiating process' and 'addressing remaining gaps' such as adaptation, forestry, technology and emissions MRV. The conclusions also mentioned other settings that could complement the UNFCCC process on specific issues as well as possible partnerships with other industrialized countries.

On 12 May 2010 EU finance ministers adopted a progress report on the state of play on EU and Member States fast-start financing climate change, as prepared by the Economic Policy Committee and the Economic and Financial Committee and approved by COREPER. The progress report confirmed that the EU has so far delivered €2.18 billion to developing countries for the period 2010–2012 and is therefore well on track to meet the agreed €2.4 billion pledge. Details of the division of financial commitments among the Member States were not provided. The main part of the funds will be dedicated to mitigation projects and distributed through bilateral agreements.

On 26 May 2010 the European Commission released Communication COM(2010)265 Analysis of options to move beyond 20 per cent GHG emission reductions and assessing the risk of carbon leakage, which showed that it is economically and technically affordable for the EU to move to a 30 per cent cut of

GHG emissions by 2020.[11] The policy paper of the Commission indicated the additional cost of such a move as equivalent to approximately €33 billion, or 0.2 per cent of GDP. Amongst others, the paper identified the following measures to be introduced to achieve the 30 per cent target: a carbon tax set at €30 per tonne of CO_2 equivalent for non-EU ETS sectors together with a 'more targeted use' of funds linked to climate-related programmes; the cancellation of 1.4 billion EUAs (to be decided by co-decision) from the total cap for the period 2013–2020 for EU ETS sectors; new funds for green investments in Central and Eastern Europe; and increases in the value of domestic reductions compared to credits generated by CDM projects. However, according to Communication COM(2010)265 the conditions for scaling up the EU target to 30 per cent have not yet been met since the level of ambition of the new climate change regime is still unclear and the economic crisis in the EU has reduced the capacity of the business sector to make new investments.

On 30 June 2010 the Directorate General climate commissioner circulated the document 'A vision for an enhanced Monitoring, Reporting and Verification (MRV) system' at the meeting of the Major Economies Forum.

On 11 June 2010 the Environment Council adopted Council conclusions on Climate change: Follow-up to the Copenhagen Conference[12] that, on the issue of the unilateral move to a 30 per cent GHG emission reduction target, requested further analysis on the effects of such a decision on individual Member States. The Environment Council concluded that the speed and evolution of the international climate change negotiations under the UNFCCC 'made a definitive assessment of the risk of carbon leakage difficult for the time being' and that no policy options should be excluded in the future.

The Presidency Conclusions of the European Council adopted on 17 June 2010 acknowledged the Commission's communication, analysing options to move beyond 20 per cent GHG emission reductions and assessing the risk of carbon leakage. The Council also recognized that the EU and its Member States 'have advanced in the implementation of their fast start commitments for 2010 and will report at the Cancun conference in a coordinated manner on progress achieved'.

16.1   INTEGRATED CLIMATE AND ENERGY PACKAGE

On 23 January 2008 the European Commission presented the integrated package of legislative proposals on climate and energy, providing for the long-term strategy of the future EU climate and clean energy policy. The details of the package were explained in Communication COM(2008)30, 20 20 by 2020: Europe's climate change opportunity.[13] Communication COM(2008)30 included several legislative proposals:

* Proposal for a Decision on the effort of Member States to reduce their GHG emissions to meet the Community's GHG emission reduction commitments up to 2020 COM(2008)17 (Effort Sharing);[14]

- Proposal for a Directive amending Directive 2003/87/EC so as to improve and extend the GHG emission allowance trading system of the Community COM(2008)16 (EU ETS Review);[15]
- Proposal for a Directive on the promotion of the use of energy from renewable sources COM(2008)19;[16]
- Proposal for a Directive on the geological storage of carbon dioxide and amending Council Directives 85/337/EEC, 96/61/EC, Directives 2000/60/EC, 2001/80/EC, 2004/35/EC, 2006/12/EC and Regulation (EC) No 1013/2006 COM(2008)18;[17]
- Communication on a First assessment of national energy efficiency action plans as required by Directive 2006/32/EC on energy end-use efficiency and energy services COM(2008)11;
- Communication on Supporting early demonstration of sustainable power generation from fossil fuels COM(2008)13 (CO_2 capture and storage);[18]
- Community Guidelines on State Aid for Environmental Protection.[19]

The basic principles behind the legislative proposals included in the integrated package on climate and energy were:

- Synergy between energy, energy security and climate issues;
- Fair distribution and cost effectiveness of the burden and costs of the implementation of the measures among the Member States;
- EU aiming at a leading role at the international level in the fight against climate change and in the promotion of renewable energy.

The package was based on the 20 20 (20) targets by 2020 updated to 2005, as shown in Table 16.2.

On the fair distribution of the efforts among the Member States for the implementation of the new measures, the European Commission's goal was the equal treatment of companies in the EU based on the differentiation of efforts according to GDP/capita. Accordingly, the following targets were established:

- National targets for non-EU ETS sectors;
- National targets for renewable energy;
- Redistribution of auctioning rights under the EU ETS.

Cost effectiveness in the package of climate and clean energy legislation was ensured by the introduction of more flexibility in the use of market-based

Table 16.2 *20 20 targets by 2020 updated to 2005*

Target for EU27	Compared to 1990 levels	Compared to 2005 levels
Reduction of GHG emissions in the EU by 2020	−20%	−14%
Share of renewable energies in the EU final energy consumption by 2020	+20%	+11.5%

instruments, both the EU ETS and the system of transferability of GOs for the promotion of renewable energy. While more flexibility in the EU ETS was introduced thanks to pressure from some EU Member States that perceived the new targets on the reduction of GHG emissions as too rigid, the system of GOs, namely a uniform voluntary system at the Member State or company level, was criticized by many Member States that expressed the fear that this mechanism could undermine the value of the existing supporting schemes.

The European Commission decided to differentiate the target for the reduction of GHG emissions between EU ETS and non-EU ETS sectors (transport, housing, agriculture, waste and non-CO_2 gases) in order to preserve the balance among the different EU industrial sectors affected by the measures as well as among the Member States. Therefore, on the basis of the 2005 levels, the EU ETS sectors and the non-EU ETS sectors were required to reduce their GHG emissions by 2020, by 21 per cent and 10 per cent, respectively. In the latter case, the European Commission proposed differentiated targets for the EU27, with targets stretching from −20 per cent to +20 per cent. The choice of 2005 as a base year for the calculation of new GHG emission reduction targets for the EU Member States created discontent, especially among some of the EU new Member States that have assumed in the current international climate regime base years different from 1990. The details of the effort sharing among Member States in respect of the non-EU ETS sectors were highlighted in the Proposal for a Decision on the effort of Member States to reduce their GHG emissions to meet the Community's GHG emission reduction commitments up to 2020 COM(2008)17 (Effort Sharing), and are also shown in Table 16.3.

On 26 February 2008 the Environment Committee of the European Parliament held its first exchange of views on the integrated climate and energy package. MEPs' main concerns were about the competitiveness of EU energy-intensive industries, and to this end support for the introduction of a carbon import tax was expressed by the European Commission's Environmental Policy Directorate. Furthermore, the focus was put on CCS, CDM projects, aviation and on the necessity to strive for an ambitious international agreement.

EU energy ministers welcomed the integrated climate and energy package at the Council Transport, Telecommunications and Energy of 28 February 2008. The system for the trading of GOs was one of the main issues that remained unresolved, as well as the need to introduce sustainability criteria in the proposal for the promotion of biofuels and the importance of protecting energy-intensive industries from external competition loss.

At the Environment Council of 3 March 2008 EU ministers discussed the integrated climate and energy package and adopted a draft resolution including political guidelines to be forwarded to the European Council on 13 and 14 March 2008. Key aspects discussed at the Environment Council included:

- More flexibility in implementing the climate and energy legislative proposals;
- Greater use of CERs and ERUs from CDM and JI projects (the UK, Spain and Austria);

Table 16.3 *Member States limits by 2020 for non-EU ETS sectors (proposed by the Commission)*

	Member State greenhouse gas emission limits by 2020 compared to 2005 greenhouse gas emission levels for sources not covered under Directive 2003/87/EC	Member State greenhouse gas emissions in 2020 resulting from the implementation of Article 3 (in tonnes of CO_2 equivalent)
Belgium	− 15%	70,954,356
Bulgaria	20%	35,161,279
Czech Republic	9%	68,739,717
Denmark	− 20%	29,868,050
Germany	− 14%	438,917,769
Estonia	11%	8,886,125
Ireland	− 20%	37,916,451
Greece	− 4%	64,052,250
Spain	− 10%	219,018,864
France	− 14%	354,448,112
Italy	− 13%	305,319,498
Cyprus	− 5%	4,633,210
Latvia	17%	9,386,920
Lithuania	15%	18,429,024
Luxembourg	− 20%	8,522,041
Hungary	10%	58,024,562
Malta	5%	1,532,621
Netherlands	− 16%	107,302,767
Austria	− 16%	49,842,602
Poland	14%	216,592,037

- Spreading of the reduction targets to all sectors (main Central and Eastern European countries);
- Opposition to the adoption of 2005 as base year (main Central and Eastern European countries);
- More efforts to address 'carbon leakage' of the energy-intensive industries;
- Adoption of burden sharing figures in a 30 per cent reduction scenario (Germany, Sweden and the Netherlands);
- Promotion of mineral and chemical sequestration of carbon in developing rules for CCS;
- Need for more measures to promote energy efficiency (Germany, Italy and Austria);
- Political agreement on the entire package by the end of 2008.

The Presidency Conclusions of the Brussels European Council held on 13 and 14 March 2008 included a specific reference to the commitments of the EU leaders to reach an agreement on the proposal for the integrated climate and energy package by the end of 2008 in order to have the formal adoption early in 2009. Furthermore, the EU summit decided that actions to preserve economic integrity of the European energy-intensive industries will be taken only in the case of the failure of international negotiations for the post-2012 regime under the Kyoto

Protocol. The proposal for the elimination of VAT on green products did not reach unanimity among the EU leaders.

On 8 May 2008 the Environment Committee of the European Parliament debated for the first time plans to review proposals COM(2008)16 and COM(2008)17 on the review of the EU ETS and on effort sharing, respectively. The major issues of debate were: the inclusion of forestry projects (avoided deforestation) and shipping in the EU ETS, and limits on the use of JI and CDM reduction credits in the EU ETS after 2013.

On 5 June 2008 Environment Ministers in the Council Environment met to discuss the 2008 integrated package on climate and energy. Amendments to the Commission's proposals proposed by the Environment Council included: base year for legislation aimed at the reduction of GHG emissions to be shifted from 2005 to 1990 (group of Central and Eastern European states); more flexibility in burden sharing legislation (Sweden); trading of emissions credits between EU ETS and non-EU ETS sectors (finally rejected). On 16 and 18 June 2008 the package was discussed in the Parliament and the Doyle amendments and the Hassi amendments were presented.

On 3 and 4 July 2008 at the first informal meeting of the French presidency of the Council of the EU, the EU environment and energy ministers agreed to subordinate the adoption of the integrated climate and energy package to have flexibility over the emission reduction targets and financial assistance mechanisms to be provided to the Member States with a lower GDP. Also, the EU ministers requested the introduction of 'carbon leakage' safeguards aimed at the protection of the competitiveness of European energy-intensive industries vis-à-vis similar companies in developing countries in case a global agreement was not reached. Other issues of concern among the Member States were: shifting the baseline for 2020 targets from 2005 to 1990 (Hungary, Poland and others), and opposition to the use of at least 20 per cent of revenues from auctioning for climate-related activities (Germany, the UK and others).

Following the discussions in the COREPER of 17 September 2008, on 23 September 2008 the EU presidency (France) presented a number of compromise amendments on the climate and energy legislative package addressing, amongst others, the following points:

- Carbon leakage and the EU ETS directive;
- Adjustments in the event of an international agreement in connection with the EU ETS directive and the effort sharing decision;
- A general rendez-vous clause in connection with the renewable energy directive.

On 6 October 2008 the Environment Committee voted on three major legislative proposals included in the package: the EU ETS review (Doyle's amendments), the efforts sharing (Hassi's amendments) directive and the CCS (Davies's amendments) directives.

The Presidency Conclusions adopted by the EU heads of state and government on 16 October 2008 contained a chapter on climate change and energy where the deadline of December 2008 for the adoption of the integrated climate and energy package was confirmed. During the EU summit eight Eastern European countries and Italy expressed their concerns about the package in a joint statement on the impact of some parts of the new legislative measures on the national economies. As a consequence, the presidency conclusions mentioned the 'Member State's specific situation' and the need to apply the package in a 'cost-effective manner to all sectors of the European economy and all Member States'.

EU environment ministers met in Luxembourg on 20 October 2008 to discuss the integrated climate and energy package but no consensus was reached on several key issues. Disagreement among the Member States centred on how to identify the industrial sectors at risk of carbon leakage and on the nature and scope of a solidarity mechanism recognizing the reductions of GHG emissions achieved in the Eastern European countries since 1990. In particular, the German proposal to allow free allocation of allowances to the most exposed industries under the EU ETS was backed by almost all the Member States. The new Member States asked for less space for JI and CDM credits in the EU ETS and the introduction of intra-Member States trading that could easily allow these countries to sell part of their GHG emissions surplus.

On 12 December 2008 the European Council reached a unanimous agreement on the integrated climate and energy package and urged the Council and the Parliament to adopt the measures including the details of the final compromise to which paragraph 20 of the European Council conclusions referred. The issues discussed in the final compromise were:

- Industrial sectors not exposed to the risk of carbon leakage;
- Industrial sectors exposed to the risk of carbon leakage;
- Possible derogations to setting the auctioning rate at 100 per cent in 2013 in the electricity sector;
- Allocation of allowances;
- Correction of the linear trajectory for Member States authorized to increase their emissions (effort sharing decision);
- Funding for innovative CCS technologies and renewable energy sources;
- CDM and JI (effort sharing decision);
- Voluntary pre-allocation of part of auctioning revenues;
- Commission report on the feasibility of an option related to auctioning;
- Basis for calculating the allocation of emission rights;
- Permission to exceed the carry-forward rate in the event of extreme meteorological conditions (effort sharing decision);
- Use of specific credits from project types (ETS directive);
- Closure of Ignalina;
- Energy security;
- Carbon leakage.

On 17 December 2008 the European Parliament approved with a big majority the integrated climate and energy package including four different EU laws and formalizing the reduction of GHG emissions in the EU by 20 per cent by 2020 compared with 1990 levels. The compromise deals with the Council voted on by the MEPs in a first-reading agreement concerned the following texts:

- European Parliament legislative resolution of 17 December 2008 on the Proposal for a Directive of the European Parliament and of the Council amending Directive 2003/87/EC so as to improve and extend the GHG emission allowance trading system of the Community (COM(2008)16) (EU ETS review);[20]
- European Parliament legislative resolution of 17 December 2008 on the Proposal for a Decision of the European Parliament and of the Council on the effort of Member States to reduce their GHG emissions to meet the Community's GHG emission reduction commitments up to 2020 (COM(2008)17) (Effort Sharing);[21]
- European Parliament legislative resolution of 17 December 2008 on the Proposal for a Directive of the European Parliament and of the Council on the promotion of the use of energy from renewable sources (COM(2008)19);[22]
- European Parliament legislative resolution of 17 December 2008 on the Proposal for a Directive of the European Parliament and of the Council on the geological storage of carbon dioxide and amending Council Directives 85/337/EEC, 96/61/EC, Directives 2000/60/EC, 2001/80/EC, 2004/35/EC, 2006/12/EC and Regulation (EC) No 1013/2006 (COM(2008)18).[23]

The integrated climate and energy package will be a key instrument for the implementation of the clean energy and climate commitments undertaken by the EU in 2007 and 2008, notably the 20-20-20 + 10 targets, in particular the reduction of GHG emissions of 20 per cent by 2020. In this sense the agreement represented a message to the international community negotiating the details of the post-2012 climate regime. The agreement was possible thanks to concessions made to countries that opposed the package in order to protect national economies from the impact of the new measures on some industrial sectors. In this respect, installations covered by the EU ETS will get at least 30 per cent of their EUAs for free until at least 2020 and some concessions were granted to some installations in the power sector. Furthermore, a deal was reached at the last moment on access to carbon credits from JI and CDM and on the earmarking of carbon auction revenues towards the new Member States and the funding of CCS activities.

On the same day, the Parliament also approved two important new regulations on road transport, notably: European Parliament legislative resolution of 17 December 2008 on the Proposal for a Regulation of the European Parliament and of the Council setting emission performance standards for new passenger cars as part of the Community's integrated approach to reduce CO_2 emissions from light-duty vehicles (COM(2007)856;[24] and European Parliament legislative resolution of 17 December 2008 on the Proposal for a Directive of the European Parliament and of the Council amending Directive 98/70/EC as regards the specification of petrol, diesel and gas-oil and introducing a mechanism to monitor and reduce GHG emissions from the use of road transport fuels and amending Council

Directive 1999/32/EC, as regards the specification of fuel used by inland waterway vessels and repealing Directive 93/12/EEC (COM(2007)18.[25]

In accordance with the decision on efforts sharing COM(2008)17 endorsed by the European Parliament on 17 December 2008, the EU Member States are required to reduce GHG emissions from sources not covered by the EU ETS by 10 per cent by 2020 compared with 2005 levels. These sectors correspond to transport, buildings, services, agriculture and small industrial plants. The 10 per cent target has been distributed into binding national targets as proposed by the European Commission in January 2008. National targets are divided into annual binding targets for the period 2013–2020. The final compromise has increased flexibility in the efforts sharing decision, notably by allowing Member States to bank up to 5 per cent of their GHG emission limit for the subsequent year and to transfer the same amount to another Member State in a 'manner that is mutually convenient' (auctioning, market intermediaries, bilateral arrangement). Furthermore, flexibility is ensured by allowing Member States to purchase JI and CDM credits up to 3 per cent of the national emissions in 2005. This limit may be raised by 1 per cent annually by states facing tough emission reduction obligations (Annex III); in this case only CERs from projects in LDCs and small island developing states can be accepted. Finally, the efforts sharing decision introduced specific yearly compliance sanctions for states exceeding emissions, and specific conditions and consequences in the event that an international climate treaty is achieved (and vice versa).

The final text of the six new provisions included in the integrated climate and energy package was adopted on 6 April 2009 by the EU Council of Justice and Home Affairs to cover:

- Directive 2009/28/EC of the European Parliament and of the Council of 23 April 2009 on the promotion of the use of energy from renewable sources and amending and subsequently repealing Directives 2001/77/EC and 2003/30/EC;[26]
- Directive 2009/29/EC of the European Parliament and of the Council of 23 April 2009 amending Directive 2003/87/EC so as to improve and extend the GHG emission allowance trading scheme of the Community;[27]
- Directive 2009/30/EC of the European Parliament and of the Council of 23 April 2009 amending Directive 98/70/EC as regards the specification of petrol, diesel and gas-oil and introducing a mechanism to monitor and reduce GHG emissions and amending Council Directive 1999/32/EC as regards the specification of fuel used by inland waterway vessels and repealing Directive 93/12/EEC;[28]
- Directive 2009/31/EC of the European Parliament and of the Council of 23 April 2009 on the geological storage of carbon dioxide and amending Council Directive 85/337/EEC, Directives 2000/60/EC, 2001/80/EC, 2004/35/EC, 2006/12/EC, 2008/1/EC and Regulation (EC) No 1013/2006;[29]
- Decision No 406/2009/EC of the European Parliament and of the Council of 23 April 2009 on the effort of Member States to reduce their GHG emissions to meet the Community's GHG emission reduction commitment up to 2020;[30]

- Regulation (EC) No 443/2009 of the European Parliament and of the Council of 23 April 2009 setting emission performance standards for new passenger cars as part of the Community's integrated approach to reduce CO_2 emissions from light-duty vehicles.[31]

On 5 June 2009 five of the six laws included in the integrated climate and energy package presented in 2008 were published in the *Official Journal* of the EU. These are the renewable energy directive (2009/28/EC), the EU ETS review directive (2009/29/EC), the passenger car emissions regulation (443/2009/EC), the CCS directive (2009/31/EC) and the fuel quality directive (2009/30/EC). The regulation on CO_2 emissions from new cars entered into force on 8 June 2009, with the remaining four laws on 25 June 2009. The sixth law, the decision on efforts sharing (406/2009/EC), was published on 5 June 2009 and entered into force on 25 June 2009. Member States were required to transpose the new renewable energy directive by 5 December 2010, while the EU ETS review directive shall be implemented by 31 December 2012.

NOTES

1. Communication from the Commission to the Council, the European Parliament, the European Economic and Social Committee and the Committee of the Regions Winning the Battle Against Global Climate Change, COM(2005)35, Brussels, 9 February 2005.

2. Communication from the Commission to the Council, the European Parliament, the European Economic and Social Committee and the Committee of the Regions, Limiting Global Climate Change to 2 degrees Celsius - The way ahead for 2020 and beyond, COM(2007)2, Brussels, 10 January 2007.

3. Commission Staff Working Document, Limiting Global Climate Change to 2° Celsius – The way ahead for 2020 and beyond, SEC(2007)7, Brussels, 10 January 2007.

4. Communication from the Commission to the Council, the European Parliament, the European Economic and Social Committee and the Committee of the Regions Annual Policy Strategy for 2008, COM(2007)65, Brussels, 21 February 2007.

5. Communication from the Commission to the Council, the European Parliament, the European Economic and Social Committee and the Committee of the Regions on the European Strategic Energy Technology Plan (SET-PLAN), COM(2007)723, Brussels, 22 November 2007.

6. Communication from the Commission to the Council, the European Parliament, the European Economic and Social Committee and the Committee of the Regions, Towards a comprehensive climate change agreement in Copenhagen, COM(2009)39, Brussels, 28 January 2009.

7. Council Conclusions on EU position for the Copenhagen Climate Conference (7–18 December 2009), 2968th Environment Council meeting, Luxembourg, 21 October 2009.

8. Communication from the Commission to the Council, the European Parliament, the European Economic and Social Committee and the Committee of the Regions Communication Stepping up international climate finance: A European blueprint for the Copenhagen deal, COM(2009)475/3, Brussels, 10 September 2009.

9. Communication from the Commission to the Council, the European Parliament, the European Economic and Social Committee and the Committee of the Regions, International climate policy post-Copenhagen: Acting now to reinvigorate global action on climate change, COM(2010)86, Brussels, 9 March 2010.

10. 3002nd Council meeting, Council Environment, Council conclusions on Climate change: Follow-up to the Copenhagen Conference, Brussels, 15 March 2010.

11. Communication from the Commission to the Council, the European Parliament, the European Economic and Social Committee and the Committee of the Regions, Analysis of options to move beyond 20 per cent greenhouse gas emission reductions and assessing the risk of carbon leakage, COM(2010)265, Brussels, 26 May 2010.

12. 3021st Council meeting, Council Environment Conclusions, Brussels, 11 June 2010.

13. Communication from the Commission to the Council, the European Parliament, the European Economic and Social Committee and the Committee of the Regions, 20 20 by 2020: Europe's climate change opportunity, COM(2008)30, Brussels, 23 January 2008.

14. Proposal for a Decision on the effort of Member States to reduce their GHG emissions to meet the Community's GHG emission reduction commitments up to 2020, COM(2008)17 (Effort Sharing), Brussels, 23 January 2008.

15. Proposal for a Directive amending Directive 2003/87/EC so as to improve and extend the GHG emission allowance trading system of the Community, COM(2008)16 (EU ETS review), Brussels, 23 January 2008.

16. Proposal for a Directive on the promotion of the use of energy from renewable sources, COM(2008)19, Brussels, 23 January 2008.

17. Proposal for a Directive on the geological storage of carbon dioxide and amending Council Directives 85/337/EEC, 96/61/EC, Directives 2000/60/EC, 2001/80/EC, 2004/35/EC, 2006/12/EC and Regulation (EC) No 1013/2006, COM(2008)18, Brussels, 23 January 2008.

18. Communication on Supporting early demonstration of sustainable power generation from fossil fuels, COM(2008)13 (CO_2 capture and storage), Brussels, 23 January 2008.

19. Community Guidelines on State Aid for Environmental Protection, Brussels, 23 January 2008.

20. European Parliament legislative resolution of 17 December 2008 on the Proposal for a Directive of the European Parliament and of the Council amending Directive 2003/87/EC so as to improve and extend the GHG emission allowance trading system of the Community (COM(2008)16) (EU ETS review), Brussels, 17 December 2008.

21. European Parliament legislative resolution of 17 December 2008 on the Proposal for a Decision of the European Parliament and of the Council on the effort of Member States to reduce their GHG emissions to meet the Community's GHG emission reduction commitments up to 2020 (COM(2008)17) (Effort Sharing), Brussels, 17 December 2008.

22. European Parliament legislative resolution of 17 December 2008 on the Proposal for a Directive of the European Parliament and of the Council on the promotion of the use of energy from renewable sources (COM(2008)19), Brussels, 17 December 2008.

23. European Parliament legislative resolution of 17 December 2008 on the Proposal for a Directive of the European Parliament and of the Council on the geological storage of carbon dioxide and amending Council Directives 85/337/EEC, 96/61/EC, Directives 2000/60/EC, 2001/80/EC, 2004/35/EC, 2006/12/EC and Regulation (EC) No 1013/2006 (COM(2008)18), Brussels, 17 December 2008.

24. European Parliament legislative resolution of 17 December 2008 on the Proposal for a Regulation of the European Parliament and of the Council setting emission performance

standards for new passenger cars as part of the Community's integrated approach to reduce CO_2 emissions from light-duty vehicles (COM(2007)856, Brussels, 17 December 2008.

25. European Parliament legislative resolution of 17 December 2008 on the Proposal for a Directive of the European Parliament and of the Council amending Directive 98/70/EC as regards the specification of petrol, diesel and gas-oil and introducing a mechanism to monitor and reduce GHG emissions from the use of road transport fuels and amending Council Directive 1999/32/EC, as regards the specification of fuel used by inland waterway vessels and repealing Directive 93/12/EEC (COM(2007)18, Brussels, 17 December 2008.

26. Directive 2009/28/EC of the European Parliament and of the Council of 23 April 2009 on the promotion of the use of energy from renewable sources and amending and subsequently repealing Directives 2001/77/EC and 2003/30/EC, OJ L 140 of 5 June 2009, pp16–62.

27. Directive 2009/29/EC of the European Parliament and of the Council of 23 April 2009 amending Directive 2003/87/EC so as to improve and extend the GHG emission allowance trading scheme of the Community, OJ L 140 of 5 June 2009, pp63–87.

28. Directive 2009/30/EC of the European Parliament and of the Council of 23 April 2009 amending Directive 98/70/EC as regards the specification of petrol, diesel and gas-oil and introducing a mechanism to monitor and reduce GHG emissions and amending Council Directive 1999/32/EC as regards the specification of fuel used by inland waterway vessels and repealing Directive 93/12/EEC, OJ L 140 of 5 June 2009, pp88–113.

29. Directive 2009/31/EC of the European Parliament and of the Council of 23 April 2009 on the geological storage of carbon dioxide and amending Council Directive 85/337/EEC, Directives 2000/60/EC, 2001/80/EC, 2004/35/EC, 2006/12/EC, 2008/1/EC and Regulation (EC) No 1013/2006, OJ L 140 of 5 June 2009, pp114–135.

30. Decision No 406/2009/EC of the European Parliament and of the Council of 23 April 2009 on the effort of Member States to reduce their GHG emissions to meet the Community's GHG emission reduction commitment up to 2020, OJ L 140 of 5 June 2009, pp136–148.

31. Regulation (EC) No 443/2009 of the European Parliament and of the Council of 23 April 2009 setting emission performance standards for new passenger cars as part of the Community's integrated approach to reduce CO_2 emissions from light-duty vehicles, OJ L 140 of 5 June 2009, pp1–15.

CHAPTER 17

Impacts and Adaptation

The second pillar of the international climate regime is adaptation to the adverse effects of climate change. This is one of the priorities of many developing countries and, recently, of developed states too. The EU and its Member States have therefore initiated a proper and adequate adaptation policy as a complement to reducing GHG emissions.

The EU is providing financial and technical support to developing countries in the setting up and implementation of adaptation measures such as developing crops that can tolerate drought and strengthening coastal flood defences against sea level rise. In addition, the European Commission is committed to the strengthening of Europe's resilience to climate change with a view to establishing a comprehensive EU adaptation strategy by 2013.

In particular, adaptation is one of the main topics for the establishment of a solid and durable future commitment period after 2012. The EU has started to work on the definition of an advanced adaptation policy with the establishment of the ECCP II working group on impacts and adaptation. The activities of this working group cover:

- Water resource management and extreme events;
- Marine resources and coastal zones and tourism;
- Human health;
- Agriculture and forestry;
- Biodiversity;
- Regional and urban planning.

On 18 January 2006 the Commission tabled the Proposal for a Directive on the assessment and management of flood risks COM(2006)15 according to which the

EU Member States were required to identify river basins and coastal areas subject to flooding and adopt adequate policies and measures to reduce risks.[1]

On 22 September 2006 the European Commission adopted Proposal COM(2006)232 for a Directive of the European Parliament and of the Council establishing a framework for the protection of soil and amending Directive 2004/35/EC.[2]

On 25 April 2007 the European Parliament adopted at its second reading a compromised text on the Council Common Position 33/2006 of 23 November 2006 referring to the Commission proposal COM(2006)15.

On 2 July 2007, the European Commission published the Green Paper 'Adapting to climate change in Europe – options for EU action' COM(2007)354 focusing on the introduction of policies, measures and strategies on adaptation in the European and national climate policy.[3] The actions identified by the Green Paper were:

• Development and integration of adaptation strategies into EU policies;
• Cooperation on adaptation policy with third countries;
• A greater focus on research;
• Establishment of a European advisory group for the coordination of action.

On 18 September 2007 the European Commission released Communication COM(2007)540 Building a global climate change alliance between the European Union and poor developing countries most vulnerable to climate change, including a proposal for a new initiative worth €350 million to assist developing countries to minimize the adverse effects of climate change. The plan addressed five areas: adaptation measures, reduction of emissions from deforestation, carbon markets, natural disasters, and integration of climate change into poverty reduction.[4]

On the same day Directive 2007/60/EC of 23 October 2007 on the assessment and management of flood risks was adopted by Council on the basis of the agreement reached with the European Parliament in April 2007.[5] Directive 2007/60/EC entered into force on 26 November 2007 and it required Member States to identify river basins and coastal areas vulnerable to flooding and to adopt finalized measures for the reduction of risks. According to the directive, Member States are required to provide preliminary flood assessments by 2011, as well as flood risk maps and management plans by 2013 and 2015. The directive was supposed to be transposed in the national systems of the Member States by November 2009.

On 14 November 2007 the European Parliament adopted its first-reading report on the Commission proposal for an EU directive on soil protection. The Parliament's vote would give EU Member States more space of manoeuvre in deciding how to protect areas at risk of soil degradation.

The Environment Council of 21 December 2007 failed to obtain the qualified majority required to adopt a Directive on the protection of soil, amending Directive 2004/35/EC. The proposal was rejected by Germany, France, the Netherlands, Austria and the UK, which justified their position by considering the proposed new directive as an extra burden and cost for national public administration.

Directive 2007/60/EC of 23 October 2007 on the assessment and management of flood risks entered into force on 26 November 2007. Directive 2007/60/EC had to be transposed into national law by November 2009 and requires Member States to prepare preliminary flood assessments by 2011 as well as flood risk maps and management plans by 2013 and 2015.

On 1 April 2008 the Temporary Committee on Climate Change of the European Parliament, whose mandate was extended until February 2009, released its first report since it was established. The Interim Report on the scientific facts of climate change: findings and recommendations for decision-making confirmed the findings of the IPCC 4th Report that climate change is 'settled' and that it is time to act. Furthermore, the report gives support to the EU objective of limiting global warming to 2°C and urges the IPCC to investigate new arguments seriously.

In response to the European Commission Green Paper on 'Adapting to climate change in Europe: Options for EU action' in 2007, the European Parliament adopted on 10 April 2008 a non-binding resolution urging the Commission to investigate costs and benefits of adaptation to climate change and promote economic investment to support measures on adaptation. The Parliament proposed including funds for adaptation in regional policies and invited Member States to utilize rural development funds to promote adaptation in agriculture.

On 29 September 2008 the EEA published Report No 4/2008 Impacts of Europe's changing climate – 2008 indicator based assessment.[6] The report focused on the impact of climate change in Europe on the basis of 40 key indicators and included research on biodiversity and marine, terrestrial and freshwater ecosystems. The main finding of the report was the necessity for the EU to intensify action on adaptation and to improve the quality and quantity of monitoring and reporting of observed impacts.

The proposal of 15 September 2008 of the EU presidency (France) to amend the Proposal for a Directive establishing a framework for the protection of soil did not get the support of the Member States and European Commission. The opposition to the original proposal of the Commission by Germany, the UK, Austria, the Netherlands and France expressed in December 2007 was confirmed, although the French presidency amendments would have foreseen the removal of a binding obligation on the Member States to identify sites at risk of soil degradation and would have introduced more flexibility in deciding the actions to be implemented to clean the contaminated sites.

On 9 October 2008 the Parliament adopted a non-legislative resolution on Arctic governance urging the Commission to start negotiations on an international treaty for the protection of the Arctic and to address the issue of ice melting in that region.

On 1 April 2009 the European Commission produced the White Paper COM(2009)147 Adapting to climate change: Towards a European framework for action, addressing the climate change impacts in Europe and including a coordinated strategy and proposals to reduce the negative effects of climate change in the region. The White Paper presented an adaptation strategy divided into two phases: (1) research and potential actions (2009–2012) and (2) implementation (from 2013). Furthermore, the Commission proposed the establishment of a European information exchange platform, plus an Impact and Adaptation Steering Group (IASG) composed of Member States representatives. The White Paper considered the costs of adaptation and possible means of finance, i.e. general budget, auctioning revenues from the EU ETS and insurance funds.

On 14 and 15 April 2009 an Informal Meeting of the EU Ministers of Environment was held in Prague to debate the role of the EU in adaptation towards climate change and to discuss the Commission policy paper released on 1 April. The Ministers questioned the proposal of the Commission to monitor the adaptation strategies of the Member States and the mandate of the proposed IASG. Criticism was expressed by some Member States over the proposal of the Commission to share adaptation costs. Ministers demanded more integration of adaptation issues into existing policies, more coordination and cooperation among countries in data collection, cross-border responses and mutual support.

The list of National Adaptation Strategies (NASs) adopted by the Member States and included in the fourth National Communication to the UNFCCC of 2005 are collected by the EEA NASs. At the moment of writing, the following NASs have been adopted by the Member States:

- Denmark (2008)
- Finland (2005)
- Germany (2008)
- France (2007)
- Hungary (2008)
- Netherlands (2008)
- Spain (2006)
- Sweden (2009)
- UK (2008)

The 2009 Environmental Policy Review, released by the European Commission on 10 August 2010, showed that more than 360 natural disasters such as floods and droughts were linked to climate change from 2000 to 2009, up from 100 between 1980 and 1989, according to the report. Progress in protecting EU biodiversity remained mixed, with positive developments for some species and habitats but not for others.

NOTES

1. European Commission, Proposal for a Directive on the assessment and management of flood risks, COM(2006)15, Brussels, 18 January 2006.

2. European Commission, Proposal for a Directive of the European Parliament and of the Council establishing a framework for the protection of soil and amending Directive 2004/35/EC, COM(2006)232, Brussels, 22 September 2006.

3. European Commission, Green Paper 'Adapting to climate change in Europe – options for EU action', COM(2007)354, Brussels, 2 July 2007.

4. European Commission Communication Building a global climate change alliance between the European Union and poor developing countries most vulnerable to climate change, COM(2007)540, Brussels, 18 September 2007.

5. Directive 2007/60/EC of the European Parliament and of the Council of 23 October 2007 on the assessment and management of flood risks, *OJ* L 288 of 6 November 2007, pp27–34.

6. Report No 4/2008 Impacts of Europe's changing climate – 2008 indicator based assessment, EEA, 29 September 2008.

CHAPTER 18

Climate Change Litigation

*W*ith the term 'climate change litigation', we refer to any legal proceeding initiated in front of international, regional and national courts for a dispute directly related to climate change and its effects. The concept and the practice of climate change litigation, namely the frequency of legal cases dealing with climate change, are relatively new and scarce if compared with other environment-related litigation. In this book, the analysis of climate change litigation is mainly limited to the litigation in front of the EU courts (ECJ and CFI) that has developed since the establishment of the EU ETS and in respect of issues related with the implementation and application of Directive 2003/87/EC in the Member States. A number of other examples of cases related to the compliance by the Member States with other legislation on climate change adopted by the EU are also provided. Outside the EU, in particular in the US and Australian courts, climate change litigation refers mostly to proceedings dealing with damages caused by the adverse effects of climate change.

The success and development of climate change litigation in the EU is mainly related to the power of the European Commission to approve or reject NAPs submitted by the Member States. According to Article 9(3) of Directive 2003/87/EC:

> within three months of notification of a national allocation plan by a Member State under paragraph 1, the Commission may reject that plan, or any aspect thereof, on the basis that it is incompatible with the criteria listed in Annex III or with Article 10. The Member State shall only take a decision under Article 11(1) or (2) if proposed amendments are accepted by the Commission. Reasons shall be given for any rejection decision by the Commission.

CASE T-27/07 *US STEEL KOŠICE V. COMMISSION*[1]

The Commission's hard line on second phase NAPs for the EU ETS was challenged for the first time by Slovakia on 7 February 2007. Slovakia was among those Member States whose draft NAP had been rejected by the Commission. According to the Slovakian government, the 25 per cent cut requested by the Commission on the NAP will affect national economic growth. Moreover, Slovakia claimed that the Commission was acting beyond the scope of its competences. For all these reasons Slovakia sought the annulment of Commission decision of 29 November 2006. According to the order of the CFI of 1 October 2007 the application was dismissed as inadmissible because of lack of direct concern and US Steel Košice s.r.o was ordered to pay the costs.

CASE T-387/04 *ENBW ENERGY BADEN WÜRTTEMBERG AG V. COMMISSION*

On 17 June 2004 one of the main German energy producers requested the annulment of Commission decision of 7 July 2004 on the German NAP concerning the NAP notified by Germany under Directive 2003/87/EC. The company EnBW Energy Baden Württemberg AG did not agree with the allocation of EUAs for power stations decommissioning nuclear energy installations (such as one of its main competitors RWE) and claimed that this method of allocation should be considered as state aid. EnBW claimed that the Commission failed to start the state aid procedure under Community law, breaching Article 88(2) of the EC Treaty. The Order of the Court of First Instance (Third Chamber) of 30 April 2007 considered the actions for annulment inadmissible for the lack interest in bringing proceedings in front of the Court by the applicant.

On 25 May 2007 Poland and the Czech Republic announced their intention to take the Commission to court because the CO_2 emission allowances distributed under EU ETS were considered too stringent. In March 2007 the Polish and Czech NAPs were rejected by the European Commission under the condition that the quantity of EUAs was going to be reduced significantly. According to the two countries, the Commission's allowances would prevent the countries from catching up with the Western European countries from an economic perspective. The decisions of the Commission would hamper industrial development. Slovakia had already started a formal complaint to the CFI against the Commission in February 2007.

CASE T-182/06 *NETHERLANDS V. COMMISSION*

Euro 4 norms laid down in Directive 98/69/EC of 13 October 1998 relating to measures to be taken against air pollution by emissions from motor vehicles and amending Council Directive 70/220/EEC allowed for emissions of small particles of up to 25mg/km from passenger cars and small commercial vehicles with diesel engines. The Netherlands asked permission to introduce stricter norms, aimed at limiting emissions to 5mg/km as of 1 January 2007 by means of an obligatory particle filter. The Commission rejected the Dutch request, made under ex Article 95(5) EC Treaty on the approximation of laws in the field of the internal market, by means of Commission Decision 2006/372/EC of 3 May 2006. With action brought on 12 July 2006, the Kingdom of the Netherlands sought the annulment of Commission Decision 2006/372/EC claiming infringement by the Commission of 'the appraisal criteria of Article 95(5) EC, the duty of care and the duty to state reasons under Article 253 EC' and the breach by the Commission of its appraisal of alternative measures pursuant to Article 95(6) EC. Judgment of the Court of First Instance of 27 June 2007 confirmed that while accepting that the emissions of particulate matter produced by diesel vehicles is acute and leads to not meeting quality norms laid down in Directive 1999/30/EC, adoption of national measures that disrupt the functioning of the internal market is only allowed if the environmental problem is specific to the notifying state. The Dutch ambient air quality problems were not considered significantly different from those in other regions in the EU, according to the Court. The action was dismissed and the Kingdom of the Netherlands obliged to pay the costs.

CASE T-28/07 FELS-WERKE GMBH AND OTHERS V. COMMISSION

On 7 February 2007 Fels-Werke GmbH and Others brought an action against the European Commission seeking the annulment of Articles 1(2) and 2(2) of the Commission's decision of 29 November 2006 concerning the national allocation plan for Germany, namely the incompatibility of the German NAP with Annex III to Directive 2003/87/EC. The group of German companies asked the Court to annul the Commission's decision that provided EUAs for free for 12 years for new installations, therefore breaching EU state aid rules by giving unfair advantage to those installations. The applicants claim to be directly and individually concerned by the contested decision. In its order of 11 September 2007, the CFI declared the action inadmissible since the decision of the Commission was addressed to Germany (state) and therefore applicants were not considered individually concerned.

CASE C-503/07 SAINT-GOBAIN V. EUROPEAN COMMISSION

On 19 November 2007 Saint-Gobain Glass Deutschland GmbH appealed to the ECJ against the order delivered by the Court of First Instance (Third Chamber) on 11 September 2007 on the above mentioned case on the basis of the right to be heard by a court and right to a fair hearing. The order of the Court of 8 April 2008 dismissed the appeal of Saint-Gobain.

CASE T-374/04 *GERMANY V. COMMISSION*

On 7 November 2007 the Court of First Instance (CFI) annulled part of the Commission Decision C(2004) 2515/2 final of 7 July 2004 concerning Directive 2003/87/EC on the German NAP for the period 2005–2007, which prohibited ex-post adjustments in the distribution of EU Allowances. In the action brought on 20 September 2004 against the Commission, Germany sustained that it maintained the right to reduce, in certain specific cases, the amount of allowances distributed to the single installations and to transfer EUAs from closed to new installations. Commission Decision of 7 July 2004 refused the ex-post adjustments to a NAP on the basis of the infringement of criteria 5 and 10 laid down in Annex III to the directive. The CFI declared the admissibility of downward ex-post adjustments in the German NAP. According to the Court, it would be contrary to the nature of directives to limit the room for manoeuvre for Member States to comply with the aim to be achieved, as Directive 2003/87/EC itself did not contain an explicit prohibition for downward ex-post adjustments. The decision of the CFI challenged one of the main pillars of the reasoning of the Commission in the judgment of the various NAPs of the EU Member States, namely the refusal to authorize ex-post adjustments on the level of EUAs distribtuted by the national authorities.

CASE T-183/07 *POLAND V. COMMISSION*

The judgment of the Court of First Instance of 9 November 2007 refused the action brought by Poland on 28 May 2007 to suspend Commission Decision C(2007) 1295 of 26 March 2007 on the Polish NAP for the period 2008–2012. The action initiated by Poland and supported by Hungary, Lithuania and the Slovak Republic challenged the Commission decision to lower the CO_2 emission limit by 27 per cent. The Commission was supported by the UK. The CFI annulled Commission Decision C(2007) 12395 in part on the basis of the lack of competence of the Commission provided by Directive 2003/87/EC in the establishment of a maximum cap, above which the NAP is considered incompatible with the directive.

On 1 November 2007 the Norwegian employers' organization (NHO) filed a formal complaint to the EFTA Association on the draft NAP for the period 2008–2012. The reason behind this action was the lack of credit reserve for new entrants.

On 21 and 28 December 2007 Romania and Bulgaria respectively initiated a legal proceeding against the European Commission before the ECJ, challenging the decisions of the Commission to reduce the level of EUAs included in their NAPs for 2008–2012. For Bulgaria, the Commission did not take into consideration a fundamental switch from nuclear power to thermal power plants due to the shutting down of units 3 and 4 of the Kozloduy nuclear power plant.

After the decision of the European Commission of 7 December 2007 concerning the amendment to the NAP of Slovakia for 2008–2012, the Slovakian government decided on 17 January 2008 to withdraw the legal action filed against the European Commission challenging the requested cut of its EUAs.

On 7 February 2008 a German Court ruled that the per-tonne administrative fee for allowances charged to national installations in the first period of the EU ETS by the German Emission Trading Authority (DEHSt) was illegitimate. DEHSt charged the German installations €0.25 per allowance and this decision was challenged by more than 100 companies.

CASE C-127/07 SOCIÉTÉ ARCELOR ATLANTIQUE ET LORRAINE AND OTHERS V. PRIME MINISTER AND OTHERS (FRANCE)

On 5 March 2007 the Conseil d'État (France's highest administrative court) lodged a reference for a preliminary ruling to the ECJ in reference to validity of Directive 2003/87/EC in the light of the principle of equal treatment, in so far as the EU ETS imposes obligations on installations in the steel sector without including in its scope the aluminium and plastic industries. The reference referred to the proceedings between Société Arcelor Atlantique et Lorraine and Others and the French prime minister, the minister for ecology and sustainable development and the minister for the economy, finance and industry concerning the implementation of Directive 2003/87 in the French legal order. The French court asked for the annulment of the 2004 decree transposing the EU emission trading directive into French law, in particular in reference to the fact that the EU ETS was applicable to installations in the steel sector without including in its scope the aluminium and plastic industries. On 16 April 2008 the ECJ decided to suspend the legal challenge launched by Arcelor and on 21 May 2008 the conclusions of Advocate General Maduro considered the nature of the relation between the national constitutions and Community law and clarified the absence of a potential conflict between them. Maduro concluded that the assessment of the preliminary ruling had clarified that Directive 2003/87/EC was valid and there was not any element to justify its annulment on the basis of the principle of equal treatment. He concluded for the inadmissibility of the annulment challenge thus rejecting Arcelor's argument on the unfair advantage that Directive 2003/87/EC would have conferred to sectors not falling within the scheme, such as the aluminium and plastic industries. On 16 December 2008 the judgment of the Court supported Maduro's advisory opinion and rejected the mentioned reference for a preliminary ruling on the basis that Directive 2003/87/EC did not breach the principle of equal treatment under EC law. According to the Court the gradual expansion of the EU ETS was lawful.

CASE T-16/04 ARCELOR V. PARLIAMENT AND COUNCIL

In the action brought on 15 January 2004 by Arcelor S.A. before the CFI against the Parliament and the Council, the French company Arcelor challenged the EU Emissions Trading Directive and asked for the annulment of several articles of that instrument on the basis of the breach of four fundamental principles guaranteed by the EU Treaty: equality, proportionality, legal certainty and the freedom of establishment. Furthermore, Arcelor's argument relied on the little potential for steel producers to reduce further their CO_2 emissions. The judgment of the Court of 2 March 2010 dismissed the action.

CASE T-241/07 *BUZZI UNICEM V. COMMISSION*

On 27 October 2008 the CFI ruled against the plea of the Italian cement producer Buzzi Unicem, which had challenged the Commission Decision of 15 May 2007 on the Italian NAP for 2008–2012. According to the Court, the Commission had the right to object to the Italian NAP in the event that the clause allowing firms to shut down an installation due to the rationalization of production to keep their allowances would not be removed, as indicated in its decision. Additionally, the Court also made clear that the Italian producer was not 'directly and individually affected' by the decision of the Commission since other firms could be in a similar situation.

On 23 September 2009 the CFI condemned the European Commission in cases T-183/07 and T-263/07 against Poland and Estonia respectively for the failure to exercise its competences correctly under the limit of EC law in the implementation of Directive 2003/87/EC. According to the Court, the Commission exceeded its powers by revising downwards the GHG emission caps imposed by Poland and Estonia on their national industries under the NAPs for the period 2008–2012. Despite the fact that Article 9 of Directive 2003/87/EC provided the Commission with the power to review the NAPs and to verify the conformity of those with the criteria set out in the directive and to reject the plans on the basis of incompatibility with the specific criteria and provisions, the Court condemned the Commission for the replacement of data in the NAP with data based on its own assessment method. In case T-183/07 *Poland* v. *Commission*, the Court contested the Commission's failure to provide sufficient reasons in the decision rejecting the NAP and the extent to which the choice of the methodology for the economic analysis and the data used by Poland were contrary to Community law. In case T-263/07 *Estonia* v. *Commission*, the Court contested Estonia's infringement of the principle of sound administration by not examining in the proper manner the Estonian NAP.

Directive 2009/29/EC (EU ETS review) established that in the period 2013–2020 allowances will be allocated directly to industrial sectors rather than by the Member States in the NAPs. European installations will have the right to challenge the decision of the Commission on the allocation of EUAs before the EU courts.

NOTE

1. Action brought on 7 February 2007 – *US Steel Košice* v. *Commission* (Case T-27/07) (2007/C 69/54), *OJ* C 69 of 24 March 2007, pp25–26.

CHAPTER 19

Compliance

T he non-compliance system of the Kyoto Protocol is based on Article 18 of the Kyoto Protocol and is regulated by the Marrakech Accords adopted by the Parties at COP7 in Marrakech, namely the procedures and mechanisms relating to compliance contained in the annex to Decision 27/CMP.1.

The non-compliance system of the Kyoto Protocol is composed of the Compliance Committee and two branches: the facilitative branch and the enforcement branch. While the first is aimed at assisting parties to ensure compliance with the Kyoto Protocol obligations, the second is a quasi-judicial body designed to determine and apply consequences for cases of non-compliance.

At the European Community law level, the compliance of the Member States with the obligations created by the Kyoto Protocol is ensured through the adoption of EC secondary legislation and by the consequent infringement procedure established under the EC Treaty.

At the time of writing, the enforcement branch of the Compliance Committee had addressed the following four on questions of implementation: Greece, Canada, Croatia and Bulgaria.

On 17 April 2008 the enforcement branch of the Compliance Committee of the Kyoto Protocol adopted its final decision declaring the ineligibility of Greece to participate in the Protocol's flexible mechanisms (ET, JI and CDM). Greece was not in compliance with the eligibility criteria, namely it failed to submit adequate information on how GHG emissions have been calculated since 1990. In particular, Greece was found to be not in compliance with the annual inventory submission and specifically with the guidelines for national systems under Article 5(1) of the Kyoto Protocol, in particular paragraph 10 of the annex to Decision 19/CMP.1, and the guidelines for the preparation of the information required under Article 7 of the Kyoto Protocol contained in the annex to Decision

15/CMP.1. This decision of the enforcement branch does not affect the Greek installations falling under the EU ETS, but instead only refers to Greece's eligibility as a state.

On 16 August 2010 the enforcement branch adopted its final decision on the case of Bulgaria. Bulgaria was found to be in non-compliance with national system requirements for countries with 2012 targets (Annex B parties). On the basis of this decision, Bulgaria was declared to be in non-compliance and thus required to submit a plan to address its non-compliance within three months. It was declared to be not eligible to participate in the flexible mechanisms.

Bibliography

EEA (European Environment Agency) (2009a) 'Annual EC GHG inventory 1990–2007 and inventory report 2009', Technical Report No 4/2009, EEA, 29 May.

EEA (2009b) 'GHG emission trends and projections in Europe 2009: Tracking progress towards Kyoto targets', EEA Report No 9/2009, EEA, 12 November.

EEA (2010) 'Annual EU GHG inventory 1990–2008 and inventory report 2010', Technical Report No 6/2010, EEA, 2 June.

Epiney A. (2003) 'Division of competence between Member States and the EC', in J. H. Jans (ed) *The European Convention and the Future of European Environmental Law*, Europa Law Publishing, Groningen, pp43–54.

Eritja, M. C. (2006) 'Reviewing the challenging task faced by Member States in implementing the Emissions Trading Directive: Issues of Member States liability', in M. Peeters and K. Deketelaere (eds) *EU Climate Change Policy*, Edward Elgar, Cheltenham.

European Commission (2009) 'Report from the Commission to the European Parliament and the Council on the progress towards achieving the Kyoto objectives, required under Article 5 of Decision 280/2004/EC of the European Parliament and of the Council concerning a mechanism for monitoring Community GHG emissions and for implementing the Kyoto Protocol', COM(2009)630, Brussels, 12 November.

Eurostat (2010) 'Taxation trends in the European Union: Data for the EU Member States, Iceland and Norway', European Commission, Taxation and Customs Union Programme and the World Trade Organization, Publications Office of the European Union, Luxembourg.

Haigh, N. (1996), 'EC climate Change and Politics', in T. O'Riordan and J. Jäger (eds) *Politics of Climate Change: A European Perspective*, Routledge, London.

Heliskoski, J. (2001) *Mixed Agreements as a Technique for Organizing the International Relations of the European Community and its Member States*, Kluwer Law International, The Hague.

Macleod, I., Hendry, I. D. and Hyett, S. (1996) *The External Relations of the European Communities*, Clarendon Press, Oxford.

Massai, L. (2010a) 'The long way to the Copenhagen Accord: Climate change negotiations in 2009', *Review of European Community and International Environmental Law*, vol 19, no 1, pp102–119.

Massai, L. (2010b) 'The revision of the EU Emissions Trading System', in M. Roggenkamp and U. Hammer (eds) *European Energy Law Report VII*, Intersentia Publishing, Antwerp, pp3–25.

O'Keeffe, D. and Schermers, H. G. (eds) (1983) *Mixed Agreements*, Kluwer, Deventer.

Roggenkamp, M., Redgwell, C., Rønne, A. and Guayo, I. (2007) *Energy Law in Europe*, Oxford University Press, Oxford.

Rosas, A. (2000) 'The European Union and mixed agreements', in A. Dashwood and C. Hillion (eds) *The General Law of EC External Relations*, Sweet and Maxwell, London, pp200–220.

Schermers, H. G. (1983), 'A typology of mixed agreements', in D. O'Keeffe and H. G. Schermers (eds) *Mixed Agreements*, Kluwer, Deventer, pp23–33.

Torvanger, A. and Ringius, L. (2000) 'Burden differentiation: Criteria for evaluation and development of burden sharing rules', CICERO (Center for International Climate and Environmental Research), Oslo, Norway.

UNEP and UNFCCC (2002a) *Understanding Climate Change: A Beginners' Guide to the UN Framework Convention and its Kyoto Protocol*, Bonn Climate Change Secretariat.

UNEP and UNFCCC (2002b) *Climate Change Information Kit*, Bonn Climate Change Secretariat.

UNFCCC (United Nations Framework Convention on Climate Change) (2000) 'Tracing the origins of the Kyoto Protocol: An article-by-article textual history', Technical paper FCCC/TP/2000/2 prepared under contract to UNFCCC by Joanna Depledge.

UNFCCC (2002) '"Good practices" in policies and measures among Parties included in Annex I to the Convention. Policies and measures of Parties included in Annex I to the Convention reported in third national communications', Report by the Secretariat, FCCC/SBSTA/2002/INF.13, 9 October.

UNFCCC (2006) 'Highlights from greenhouse gas (GHG) emissions data for 1990–2004 for Annex I Parties, UNFCCC, http://unfccc.int/files/essential_background/background_publications_htmlpdf/application/pdf/ghg_booklet_06.pdf (accessed 2 May 2011).

Van Schaik, L. and Egenhofer, C. (2003) 'Reform of the EU Institutions: Implications for the EU's performance in climate negotiations', Policy Brief Series No. 40, Centre for European Policy Studies (CEPS).

WTO/UNEP (World Trade Organization/United Nations Environment Programme) (2009) 'Trade and climate change: A report by the United Nations Environment Programme and the World Trade Organization', WTO Publications, Geneva.

Index

Entertainment Rigging
for the 21st Century

Entertainment Rigging for the 21st Century

Compilation of Work on Rigging Practices, Safety, and Related Topics

Edited by Bill Sapsis

Focal Press
Taylor & Francis Group

NEW YORK AND LONDON

First published 2015
by Focal Press
70 Blanchard Road, Suite 402, Burlington, MA 01803

and by Focal Press
2 Park Square, Milton Park, Abingdon, Oxon OX14 4RN

Focal Press is an imprint of the Taylor & Francis Group, an informa business

Library of Congress Cataloging in Publication Data
   21st century : compilation of rigging practices, safety, automation, and
   related issues / edited by Bill Sapsis.
     pages cm
   Includes index.
   ISBN 978-0-415-70274-4 (pbk) – ISBN 978-0-203-79541-5 (ebk)
   1. Hoisting machinery–Rigging. 2. Theaters–Stage-setting and scenery–
   Safety measures. I. Sapsis, Bill, editor.
   TA660.C3E58 2015
   790.028'4–dc23   2014018242

ISBN: 978-0-415-70274-4 (pbk)
ISBN: 978-0-203-79541-5 (ebk)

Typeset in Minion
By Keystroke, Station Road, Codsall, Wolverhampton

Contents

List of Contributors

ROCKY PAULSON
Rocky Paulson started rigging in 1969 at the Cow Palace in San Francisco. In 1973 he became a member of IATSE. That same year he went to work for NBC as a rigger for the *Disney on Parade* shows, as well as working as a production rigger on NBC's production of *Peter Pan*. After traveling on several continents with these shows, Rocky did many rock tours. In 1977 he formed the first entertainment rigging production company. Since that time Stage Rigging has maintained its position as one of the most respected rigging companies in the US. At the end of 2005, Rocky retired from Stage Rigging, but has continued teaching activities.

BILL GORLIN
Bill Gorlin serves as McLaren Engineering Group's vice-president of the Entertainment Division. A graduate of Cornell University in engineering, he is registered as a professional engineer in ten states and is a board-certified structural engineer. His more than 27 years of experience include engineering of scenic, entertainment, and amusement structures, staging, rigging, buildings, show action equipment, architectural theming, sculptures, and other frameworks, nationwide and worldwide.

Bill is a member of the PLASA NA Rigging Working Group, the Performer Flying and Temporary Structures Task Groups; American Society of Civil Engineers; Cornell Society of Engineers; and Structural Engineers Association of New York. He has published articles in *Architecture Week* and *Structural Engineering Forum*, and is a frequent lecturer at various universities and industry conventions.

TRAY ALLEN
Tray Allen became involved with theatre while attending David Lipscomb University in Nashville, Tennessee. His first lighting truss hang was at this university and involved 20' of truss and two crank-up lifts. After graduating with a BSc in Engineering Science he worked for Bradfield Stage Lighting in Nashville, then as a master electrician for Opryland USA. After getting married, Tray moved to Knoxville, Tennessee, and, in 1992, went to work for James Thomas Engineering.

ROY BICKEL

Roy Bickel's career started more than 50 years ago as a circus performer whose skills included trapeze, trampoline, and human cannonball. He started rigging with the circus and then moved on to become the first rigger for *Disney on Parade*. He was also the first large arena tour rigger in the USA and was the first to introduce fixed length wire rope slings and sling color-coding. Roy also worked on a number of Broadway show including *Chicago*, *The Wiz*, and *Truckloads*. Roy introduced to the industry many of the practices still in use by riggers today. After 50 years in the industry, Roy is still rigging for conventions, corporate events, and tours.

KEITH BOHN

Keith Bohn has been in the entertainment industry for more than 25 years. During this time he has been involved in the use, manufacturing, and design of structural rigging solutions ranging from simple truss to complex, permanently installed structures. Keith has also served the industry through PLASA as a principle voting member for the Rigging Working Group since 1998, and has chaired the task group assigned to create and revise ANSI E1.21-2013, Temporary Structures Used for Technical Production of Outdoor Entertainment Events. A founding contributor of the Event Safety Alliance, he has contributed on a number of topics contained in *The Event Safety Guide*. Additionally, he has taught classes on the safe use of truss and outdoor structures worldwide.

KAREN BUTLER

Karen Butler lives in Phoenix, Arizona, and has been working in the entertainment industry since 1983. She has had the opportunity to work in all facets and departments for music, dance, movies, commercials, industrials, trade shows, and—her first love—theatre. In 1986, Karen became a proud member of IATSE Local 336. In 1990 she created Suddenly Scenic and has worked as a scenic artist for 25 years. She is the master scenic for Phoenix Theater and Childsplay Inc.

In 1996, Karen became part of the team to reopen the Orpheum Theater in Phoenix as the house steward and head flyman. It was during this period that Karen became part of the ETCP Theatrical Rigging committee; first as an SME, then as chair.

STU COX

Stu Cox has traveled the planet as a ZFX flying director, rigging and choreographing performer flying effects, along with aerialists, oversized inflatable scenery, 500' motorized zip lines, and no shortage of flying monkeys, ghosts, angels, and Peter Pans. He received a BFA in Theatre Design and Production from the University of Louisville. He has worked with *Wicked*, the Vancouver Winter Olympics, Green Day's *American Idiot*, Fox Sports, and FIFA. Stu is an

ETCP theatre and arena rigger, and is an ETCP recognized trainer. He is a member of USITT, CITT, and PLASA. In his downtime, Stu snowboards and spends time with his family in Ontario, Canada.

JOE MCGEOUGH

Joe McGeough is director of operations for Foy Inventerprises, Inc., based in Las Vegas. Over the years, Joe has worked on the development of new flying systems for productions worldwide, and has collaborated on hundreds of productions for Foy, including shows on Broadway (*American Idiot, The Lion King, Mary Poppins, Tarzan the Musical*), concert tours (The Backstreet Boys' *Into the Millennium World Tour*), television (*The Drew Carey Show, The Grammy Awards, The American Music Awards*), international productions (*Wicked* in Tokyo), seasonal shows (*The Flying Angels* at the Crystal Cathedral and Phoenix First Assembly of God, *The Radio City Music Hall Christmas Spectacular, The Shoji Tabuchi Show* in Branson, MO), theme parks (*Finding Nemo* at Disney's Animal Kingdom), industrials (*The Microsoft Global Summit*), special events (*Olympic Torch Relay in Times Square, Super Bowl XLV Halftime Show*) and more than a dozen Royal Caribbean cruise ships.

SCOTT FISHER

Scott Fisher is the founder of Fisher Technical Services and a pioneer in the development of theatrical and rigging automation systems. A 25-year veteran of the entertainment industry, Scott and the Fisher Technical team have provided cutting-edge automation systems to hundreds of theaters, theme parks, motion pictures, and attractions, and the systems and techniques developed at Fisher Technical continue to be used throughout the entertainment world today.

DAN CULHANE

Dan Culhane is currently the technical business development manager at SECOA Inc. Prior to this he spent more than 11 years as SECOA's engineering manager. He is an ETCP-certified rigger for theater and a subject matter expert. Dan has spent 15 years as a technical director working for theatres across the country including the Guthrie Theater and the Children's Theatre Company, in Minneapolis, MN.

He serves on the PLASA Technical Standards Program, Rigging Working Group and chairs the task group revising the standard for fire safety curtains. He is a member of the Stage Lift Working Group. Dan also is on the board of directors for USITT and serves as treasurer. He is a member of the UL Standards Technical Panel for Fire Doors (STP 10), and serves as an alternate committee member to the NFPA Technical Committee on Fire Doors and Windows.

EDDIE RAYMOND

Eddie Raymond is a lifelong San Francisco Bay Area resident, graduating from and attending postgraduate work in education at UC Berkeley. He has been a stagehand with Local 16 of the IATSE since 1975. After graduating the Apprentice Program at Local 16, Eddie became involved as a member of the examining board in 1981 and served as the chair of that committee from 1984 until February of 2014. Since 1981 he has been a leader in the progressive improvement of stagehand training in Local 16 as well as in the International Alliance. Currently he serves the IATSE as a member of their Career Advancement Program, providing training and advising the IA's Exhibition and Entertainment Joint Training Trust.

Eddie was a co-chair of PLASA's ETCP rigging certification program and sits as a member of the ETCP Council. He is the second term chair of PLASA's North American Regional Board and sits on PLASA's governing body.

CHRIS HIGGS

Chris Higgs provided rigging for theatre, corporates, television, and concert touring from the early 1970s to the mid-1990s and is one of the founders of entertainment rigging training in the UK. Total Training started in 1998, as part of the Total Solutions Group, and delivers training courses in the UK and overseas in rigging, work at height, rescue, and inspection, amongst other associated subjects. The Total Training three-day rigging course is unique in the world, being held at least three times monthly throughout the year.

BILL SAPSIS

Bill Sapsis, president of Sapsis Rigging, Inc., has been involved in the entertainment industry since 1972. His work on Broadway includes the original productions of *A Chorus Line* and *The Runner Stumbles*. In 1981, Bill began Sapsis Rigging and has grown the company into a multifaceted installation/production/service company.

Bill has written and lectured on safety related issues on an international basis. Bill is a member of the ETCP council and chair of the Rigging Subject Matter Experts. He serves on PLASA's Technical Standards Committee and is the chair of the Rigging Working Group. Bill is a member of the ESTA Foundation's board of directors.

Bill is a USITT fellow. He is a founding member of the Long Reach Long Riders, an industry-based charity motorcycle group, and he was the 2010 recipient of the Eva Swan Award, PLASA NA's highest honor.

CARLA D. RICHTERS

Carla Richters is a 30-year member of IATSE TWU local 805. She was a road wardrobe supervisor and received her MFA from the University of Texas at Austin in Theatrical Production Design. She was the costume shop manager and

Theatre Department safety officer at Dartmouth College for 24 years. She was an EMT at Upper Valley Ambulance in Fairlee, Vermont from 1995 to 2010, and a paramedic from 2001 to 2010. An ambulance squad training officer, she was the medical director of the Vermont Special Olympics Winter Games. She was twice awarded the Jack Seusse Memorial Stump the Rigger trophy. Her interest in rigging safety happened because, as she puts it, "they hang heavy stuff over our heads. I wanted to know what they were doing and how to help when things went wrong."

FOREWORD BY MONONA ROSSOL
Monona Rossol was born into a theatrical family and began working as a professional entertainer at three years of age. She holds a BS in Chemistry with a Math minor, and MS and MFA degrees in Art. Currently, she is an industrial hygienist, chemist, and the president and founder of Arts, Crafts & Theater Safety, Inc., a nonprofit organization providing health and safety services to the arts. She also is the safety officer for IATSE Local USA829 and for the New York Production Locals.

Foreword

As both a former performer and current safety professional, I can heartily endorse the publication of this book. There have been several good basic rigging texts published. But *Entertainment Rigging for the 21st Century* addresses basic information plus what is happening right now in this field. And a lot is happening right now.

For example, the Occupational Safety and Health Administration finally recognizes that theatre and arena rigging are every bit as complex and potentially life-threatening as rigging on major construction sites. Today, OSHA applies the same Construction and General Industry rules to entertainment workplaces. And if there is some new rigging equipment or process for which there is no written OSHA regulation, OSHA can cite them anyway under their General Duty Clause.

The General Duty Clause allows OSHA to cite or fine an employer for failure to address any "recognized hazard." Well, hitting the deck at a rapid rate of speed, clearly, is a recognized hazard. So the producers of *Spider-Man* were cited under this clause (29 CFR 1910.5(a)(1)) when mistakes made in flying some of the various actors playing bits of the Spider-Man role led to accidents.

Today even the general public is aware of rigging perils when the press covers entertainment and theatrical accidents. In 2013, two rigging accidents resulted in fatalities from falls in the range of 100' above the stage. On April 5, a rigger fell after a Romeo Santos performance at the AT&T arena in San Antonio, Texas. Then on June 29, a Las Vegas Cirque du Soleil acrobat in harness fell to her death in full view of an audience.

Chapter 13 by Bill Sapsis on fall arrest systems will make clear how riggers should be protected by fall protection gear in order to prevent tragedies. Keith Bohn's Chapter 5 on outdoor structures should help readers understand what is needed to avoid accidents such as the 2011 Indiana State Fair stage collapse that killed seven and injured 58 people. Chapter 3 by Tray Allen on lighting trusses explains how these should be used and should remind us all of the many spectacular photos and videos of arena lighting truss collapses on the Internet where we can watch millions of dollars in damages occurring before our eyes.

While the general public certainly can understand the results of rigging accidents, they are not as likely to understand the technical causes. It is also likely they wouldn't even understand the language used by rigging experts to explain the issues involved. One reason is that riggers have their own language. This language is a collection of terms ranging in origin over centuries of time.

The roots of rigger-speak are found in the colorful words used by the very first riggers who were the crews on ancient sailing ships. More terminology was added to describe developments in rigging gear developed during the Industrial Revolution. And the final confusion arises from the geek-generated language applicable to complex computer-driven rigging systems.

A sampling of these words include: belay, bo'sun, cleat, clew, crew, deck, hitch, lanyard, pinrail, locking rail, purchase, trapeze, trim, trim clamp (or knuckle-buster), fly, motor-assist and dead-haul fly systems, fly loft, fly gallery, grid deck, loading bridge, arbor, standard pipe battens, truss battens, electric battens, light ladder battens, tab battens, lines, line-sets, jack lines, hemp lines, nylon lines, cables (wire ropes), steel bands, proof coil chains, rope locks, swage (compression) fittings or cable clips, trim chain, shackles or turnbuckles (which can be moused), pipe clamps, counterweight systems, electrical hoists (or winches), drum winches, tension blocks, head blocks, loftblocks, mule blocks, and programmable logic controllers (PLCs).

And the term "runaway" does *not* refer to a recalcitrant teenager.

And when this language is used in *Entertainment Rigging for the 21st Century*, each author makes sure that the definitions are clear to us all.

Today's riggers not only need to understand this strange language, they also need to provide proof to employers that they understand and rig competently. In other words, many employers today want to see evidence that the riggers they hire have had some formal training. The simple résumé of past jobs may not be enough in the 21st century.

Theatrical riggers used to learn their trade by the seat of their pants, working with older riggers who also learned by the seat of *their* pants. Today, there are formal training programs, the most important of which is the certified training offered by the Entertainment Technicians Certification Program (ETCP). These programs, originated by the Entertainment Services Technology Association and run now by PLASA, require their candidates to pass a rather tough test. In other words, today's riggers should have professional certifications in addition to experience.

Employers like certification programs because it helps protect their liability after an accident. The human toll and damages caused by major rigging accidents almost always result in lawsuits. If employers can show that they hired people with rigging credentials backed up by certifying agencies, they at least cannot be accused of having hired incompetent labor.

Entertainment Rigging for the 21st Century covers these kinds of training programs in Chapters 11 and 12, by Eddie Raymond and Chris Higgs. They cover training resources and issues in both the USA and the UK. And the fact that there are two chapters, one for North America and the other for our pals across the pond, illustrates that today's entertainment riggers are conforming to, and aware of, both local and international standards.

Standards for various types of equipment and the development of protocols for use of the equipment are now being developed under the watchful eye of PLASA. These standards will be referenced repeatedly by authors of the various chapters in *Entertainment Rigging for the 21st Century*.

In fact, PLASA itself is the result of a merger between UK and USA organizations. And their standards are accredited here and in Europe. In North America, PLASA standards usually receive American National Standards Institute (ANSI) accreditation. In Europe, PLASA works closely with the British Standards Institute in the UK and the European Committee for Electrotechnical Standardization (CENELEC). PLASA is also a major contributor to the development of international standards through the International Standards Organization.

While all this sounds technical, the writers have made it easy to understand by providing definitions of their terms, diagrams and drawings, and formulas. Even non-riggers like myself, can follow these chapters and learn. And to keep our interest, the book is peppered with personal observations and stories of actual incidents experienced on the job by those authors who are also riggers.

Entertainment Rigging for the 21st Century is not just for riggers and rigger wannabes. I personally think it also belongs in the library of theatrical safety professionals, regulatory personnel, theatre administrators, writers who address theatre and entertainment subjects, or anyone who wants to know what's really going on in theatres and arenas to make all those wonderful things happen on stage and above.

Monona Rossol

1

Forces and Formulas

ROCKY PAULSON

Introduction

Most riggers choose their profession because they enjoy working at height, the challenges and responsibilities of the job, and having a job that keeps them in shape. When we head down the road to becoming a rigger, most of us never dream that we would need to hone our math skills along with our climbing and knot-tying skills. However, the most important part of a rigger's job is to make sure that the rigging system, the attachment points, and the support structure as a whole are able to support the forces imposed by the rigging load. In order to know whether or not the rigging system and attachment points (anchorages) can support the loads, the rigger must know the strengths of the support elements and how the rigging load distributes the forces produced by it to these elements. This chapter will give the reader the techniques to do much of the force-estimating required to prevent overloading in the rigging system or anchorages.

Much of the estimating we do requires the use of math formulas. To keep the formulas concise, we abbreviate the forces and distances by creating symbols. The symbol system used in this chapter conforms to a system developed over a few years by several teachers and was first published in 2009. Since all of our formulas use forces and distances, the system uses F to indicate a force and D to indicate a distance. S will be used to indicate span, the distance between support points or anchorages. When there are multiple forces or distances within a formula, which happens in most cases, subscripts are added to the symbol. The subscript will consist of a number or a letter, and in some cases a number followed by a letter. The letters are used to indicate the direction of the force or distance as well as being a designator for the force being analyzed. Table 1.1 below shows many of the symbols and their definitions.

By reviewing the symbols in the table, the reader will become familiar with the meaning of the symbols including the use of the subscripts.

In addition to creating symbols for use in the formulas, it is useful to have symbols to denote the points of analysis in the rigging system. The end points of

TABLE 1.1 Common symbols used in formulas

F_A	The force applied to the rigging system
F_1	The force at the near support or anchorage
F_2	The force at the far support or anchorage
D_1	The distance from applied force to the near support
D_2	The distance from applied force to the far support
S	The distance between supports
F_H	The horizontal force
F_L	The force in line with an angled leg
F_V	The vertical force
D_H	The horizontal distance from the anchorage to the applied force
D_L	The length of the angled leg
D_V	The vertical distance from the anchorage to the applied force
F_{1H}	The horizontal force associated with leg 1 of a bridle
F_{1L}	The force in leg 1 of a bridle
F_{1V}	The vertical force associated with leg 1 of a bridle
D_{1H}	The horizontal distance from the leg 1 anchorage to the bridle point
D_{1L}	The leg length of leg 1 of a bridle
D_{1V}	The vertical distance from the leg 1 anchorage to the bridle point
F_{2H}	The horizontal force associated with leg 2 of a bridle
F_{2L}	The force in leg 2 of a bridle
F_{2V}	The vertical force associated with leg 2 of a bridle
D_{2H}	The horizontal distance from the leg 2 anchorage to the bridle point
D_{2L}	The leg length of leg 2 of a bridle
D_{2V}	The vertical distance from the leg 2 anchorage to the bridle point

TABLE 1.2 Analysis point symbols

A	An anchorage or support point
A_1	An anchorage for a truss or bridle leg
A_2	An anchorage for a truss or bridle leg
A_3	The third anchorage for a three-leg bridle
P	The point of attachment for the applied force

the span being analyzed or, in the case of bridles, the anchor points of the bridle legs will be labeled A_1, A_2 or A_3 as required. For the point of attachment of the applied force, the symbol P will be used (see Table 1.2).

Force Distribution

Force distribution analysis is arguably the most important activity performed by a production rigger. This analysis must be done in some form to determine the force at each hang point for creating a rig plot and, in addition, many times this analysis must be performed to determine whether or not the support structures are being overloaded. Although the Internet and smartphones have made the process of preplanning much easier, there is no app for force distribution estimation. In many cases, the time required for this activity is a very significant portion of the total preplanning of the show for the rigger. In this section, we will examine the formulas and techniques for estimating the resulting forces at the supports caused by applying loads to a truss or beam.

There are several important concepts that need to be put into context prior to any calculation.

- **Force, weight and load:** *Force* is an action on a body caused by another body that tends to cause motion. It can be direct from physical contact or indirect such as the force of gravity. When a motion causing force is resisted by an equal and opposite force a static force exists. It is these static forces which are the subject of our analysis in this chapter. The US customary unit for force is the pound-force (lb). The SI (metric) unit is the Newton (N). *Weight* is simply a measure of the force of gravity on an object. The word *load*, although widely used by riggers, is more general and less succinct in its definition. In general terms it is the sum of all forces acting on an object.
- **Simply supported span:** A *simply supported span* is a beam or truss supported by two supports where the ends of the beam or truss are free to rotate. Imagine a truss supported by a chain hoist at either end. As the load increases between the hoists, the truss is free to deflect more than if the beam were fixed at the ends to prevent free rotation. See Figure 1.1 for examples of simply supported spans.
- **Cantilever:** In a *cantilever* at least one of the supports is not at the end of the beam and there is loading outside of the two supports such that both supports are on the same side as the load. See Figure 1.1 for an example.
- **Point load (PL):** A *point load* creates a force that is concentrated in a single point or very small area along the length of the beam or truss.
- **Uniformly distributed load (UDL).** As opposed to a point load, a *uniformly distributed load* creates a force that is evenly spread over a significant portion of the length of the beam or truss.
- **Center of Gravity (CG).** The *center of gravity* is the single point at which, if an object is supported, it will remain in equilibrium. The CG of a UDL, which is located at the center of the UDL, will be used to calculate the forces distributed to the supports resulting from the force applied by the UDL.

Figure 1.1

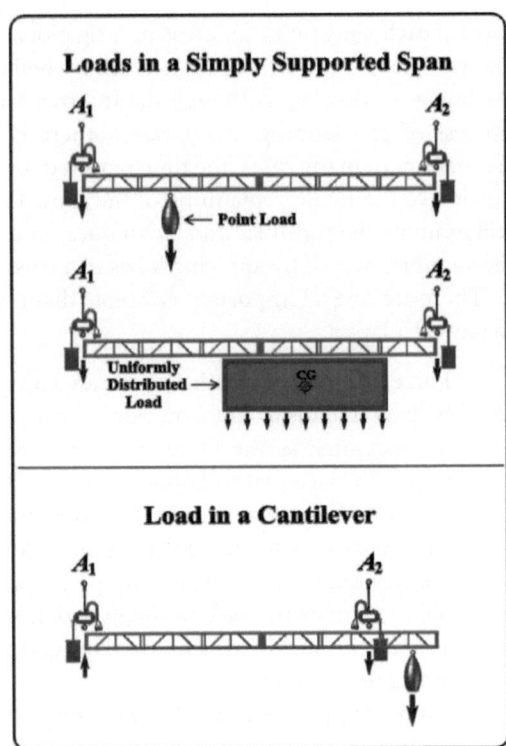

Loads in a Simply Supported Span

A_1 A_2

← Point Load

A_1 A_2

Uniformly Distributed → Load   CG

Load in a Cantilever

A_1 A_2

Sample Problems

Problem 1: Force Distribution in a Simply Supported Span resulting from a PL

On a 40′ beam supported at each end, a 1,250lb PL is placed 15′ away from the left-hand support. The top portion of Figure 1.2 illustrates the problem to solve.

To solve for forces F_1 and F_2, the following equations will be used:

$$F_1 = \frac{F_A D_2}{S} \text{ and } F_2 = \frac{F_A D_1}{S}$$

These are the basic formulas that form the building blocks for much of the rest of the equations in this chapter. These equations become a part of the rigger's bag of skills and are used as much, or more, than a bowline is used over the career of a production rigger.

Solving math problems accurately requires a combination of organizational skills and math skills. Certainly, those with less of the latter would do well to use as much of the former as possible while the math skills are developed. For this reason, the first steps to all the solutions in this chapter will begin with organizing the data in a fashion that will minimize the chance of making a mathematical error.

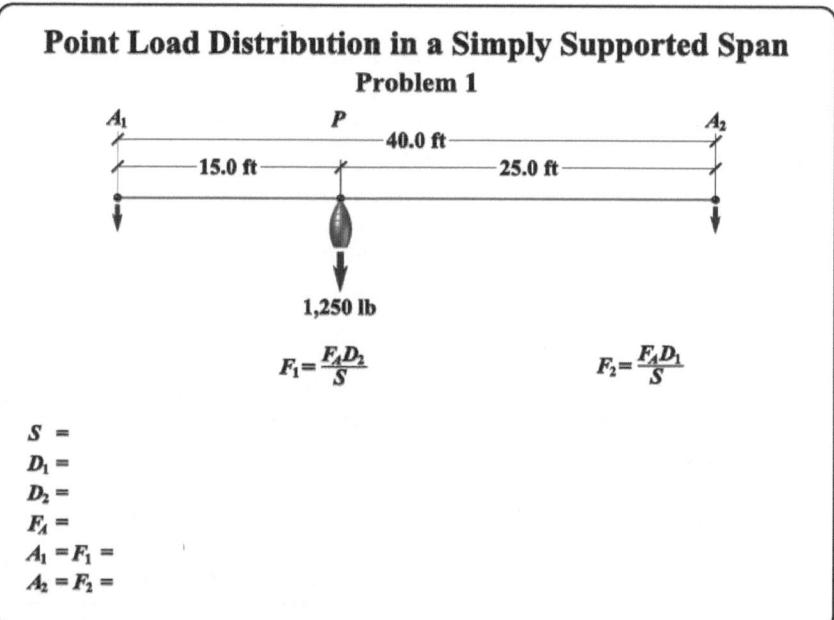

Point Load Distribution in a Simply Supported Span
Problem 1

$$F_1 = \frac{F_A D_2}{S} \qquad F_2 = \frac{F_A D_1}{S}$$

$S =$
$D_1 =$
$D_2 =$
$F_A =$
$A_1 = F_1 =$
$A_2 = F_2 =$

Figure 1.2

The lower half of Figure 1.2 is the structure which will help keep the data organized. It is well worth setting up this structure or a similar one. First, the equations for the unknowns are positioned so that there is room below to perform the required mathematics. On the left side of the figure is an informal table of the known forces and dimensions. Below, the locations A_1 and A_2 of the unknown forces, F_1 and F_2, are listed. For most of this chapter F_1 will be associated with the nearest support or anchorage, so a decision will have to be made initially as to with which supports F_1 and F_2 will be associated.

This and all the following problems will be solved in a step-by-step fashion.

Rounding

Rounding is useful in some cases, but must be done only after evaluating the benefits. For a rigger, a force of 781.25 may convey a false sense of accuracy, since most of our measurements of weight lack precision or are estimates to begin with.

- In this chapter the rounding convention used will be to look to the right of the digit being rounded. If that number is less than five, do nothing to the rounded digit and drop all digits to the right. If the number to the right of the rounding digit is five or greater, increase the rounding digit by one and drop all digits to the right. If changing the rounding digit changes it to zero, then increase the digit to the left by one.

- Be very careful when rounding numbers and then using the rounded numbers for further computations. Rounding a distance to the nearest foot and then using the rounded number to calculate a force can change the force by tens if not hundreds of pounds in some cases. In this chapter, any calculation that results in a number with more than three digits to the right of the decimal will be rounded to three digits. The rounded number will be used in any future calculations.
- If a rounded number is used in a calculation, notice will be given to the reader so that both our calculations will result in the same answer.

SOLUTION

STEP 1: ORGANIZE THE DATA

- Write the equations with space below for solving:

$$F_1 = \frac{F_A D_2}{S} \text{ and } F_2 = \frac{F_A D_1}{S}$$

- Make a table or list of the symbols for the known values used in the equations (S, D_1, D_2, F_A).
- Below these, add the anchorage labels, in this case A_1 and A_2. Since P, the location of the PL, is closer to A_1, A_1 will be the location of F_1 and A_2 will be the location of F_2.

STEP 2: INSERT VALUES AND SOLVE

- Using Figure 1.2, insert the values next to the symbols on the list. Until you are familiar with the meaning of the symbols, refer to Table 1.1 for their definitions.
- Substitute the values from your adjacent list into the formulas for F_1 and F_2 and solve for F_1 and F_2.
- Transfer the values for F_1 and F_2 to the appropriate spaces on your table of symbols, rounding as desired once all calculations are complete.

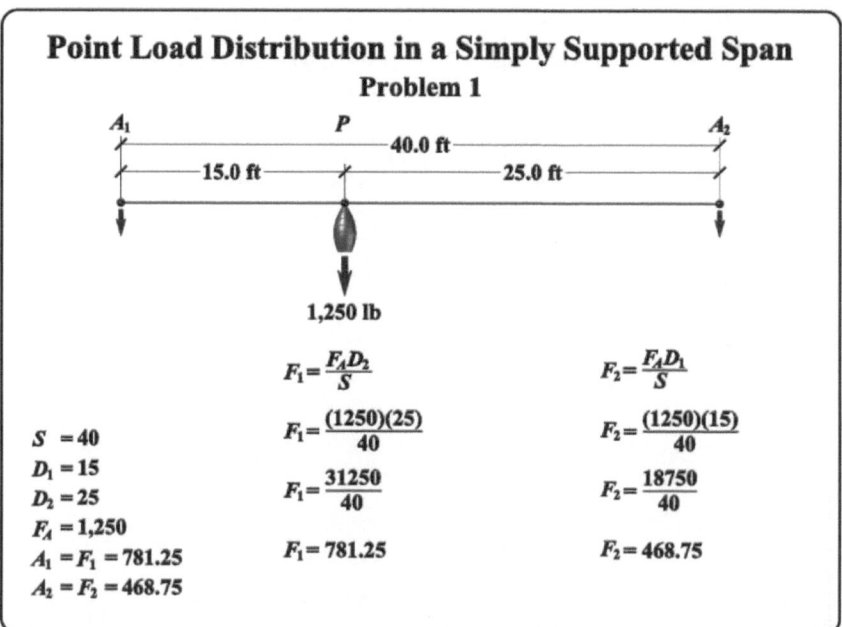

Point Load Distribution in a Simply Supported Span
Problem 1

$F_1 = \frac{F_A D_2}{S}$ $F_2 = \frac{F_A D_1}{S}$

$S = 40$

$D_1 = 15$

$D_2 = 25$

$F_A = 1,250$

$A_1 = F_1 = 781.25$

$A_2 = F_2 = 468.75$

$F_1 = \frac{(1250)(25)}{40}$ $F_2 = \frac{(1250)(15)}{40}$

$F_1 = \frac{31250}{40}$ $F_2 = \frac{18750}{40}$

$F_1 = 781.25$ $F_2 = 468.75$

Figure 1.3

Tips

- Double-check your math. $F_1 + F_2 = F_A$
- The value of F_1 should be greater than the value of F_2. The near support will always have more force on it than the far support.

Problem 2

Force Distribution in a Simply Supported Span resulting from multiple PLs, each being the same weight and equally spaced apart. The next problem will illustrate a slightly different way to use these equations. In this case there are four PLs of equal weight and evenly spaced within the span (see Figure 1.4).

SOLUTION

STEP 1: ORGANIZE THE DATA

- Write the equations with space below for solving:

$$F_1 = \frac{F_A D_2}{S} \text{ and } F_2 = \frac{F_A D_1}{S}$$

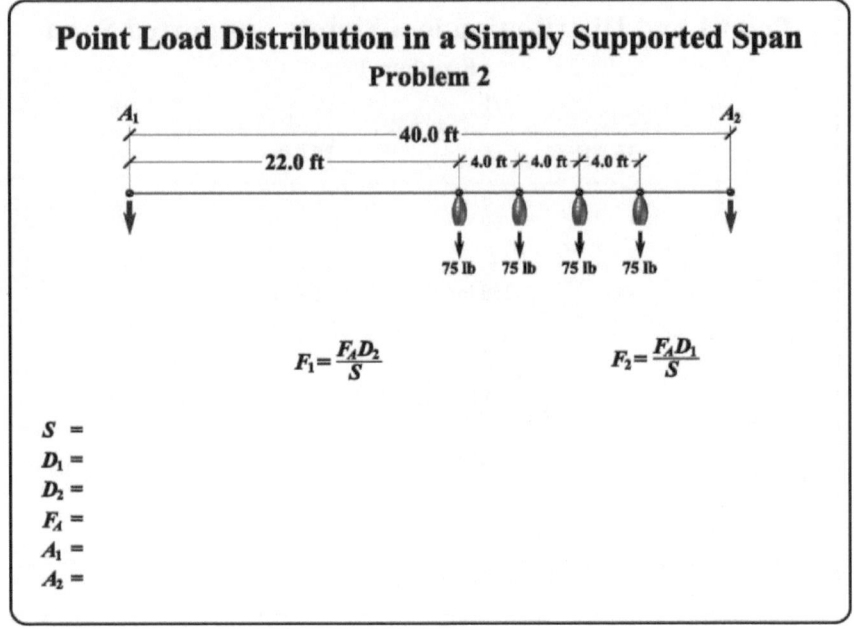

Point Load Distribution in a Simply Supported Span
Problem 2

A_1

—40.0 ft—

—22.0 ft—   4.0 ft ⁄ 4.0 ft ⁄ 4.0 ft ⁄

A_2

75 lb   75 lb   75 lb   75 lb

$$F_1 = \frac{F_A D_2}{S}$$

$$F_2 = \frac{F_A D_1}{S}$$

S =
D_1 =
D_2 =
F_A =
A_1 =
A_2 =

Figure 1.4

- Make a table or list of the symbols for the known values used in the equations (S, D_1, D_2, F_A).
- Below add the anchorage labels, in this case A_1 and A_2.

STEP 2: FIND P

- In order to solve the problem using our previous equations, we must first locate P, so we can determine the values for D_1 and D_2. When there are multiple PLs supported by the same two support points, and they are of equal weight and spacing, they can be converted into one PL with the weight being the sum of the PLs. The location of P is centered between the two outside PLs (see Figure 1.5).

STEP 3: INSERT VALUES AND SOLVE

- Using Figure 1.5, insert the values next to the symbols on the list.
- Substitute the values from your adjacent list into the formulas for F_1 and F_2 and solve for F_1 and F_2.
- Transfer the values for F_1 and F_2 to the appropriate spaces on your table of symbols, rounding as desired once all calculations are complete (see Figure 1.6).

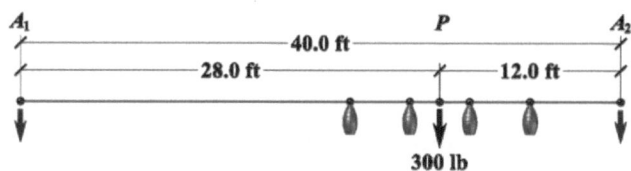

Point Load Distribution in a Simply Supported Span
Problem 2

$$F_1 = \frac{F_A D_2}{S}$$

$$F_2 = \frac{F_A D_1}{S}$$

$S =$
$D_1 =$
$D_2 =$
$F_A =$
$A_1 = F_2 =$
$A_2 = F_1 =$

Figure 1.5

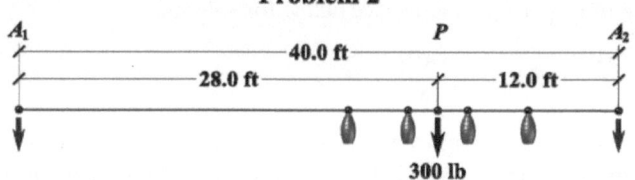

Point Load Distribution in a Simply Supported Span
Problem 2

$$F_1 = \frac{F_A D_2}{S} \qquad F_2 = \frac{F_A D_1}{S}$$

$$F_1 = \frac{(300)(28)}{40} \qquad F_2 = \frac{(300)(12)}{40}$$

$$F_1 = \frac{8400}{40} \qquad F_2 = \frac{3600}{40}$$

$$F_1 = 210 \qquad F_2 = 90$$

$S = 40$
$D_1 = 12$
$D_2 = 28$
$F_A = 300$
$A_1 = F_2 = 90$
$A_2 = F_1 = 210$

Figure 1.6

Tips

- Note that in this problem P is closer to A_2, so F_1 is located at A_2 and F_2 is located at A_1.
- A critical look at the problem would reveal that point P is located three-tenths of the way across the span. Note that the solution for the force at the far support F_1 is 90lb, or three-tenths of the applied force F_A. Had the location of P been one-quarter of the way across the span, then one-quarter of the applied force or 75lb would be at the far support. If the applied force is easily divisible by the fractional distance across the span, the need for solving the equations becomes unnecessary.

Problem 3: Force Distribution in a Cantilever resulting from a PL outside of A_2

For Problem 3, a 40ft beam is supported at the left end by anchorage A_1 and supported 8′ from the right end of the beam by anchorage A_2; 4′ to the right of anchorage A_2 is a PL of 750lb.

SOLUTION

STEP 1: ORGANIZE THE DATA

- Write the equations with space below for solving:

$$F_1 = \frac{F_A D_2}{S} \text{ and } F_2 = -\frac{F_A D_1}{S}$$

Note: these are the same equations as are used in a simply supported span except for the minus sign in the equation for F_2. The reason for the minus sign is that in a cantilever the force on the far support, F_2, is always in the opposite direction to the force on F_1.

- Make a table or list of the symbols for the known values used in the equations (S, D_1, D_2, F_A).
- Below add the anchorage labels, in this case A_1 and A_2. In this case A_1 is the far support, so F_2 will be located at A_1 and F_1 will be located at A_2 (see Figure 1.7).

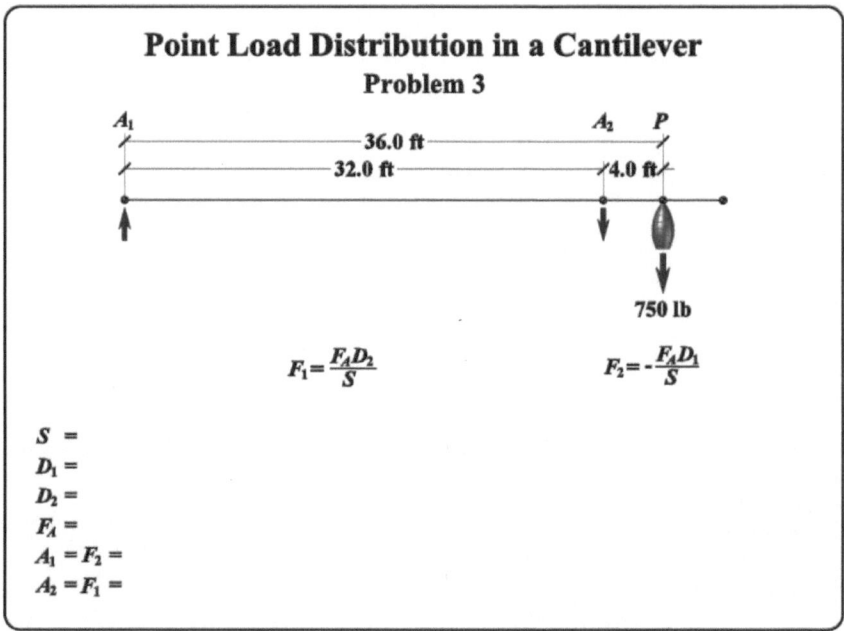

Figure 1.7

STEP 2: INSERT VALUES AND SOLVE

- Using Figure 1.7, insert the values next to the symbols on the list. Until you are familiar with the meaning of the symbols, refer to Table 1.1 for their definitions.
- Substitute the values from your adjacent list into the formulas for F_1 and F_2 and solve for F_1 and F_2.
- Transfer the values for F_1 and F_2 to the appropriate spaces on your table of symbols, rounding as desired once all calculations are complete (see Figure 1.8).

Tips

- The force at the fulcrum, F_1 will always be greater than F_A.
- As in simply supported span problems, the sum of F_1 and F_2 will always equal F_A.

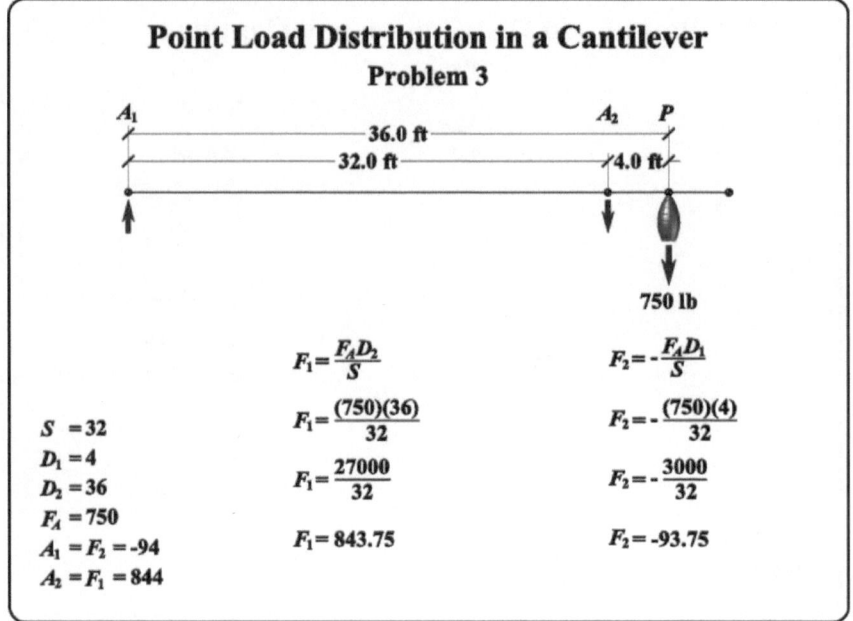

Point Load Distribution in a Cantilever
Problem 3

A_1 A_2 P

——36.0 ft——
——32.0 ft——   4.0 ft

750 lb

$$F_1 = \frac{F_A D_2}{S}$$ $$F_2 = -\frac{F_A D_1}{S}$$

$S = 32$

$D_1 = 4$

$D_2 = 36$

$F_A = 750$

$A_1 = F_2 = -94$

$A_2 = F_1 = 844$

$$F_1 = \frac{(750)(36)}{32}$$ $$F_2 = -\frac{(750)(4)}{32}$$

$$F_1 = \frac{27000}{32}$$ $$F_2 = -\frac{3000}{32}$$

$$F_1 = 843.75$$ $$F_2 = -93.75$$

Figure 1.8

Problem 4

Force Distribution in a Cantilever resulting from a PL outside of A_1. For problem 4, a 40′ beam is supported 8′ from the left end and also at the right end of the beam. The PL of 750lb is located at the left end of the beam.

SOLUTION

STEP 1: ORGANIZE THE DATA

- Write the equations with space below for solving:

$$F_1 = \frac{F_A D_2}{S} \text{ and } F_2 = - \frac{F_A D_1}{S}$$

- Make a table or list of the symbols for the known values used in the equations (S, D_1, D_2, F_A).
- Below add the anchorage labels, in this case A_1 and A_2. In this case A_1 is the near support, so F_1 will be located at A_1 and F_2 will be located at A_2. See Figure 1.9 overleaf.

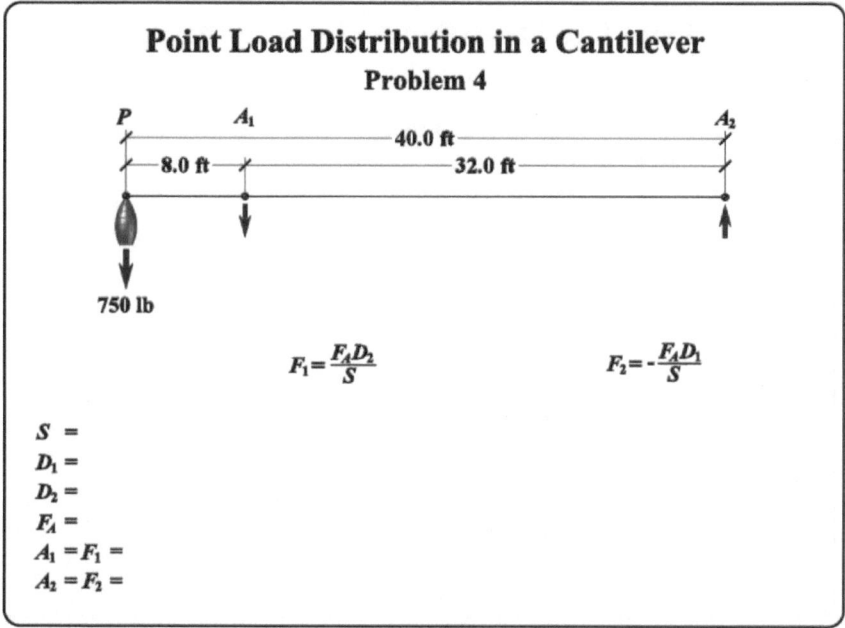

Point Load Distribution in a Cantilever
Problem 4

$$F_1 = \frac{F_A D_2}{S}$$

$$F_2 = -\frac{F_A D_1}{S}$$

$S =$
$D_1 =$
$D_2 =$
$F_A =$
$A_1 = F_1 =$
$A_2 = F_2 =$

Figure 1.9

STEP 2: INSERT VALUES AND SOLVE

- Using Figure 1.9, insert the values next to the symbols on the list.
- Substitute the values from your adjacent list into the formulas for F_1 and F_2 and solve for F_1 and F_2.
- Transfer the values for F_1 and F_2 to the appropriate spaces on your table of symbols, rounding as desired once all calculations are complete (see Figure 1.10).

Tips

- The force at the fulcrum, F_1 will always be greater than F_A.
- By comparing Figure 1.8 and Figure 1.10, you will see that S and F_A are the same in both figures, but the application point P of F_A was moved away from the fulcrum by 4′ in Problem 3 to 8′ in Problem 4. Note that this change caused nearly a 100lb increase in F_1 in Problem 4. Whenever possible, riggers should attempt to minimize any cantilevered loads by minimizing the weight or by minimizing the distance to the fulcrum D_1.

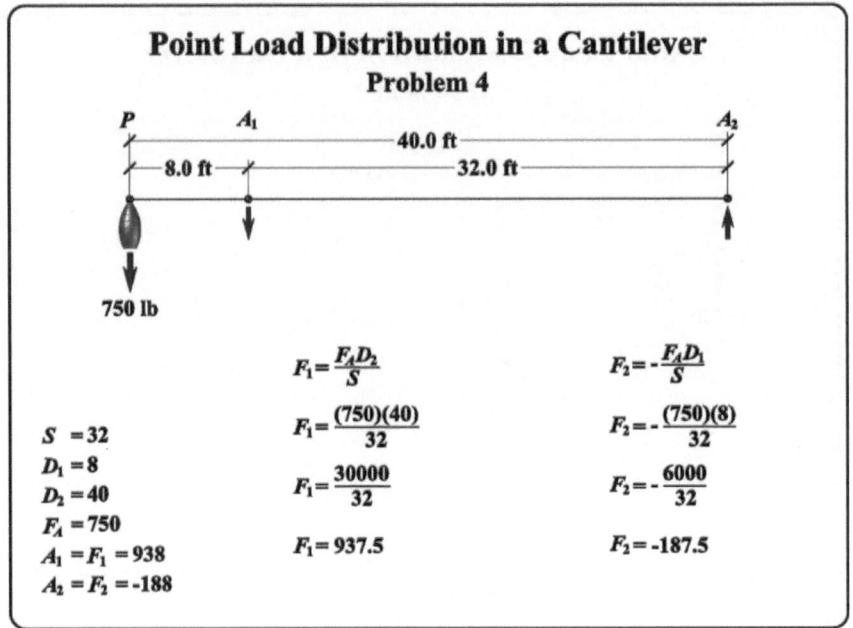

Figure 1.10

Problem 5: Force Distribution in a Simply Supported Span resulting from a UDL

Problem 5 illustrates this issue. In this problem a 16′ wide UDL weighing 1,600lb is attached to a 40′ beam that is supported by anchorages A_1 and A_2 at either end of the beam. The UDL is located such that the right end of it is 22′ from the right-hand support A_2.

SOLUTION

STEP 1: ORGANIZE THE DATA

- Write the equations with space below for solving:

$$F_1 = \frac{F_A D_2}{S} \text{ and } F_2 = \frac{F_A D_1}{S}$$

- Make a table or list of the symbols for the known values used in the equations (S, D_1, D_2, F_A).
- Below add the anchorage labels A_1 and A_2. See Figure 1.11 below.

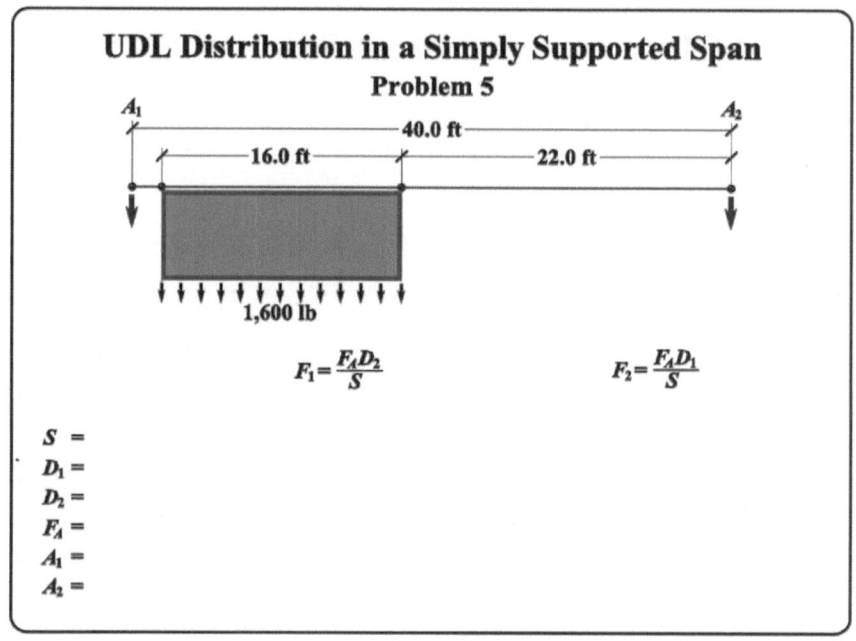

UDL Distribution in a Simply Supported Span
Problem 5

$$F_1 = \frac{F_A D_2}{S} \qquad\qquad F_2 = \frac{F_A D_1}{S}$$

$S =$
$D_1 =$
$D_2 =$
$F_A =$
$A_1 =$
$A_2 =$

Figure 1.11

STEP 2: FIND P

- In order to fill in values for D_1 and D_2 we must first locate P. P is located by finding the center of gravity of the UDL, which would be at its center. In this case it is 22′ from A_2 to the edge of the UDL plus half the width of it, or 8′. This places P at 30′ to the left of A_2 (see Figure 1.12).

STEP 3: INSERT VALUES AND SOLVE

- Using Figure 1.12, insert the values next to the symbols on the list.
- Substitute the values from your adjacent list into the formulas for F_1 and F_2 and solve for F_1 and F_2.
- Transfer the values for F_1 and F_2 to the appropriate spaces on your table of symbols, rounding as desired once all calculations are complete. In this case A_1 is the closest support to P, so F_1 would be located at it and F_2 would be located at A_2 (see Figure 1.13).

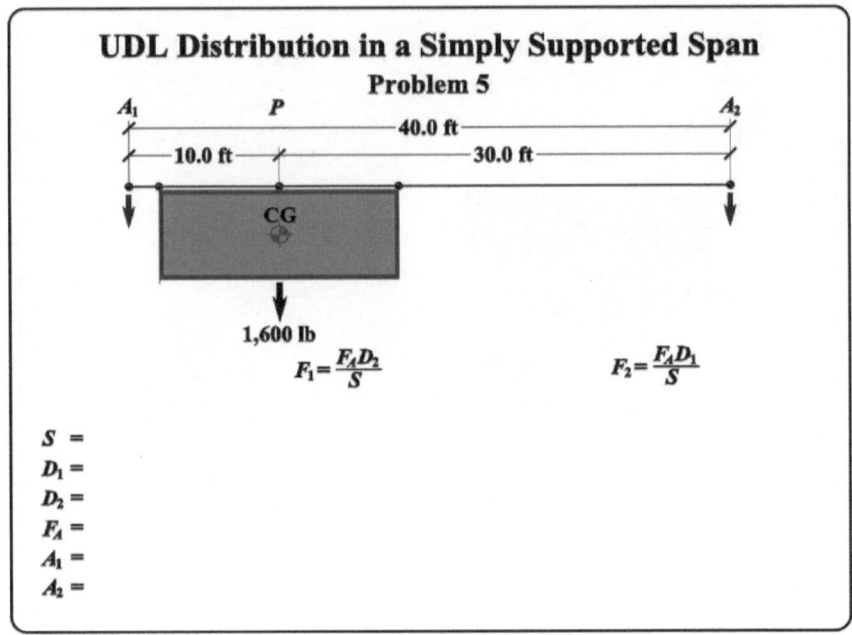

Figure 1.12

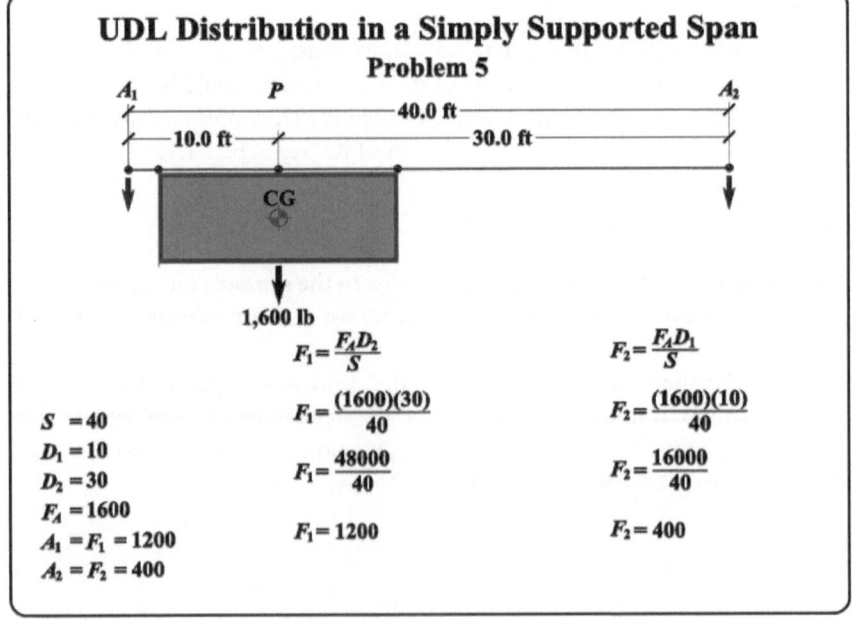

Figure 1.13

Tips

- On all force distribution problems where the applied force is a UDL, *P* must be located to find D_1 and D_2 by finding the center of the UDL.
- By looking at the values of D_1 and D_2 and comparing them to the value of *S* in Figure 1.12, *P* is located at a quarter point along the beam. This means that three-quarters of the force is distributed to the near support A_1 (1,200lb) and one-quarter is distributed to A_2 (400lb). This is the same result as obtained by using the equations.

Problem 6: Force Distribution in a Cantilever resulting from a UDL

Problem 6 consists of a 40′ beam supported by two points: A_1, which is 10′ in from the left end of the beam, and A_2 at the right end of the beam. An 8′ long UDL weighing 1,200lb is attached flush to the left end of the beam.

SOLUTION

STEP 1: ORGANIZE THE DATA

- Write the equations with space below for solving:

$$F_1 = \frac{F_A D_2}{S} \text{ and } F_2 = -\frac{F_A D_1}{S}$$

- Make a table or list of the symbols for the known values used in the equations (S, D_1, D_2, F_A).
- Below add the anchorage labels, in this case A_1 and A_2. In this case A_1 is the near support, so F_1 will be located at A_1 and F_2 will be located at A_2 (see Figure 1.14).

STEP 2: FIND P

- In order to fill in values for D_1 and D_2 we must first locate *P*. *P* is located by finding the center of gravity of the UDL, which would be in its center. In this case the distance from A_1 is 2 + 4, or 6′ (see Figure 1.15).

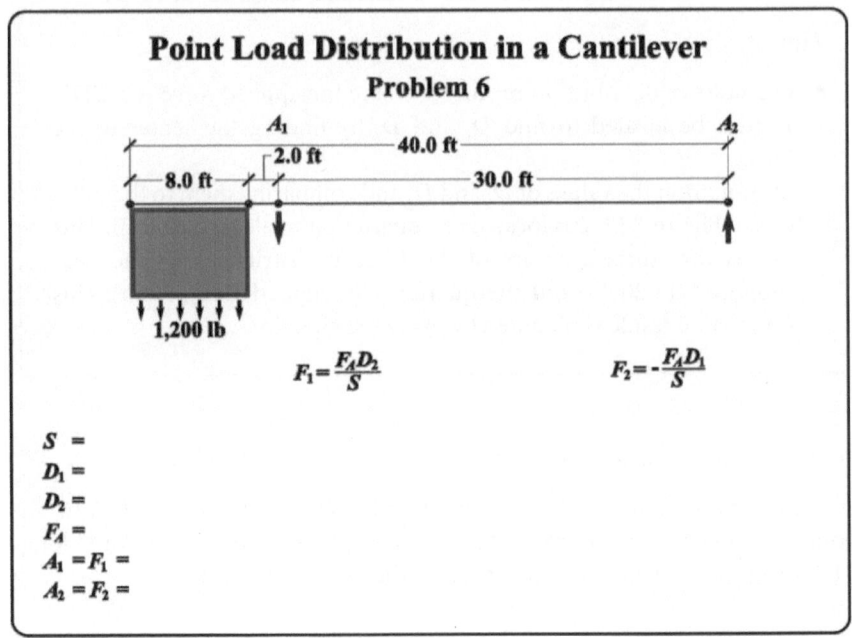

Figure 1.14

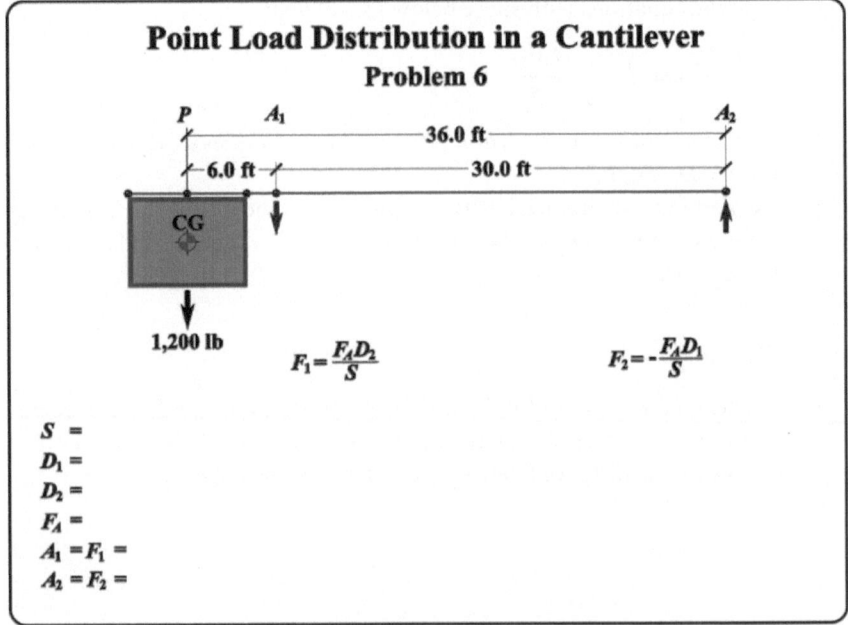

Figure 1.15

- Using Figure 1.15, insert the values next to the symbols on the list.
- Substitute the values from your adjacent list into the formulas for F_1 and F_2 and solve for F_1 and F_2.
- Transfer the values for F_1 and F_2 to the appropriate spaces on your table of symbols, rounding as desired once all calculations are complete. In this case A_1 is the closest support to P, so F_1 would be located at it and F_2 would be located at A_2 (see Figure 1.16).

Tip

- As in all cantilever problems, the force on the far support, in this case A_2, is in the opposite direction of the force on the near support, in this case A_1.

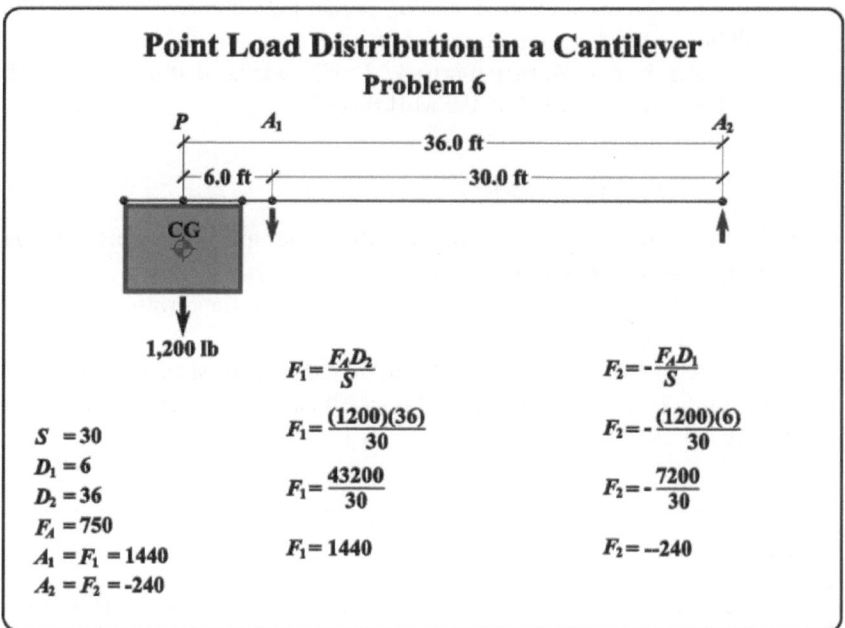

Figure 1.16

Problem 7: Complex Use of Force Distribution Equations

Problem 7 is an example of a 16' wide UDL that is partially in a simply supported span and partially in a cantilever. For this problem the UDL is attached to a 40' beam that is supported 6' from the left end and at the right end of the beam. The 1,350lb UDL is positioned such that its left end is 4' to the left of anchorage A_1 and extends past A_1 to the right 12' towards A_2. Figure 1.17 is an illustration of the problem.

Note that the equations can't be laid out until the position of P is located.

SOLUTION

STEP 1: FIND P

- P is located by finding the CG of the UDL which would be 4' to the right of A_1 (see Figure 1.18).

STEP 2: ORGANIZE THE DATA

- Since P is located between A_1 and A_2, the simply supported span equations will be used and written on the worksheet.
- Since P is closer to A_1 than it is to A_2, F_1 will be located at A_1 and F_2 will be located at A_2 and noted on the worksheet.

STEP 3: INSERT VALUES AND SOLVE

- From the data on Figure 1.18, fill in the values for the knowns S, D_1, D_2, and F_A on the worksheet.
- Substitute the values from your adjacent list into the formulas for F_1 and F_2 and solve for F_1 and F_2.
- Transfer the values for F_1 and F_2 to the appropriate spaces on your table of symbols, rounding as desired once all calculations are complete (see Figure 1.19).

Tip

- The method used for Problem 7 will also work for solving problems with equally spaced point loads of the same weight when they are located partially in a simply supported span and in a cantilever. As long as the PLs are all being supported by the same two supports, they can be grouped as was done in Problem 2.

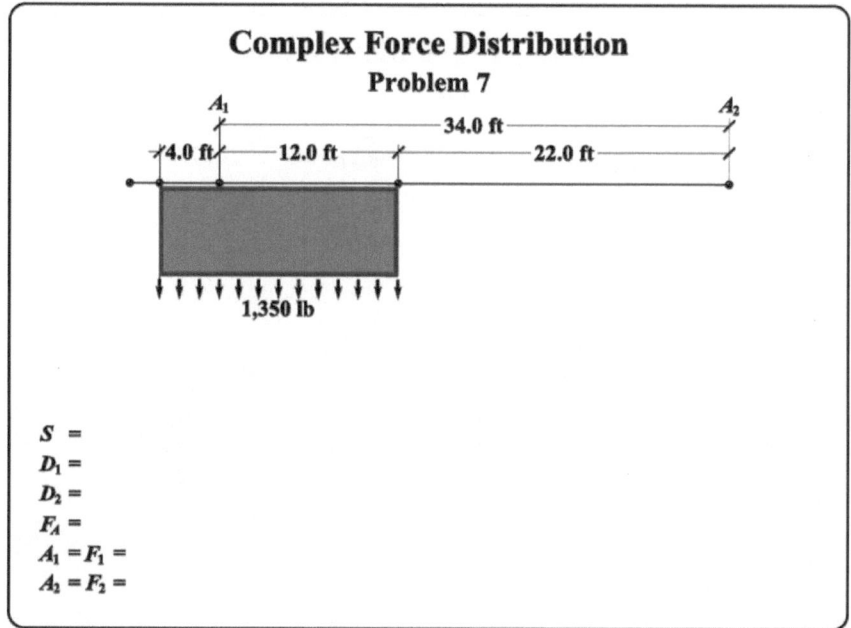

Figure 1.17

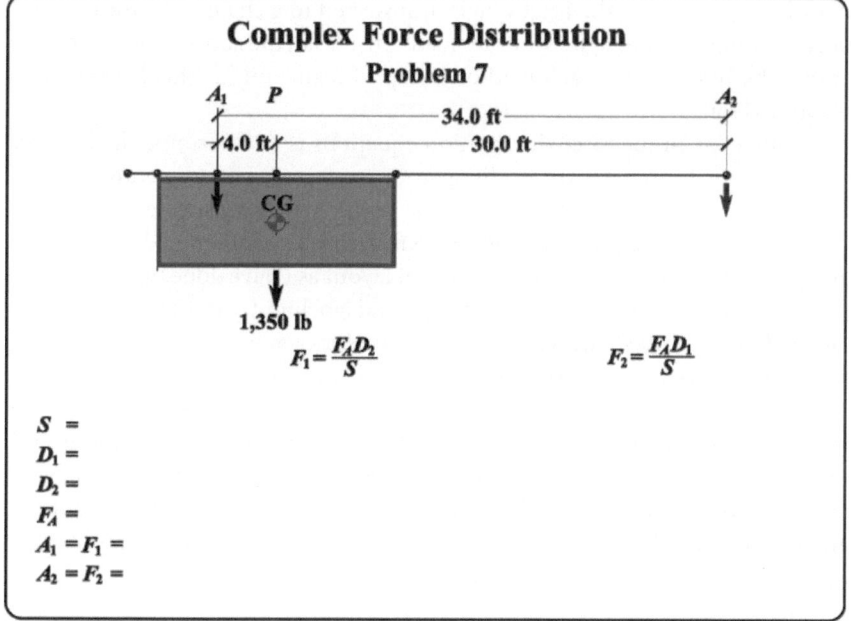

Figure 1.18

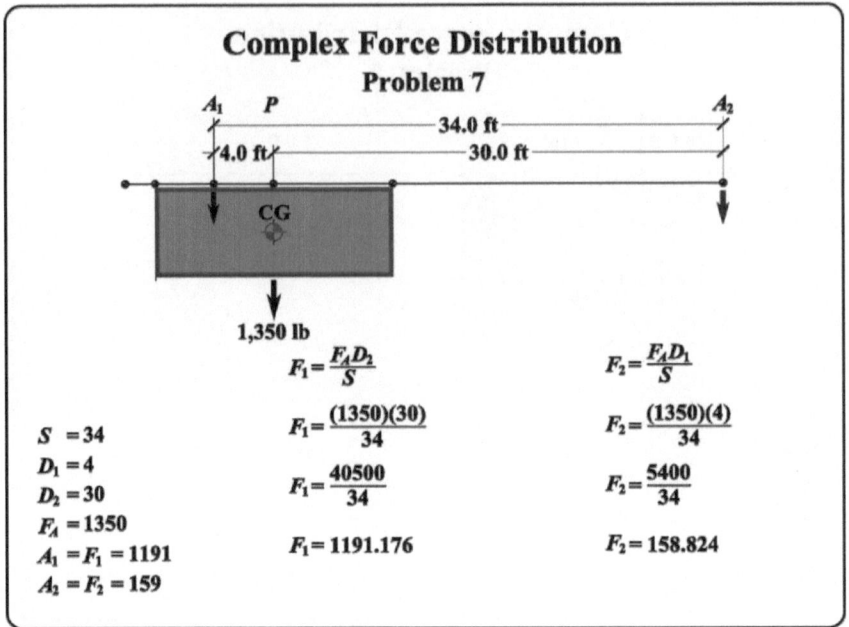

Figure 1.19

In order to determine the total weight transferred to a chain hoist from a particular arrangement of lights or other loads on a truss, the above procedures will have to be repeated for each load or group of loads until all loads have been accounted for.

Another use of the force distribution equations is to determine the load on the trusses or beams we are bridling from on the support structure. This is particularly important on low capacity support structures such as temporary structures or rigging grids, sometimes referred to as mother grids. By superimposing a rig plot onto a structure's beam layout as I have done in Figure 1.20, using the distance between the beams as S and labeling D_1 and D_2 according to the symbol definitions, the weight distribution for each point can be calculated. After the calculations are complete for all four PLs, the distributions for each beam can be summed to arrive at the total imposed load on each beam.

I have discussed many of the uses for the force distribution equations and solved several problems using them. Remember, organization of data is very important to make sure that the correct numbers are entered into the appropriate equations. After solving 15–20 of this type of problem the symbols and equations will probably be committed to your memory.

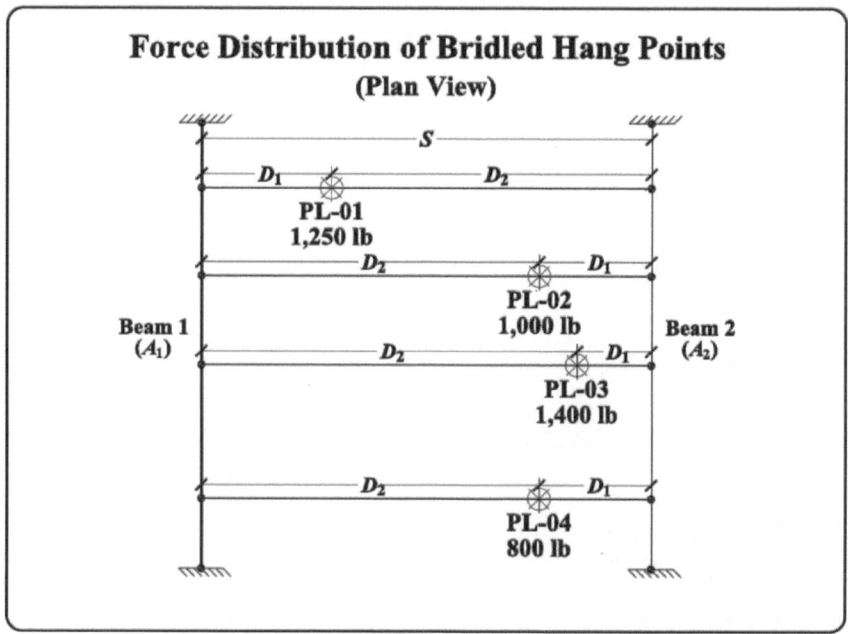

Figure 1.20

Angle Loading

In the previous section, all the forces discussed were vertical forces and could be added together if they were in the same direction or subtracted if they are in the opposite direction. However, what happens when we combine a vertical and a horizontal force, forces that are 90 degrees apart instead of 0 or 180 degrees apart? To visualize angle loading, imagine a point being deadhung and then breasted off to the side a little with a hand line. This situation will not usually make us break out our calculators though, as how much force could we be adding to the system with a hand line? Another common situation that occurs often is of more concern and can cause increases in tension worthy of a quick calculation. This situation occurs when anchorages in the support structure are not directly above their attachment points on the truss being raised. Figure 1.21 is an example of this situation. If the truss was being raised using half-ton hoists, would they be overloaded by the time the truss reached its trim height 10' below the anchorages?

Our objective will be to calculate the force on the angled support line, but first let's explain some new concepts.

- **Force as a vector quantity:** In order to accurately describe force, it must be stated in terms of both magnitude and direction as well as its origin.

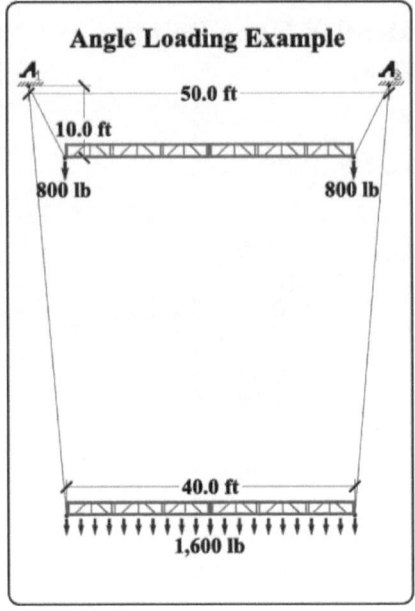

Figure 1.21

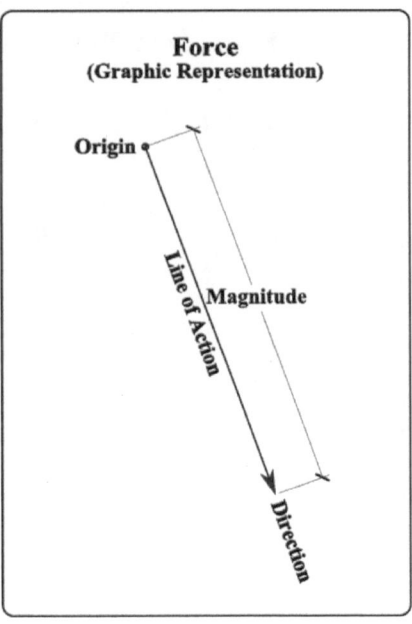

Figure 1.22

When we talk about weight, which is a special case of force, we tend not to describe the direction because gravity always works in the same direction. Force as a vector can best be viewed graphically as an arrow. The beginning of the arrow shaft is the *origin* of the force. The length of the arrow is a scale representation of the *magnitude*. And finally, the point of the arrow indicates the *direction* of the force (see Figure 1.22).

- **Parallelogram rule for combining forces:** We have noted that when forces are not in the same or opposite directions, they cannot be added or subtracted to arrive at a total force. The *parallelogram rule* must be used in cases where addition and subtraction won't work. The rule states that if two forces with the same origin are drawn as sides of a parallelogram, the diagonal of the parallelogram will represent the resultant force. The converse of the rule is also useful when thinking about vertical and horizontal component forces resulting from an angled support or bridle leg (see Figure 1.23).

- **Force triangles:** A *force triangle* is simply one-half of a force parallelogram. Looking at the rectangle at the bottom of Figure 1.23, if the arrow representing the horizontal force, Force B, were moved to the bottom side of the rectangle, a triangle would be formed by Forces A, B, and C.

- **Similar triangles:** *Similar triangles* are triangles that have the same shape, but not necessarily the same size. The unique property of similar triangles

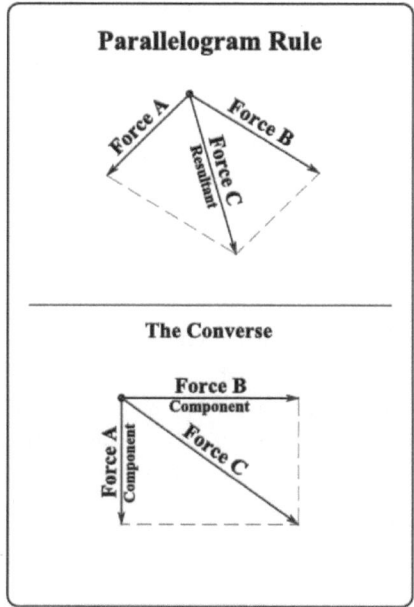

Figure 1.23

is that the ratio of the lengths of all the sides is the same for both triangles. In other words, if side A of triangle 1 is twice the length of side A of triangle 2, then sides B and C of triangle 1 will also be twice the length of sides B and C of triangle 2. This property can be used to solve force problems, since the force in the angled leg forms the same shaped triangle when combined with the vertical and horizontal forces as the triangle formed by the vertical and horizontal distances measuring from *A* to *P*. In fact, since vertical and horizontal form a 90-degree angle, the similar triangles are *right triangles.*

- **Right triangles:** Right triangles are triangles that contain a 90-degree angle as one of the three angles that make up the triangle.
- **Pythagorean Theorem:** If the length of two sides of a right triangle are known, the third length can be found use the *Pythagorean Theorem.* The theorem states that the hypotenuse, the side opposite the 90-degree angle, is the square root of the sum of the squares of the other two sides. Expressed as an equation it is:

$$a^2 + b^2 = c^2 \text{ or } c = \sqrt{a^2 + b^2}$$

Using it with a force triangle and the symbols used in this chapter the equation becomes:

$$F_L^2 = F_H^2 + F_V^2 \text{ or } F_L = \sqrt{F_H^2 + F_V^2}$$

Solving Angled Leg Force Problems

Problem 8

Finding the angled leg force when the vertical force is known, method one. For Problem 8, the example shown in Figure 1.21 will be used. In this case we have an anchorage (A) that is offset 10′ vertically and 5′ horizontally from the point (P), at which the applied force (F_A) is attached. Since the truss is uniformly loaded and weighs 1,600lb, there is an 800lb vertical force (F_V) at each end. We want to know the value of the force in the angled leg (F_L) to determine whether or not we would be overloading the half-ton hoists that are lifting the truss.

SOLUTION

STEP 1: ORGANIZE THE DATA

- To better visualize the problem, we will draw two triangles. The first will represent the distances. Starting at the anchorage a vertical line is drawn representing D_V, the 10′ height difference between A and P. From the bottom of that line we will draw a line horizontally representing D_H, the 5′ offset between A and P. The line connecting A and P represents D_L, the slope and length of the angled leg.
- Adjacent to the triangle, make a table of the three distances and fill in the two that are known.
- Below the distance triangle, draw a similar triangle representing the forces. For this triangle we will start at A and draw a vertical line F_V, representing the 800lb vertical force. We know this triangle is similar to the distance triangle so the angles will be the same, but the forces F_H and F_L are unknown at this point.
- Next to this triangle make a table representing the three forces F_H, F_L, and F_V, filling in the value for F_V.
- Finding F_L will be a two-step process and there is more than one way to come to a solution. We could begin by either finding the length of D_L or by finding F_H. For this problem we wouldn't care what the length of D_L is as there is a chain hoist in line with it. F_H might be more useful, so we will start by finding its value. Since the force and distance triangles are similar, the ratio of F_H to F_V will be the same as the ratio of D_H to D_V. The equation for this is:

$$F_H = F_V \frac{D_H}{D_V}$$

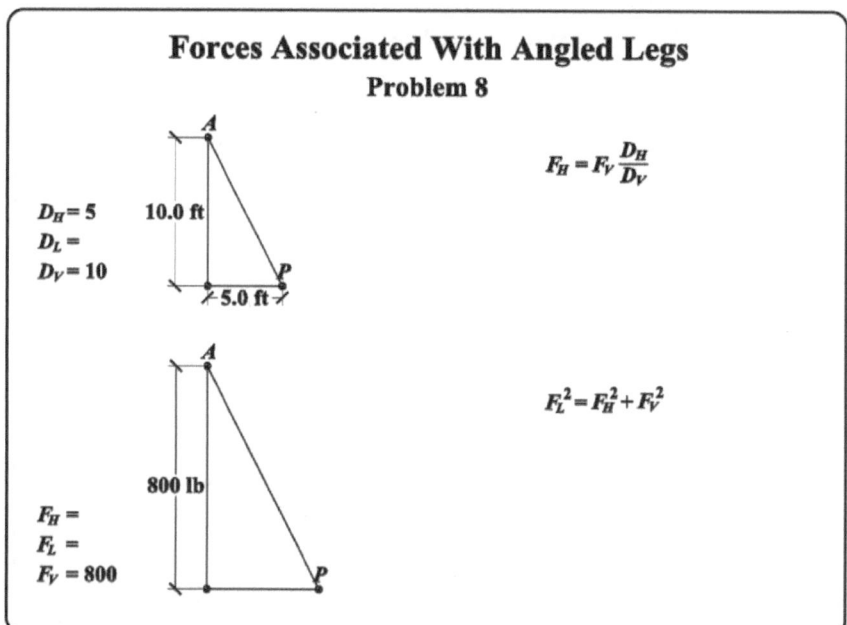

Forces Associated With Angled Legs
Problem 8

$$F_H = F_V \frac{D_H}{D_V}$$

$D_H = 5$ 10.0 ft
$D_L =$
$D_V = 10$ 5.0 ft

$$F_L^2 = F_H^2 + F_V^2$$

800 lb

$F_H =$
$F_L =$
$F_V = 800$

Figure 1.24

This formula is the first of two needed and will be useful to remember for use later in solving bridle problems as well. It should be written down with space below for solving.

Once we have the value for F_H, we can use the Pythagorean Theorem to find F_L, the hypotenuse of the force triangle (see Figure 1.24). For this triangle it would be written down as follows:

$$F_L^2 = F_H^2 + F_V^2$$

STEP 2: INSERT VALUES AND SOLVE

- Solving for F_H first by substituting the values for F_V, D_H, and D_V, we find the value of 400lb for F_H.
- After writing this in the table, we can then substitute the values into our Pythagorean Theorem equation and arrive at an answer of approximately 894.427lb. So, by doing this calculation we see that we would not be overloading the half-ton chain hoists.
- Finally, transfer the value of F_L to the table and round as desired once all calculations are complete (see Figure 1.25).

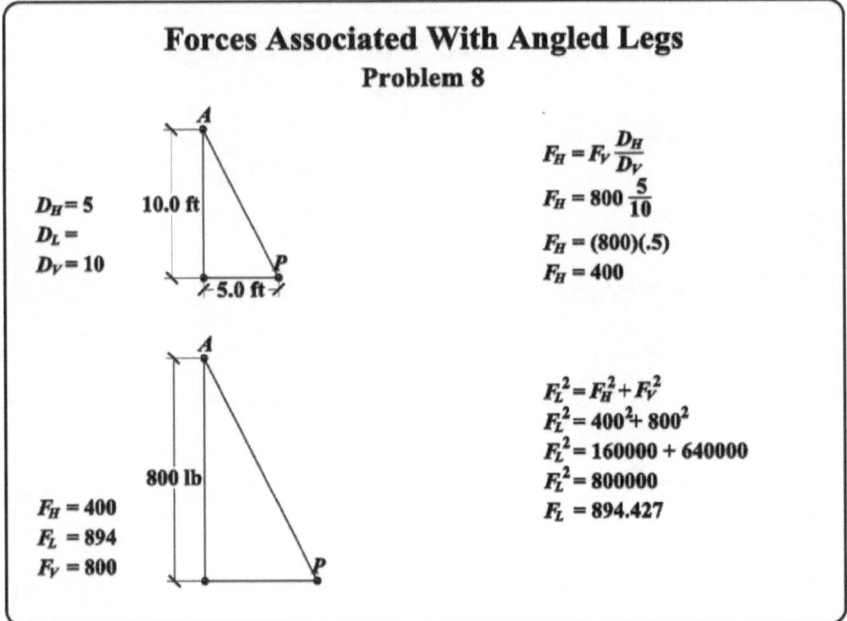

Forces Associated With Angled Legs
Problem 8

$D_H = 5$
$D_L =$
$D_V = 10$

10.0 ft

A

P

5.0 ft

$F_H = F_V \dfrac{D_H}{D_V}$

$F_H = 800 \dfrac{5}{10}$

$F_H = (800)(.5)$

$F_H = 400$

A

800 lb

P

$F_H = 400$
$F_L = 894$
$F_V = 800$

$F_L^2 = F_H^2 + F_V^2$

$F_L^2 = 400^2 + 800^2$

$F_L^2 = 160000 + 640000$

$F_L^2 = 800000$

$F_L = 894.427$

Figure 1.25

Problem 9

Finding the angled leg force when the vertical force is known, method two. For Problem 9, we will create a little more angle loading and solve the problem using the Pythagorean Theorem first and use the similar triangle ratio second. For this problem we will keep D_V and F_V the same as in Problem 8, with values of $10'$ and 800lb respectively. We will increase the horizontal offset to $7.5'$ to see how this affects F_L.

SOLUTION

STEP 1: ORGANIZE THE DATA

- As in Problem 8 we will draw the distance triangle using the dimensions given for Problem 9.
- Create a table of the distances adjacent to the triangle and fill in the known distances.
- Create a force triangle using F_V as the only known side, approximating the angles of the distance triangle.
- Create a table of the forces adjacent to the triangle and fill in the value for F_V.
- This time, to find F_L we will start off by finding the value of D_L using the Pythagorean Theorem using this equation:

$$D_L^2 = D_H^2 + D_V^2$$

This equation should be copied with space below for solving.

- Once we have a value for D_L, we can use the ratio of sides in the distance triangle to find F_L in the force triangle (see Figure 1.26). Below is the equation we will use. It should be copied with space for solving:

$$F_L = F_V \frac{D_L}{D_V}$$

STEP 2: INSERT VALUES AND SOLVE

- In this case we will substitute the values into our Pythagorean Theorem equation to find the value of D_L.
- Once the value of D_L has been found the second equation can be solved to find F_L (see Figure 1.27). Note that F_L has increased from the previous problem by increasing D_H.

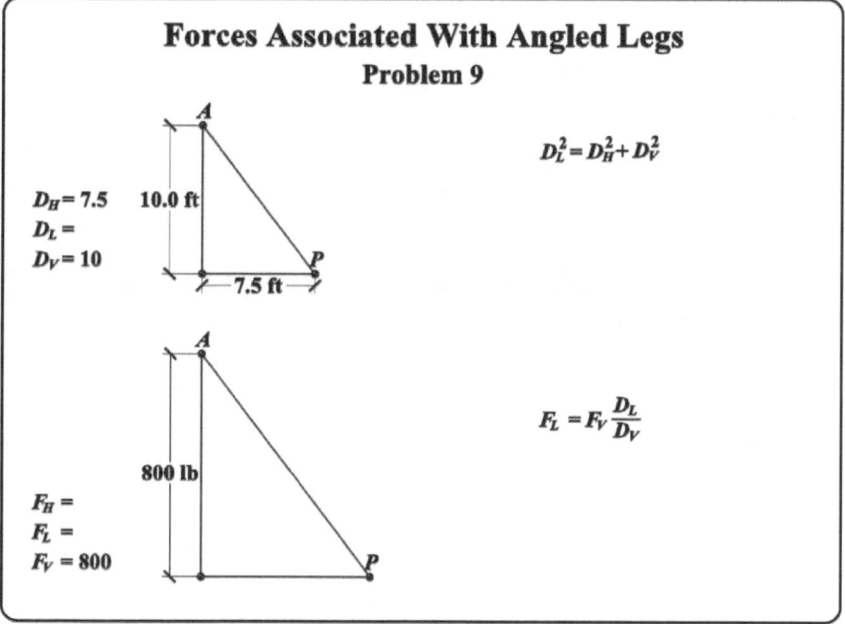

Figure 1.26

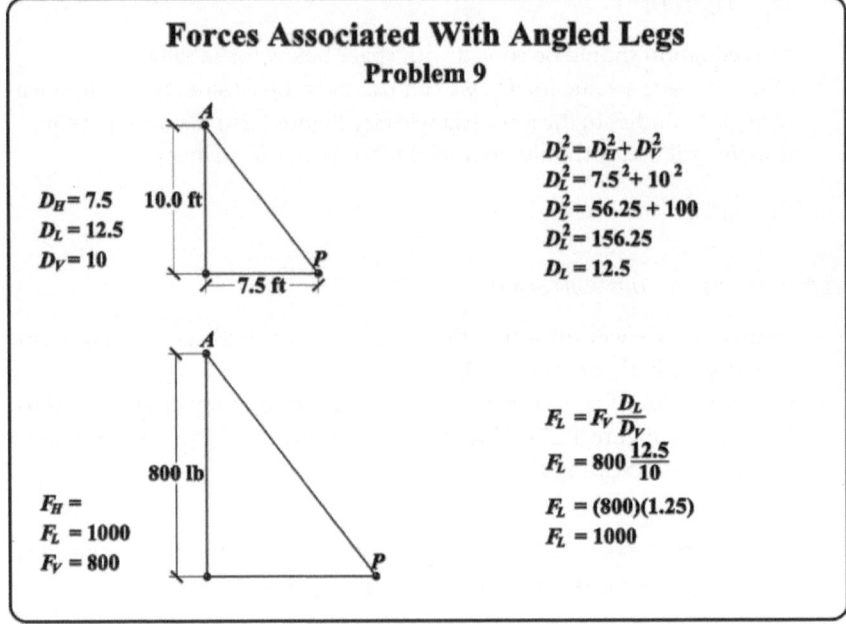

Forces Associated With Angled Legs
Problem 9

$D_H = 7.5$
$D_L = 12.5$
$D_V = 10$

$$D_L^2 = D_H^2 + D_V^2$$
$$D_L^2 = 7.5^2 + 10^2$$
$$D_L^2 = 56.25 + 100$$
$$D_L^2 = 156.25$$
$$D_L = 12.5$$

$F_H =$
$F_L = 1000$
$F_V = 800$

$$F_L = F_V \frac{D_L}{D_V}$$
$$F_L = 800 \frac{12.5}{10}$$
$$F_L = (800)(1.25)$$
$$F_L = 1000$$

Figure 1.27

Tips

- If you look at both ratio equations side by side, it may make them easier to remember:

$$F_L = F_V \frac{D_L}{D_V} \text{ and } F_H = F_V \frac{D_H}{D_V}$$

  - Whether you are solving for F_H or F_L, the vertical force is multiplied by a fraction.
  - The fraction always has D_V as the denominator.
  - The numerator uses D_H when solving for F_H, and D_L when solving for F_L.

- The distance triangle in Problem 9 is a *3-4-5 triangle*. That means that the short side is a multiple of 3, the long side is a multiple of 4, and the hypotenuse is a multiple of 5. In this case the multiple is 2.5. If this had been recognized, the ratio equation need not have been solved as 2.5 × 5 would have yielded the correct value for D_L of 12.5. Since the force triangle is similar to the distance triangle, it also must be a 3-4-5 triangle with a long side of 800. 4 × 200 equals 800, so F_H = 3 × 200 or 600, and the hypotenuse F_L = 5 × 200 or 1,000. Recognizing things like this will make you a more accurate and more efficient rigger.

The main concern about angle loading is that as the angled leg gets flatter and flatter, F_H and F_L dramatically increase. The angle riggers commonly refer to when talking about bridle or angled legs is the acute angle formed at the anchorage from horizontal down to the angled leg. I will refer to it as *angle-a* ($\angle a$) because it is associated with the anchorage *A*. Some of the key approximate relationships to remember are:

If $\angle a$ = 45°, then F_H = F_V
If $\angle a$ = 45°, then F_L = 1.5F_V
If $\angle a$ = 30°, then F_L = 2F_V
If $\angle a$ = 15°, then F_L = 4F_V

See Figure 1.28 for an illustration.

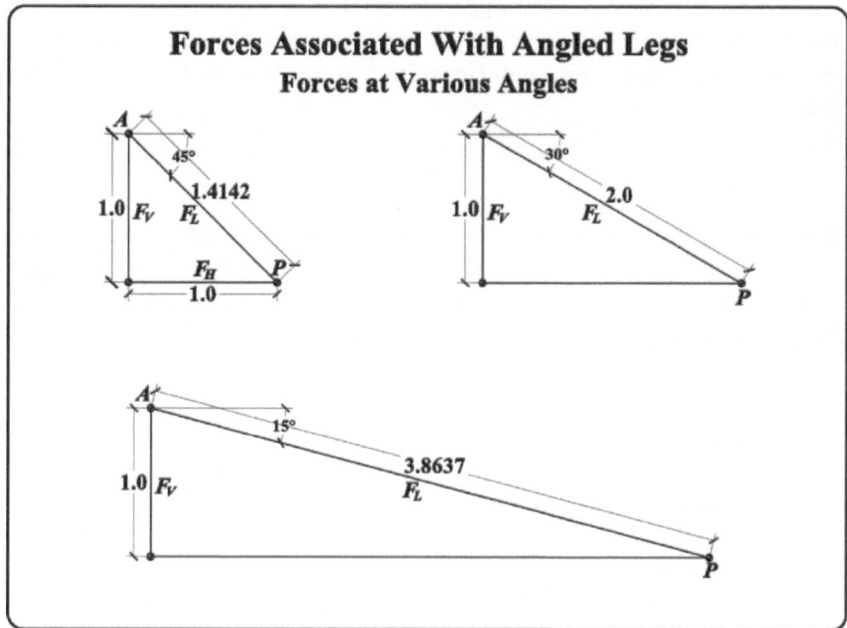

Forces Associated With Angled Legs
Forces at Various Angles

Figure 1.28

Two-Leg Bridles—Equal Height Anchorages

Two-leg, equal height bridles are the most common type of bridle done by entertainment riggers. There are, however, many buildings around the world where the anchorage beams for the bridles are at different heights, as well as a few where three-leg bridles are common. Not only are the two-leg, equal height bridles more common, they are the simplest to install accurately and the simplest to solve mathematically. To differentiate them from the other types I will call them *simple bridles* in this chapter. In this section, I will present methods for determining all the forces associated with simple bridles.

Solving bridle force problems can be accomplished several different ways. When I was first reading about forces in bridle legs, I stumbled upon a chart of "multipliers" used for calculating the force in a bridle leg when the angle-a ($\angle a$) of the leg was known. It was useful, but I didn't want to carry the chart around with me. The best I could do was to memorize a couple of the key multipliers. One day I discovered that the chart of multipliers was in fact the table of values for a trigonometric function. Since this was long before the days of smartphones and even the Internet, my option was to buy a calculator that had the functions programmed in it. For me back then, it was the equivalent of getting a smartphone and downloading a bridle calculation app. As useful as that was back then, I found it very challenging to teach the use of trigonometry in my rigging classes. Over the years I have found that instead of thinking about angles and trigonometric functions, most riggers found that learning to solve bridle problems by directly applying the ratios involved was easier. However, no discussion of rigging math would be complete without at least the briefest description of the various functions.

Trigonometric Functions

Trigonometric or *trig functions* are the ratios of the various sides that contain the angle being evaluated. With a given angle in a right triangle, the sides have a specific ratio. These ratios are described by three primary and three secondary functions listed below:

- Primary functions

  - *Sine (sin)* $= \frac{Opposite\ Side}{Hypotenuse}$

  - *Cosine (cos)* $= \frac{Adjacent\ Side}{Hypotenuse}$

  - *Tangent (tan)* $= \frac{Opposite\ Side}{Adjacent\ Side}$

- Secondary functions

  - $Cosecant\ (csc) = \frac{Hypotenuse}{Opposite\ Side}$

  - $Secant\ (sec) = \frac{Hypotenuse}{Adjacent\ Side}$

  - $Cotangent\ (cot) = \frac{Adjacent\ Side}{Opposite\ Side}$

The secondary functions are the reciprocal of their primary counterparts. For instance, *csc* is the reciprocal of **sin**. That means that the fractions are reversed. On a calculator the reciprocal can be found by dividing the number into one, or in some cases pressing the "1/x" button.

On a calculator, entering the angle and pressing the function button will yield the value for that function. The reverse can be done although all calculators are not labeled the same. On an iPhone, pressing the "2nd" button, then typing in the ratio, followed by the pressing the desired function button will yield the angle.

The problem with using trig functions is that we either have to know what the angle is to find the appropriate ratio, or we have to find the angle by using a different ratio in order to find the ratio we need. In many ways ratios are easier to understand and measure since we generally have tape measures available but not protractors. Most of us have learned about fleet angles and that, for a smooth drum, the maximum fleet angle is 1.5 degrees. It seems more useful to me to know the ratio of 1:40 instead of the angle. This tells us that for every 40″ of distance away from the drum that the wire rope can be 1″ off-center.

Span to Height Ratio ($S:D_V$)

All riggers should be aware that tremendous forces are produced in very flat bridles, and many rigging sources teach different rules of thumb for avoiding potentially dangerous bridles. Most riggers have heard of the 90-degree rule, or the not quite as conservative 120-degree rule, about never exceeding these values for the bridle angle. For us, the bridle is in the middle of space and estimating or measuring it is difficult at best. Instead of thinking about a bridle angle, think about the ratio of bridle height to span. If the S to D_V ratio is 3 to 1 or less you will avoid overloading the rigging slings and hardware as long as they would safely support the load in a dead hang.

Problem 10: Bridle labels

Figure 1.29 illustrates the labels and symbols used for discussion and solving bridle problems. Finding bridle leg forces when the anchorage heights of the legs are equal. In Problem 10 we will be solving for the three forces associated with

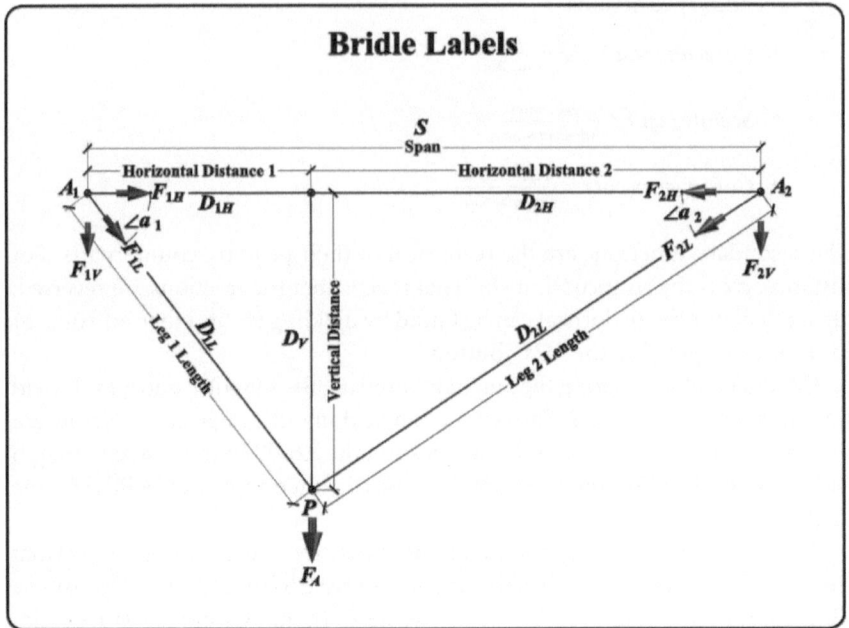

Figure 1.29

each leg of the bridle as well as the bridle leg lengths. In order to solve for all six unknown forces we will use the force distribution equations along with the ratio formulas learned in the previous section. The bridle leg lengths will be found using the Pythagorean Theorem. For this problem the distance between the beams is 45′, and the bridle point is 15′ away from the near beam. The point weighs 1,800lb, and the bridle point is 10′ below the beams.

SOLUTION

STEP 1: ORGANIZE THE DATA

- Sketch a drawing of the bridle with the dimensions given above, showing the weight applied.
- Below the drawing make a table of the known values and their symbols. Also make a table of the symbols for the eight unknown values (see Figure 1.30).
- On a second sheet of paper, divide it into two columns with a vertical line and then divide the columns into four sections each. The four unknown values for leg 1 will be solved on the left side and the unknowns for leg 2 on the right side.

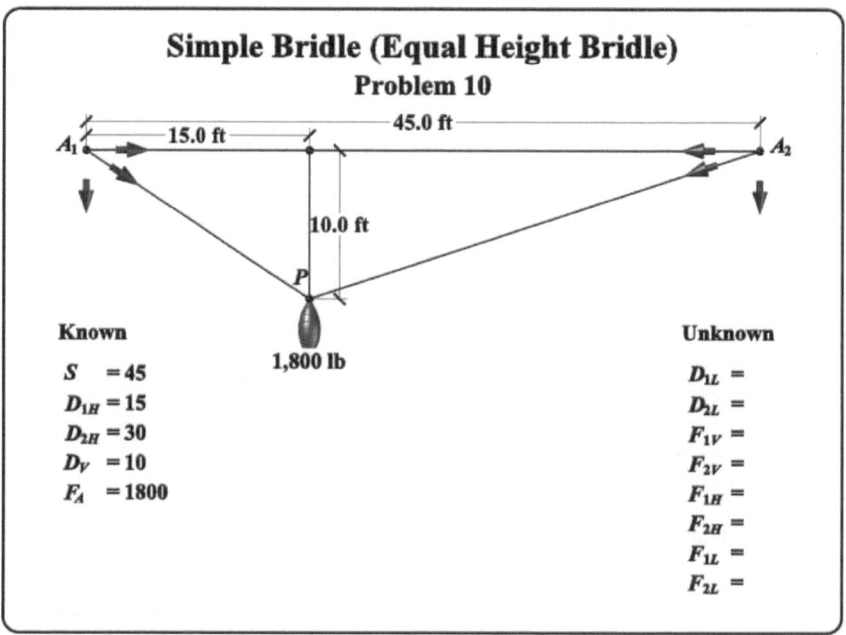

Simple Bridle (Equal Height Bridle)
Problem 10

Known	Unknown
S = 45	D_{1L} =
D_{1H} = 15	D_{2L} =
D_{2H} = 30	F_{1V} =
D_V = 10	F_{2V} =
F_A = 1800	F_{1H} =
	F_{2H} =
	F_{1L} =
	F_{2L} =

Figure 1.30

STEP 2: SOLVE FOR BRIDLE LEG LENGTH

It may help when solving bridle problems to think of a bridle as two back-to-back angled leg triangles that in this case share a common side. The vertical distance, D_V, is the common side. By looking at Figure 1.30, we know the values for the short sides of both triangles, D_{1H} and D_V for bridle leg 1, and D_{2H} and D_V for bridle leg 2. Using the Pythagorean Theorem, the hypotenuses of both legs, D_{1L} and D_{2L}, can be calculated.

- Using the top sections of the two columns, find the bridle leg lengths D_{1L} and D_{2L}. At the top of the left-hand section write either equation:

$$D_{1L} = \sqrt{D_{1H}^2 + D_{1V}^2} \text{ or } D_{1L}^2 = D_{1H}^2 + D_{1V}^2$$

- At the top of the right-hand upper section write either of the following forms of the Pythagorean Theorem for D_{2L}:

$$D_{2L} = \sqrt{D_{2H}^2 + D_{2V}^2} \text{ or } D_{2L}^2 = D_{2H}^2 + D_{2V}^2$$

- Substitute values into the equations and solve for D_{1L} and D_{2L}. Transfer your answers to the table of unknowns, rounding if desired once all calculations are complete.

Note: The solutions for both D_{1L} and D_{2L} resulted in numbers with many decimal places. The values of D_{1L} and D_{2L} rounded to three places will be used later in this problem.

STEP 3: SOLVE FOR VERTICAL FORCE

We now have values for all three sides of the distance triangles associated with this problem. We need to find the values for the sides of the force triangles. Looking back at the angled leg problem, we were given the vertical force, which we used in finding the horizontal and leg force. So finding the vertical force first is a useful next step. Finding the vertical force is a force distribution problem for which we know the distances S, D_{1H} and D_{2H}, along with F_A.

- At the top of the second section down on the scratch paper, write the force distribution formulas below:

  For Leg 1 $F_{1V} = \frac{F_A D_{2H}}{S}$, and for Leg 2, $F_{2V} = \frac{F_A D_{1H}}{S}$

- Substitute the values into the equations and solve, transferring the values for F_{1V} and F_{2V} to the table.

STEP 4: SOLVE FOR HORIZONTAL FORCE

Once we have the values for F_{1V} and F_{2V}, we can solve for the horizontal and bridle leg forces. Starting with the horizontal forces F_{1H} and F_{2H} we can use the ratio formulas we used in the angled leg problems. We will multiply the vertical force by the horizontal distance divided by the vertical distance.

- At the top of the next section down on your worksheet write the formulas below for F_{1H} and F_{2H}:

  $F_{1H} = F_{1V} \frac{D_{1H}}{D_{1V}}$ and $F_{2H} = F_{2V} \frac{D_{2H}}{D_{2V}}$

- Substitute the values into the equations and solve, transferring the values for F_{1H} and F_{2H} to the table.

STEP 5: SOLVE FOR BRIDLE LEG FORCE

Finally we can solve for the bridle leg forces. This time the ratio will be D_L over D_V since we are trying to find F_L.

- Write the ratio equations for finding F_{1L} and F_{2L} at the top of the bottom section of the worksheet:

  $F_{1L} = F_{1V} \frac{D_{1L}}{D_{1V}}$ and $F_{2L} = F_{2V} \frac{D_{2L}}{D_{2V}}$

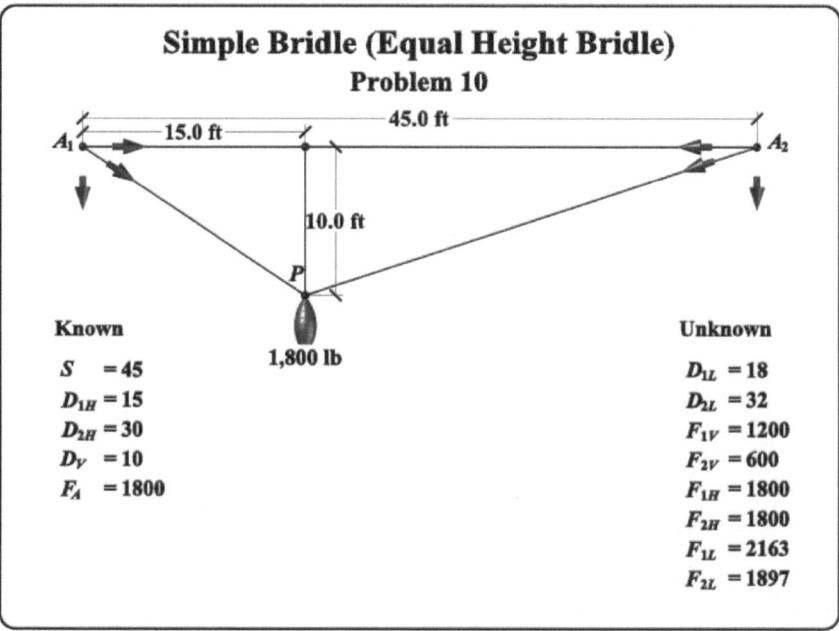

Simple Bridle (Equal Height Bridle)
Problem 10

Known

$S = 45$
$D_{1H} = 15$
$D_{2H} = 30$
$D_V = 10$
$F_A = 1800$

1,800 lb

Unknown

$D_{1L} = 18$
$D_{2L} = 32$
$F_{1V} = 1200$
$F_{2V} = 600$
$F_{1H} = 1800$
$F_{2H} = 1800$
$F_{1L} = 2163$
$F_{2L} = 1897$

Figure 1.31

- Substitute the values into the equations and solve, transferring the values for F_{1L} and F_{2L} to the table. To obtain the same result as the book does, you must round D_{1L} and D_{2L} to three places to the right of the decimal point in the above equations.

Review Figures 1.31 and 1.32 to check your results.

Tips

- When calculating the leg lengths, the square roots yielded many places after the decimal. I only kept three places to the right of the decimal point for future use in equations. When transferring the data to a rig plot, the data needs not be that accurate and can be rounded further.
- Note that P is one-third of the way across the span in this problem, therefore two-thirds of the force is transferred to the near support and one-third to the far support. Using this method works on simple bridles, but not on bridles with different height anchorages. They will be discussed later in the chapter.

Bridle Leg Length

$$D_{1L}^2 = D_{1H}^2 + D_{1V}^2$$
$$D_{1L}^2 = 15^2 + 10^2$$
$$D_{1L}^2 = 225 + 100$$
$$D_{1L}^2 = 325$$
$$D_{1L} = 18.028$$

$$D_{2L}^2 = D_{2H}^2 + D_{2V}^2$$
$$D_{2L}^2 = 30^2 + 10^2$$
$$D_{2L}^2 = 900 + 100$$
$$D_{2L}^2 = 1000$$
$$D_{2L} = 31.623$$

Vertical Force

$$F_{1V} = \frac{F_A \, D_{2H}}{S}$$
$$F_{1V} = \frac{(1800)(30)}{45}$$
$$F_{1V} = \frac{54000}{45}$$
$$F_{1V} = 1200$$

$$F_{2V} = \frac{F_A \, D_{1H}}{S}$$
$$F_{2V} = \frac{(1800)(15)}{45}$$
$$F_{2V} = \frac{27000}{45}$$
$$F_{2V} = 600$$

Horizontal Force

$$F_{1H} = F_{1V} \frac{D_{1H}}{D_V}$$
$$F_{1H} = 1200 \frac{15}{10}$$
$$F_{1H} = (1200)(1.5)$$
$$F_{1H} = 1800$$

$$F_{2H} = F_{2V} \frac{D_{2H}}{D_V}$$
$$F_{2H} = 600 \frac{30}{10}$$
$$F_{2H} = (600)(3)$$
$$F_{2H} = 1800$$

Bridle Leg Force

$$F_{1L} = F_{1V} \frac{D_{1L}}{D_V}$$
$$F_{1L} = 1200 \frac{18.028}{10}$$
$$F_{1L} = (1200)(1.8028)$$
$$F_{1L} = 2163.36$$

$$F_{2L} = F_{2V} \frac{D_{2L}}{D_V}$$
$$F_{2L} = 600 \frac{31.623}{10}$$
$$F_{2L} = (600)(3.1623)$$
$$F_{2L} = 1897.38$$

Figure 1.32

- When you compare the horizontal force and the bridle leg force equations, the vertical force for the triangle is multiplied by a ratio with the denominator being the vertical distance. The only change to the numerator is that it is the horizontal distance when finding the horizontal force, and the leg length when finding the leg force. By studying these equations, they can be committed to memory.
- In this problem, the bridle leg forces were the last sides of the force triangles that were unknown. Instead of using the ratio equations we could have found them using the Pythagorean Theorem on the force triangles. See the equations below:

$$F_{1L}^2 = F_{1H}^2 + F_{1V}^2 \text{ and } F_{2L}^2 = F_{2H}^2 + F_{2V}^2$$

- The forces in both legs are significantly more than 1,800lb, the value of F_A. This is because the $S{:}D_V$ ratio is 4.5, well above 3. If possible, when S is 45', D_V should be at least 15'.

Two-Leg Bridles—Unequal Height Anchorages

Unequal height bridles cannot be solved exactly the same way as simple bridles because the anchorages are at different heights. However, these problems can be solved using the concepts we have already learned. We just need to apply the concept of similar triangles in a different application when trying to find forces. Using the concept of similar triangles we can convert the bridle into a simple bridle in order to apply the force distribution equation to find the vertical forces.

First, think of the legs in a bridle as representing the lines of actions for the forces in the legs and realize that the force is the same anywhere along the line of action (see Figure 1.33).

One side of the bridle can be changed to a similar triangle with the same D_V as the other side (see Figure 1.34).

Since we are creating a new leg 2 similar triangle with a D_V equal to D_{1V}, the new horizontal distance, D_{2Hnew}, can be calculated by multiplying D_{2H} by the ratio of D_{1V} divided by D_{2V}. Once we have a value for D_{2Hnew}, it can be added to D_{1H} to arrive at the new span S_{new}. These values can then be put into the force distribution formulas as follows:

$$F_{1V} = F_A \frac{D_{2Hnew}}{S_{new}} \text{ and } F_{2V} = F_A \frac{D_{1H}}{S_{new}}$$

Once the vertical forces are obtained, the other forces can be found by the same methods as were done with the simple bridle.

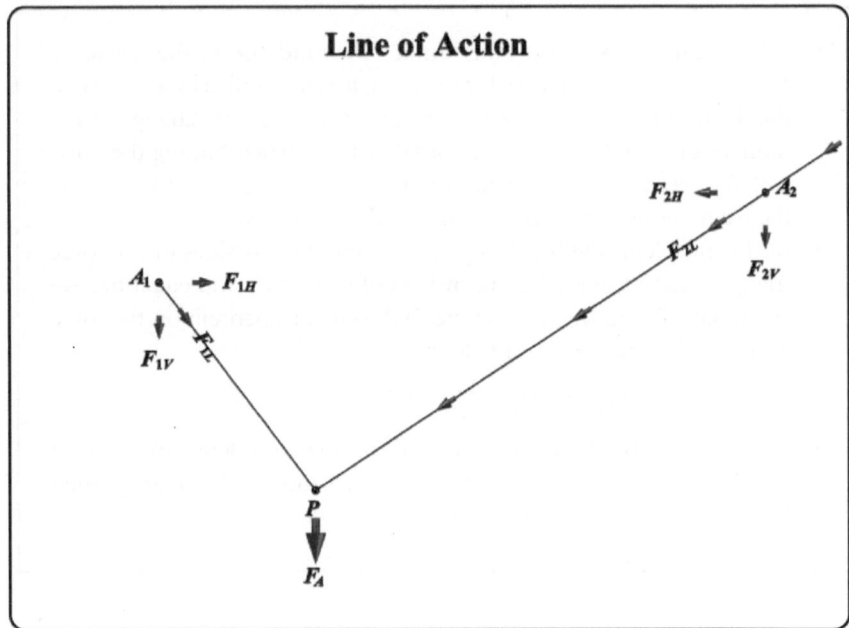

Figure 1.33

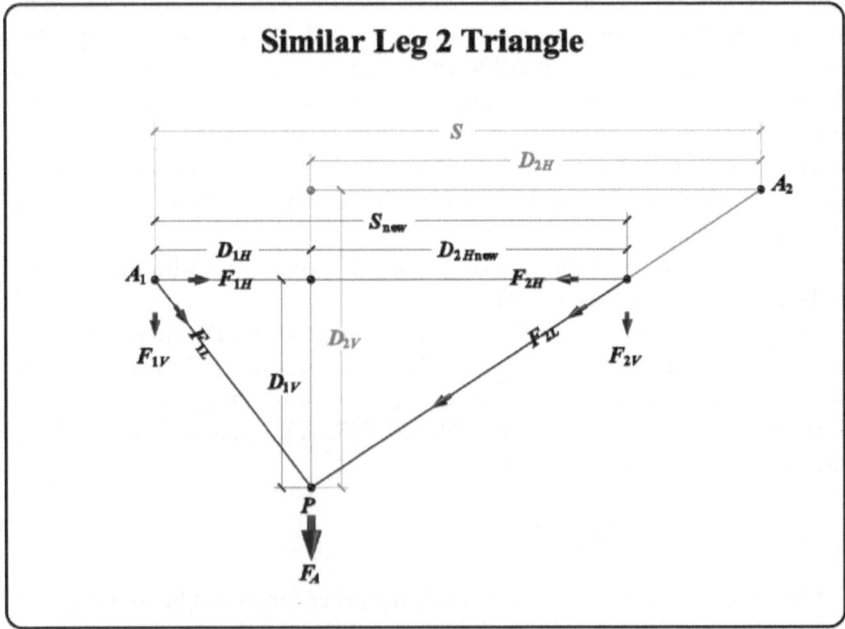

Figure 1.34

Problem 11

Finding bridle leg forces when the anchorage heights of the legs are not equal. For problem 11, the span between the beams is 40.5′. The bridle point, *P*, is 10.5′ from anchorage A_1. The vertical distance, D_{1V}, from A_1 to *P* is 14′, while the vertical distance, D_{2V}, from A_2 to *P* is 20′. The applied force, F_A is 1,600lb as in the simple bridle problem, Problem 10.

SOLUTION

STEP 1: ORGANIZE THE DATA

- Sketch a drawing of the bridle with the dimensions given above, showing the weight applied.
- Below the drawing make a table of the known values and their symbols. Also make a table of the symbols for the ten unknown values (see Figure 1.35).
- On a second sheet of paper, divide it into two columns with a vertical line and then divide the columns into five sections each. The four unknown values for leg 1 will be solved on the left side and the unknowns for leg 2 on the right side. The fifth section will be used to find the new span by creating a similar triangle for leg 2 with the same height as leg 1.

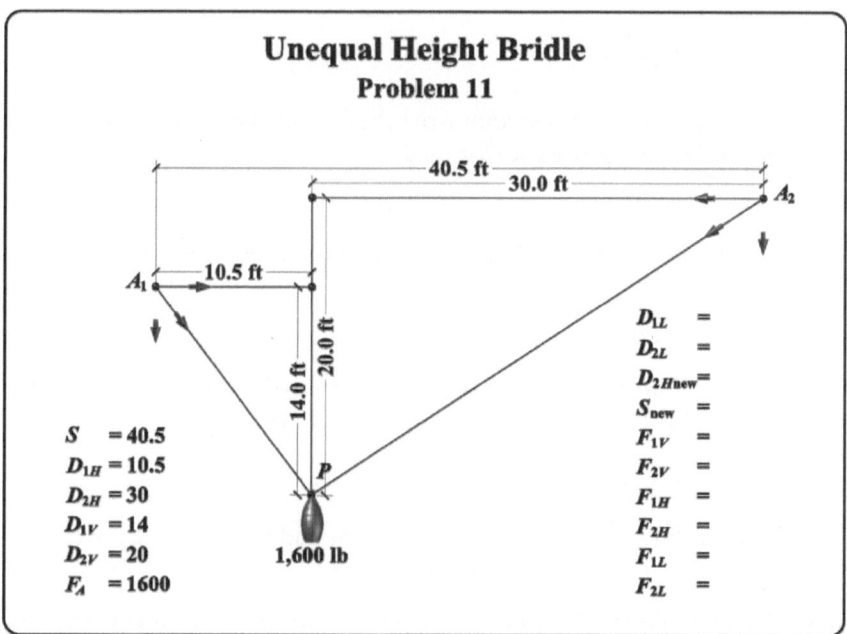

Unequal Height Bridle
Problem 11

$S = 40.5$
$D_{1H} = 10.5$
$D_{2H} = 30$
$D_{1V} = 14$
$D_{2V} = 20$
$F_A = 1600$

1,600 lb

$D_{1L} =$
$D_{2L} =$
$D_{2Hnew} =$
$S_{new} =$
$F_{1V} =$
$F_{2V} =$
$F_{1H} =$
$F_{2H} =$
$F_{1L} =$
$F_{2L} =$

Figure 1.35

STEP 2: SOLVE FOR BRIDLE LEG LENGTH

Since the short sides of both the distance triangles are known, the Pythagorean Theorem can be used to find the bridle leg lengths. The short sides of the triangles would be the heights, D_{1V} and D_{2V}, and the horizontal distances D_{1H} and D_{2H}. The hypotenuse of the leg 1 triangle would be represented by D_{1L}, and for leg 2 the hypotenuse would be D_{2L}.

- Using the top section of your worksheet, in the left side add a form of the Pythagorean Theorem for Leg 1 such as:

$$D_{1L}^2 = D_{1H}^2 + D_{1V}^2$$

- On the right side of the top section add the equation for Leg 2:

$$D_{2L}^2 = D_{2H}^2 + D_{2V}^2$$

- Substitute the known values into the equations and solve for D_{1L} and D_{2L}.
- Transfer the values to the list of unknowns, rounding as desired once all calculations are complete.

STEP 3: FIND THE NEW SPAN

- In the next section down, on the left side, write the ratio equation to find the horizontal distance for a new leg 2 triangle that has the same height as leg1:

$$D_{2Hnew} = D_{2H} \frac{D_{1V}}{D_{2V}}$$

- In the right side of this section write the formula for the new span using the new horizontal distance for leg 2:

$$S_{new} = D_{1H} + D_{2Hnew}$$

- Solve both equations, adding the solutions to the list of unknowns.

STEP 4: FIND THE VERTICAL FORCE

- Using the information from the newly equalized height bridle write the formulas for F_{1V} and F_{2V} in the next section down:

$$F_{1V} = \frac{F_A D_{2Hnew}}{S_{new}} \text{ and } F_{2V} = \frac{F_A D_{1H}}{S_{new}}$$

- Insert the known values into the equations and solve, adding the results to the list of unknowns rounded as desired.

Note: The solutions for F_{1V} and F_{2V} resulted in numbers with many decimal places. I rounded both to three decimal places on the worksheet for use later in the problem.

STEP 5: FIND THE HORIZONTAL FORCE

- In the next section down, write the ratio equations for finding F_{1H} and F_{2H}:

$$F_{1H} = F_{1V}\frac{D_{1H}}{D_{1V}} \text{ and } F_{2H} = F_{2V}\frac{D_{2H}}{D_{2V}}$$

- Insert the known values into the equations and solve, adding the results to the list of unknowns rounded as desired.

Note: The solutions for F_{1H} and F_{2H} resulted in numbers with many decimal places. I rounded both to three decimal places on the worksheet for use later in the problem.

STEP 6: FIND THE BRIDLE LEG FORCE

In Problem 10, we used ratio equations to find the values for F_{1L} and F_{2L}. Instead of doing that, this time we will use the Pythagorean Theorem to find the forces in the bridle legs. Either way will result in similar results.

- In the bottom section, write the Pythagorean Theorems for F_{1L} and F_{2L}:

$$F_{1L}^2 = F_{1H}^2 + F_{1V}^2 \text{ and } F_{2L}^2 = F_{2H}^2 + F_{2V}^2$$

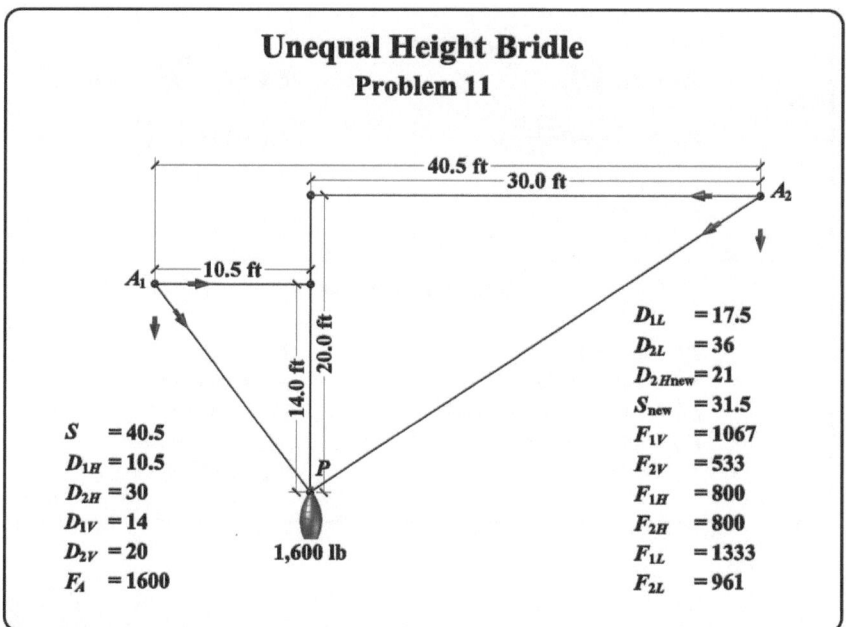

Figure 1.36

Bridle Leg Length

$$D_{1L}^2 = D_{1H}^2 + D_{1V}^2$$
$$D_{1L}^2 = 10.5^2 + 14^2$$
$$D_{1L}^2 = 110.25 + 196$$
$$D_{1L}^2 = 306.25$$
$$D_{1L} = 17.5$$

$$D_{2L}^2 = D_{2H}^2 + D_{2V}^2$$
$$D_{2L}^2 = 30^2 + 20^2$$
$$D_{2L}^2 = 900 + 400$$
$$D_{2L}^2 = 1300$$
$$D_{2L} = 36.056$$

Find New Span

$$D_{2Hnew} = D_{2H}\frac{D_{1V}}{D_{2V}}$$

$$D_{2Hnew} = 30\frac{14}{20}$$

$$D_{2Hnew} = (30)(.7)$$
$$D_{2Hnew} = 21$$

$$S_{new} = D_{1H} + D_{2Hnew}$$

$$S_{new} = 10.5 + 21$$

$$S_{new} = 31.5$$

Vertical Force

$$F_{1V} = \frac{F_A\, D_{2Hnew}}{S_{new}}$$

$$F_{1V} = \frac{(1600)(21)}{31.5}$$

$$F_{1V} = \frac{33600}{31.5}$$

$$F_{1V} = 1066.667$$

$$F_{2V} = \frac{F_A\, D_{1H}}{S_{new}}$$

$$F_{2V} = \frac{(1600)(10.5)}{31.5}$$

$$F_{2V} = \frac{16800}{31.5}$$

$$F_{2V} = 533.333$$

Horizontal Force

$$F_{1H} = F_{1V}\frac{D_{1H}}{D_{1V}}$$

$$F_{1H} = 1066.667\frac{10.5}{14}$$

$$F_{1H} = (1066.667)(.75)$$
$$F_{1H} = 800$$

$$F_{2H} = F_{2V}\frac{D_{2H}}{D_{1V}}$$

$$F_{2H} = 533.333\frac{30}{20}$$

$$F_{2H} = (533.333)(1.5)$$
$$F_{2H} = 800$$

Bridle Leg Force

$$F_{1L}^2 = F_{1H}^2 + F_{1V}^2$$
$$F_{1L}^2 = 800^2 + 1066.667^2$$
$$F_{1L}^2 = 640000 + 1137778.489$$
$$F_{1L}^2 = 1777778.489$$
$$F_{1L} = 1333.334$$

$$F_{2L}^2 = F_{2H}^2 + F_{2V}^2$$
$$F_{2L}^2 = 800^2 + 533.333^2$$
$$F_{2L}^2 = 640000 + 284444.089$$
$$F_{2L}^2 = 924444.089$$
$$F_{2L} = 961.480$$

Figure 1.37

- Insert the known values into the equations and solve, adding the results to the list of unknowns rounded as desired.

Note: The solutions for F_{1L} and F_{2L} resulted in numbers with many decimal places during the process of squaring values and taking their square roots. I rounded both to three decimal places on the worksheet each time. To check your results, look at Figures 1.36 and 1.37.

Tip

- In order to find the vertical forces in an unequal height bridle, one side of the bridle must be changed using the principle of similar triangles so that $D_{1V} = D_{2V}$. The newly created D_H is used to find the span of the newly created bridle. The newly created bridle is used only to find F_{1V} and F_{2V}. Once the vertical forces have been determined, the distances of the original bridle are used to find the other forces.

Three-Leg Bridle Geometry

In the last section of this chapter we will learn to calculate the lengths of the legs of three-leg bridles. To solve problems with two-leg bridles all the dimensions are in the same plane, which is some ways simplifies the mathematics, especially when considering the force calculations, which will not be discussed in this chapter. The bridle leg lengths, however, are simple enough to calculate with the addition of two new concepts.

Cartesian Coordinate System

The Cartesian coordinate system was developed by a French philosopher named René Descartes. The system is used to describe unique points in space. On a two-dimensional system there are two axes separated by 90 degrees in the same plane. Generally these axes are labeled x and y. The point at which the axes cross is called the *origin*. Many people in our business call it the *zero-zero point*. Where they cross is described by the ordered pair (0,0), where the first number is the value of x and the second number is the value of y. When used on the stage, normally the *x-axis* runs cross stage and the *y-axis* runs up and down stage. See the left side of Figure 1.38 for an illustration of a two-dimensional Cartesian coordinate system.

A three-dimensional system adds an additional axis, perpendicular to the plane of the *x-y* axes. This axis is the *z-axis*, and represents the elevation above the floor or stage. In this system the values are shown as (x,y,z). See the right side

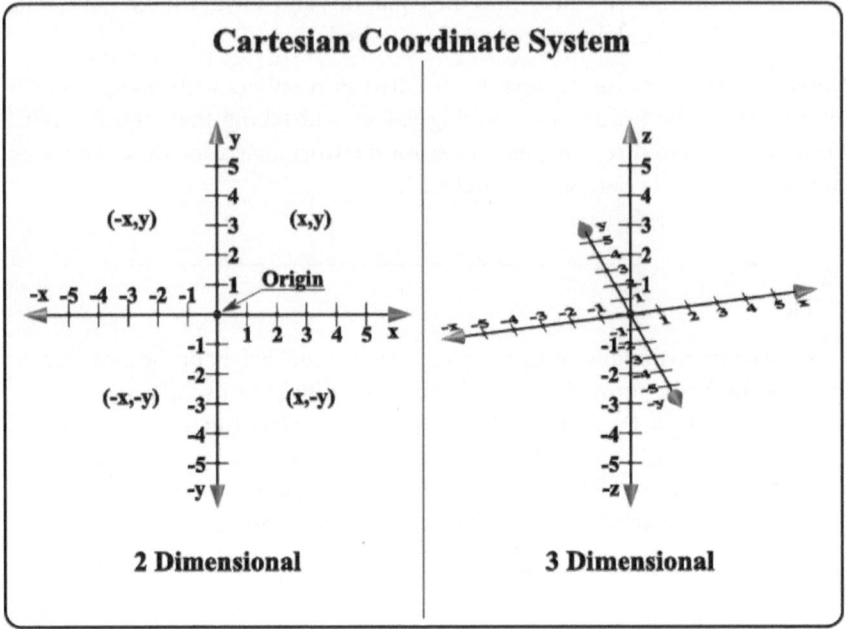

Figure 1.38

of Figure 1.38 for an illustration of a three-dimensional Cartesian coordinate system.

Delta (Δ)

In mathematics, the term *delta* (Δ) means change. It is the absolute difference in value and will be used in conjunction with the Cartesian coordinate system to solve for our bridle leg lengths. When thinking about the absolute difference in two values of *x*, or Δx, it is always a positive value no matter what the signs are of the values we are comparing (see Figure 1.39).

Three-Dimensional Pythagorean Theorem

Previously we have been using the Pythagorean Theorem on a triangle which is a two-dimensional object. We have used it to find D_L, the leg length or hypotenuse of the triangle. Because we are working in three dimensions instead of two when trying to find the leg lengths of a three-leg bridle we need to expand the Pythagorean Theorem to accommodate the change. The new theorem is:

$$D_L^2 = \Delta x^2 + \Delta y^2 + \Delta z^2$$

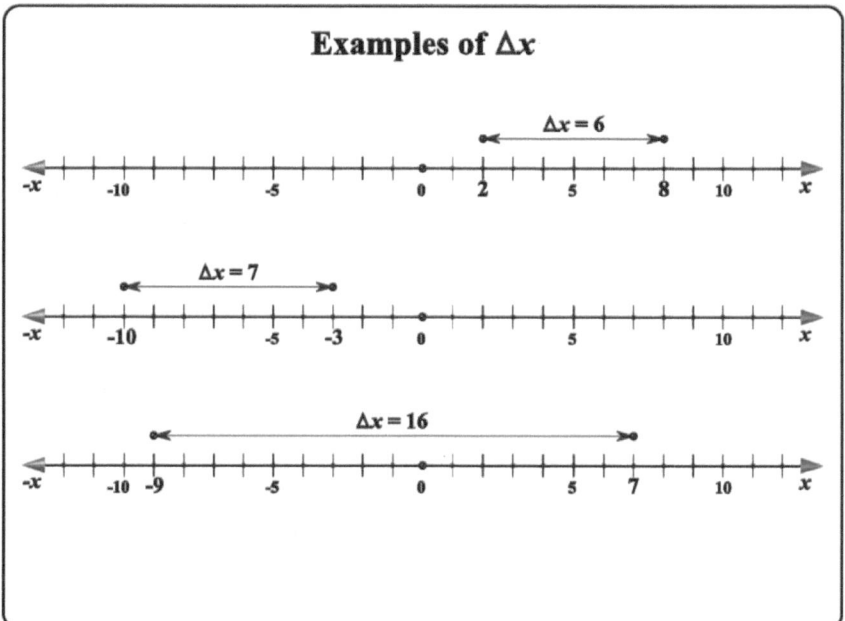

Figure 1.39

The Δ in this case is the difference in the value of the Cartesian coordinate at the anchorage of the leg in question and the respective coordinate at the bridle point *P*.

Problem 12

Finding the leg lengths of a three-legged bridle. For the final problem, we will use this new version of the Pythagorean Theorem to find the leg lengths of a three-leg bridle. For this problem the point is 28′ left of center stage, 16′ upstage of the downstage edge of the stage, and *P* needs to be 35′ above the stage. The beams we are anchoring the bridle to are 50′ above the stage, running up and downstage with attachment points every 12′ starting at the downstage edge.

To identify the anchorage points and bridle point using the Cartesian coordinate system we will place the origin at the downstage edge of the stage at center stage. The three anchorage points are: $A_1 = (36, 24, 50)$, $A_2 = (36, 12, 50)$, and $A_3 = (0, 12, 50)$. The bridle point *P* is at $(28, 16, 35)$.

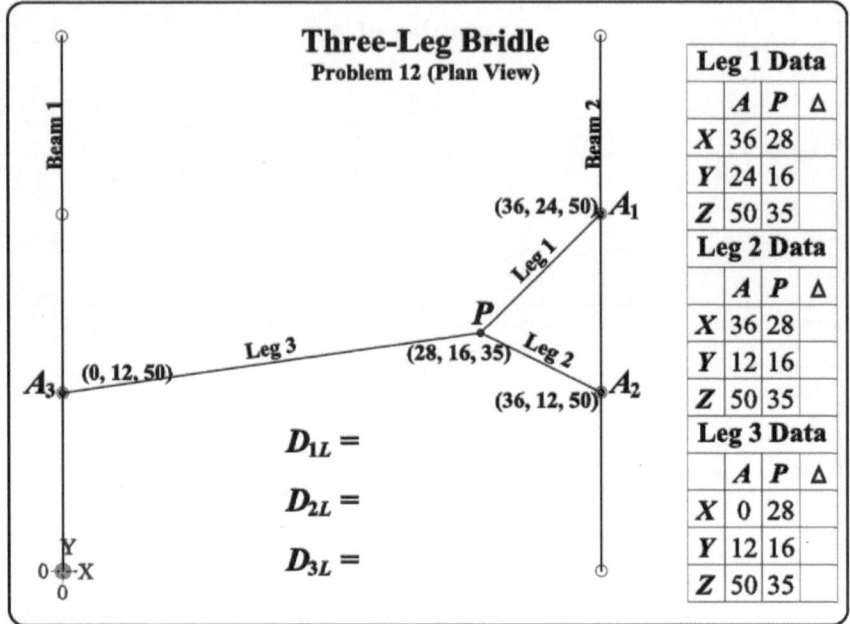

Figure 1.40

SOLUTION

STEP 1: ORGANIZE THE DATA

- Sketch a plan view drawing of the bridle with the dimensions given above, showing the beams, origin, and Cartesian coordinates.
- Make a table of data for each leg with three columns and three rows. Label the columns *A*, *P*, and Δ. Label the rows *X*, *Y*, and *Z*.
- Insert the coordinates in the tables for each leg (see Figure 1.40).
- On a second sheet of paper, write the three-dimensional Pythagorean Theorem for each leg, leaving space between for solving for leg length.

STEP 2: FIND ALL THE DELTAS

On the tables created earlier, fill in the differences between *A* and *P* for the values of *x*, *y*, and *z* for each leg.

STEP 3: SOLVE FOR LEG LENGTH

Insert the delta values into the equations and solve for leg length. Transfer the results to the drawing, rounding as desired once all calculations are complete.

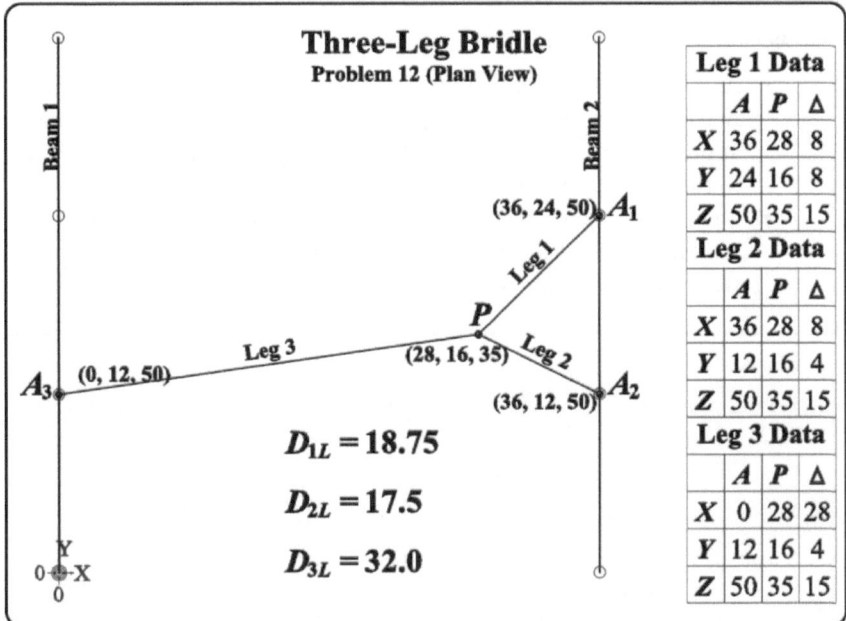

Figure 1.41

Three-Leg Bridle
Problem 12 Worksheet

$$D_{1L}^2 = \Delta x_1^2 + \Delta y_1^2 + \Delta z_1^2$$
$$D_{1L}^2 = 8^2 + 8^2 + 15^2$$
$$D_{1L}^2 = 64 + 64 + 225$$
$$D_{1L}^2 = 353$$
$$D_{1L} = 18.788$$

$$D_{2L}^2 = \Delta x_2^2 + \Delta y_2^2 + \Delta z_2^2$$
$$D_{2L}^2 = 8^2 + 4^2 + 15^2$$
$$D_{2L}^2 = 64 + 16 + 225$$
$$D_{2L}^2 = 305$$
$$D_{2L} = 17.464$$

$$D_{3L}^2 = \Delta x_3^2 + \Delta y_3^2 + \Delta z_3^2$$
$$D_{3L}^2 = 28^2 + 4^2 + 15^2$$
$$D_{3L}^2 = 784 + 16 + 225$$
$$D_{3L}^2 = 1025$$
$$D_{3L} = 32.016$$

Figure 1.42

Tips

- As a practical matter, bridle legs lengths do not need to be exact. A common adjusting method is to use STAC chain, which has approximately 3″ links. For this reason I have rounded the answers to the nearest one-quarter foot, or 3″.
- Once you become familiar with the concepts and the coordinates are known, the Δs can be found and inserted into the equations without drawing the bridle, beams, and tables.

2

Structural Behavior

BILL GORLIN

This chapter is intended to introduce structural behavior to people involved in entertainment rigging, tapping into the experiences and "feel" all of us have gained from looking at structures every day of our lives. Structures are governed by the laws of physics, namely Isaac Newton's Laws, which are all around us: buildings, bridges, box trusses, rope, fences, trees, rock formations, etc. We will explore key basic structural subjects, with the intention of improving the reader's awareness of such behaviors. This is a chapter of concepts, not equations, although I will introduce a couple of those.

Structural engineers can be helpful collaborators with designers, technical directors, riggers, and stagehands when developing entertainment designs. The engineer visualizes the load paths and stability of the system, ensuring safety and hopefully improving the ease of use and cost. It is best to involve the engineer early in the design process in order to best utilize their insights. Alert the engineer to real-world issues that are important to your project, such as the production schedule, preferred equipment and construction methods, and access and handling issues. Share insights back and forth, since the engineer and you may have very different experiences.

It is worthwhile to note that the coveted engineer's stamp on a document is more than patterned ink. To become a professional engineer, a person needs to graduate from an accredited engineering college, work several years as an engineer-in-training, apply for licensure with the state, and then take and pass a rigorous test. Once a person is licensed, he is legally responsible to the state government to comply with engineering laws and to perform engineering to a standard of care expected in the industry. In fact, an engineer is first responsible to the state to protect public safety, and then responsible to his employer and client.

Let's now get into the structural behavior.

Newton's Laws

Newton's Laws of Motion, combined with a lot of math, provide the key to most of structural engineering. These laws are as follows:

1 An object at rest will remain at rest unless acted on by an unbalanced force. An object in motion continues in motion with the same speed and in the same direction unless acted upon by an unbalanced force.
2 Acceleration is produced when a force acts on a mass. The greater the mass of the object being accelerated the greater the amount of force needed to accelerate the object.
3 For every action there is an equal and opposite reaction.

The First Law relates to stationary versus moving objects, confirming that objects will maintain their current state (moving or not) unless acted upon by an outside force. In rigging, this establishes the basis for understanding static force relationships, and it gives us a basis for understanding rigging dynamics.

The Second Law states that the force needed to move an object is greater for heavier objects.

The Third Law is simple, yet it is the most powerful tool available to a structural engineer. For example, the floor pushes up on a person's feet equal to the person's weight; a winch produces a certain amount of torque and its base must resist that same torque; a guy cable pulls on ballast and the ballast pulls back the same amount using its weight and its frictional resistance on the ground. When you visualize and apply this basic concept, you can undertake many structural systems.

Here are some examples: First, a rigger attaches a static object to a roof truss. In order for the object to remain in place, the truss needs to pull up on the hanger with a force equal to the object's weight. If the object is bridled, then the truss "pull" must resist the bridle forces. These examples utilize all three of the Laws.

How do we apply this to flexural situations, such as beams? Let's consider a playground see-saw in which the people are balancing one another so the see-saw is not moving.

Example 1

Two people of equal weight at the same distance from the fulcrum. The fulcrum supports the weight of two people, satisfying the Third Law. An engineer would say that the sum of the forces in the "Y" direction is equal to zero since there is no movement, i.e., the weight of two equal people directed down plus the reaction of the two people up at the fulcrum.

The see-saw does not rotate because the weight of the two people is balanced. An engineer would say that the sum of the moments (force times distance) about the fulcrum is equal to zero since there is no movement, i.e., weight of Person 1 times her distance to the fulcrum equals the weight of Person 2 times his distance to the fulcrum rotating in the other direction. The engineer would say that the sum of the moments is zero: $0 = (P1)(D) - (P2)(D)$, where $D =$

distance from center, and P1 and P2 = weight of each person. P1 must equal P2 for this equation to work.

Example 2

Person 2 weighs more than Person 1. The fulcrum still resists the combined weight, regardless of location on the see-saw. The engineer would say that the sum of the forces in "Y" direction is zero: $0 = P1 + P2 - R$, where R is the fulcrum reaction acting in the opposite direction from the forces.

The reader can probably visualize that the heavier person needs to be closer to the fulcrum than the lighter person in order to achieve balance, i.e., no movement. The same "sum of the moments" equation used above can be used to find the distances: $0 = (P1)(D1) - (P2)(D2)$, where D1 and D2 are the respective distances.

These relationships exist everywhere an object exists in relation to other objects, such as a cable bridle, supports for a beam, guy cables, a truss, and a group of bolts or welds in a connection.

When an externally applied moment (or force at a distance) is applied to an object, the object is anchored by a resisting moment that must equal the externally applied moment. The resisting moment is often anchored by a "force couple" consisting of equal and opposite forces separated by a distance. For example, a plated box truss connection resists flexure by a force couple consisting of the bottom bolts in tension coupled with the top plates bearing against one another in compression. A group of welds forming a shape can also provide a force couple to resist moments, e.g., a weld all around the interface of an angle and a gusset plate forms a rectangular weld group shape.

Force couples are everywhere—you just have to look for them.

Stress-Strain

Structural members of any material resist externally applied loads by pulling back (Newton's Law). Strain is the amount of movement when the force is applied and stress is the force per unit of material area. Any material will undergo strain when stress occurs, so if someone says there is "no movement" they really mean "negligible movement."

A member subjected to axial force will stretch if tension is applied or shrink if compression is applied. A member subjected to flexure will bend such that it will curve, causing one face to stretch and the other to contract. The amount of stretch/contraction decreases towards the center of the cross section, at which point the direction of stress and strain reverses.

Most common structural materials behave in an elastic fashion under normal use, which means stress and strain is directly correlated, e.g., double the pull to double the stretch. However, when the material reaches its yield stress, further

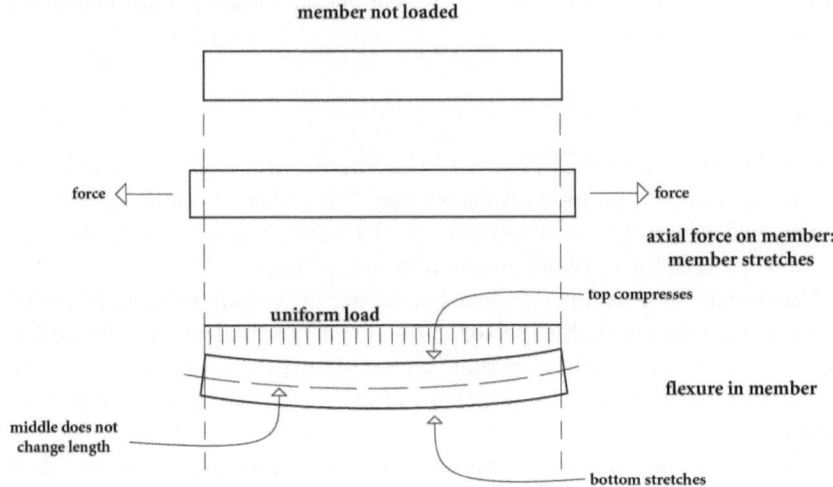

Figure 2.1 Stress – Strain

load will make it stretch like putty, unable to rebound fully when the load is removed. Then when stretched too far, the material will fracture. Each material has its own elastic-plastic behavior. For example, steel will more reliably stretch like putty before fracture, whereas aluminum is more vulnerable to sudden fracture. For rigging, always use a material in its elastic behavior range, but it is good to know how it might behave at high load.

Each material has a defined minimum yield stress above which the behavior is not elastic, as well as an ultimate stress at which it fractures. In some codes, this is expressed as strength rather than stress, and can get rather complicated.

Material Stiffness

The ratio between stress and strain defines how stiff a material is when stressed, which is called the elastic or "Young's" modulus. The higher the modulus, the

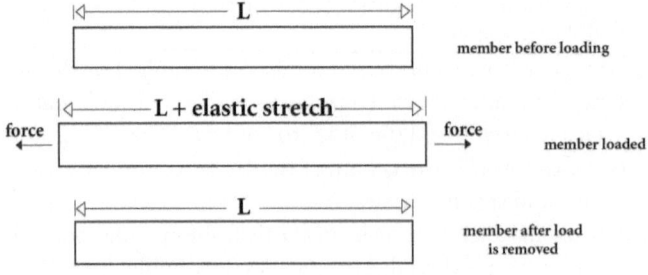

Figure 2.2 Elastic behavior

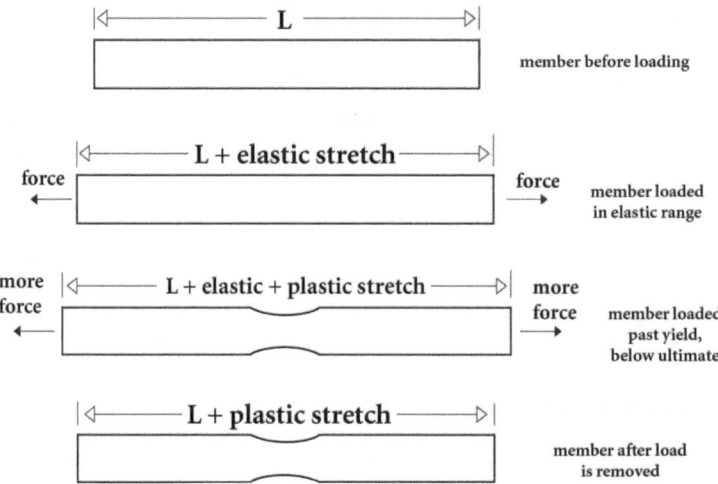

Figure 2.3 Plastic behavior

stiffer the material. Examples of common Young's modulus are listed below (psi = pounds per square inch):

- Steel: 29,000,000 psi
- Aluminum: 10,000,000 psi
- Lumber (Doug Fir North): 1,600,000 psi
- Engineered lumber: 2,000,000 psi

Member Properties

Member section properties pertain to the geometry of the member and influence allowable stresses and stiffness, and can become complicated quickly. The following is a quick overview.

Axial stiffness is related to cross sectional area. Flexural stiffness is related to moment of inertia of the cross section in the axis subject to bending. (Moment of inertia is a function of location of the material away from the cross section center axis.)

For example, a solid rod has a certain amount of material, or cross sectional area. Let's hollow out the center by stretching the material to create a pipe that is a larger diameter than the original rod, but with the same cross sectional area. This pipe is significantly stiffer in bending than the rod, using the same material. The improvement is related to the square of the distance from the center to the material.

This explains why I-beams are the most cost effective section for common building beams since most of the material in the flanges is spread far apart by the web. Of course, the very same I-beam is much less stiff when bent in the other ("weak") direction.

What if we want a member that is equally stiff when bent vertically or sideways? In this case, a pipe or tube will be a more efficient use of the material for this situation, since the material is distributed relatively uniformly in both directions.

A property listed in common structural section properties is "radius of gyration," which is calculated using moment of inertia and area, so it is solely a function of geometry. This property is the key for selecting members to resist column buckling.

Behavior of Sections Under Load

Tension

Tension is force pulling directly in line with the length of a member, i.e., axially. Members subjected to axial stress will stretch, so behavior is not dependent on length other than the amount of stretch. Pure tension is the most efficient of all load transfer methods. Be mindful that some basic building materials cannot resist much tension, such as unreinforced concrete, unreinforced masonry, and a pile of rocks—all of which can support a lot of compression.

Compression

Compression is force pushing axially on a member. Members subjected to compressive stress are at risk for buckling. This is a vital and often overlooked issue in structural systems, so please take notice. Anything under compression has a tendency to buckle and bow away from a straight line.

Let's first explore axial compression. This commonly occurs in a column or a brace. Let's consider the most basic condition where the force introduces pure axial compression with no eccentricities to complicate matters.

In the 18th century, Euler proved that slender elements will buckle suddenly when subjected to a tiny lateral force when they reach a certain stress. The definition of "slender" relates to the unbraced length of the member compared with the radius of gyration of the cross-section. The critical stress for a member is a function of these values and the Young's modulus of the material. It has no relation to the yield stress or ultimate stress. This means that the exact same size column in mild steel and high strength steel will buckle at the same critical stress.

In common applications, columns are classified as slender, compact, and in between. This means that a slender column will fail in buckling before it fails in a fashion related to cross-section stress, and a compact column will exceed an allowable stress before buckling behavior becomes a factor. Of course, there are slenderness cases that are a combination of the two.

The buckling situation becomes more complicated when considering buckling of the elements of a column failing or buckling before the overall section fails. For example, a box truss in compression needs checks of individual compression members for their own point-to-point lengths, as well as the overall truss for its full, unbraced length between supports and properties.

Let's return to basics for an example of compressive behavior:

- Imagine a common plastic drinking straw that is an inch long; it takes strong fingers to crush it. It never buckles—it just crushes and does not spring back to the original form. This is a compact column.
- Make the same straw 12″ long; this can be made to buckle pretty easily and will spring back to straight when the load is released. However, it will support load after initially buckling. If you continue to compress, it will kink and fail "plastically," i.e., it will deform and be unable to spring back to the original shape. This is a slender column, possibly with some intermediate behavior.
- Now make the same straw 200″ long; it would buckle under its own weight and you would be unable to brace it at the top. You would want to hang it from the top so that it goes into tension, or you would want to brace it laterally in the middle with a pinch of your fingers to reduce its unbraced length so it can be stabilized. This is a super-slender column.

The key to column selection when dealing with slender columns is column section geometry, not area. Considering the same force applied by one's fingers, the 200″ long unbraced straw needs a larger radius of gyration to support the same force as the shorter straws. Likewise, a really short straw could be a smaller diameter and support the same force as the 12″ straw.

The codes for various materials include a maximum slenderness that is allowed for compression members. For example, primary steel members are not permitted to have a slenderness exceeding 200, in which slenderness = (unbraced length)/(radius of gyration). Secondary members can be pushed a bit further, but the author tends to avoid slenderness above 240.

Be mindful that slender member buckling is related to Young's modulus, so a given cross section in aluminum will buckle at nearly a third of the force required to buckle the same cross section in steel.

How might one use this information?

Example 1

Say you have a steel scenic framework subjected to relatively low forces, with vertical members spaced at 6′ on center and horizontal members spaced at 8′ on center, and you want to install corner-to-corner diagonal members in bays to keep the bays square when subjected to in-plane lateral forces. The diagonal by trigonometry is 10′ long.

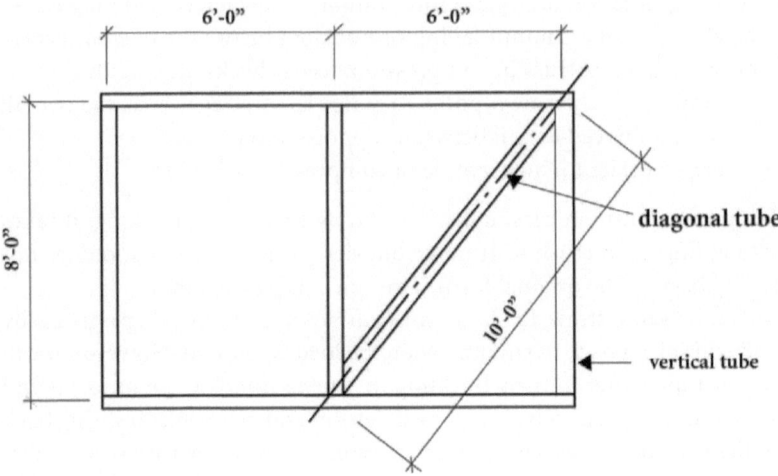

Figure 2.4 Frame braced with single tube

- Unbraced length of vertical = $(8')(12''/\text{ft})$ = 96″. Select a steel member with a radius of gyration that generates a slenderness less than 200. R ≤ 96″/200 = 0.48″. A 1.5″ × 1.5″ × 0.083″ tube has a radius of gyration of 0.58″ and a slenderness of 156, so it is a good trial choice.
- The diagonal is 10′ long, so select a steel member with a radius of gyration that generates a slenderness less than 200. R ≤ 120″/200 = 0.60″. In this case, one might try out the same 1.5″ tube or use the next size up, depending on the anticipated forces and fit-up needed.

The main point of this example is that we have defined a logical starting point for member sizing. If the designer had proposed 1″ members, this quick calculation would easily identify these as too small regardless of loads.

Example 2

In another twist on the same frame, let's say that you want the lightest weight solution. A designer could specify ⅛″ steel rod for the diagonal acting in tension instead of the heavier tube. This works for load in one direction, but not the other, since the radius of gyration is 0.031″ and the slenderness is 3,840—no good.

The appropriate solution is to install an X-brace consisting of two rods, in which one acts in tension and the other buckles out of the load path. This is a satisfactory approach as long as the taut/un-taut behavior when loads are reversed is acceptable for the frame.

A general understanding of load paths and structural behavior, particularly buckling, can help in creating a successful design. For common elements such

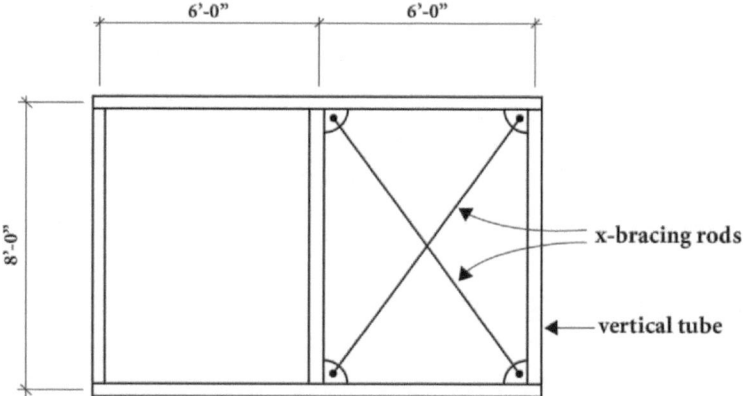

Figure 2.5 Frame braced with X-bracing

as scenic flats or trusses subjected to modest loads, member geometry is often more important than high strength.

Flexure

Flexure combines tensile and compressive behaviors. In a typical simple span beam, the deflected shape is a smile—the bottom stretches and the top compresses. The stress is greatest in the midspan and diminishes to zero at the ends. For a cantilever, the member deflects down, so the top is in tension and the bottom is in compression. The cantilever member stress is greatest at the support and diminishes at the free end. If the cantilever beam has a back span, the back end wants to deflect upwards but is pulled down by the back support.

You can imagine a cantilever beam as a see-saw with the light person at the free end and the heavy person as the "reaction" at the end of the back span, bearing on the ground. Using this image, this explains why the reaction at the first support of a cantilever with a back span is greater than the weight of the object supported on the cantilever (see Figure 2.6; refer to see-saw description above). If the cantilever is rigidly attached to a support without a back span, then there is a moment resisted by the support, rather than resisted by the bending of the back span (see Figure 2.7). Loads on the back span, or different support rigidity, further complicates the math, but the principles are the same.

Let's return to the behavior of the member.

The compressive side of the cross-section presents challenges. As noted previously, anything in compression is vulnerable to buckling. In a building, the floor framing is usually braced by the floor slab, which permits efficient use of members with slender top flanges, such as I-beams, bar joists, and 2x lumber. The exact same members subjected to the same floor loads but without the slab

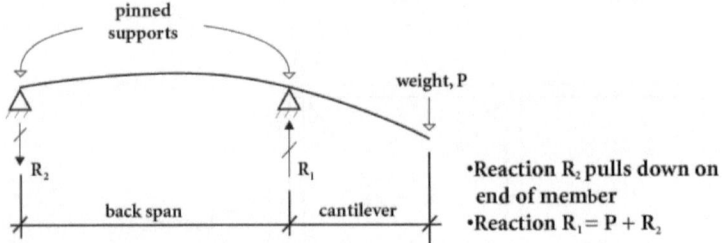

Figure 2.6 Cantilever with back span

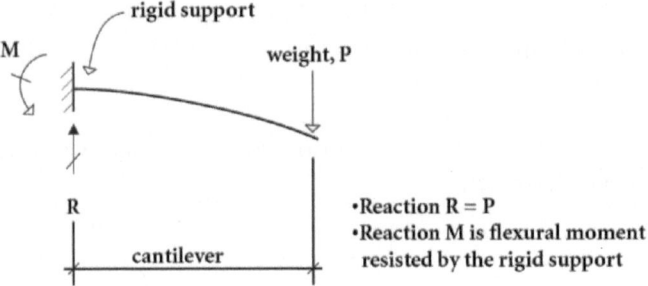

Figure 2.7 Cantilever with rigid support

or floor panel bracing the top flange will buckle laterally, or rather flop over as the top buckles and the bottom remains in tension.

While a continuous plane like a floor is the ideal system to brace compression flanges, transverse framing can also provide effective bracing, as long as it is capable. Bracing to a stiff element such as a wall, frame or major beam is very effective. Bracing that resists member twist is more effective than purely lateral framing.

Tubes and pipes have excellent resistance to lateral buckling compared with comparable size and weight open sections (I-beams, channels, angles).

Horizontal Shear

What is horizontal shear and why should I care?

Horizontal shear makes beams work. It is the difference in behavior between a stack of paper and a piece of wood the same size and shape as the stack. Support the stack of paper at its ends and subject it to beam flexure under self-weight. The center drops straight down, while the ends of the sheets of paper slip by one another, allowing the stack to deflect a lot. The amount of slip increases from the center to the end.

Now do the same to the solid piece of wood. It deflects so much less and does not exhibit the extreme slippage. What is the difference between the one thick piece of wood and the stack of really thin wood called paper? The wood fibers resist slippage in order to create effective beam behavior. This slip resistance is horizontal shear strength. The shear is the greatest at the ends, and is commonly maximum where the bending is minimum, except for cantilever supports.

The shear happens to be directly related to the reaction that accumulates at the support.

Shear resistance is provided by wood fibers naturally "glued" together or by inherent material shear resistance, for which there are values in the engineering codes.

In a truss, shear can be directly converted to axial force in a web member, and one can envision a virtual truss web occurring in the solid web of an I-beam or 2x.

In spliced or composite members, shear influences how connections are made. Imagine bolting or welding two members—one on top of the other—to create a deep composite that is stronger than the two members in parallel. The increased depth creates a much stiffer moment of inertia, thereby increasing stiffness and strength beyond the doubling of the members. Horizontal shear has to be resisted in order to mobilize this composite behavior. In this example, the bolts or welds along the interface resist the horizontal forces, and the force is greatest approaching the supports. (We will leave out equations here.)

Torsion

Torsion occurs when a member is twisted. Torsion can be evil if not handled correctly.

When twisting a plate or an open section, the resistance to twist is by warping of the shape. The thickness of the element (thickness of a web or flange) contributes much more to torsional resistance than the width. For example, if you double the width of a plate, you double its torsional stiffness. If you double the thickness of the plate, you increase the torsional stiff eight times.

Hollow sections resist torsion by creating a spiraling-type of resistance. Sections increase in torsional stiffness as they become larger. More significantly, hollow sections are orders of magnitude stiffer in torsion than open sections. For example, take a 22-gauge metal wall stud and twist it by hand; this is easy. Then take the same stud and close off the open side with a welded plate; you cannot twist this at all by hand.

When designing members to resist torsion, the supports need to reliably transfer the twist. A box truss subjected to torsion cannot simply rest on a support; it needs twist resistance.

Box trusses at first glance are excellent members to resist torsion. They are deep and wide and behave essentially as hollow sections. But remember that torsional resistance is accomplished by spiraling-type resistance. The two vertical faces have diagonal webs to facilitate the "spiral" forces around the truss centerline. However, the top and bottom are built like ladders in which resistance to parallelograming is what handles torsion; as a result, member connections are subjected to large bending moments. This is a lot less effective than the triangulated faces.

In general, it is advisable to avoid having to rely on torsion when designing structures. It is often better to arrange members so that a pair of members resists twist as a "force couple." This is how most headblock beams in theatres are constructed.

Behavior of Common Sections

The following is a brief description of the basic pros and cons of common structural shapes, meant to give the reader a general understanding of key behaviors. It is by no means an exhaustive listing of all features, and there are exceptions commonly made that may be appropriate given certain conditions.

The most common American material grades are noted in each segment. There are other materials that may be common but are not listed. It is important to note that material supply varies throughout the country. Even more importantly, material grades are quite different in other countries, including our nearest North American neighbors.

I-sections

I-beams include wide flange beams, American standard beams, and other I- or H-shaped sections. As noted previously, these sections are the most economical when bent about the strong axis (flanges top and bottom as a floor beam), because the majority of the material is kept apart by the web. They work very well when the compression flange (often the top for common beams) is braced.

These beams readily facilitate bolted and welded connections, because the flanges and web are readily accessible.

A disadvantage includes the weak axis, which is a relatively inefficient distribution of the material. If a beam has significant bending about both axes, then an I-beam could be a good choice, but not necessarily the best.

Similar to the flexural discrepancy between the strong and weak axes, there is a notable difference when these sections are used in compression. The radius of gyration can be vastly different between the strong and weak axes, so one must be careful orienting the column, providing bracing if needed. Wide flange members are often used as columns because the weak axis inefficiency is offset by the ease of bolted connections.

I-beams are relatively flexible and weak in torsion. While the wider shapes can be used successfully in torsion, one must be very careful to properly analyze the torsion, often in combination with bi-axial bending. Also, it is extremely important that the supports of torsionally loaded I-beams resist the torsion through flange connections. A web connection alone is not sufficient.

In today's market in the US, wide flange members are available in ASTM A992, which has a minimum yield stress of 50,000 psi. Other shapes are commonly available in ASTM A36, which has a minimum yield stress of 36,000 psi.

Channels

Channels have similar disadvantages as I-beams, only worse. The unbraced compression flange issues are worse, the weak axis is weaker, and the torsional behavior is very poor. The flanges are notably tapered, requiring tapered shims and making connection a bit awkward.

To make matters worse, when a channel is loaded in its strong axis, it tends to twist unless it is loaded at this magical location offset from the web.

So why do we use channels? Channels are basic shapes that are easy to use, since they are a simple geometry. They cost less to purchase per pound of steel than most other sections. When used in back-to-back pairs connected to one another, they behave more like I-beams and have some advantages in how they are handled.

Channels are typically available in ASTM A36 material, but can be purchased in ASTM A992.

Angles

The only commonly used section worse than a channel is an angle. The flexural, compressive, and torsional capabilities of angles are very poor. Single angles are almost never used as flexural members because they tend to twist so that the open side is down. Angles are so bad in torsion that they should not be allowed to twist. (Try to twist one by hand – it is not difficult.)

Similar to the channel, when angles are used back-to-back connected to one another, they behave more like tee sections. The back-to-back use allows them to be used easily with gusset plates as braces.

Like channels, angles are inexpensive per pound of steel. They can also be very easy to use in bolted connections. Angles are mostly used where the benefit of their simple shape is useful in joining framing members or for secondary members such as bracing.

Angles are typically available in ASTM A36 material.

Plates

Plates are a totally different member than the three-dimensional sections. They are great for building up a collection of plates to create the unique three dimensional object, or when used as a base or floor plate. Plates can be used alone as tension members not subject to compression.

Plates should generally not be used as beams or columns without compression flange bracing.

Hollow Sections

Hollow sections include pipes, round tubes, square tubes, and rectangular tubes. Hollow sections are much, much stronger and stiffer when twisted than any of the "open" section members, and they are better able to resist buckling. They are also strong when bent in either direction.

Tubes are often more expensive per pound of steel than open sections, whereas pipes are relatively cheap. Pipes have a rougher finish and larger tolerances than round tubes. Square and rectangular tubes in steel have rounded edges that can be of varying radii, even in the same cross-section. Aluminum tubes have either rounded or square corners.

Tubes and pipes are often more challenging than open sections when making bolted connections, since there is not a "plate" element readily available for making the attachment. Hollow sections are used when the benefits of their strength and stiffness in all directions are needed. Hollow sections often have thinner elements than comparable open sections, thereby offering weight savings.

Pipes are typically available in ASTM A53, which has a minimum yield stress of 35,000 psi. Round tubes come in a variety of materials with varying yield points, so pay careful attention to what may be available. Structural square and rectangular tubes include ⅛" and thicker tubes, and is typically available in ASTM A500 Grade B, which has a minimum yield stress of 46,000 psi. Thin gauge square and rectangular tubing up to 11 gauge is considered mechanical tubing, and is typically available in ASTM A513 Grade 1010, which has a minimum yield stress of 32,000 psi; however, check with suppliers before assuming a grade.

Types of Loads

Dead Loads

Dead loads include the weight of all permanent components of a structure, including trusses, towers, decking, cladding, etc. Such loads can be calculated using the geometry and unit weights of the materials, or they can be weighed. Structural engineers who commonly work on buildings and bridges generally consider dead loads to be more predictable and accurate than other loads.

Live Loads

Live loads include all loads that are directly caused by the weight of occupants, equipment, and other items that can be moved or repositioned. Props, furniture, scenery, soft goods, speakers, lighting, machinery, and dimmer racks are all considered live load. Building Codes have tables of required minimum live loads that are derived from ASCE-7 "Minimum Design Loads for Buildings and Other Structures." Some common live loads are as follows:

- Stage floors: 150 psf (pounds/square feet)
- Public assembly platforms, movable seats, dance halls, bleachers: 100 psf
- Fixed seating areas, fastened to the floor: 60 psf (the seats are in addition to this load)
- Catwalks for maintenance access: 40 psf
- Residential: 40 psf

These live loads are uniform loads applied to the whole area or portions of the area that may produce the maximum effects. In reality, live loads are actually the combination of many concentrated loads that average out to a uniform live load. The intent of a live load rating for a floor is to have sufficient capacity to ensure that normal use with a collection of real loads will not overload the floor, thereby avoiding an engineering reanalysis every time a new load is added.

For some uses or when equipment or other items are particularly heavy, a specific analysis is needed to confirm the structural integrity of the floor.

The building codes allow exceptions to the required minimum live loads for uses that can be managed, as long as the reduced live load is stated on a sign that is permanently posted. For some specialized entertainment structures, the live loads are specified and managed in detail in order to get the most out of a structure. Such use requires proper documentation and controls.

Another important live load that often arises is for guardrails, which is 50 psf horizontal loading along the top of the rail, but not less than a concentrated load of 200lb.

Dynamic Loads

In general terms, dynamic loads include all loads that are directly caused by movements of objects. In the entertainment industry, dynamic loads are associated with intended motions, including controlled emergency stops. Dynamic loads include revolutions of a winch drum, hydraulic cylinder actuation, fall arrest forces, rolling of wheels, mechanical braking, swinging from a rope, and dancing.

The key to a rigger is that dynamic forces are essentially related to time. The shorter the time to change movement, the greater the dynamic effect, which is often expressed in "Gs" of acceleration. For example, a motion creating 2Gs of acceleration on a body results in a force equal to twice the person's weight.

Mechanical brakes pose interesting challenges. In a typical design, it is desirable to have redundant brakes in the event of loss of one brake, and since you want either one of the brakes to be capable of holding the load, each must be sufficiently robust. Now that there are two brakes acting, there is more braking force, so the load stops more quickly, thereby increasing impact forces ("impact" refers to sudden stop). This situation is made more challenging when those brakes are on a performer flying rig, since you want to be able to reliably stop the performer, yet avoid injuring him/her during the stop.

Shock Loads

Shock loads are a type of dynamic load in which there is a sudden stop, often with a change of direction of the motion, which is typically unintended. Shock loads include chain hoist stops and category "0" uncontrolled stops (caused by the immediate removal of power to a machine causing brakes to fully engage suddenly).

Shock loads can be alleviated by incorporating features that increase the time to stop. Examples include the cushioning in a helmet and the shock absorbing fall protection lanyard. Bumpers in an automobile do the same thing.

Shock loads should not be taken lightly. If there are no components in a load path to decelerate a dropping weight, impact forces can be 20 to 50 (or more) times greater than the weight.

Environmental Loads

Theatres and arenas typically have walls and roofs that keep the environment from being a factor in rigging, with the exception of earthquakes. When you take away weather protection, weather can often be a governing loading condition and an important safety issue. Let's look at several types of environmental loads.

TEMPERATURE CHANGES

Temperature changes will cause all materials to shrink or stretch as the temperature changes. Steel and concrete move about the same amount as one another (which explains why rebar does not routinely blow out of concrete). Aluminum moves about two times more for each degree of change.

For example, a 40′ long piece of steel subjected to 40°F temperature change will change ⅛″ in length if it is free to move; if the same member is aluminum, the change is ¼″. If it is not free to move, significant stresses can develop. Some structures will adjust shape to accommodate these changes if possible, such as an outdoor amphitheater roof.

RAIN

Rain is always a potential load outdoors. The weight of water is related to the ability of a roof drain system to shed water. For the purposes of live event structures, the author generally recommends reserving 5 psf of capacity to handle rain and hail, assuming the roof is appropriately pitched to avoid ponding. Ponding is the situation where the roof deflects under the weight of water, so it collects more water and deflects more, until the roof collapses. Ponding is a serious hazard and must be avoided.

SNOW AND ICE

Snow and ice are obviously only concerns in cold weather locations and seasons. The design weight of snow as per building codes can depend on many factors, including elevation, terrain, geometry of the structure in relation to neighboring structures (sliding snow and snow drift), and local weather patterns. While snow and ice are not a concern in warm seasons, they can actually provide rigging load opportunities since roofs of permanent buildings are engineered for maximum snow loading conditions. As a result, a clever engineer can utilize the reserve capacity for snow load to allow larger roof rigging loads during warm weather, subject to local roof framing and connection limitations.

Sometimes snow removal operational procedures can be employed to take advantage of some roof snow load capacity, but this should only be done judiciously with realistic operational plans. Be very careful to avoid imbalanced loading that might result from snow removal, which in some situations can be worse than not removing any snow.

EARTHQUAKES

Earthquakes occur without warning; the risk is generally based on probability in which the longer the life of the structure, the greater the probability that it will experience an earthquake. In many locations, earthquake loading is not required for short-term temporary structures. Notwithstanding, the author generally recommends some minimal lateral load to account for earthquakes, such as 2% to 5% of the dead load and attached live load (such as lighting equipment).

The probability of earthquakes occurring and possible magnitudes are dependent on location. Major seismic regions generally occur where there is tectonic activity, such as the US West Coast, Japan, and parts of central Asia. There are other moderate seismic areas worth noting, such as the parts of the Rocky Mountains, New Madrid Fault (central Mississippi River area), South Carolina, and parts of the US Northeast.

Earthquakes produce lateral and vertical motions that go back and forth in cycles, which can last from a second to over a minute. The forces in objects from earthquakes depend on the intensity of the earthquake and the mass (weight) of the object. Heavier objects generate more force. Top-heavy objects behave badly in an earthquake, whereas stocky elements behave well.

Earthquakes tend to impose ground-shaking cycles of a second or less (sometimes much less). Overhead rigged elements that are free to swing often will take a longer time to swing when subjected to seismic forces; for these elements, engineers often neglect the mass of these items when calculating seismic effects.

For live event structures, a rigger should be careful using ballast, which itself will be subjected to seismic forces.

WIND

Wind is the nemesis of all outdoor events. Wind often occurs with some warning, but weather prediction for a very specific location is not an exact science and depends on many variables. Wind loads are influenced by region, local geographic features, local construction, seasonal conditions, temperature changes, and even the geometry of the structure under consideration.

Closed structures such as buildings will attract a lot of lateral force due to the walls. In contrast, open structures with no side walls or partial side walls will have lower lateral forces, but they can have larger roof uplift forces. This means that the amount of ballast needed for an open stage roof structure may be more or less than if the same structure has wind walls all around.

Wind prediction has improved considerably in recent years, and there are many reliable weather information services to aid in forecasting. However, these cannot pinpoint with certainty the wind speed and timing at any site. Forecasting of hurricanes often comes with a day or more of quality warning. Forecasting of tornadoes will only predict the likelihood of them occurring. The same is generally true for summer thunderstorms. In desert locations, sudden microburst wind can occur without warning.

The building codes have complex formulations to determine design forces on structures that are affected by location, structure shape, height, openness, and even the number of people in the structure. Wind on local objects and edges of the structure can be more than on the overall structure. There is different wind pressure on the windward face compared with the leeward face. It gets complicated quickly.

To make matters trickier, wind speeds are reported differently depending on where and who is reporting. US codes calculate wind forces using three-second wind gust. European codes use ten-minute average, which is a lower number than the corresponding three-second gust. News services often report one-minute average speed.

The codes include a design wind speed as shown on a map for buildings and similar structures, from which design wind forces are determined. These forces are intended for permanent structures and represent a statistical probability of that force being exceeded in 50 or 100 years. For a temporary structure, there is a lower probability. For this reason, a reduced design wind speed is often used for temporary structures. How low should the wind be reduced? The probability of the reduced wind occurring in the short time frame should be the same as for the building subjected to its code wind forces in its design lifetime.

For temporary structures, not only is it rational to use a reduced wind speed, but the nature of the wind predictability and the structure itself can help mitigate risks. For example, since hurricanes can be forecast more than a day ahead of time, some temporary structures can be partially or fully dismantled to avoid damage in a hurricane. Some structures are designed to rapidly remove items that could catch a lot of wind, such as fabric wind walls.

Weather patterns may be changing more quickly than in the past, presumably due to global warming, which means that our industry needs to continue to be diligent in addressing weather issues. Improved planning for weather is necessary, including weather monitoring and action plans that get implemented when certain weather conditions arise or become a risk. Temporary live events differ from permanent structures in that you can take advantage of people involved in managing the temporary structures. Operational actions are effective for keeping these events safe, as long as they are planned in advance, confirmed that they work as intended, and then executed when needed.

Deflections

Deflections of structures can affect the appearance, feel, and usefulness of a structure. A staging platform that is very flexible may feel overly bouncy; rolling a stage wagon over this platform may be difficult.

In addition, excessive deflections could affect the structural integrity and stability of a structure. For example, an outdoor stage roof without guys may sway laterally a foot; in doing so, the roof weight now bears on a tower that bent into a slight "S" curve, thereby bending the tower. As a result, the tower shifts over a little bit more. It keeps going like this until the system either stabilizes or falls over.

The building codes provide deflection limits than can be useful for comparison with entertainment systems. (Deflection is compared with the beam span "L" between its supports.)

- Building floor live load: Maximum deflection = L/360
- Building floor total load: Maximum deflection = L/240
- Building roof live load: Maximum deflection = L/180
- Building beam supporting elements that can crack easily (brick, glass block): Maximum deflection = L/600

For some entertainment structures, these limits can be relaxed, but a deflection of more than L/100 is not advisable. In contract, where visual or machinery alignment is needed, more stringent deflection limits may be warranted.

Determinate and Indeterminate Structures

Determinate Structures

A determinate structure is a structure in which load distributions to supports are influenced by load and support locations alone. In terms of entertainment rigging, a determinate structure is a load system supported by multiple hoists in such a fashion that small moves of one hoist do not cause large load shifting to occur between hoists in the lifting system (Section 2.6 of ANSI E1.6-3-2012). A beam is determinate if it has two supports. A rigid curved or triangular truss arrangement is determinate if it has three supports.

Determinate structures are advantageous in that their reactions are very predictable and easy to calculate, they are easy to understand, and control of hoists is rarely a problem. Load shifting does not occur, so they are reliable for entertainment truss arrangements.

A popular opinion among entertainment riggers is that determinate structures are beneficial.

However, what happens if one support fails in a determinate structure? The assembly collapses. In this case, the problem is that a single point can cause system failure.

Indeterminate Structures

An indeterminate structure is a structure in which load distributions to supports are influenced by load and support locations, as well as by structure stiffness. In terms of entertainment rigging in this document, an indeterminate structure is a load system supported by multiple hoists in such a fashion that it is not practical to calculate with accuracy the dynamic load on any one of the hoists due to load shifting (Section 2.11 of ANSI E1.6-3-2012). The structure is indeterminate if a straight line can be drawn through all of the supports carrying the load system and if it is held by more than two supports. If a straight line cannot be drawn between all supports carrying a load system, the structure is indeterminate if it is held by more than three supports.

The main advantage of indeterminate structures is that they mitigate single-point failure concerns. In a critical structure, an important objective is to avoid structure failure or progressive collapse if any one of the structural elements fails. For example, when hanging a ring-shaped structure, it may be advantageous to use at least five hangers or hoists so that if any one fails, the system remains stable. Engineers will commonly design the remaining structural system to remain safe, possibly with lower safety factors. In earthquake design, the goal is to ensure redundancy in the structural system to protect against catastrophic collapse, maximizing the use of indeterminate structures.

Indeterminate structures can also be used to improve stiffness for a given member size and span.

Indeterminate structures require the involvement of a structural engineer in order to accurately determine member forces, deflections, and support reactions. Relatively stiff members and precise construction techniques allow a design to benefit from an indeterminate structure.

When a series of hoists lifts an indeterminate structure, the hoists will run at slightly different speeds with varying start-stop characteristics. This causes the flown structure to distort. A stiff structure will redistribute the loads via its own bending resistance, wreaking havoc with the loads to hoists. In the entertainment rigging industry, flexible aluminum trusses and pipe battens are often used to rig equipment. These flexible structures are very helpful in accommodating the significant inaccuracies resulting from the use of chain hoists, cable baskets and roundslings.

Sometimes this flexibility is also a potential problem. The author is often concerned about these flexible rigging trusses when they support tracking video screens. The loads on the trusses change as the screens travel, yet the screens want to appear to be level throughout travel. This would be a condition that deserves the use of load cells to verify maximum support reactions during the installation at each venue in order to ensure the system behaves safely.

In conclusion, determinate and indeterminate structures each have pros and cons. It is important to understand these when deciding the type of support arrangement for a structure.

Closing Comment

Anyone involved in technical aspects of entertainment can contribute significantly to structural safety and stability by improving awareness of structural behavior, without needing to do any math. These behaviors are open to all to see, and many of us have a great feel for such behaviors simply by having a lifetime of observing our world.

Such awareness will enrich and improve your experience in developing solutions to entertainment technical challenges, and help avert problems. It will also empower you to productively guide your structural engineering advisor.

Now that you are more aware of structural behavior, tap into this insight and use it!

3

Lighting Truss

TRAY ALLEN

The Creation of Truss

In the beginning, there were wooden and steel beams. Steel trusses were eventually designed and of course the Great Production Company of Old took smaller versions of these trusses and started to use them in shows. The people setting up the steel truss said it was heavy and no good. So someone at the Great Production Company of Old said "What about aluminum?" So they tried it and said, "This is good." This is a simplified explanation of what could be described as an explosion and the formation of an industry.

Lighting truss is an assembly of metal, aluminum, or composite materials formed in such a way as to maximize the distance supported (the span), then maximize the amount of weight that can be applied to that assembly (the load), and finally minimize the weight and size of the assembly. In America, the 1960s saw the British invasion, not only with rock 'n' roll but also with their touring gear. The truss was steel and the technology was borrowed from antenna towers. Weight capacity and deflection were not calculated and structural engineering was still off in the future. My earliest experience with truss was at Rock Creek Park in Washington, DC. That truss had a complete mechanical system to extend the truss over the stage area. The main purpose of the steel triangle truss was to hold a curtain to allow it to open and close by parting in the middle.

So why does everyone now use aluminum? Aluminum is the most common metal found on planet Earth. It is lightweight, yet very strong pound for pound. It maintains this strength when used in normal temperatures, from 0 to 120°F. Aluminum oxidizes easily to protect itself from the elements and is also relatively inexpensive. But there are some negatives to using aluminum: it conducts electricity, and it reacts readily to alkaline materials, concrete, steel, mercury, and bromine (just to name a few).

What is Truss?

Let's start to look at a truss structure as an additive structure. Start with a pipe, load it between two points until it scares you to walk under it, or until it fails (Figures 3.1 and 3.2).

What if we took two pipes and added some vertical pipes to help the pipe assembly support more weight? So we take two pipes and add other similar pipes (spreaders) between them (Figure 3.3).

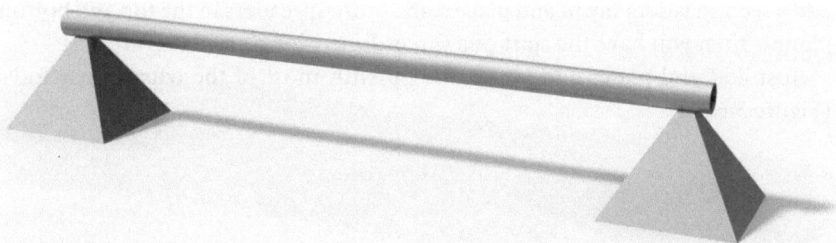

Figure 3.1

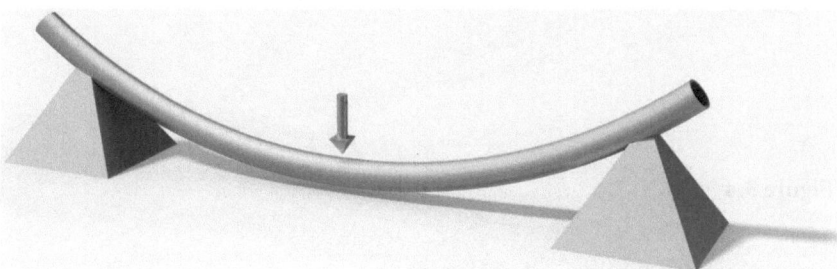

Figure 3.2

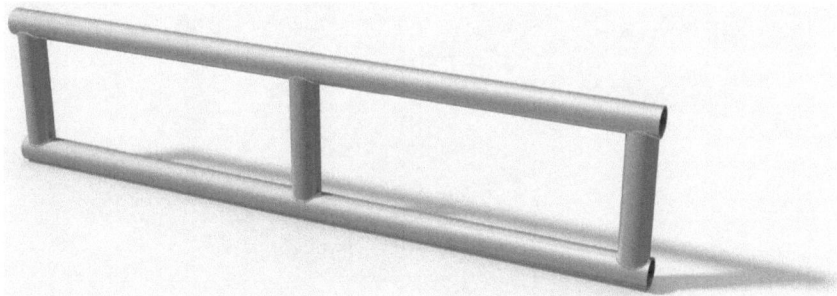

Figure 3.3

When we draw lines running through the system to represent forces we find a conundrum. These forces like to take the shortest path to the lifting or supporting device. These right angles are bad in structures. What can we do to improve this? Think Egyptian pyramids or the Eiffel Tower. Triangles are strong. To make truss stronger, add triangles aka diagonals (Figure 3.4).

This is a ladder beam or two dimensional truss. This ladder behaves kind of like a 4′ × 8′ sheet of plywood—when you are holding it vertically it is strong but when you hold it horizontally it starts to bow or deflect. If we were to add a second sheet of plywood with some 2′ × 4′s around the perimeter and through the center we would start to stiffen the structure. The same thing is truss if you add a second ladder beam and place some more spreaders in the top and bottom plane—then you have the start of a three-dimensional truss (Figure 3.5).

Just add end plates and you come up with most of the truss in use today (Figure 3.6).

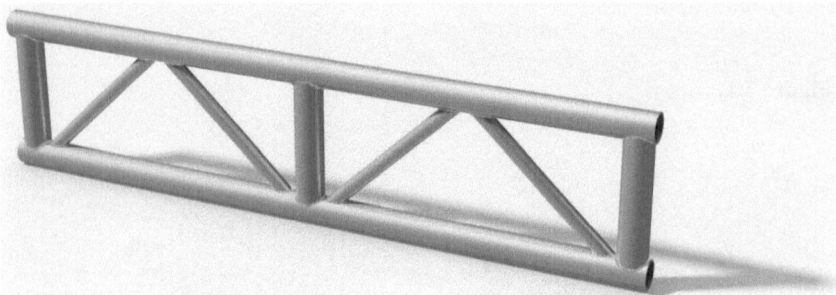

Figure 3.4

Figure 3.5

Figure 3.6

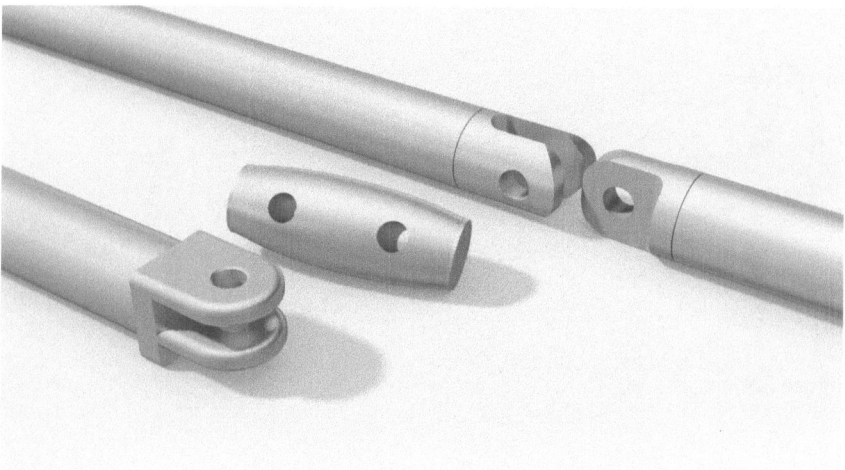

Figure 3.7

To advance the truss design further, and increase the structural capacity, you need a connection inline with the main members or chords—forks, or eggs (Figure 3.7).

That covers the design of the simple box truss. But truss comes in many forms: ladder (Figure 3.8), two mains with rungs or diagonals, triangle (Figure 3.9), square (Figure 3.10), trapezoid (Figure 3.11), hex (Figure 3.12), octagonal (Figure 3.13), even round (Figure 3.14).

Most manufacturers of truss have around 20 different box truss designs and seven or so triangle truss designs. The reason for this is simple—the customer has created the demand. "I need an inexpensive truss that will span 20′ and support 400lb of weight at the center of the span. What are my options?" The

Figure 3.8

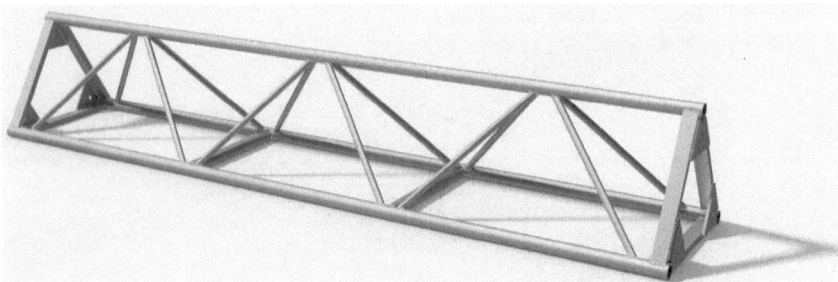

Figure 3.9

Figure 3.10

Figure 3.11

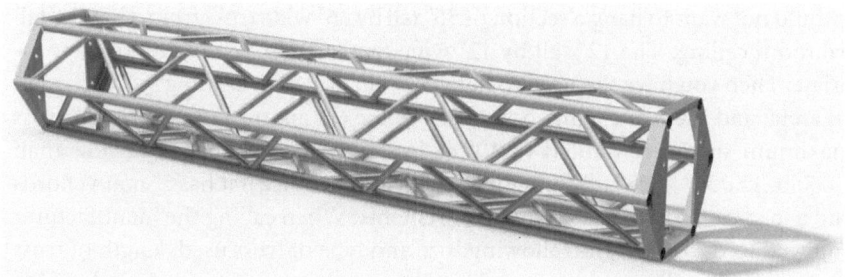

Figure 3.12

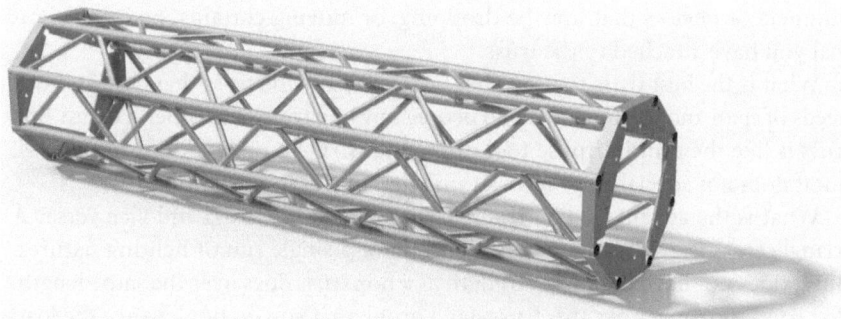

Figure 3.13

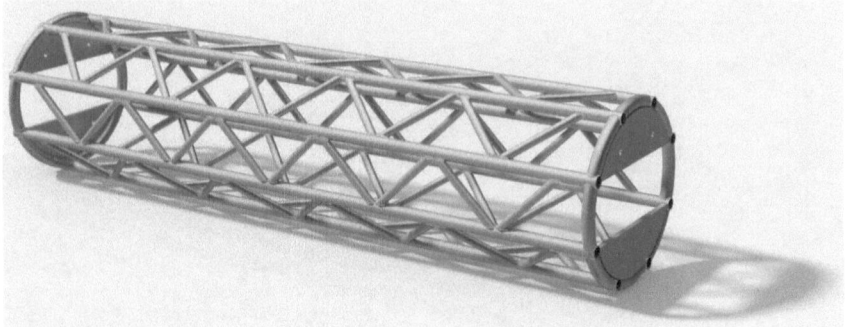

Figure 3.14

request for these options that come from various constraints given by the customers forces the manufacturers to come up with all these different varieties. I would not want to hang a section of 36″ tall by 36″ wide truss from a low 12′ tall ballroom ceiling. The 12″ tall by 12″ wide truss would be much more appropriate. Then you have the need to hang 5,000lb under an existing scoreboard in an arena and the two rigging points I have are 60' apart. Most 12″ × 12″ truss maximum spans are limited to 40' or less with a load capacity far less than 5,000lb. The 36″ × 36″ truss might be the truss of choice if it has 3″ main chords and a load capacity to match the requirements. When calling the manufacturer you will need to know the following: size and type of truss used, length of truss between the hanging points, actual weight of "whatever it is you have hanging from the truss," and where the "whatever it is you have hanging from the truss" is actually hanging on the truss. Take special care not to forget the details of where a lighting cable bundle is making its downhill run from the truss to the dimmers or devices that may be dropping, or moving curtains, or other items that you have attached to the truss.

What is the best truss to use? The short answer is the truss that best fits your needs of span and loading. Why do people buy box truss over other trusses? Box truss is like the multipurpose tool of the industry—it does a lot of things well but it does not specialize or have a thing that makes it outstanding.

What is the advantage of a triangle truss over a box truss and vice versa? A triangle truss is great for hanging a curtain or a single run of lighting fixtures. But it does not have the same strength as a box truss does over the same length. Box truss is stronger but this truss has a problem if you do not balance the load on both sets of bottom chords. Box truss tends to try and form a diamond shape if those loads are not balanced. The heavy portion of the load is on the lower set of the bottom chords of the diamond. You can correct this by hanging lighting fixtures or even ballast on the higher bottom chords. Even this diamond shape might be allowable if you consult with the manufacturer.

Making the Truss

In project design you have the triple constraint triangle with cost, scope, and schedule at the three points of the triangle and quality in the center. In truss design, the triple constraint triangle would include weight of the material, strength of the material, and quality of the material. The quality of the aluminum and thickness of the aluminum must be balanced with weight minimization. If you decrease any of these three points too much you will find that your quality (the stuff at the center) has dropped out of sight.

Figure 3.15 shows the typical bolted truss and all the members that make this truss.

The primary material in the American-made entertainment truss business is aluminum 6061-T6. The Aluminum Association describes 6061-T6 as "a system of four-digit numerical designations is used to identify wrought aluminum and wrought aluminum alloys. The 6XXX series refers to the content of magnesium and silicon present in the alloy. The last two digits identify the aluminum alloy or indicate the aluminum purity." The T6 nomenclature means the aluminum is solution heat-treated and then artificially aged. The original truss from the UK was mostly 6082-T6. While the 6082 has better machinability and base properties than 6061, most manufacturers now go for a higher quality 6061, which has a better yield and percentage of elongation. Yield is the point at which the material begins to deform plastically. Percentage elongation is the specific amount of stretch that a pipe or tube will tolerate before it fails.

The aluminum shapes used are pipes or tubes. When it comes to selecting the strongest tube, the thicker the area occupied by the aluminum, the stronger it will be. The tube could be solid aluminum, but that would be a lot of additional weight. The designer of truss tries to find the right strength-to-weight ratio.

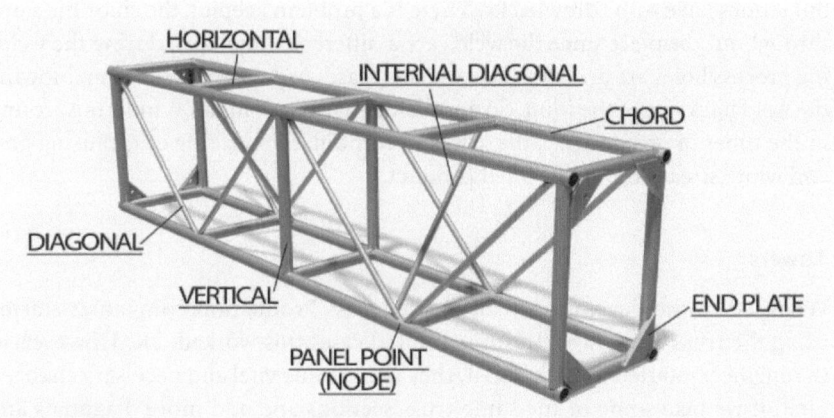

Figure 3.15

All of this tubing gets connected with two basic types of welding: TIG and MIG. TIG (Tungsten Inert Gas) is better known in the welding world as GTAW (Gas Tungsten Arc Welding). MIG (Metal Inert Gas) is better known in the welding world as GMAW (Gas Metal Arc Welding). TIG is used for holding the diagonals and other large members together. MIG is used to attach end plates to the truss and generally filling larger gaps. MIG is a faster process, but generally requires cleaning of the welded area after the fact. Currently, hybrid welds are being used with and without robotic welding. The hybrid gives the speed of MIG welding with the look and finish of the TIG welding. Even better than that, some of the welds are starting to rival that of the Friction Stir method of welding. The Friction Stir method uses great pressure and speed to mix the materials together to where the finished process has a molecular structure better than the base materials. The Friction Stir method is generally used for joining plates of aluminum together. This method has a long way to go before it breaks into the world of aluminum truss. The bad news about welding (except for the Friction Stir method) is that it weakens the strength of the base materials. Currently, for 6061-T6 weld-affected regions, the Aluminum Design Code 2010 has an allowable tension stress of 9.1ksi. This represents a 27.2% reduction from Aluminum Design Code in 2005 that had an allowable tension stress of 12.5ksi (ksi is a measure of 1,000lb per square inch). This is important to know when looking at standards, engineering, and tables from different manufacturers. When was this engineering done for the truss?

And what do you do when you want something different than the gray industrial look? Powder-coating is a process in which the powder is electrostatically charged, then the powder is sprayed on to the grounded truss, then the truss baked. This is probably the best method for changing the color of your truss. There is a limit to the number of times you can powder-coat the truss. Anodizing is a chemical and electrical etching process that changes the color of the truss, but it does have some drawbacks. There is a problem keeping the color the same throughout the piece since the welds are a different alloy. Also, during the welding process holes are drilled to allow heated gases to escape to keep from blowing the weld back out of the joint. So anodizing may result in salt wandering around in the tubes or, even worse, the salt acid combination leaking out, causing grey and white streaks on the finished product.

Towers

What led to the development of tower truss? Production companies started using the truss as a tower. In some limited cases this worked OK. However, as the engineers started to look into it, they made some vital and necessary changes. First, if we take some of the same truss sections and add more diagonals and cross-members to the section then we can then use this truss vertically. By adding internal cross-braces we help keep the truss in its box shape.

Next, if we increase the size of the main chords then these towers can support some substantial loads. (Example 12″ × 12″ towers 30′ tall supporting two tons of weight.) If we change those main chords to a 3″ diameter chord with a ¼″ wall thickness we can build towers to heights of at ranging up to 80″ supporting around 20,000lb. When towers are used outdoors they need guy wires, ballast and/or earth anchors. Ballast is generally in the form of concrete barriers. If you are attaching guy wires to these barriers they will need to be kept from sliding across the ground. As you get the barriers closer to the tower the amount and weight of the barriers will need to increase. When I use the term "earth anchors" I am not referring to tent stakes (a long solid cylinder of steel with a point at the end). I am referring to a system that is driven, pushed, or screwed into the ground. Tent stakes might keep something like a concrete barrier from sliding but they are not a solution to guying (attaching a guy wire) to a tower. These tent stakes can become ineffective if the condition of the ground changes. When it rains and is windy is a perfect example of this condition. The mainstay for outdoor use is solid ground, a solid base design, and attention to the weather when using towers. Don't forget that aluminum conducts electricity, so use proper grounding techniques if outdoors where lightning might occur. Also, keep in mind the location of power lines and other obstacles at height when erecting towers.

When indoors or even in a stadium it might give the illusion of safety and the temptation to not use guy wires or ballast will be great. Do not give in or make this mistake, if you are indoors, make sure it is not subject to the open loading door where the wind blows in and knocks your towers down. That said, the popular sizes of towers start at a cross section of 12″ × 12″, 15″ × 15″, 16″ × 16″, 20.5″ × 20.5″ and move up from there. Tower heights vary they start at 20′ and move up to around 80′ tall. The soil and base detail constraints are significantly different between these heights of towers and there weight capacities. The 80′ tower will probably require a footer detail composed of concrete and rebar and a base of solid steel or aluminum to match. The 20′ to 50′ towers may require only one to four sets of guy wires per tower (depending upon the configuration). Towers taller than 50′ may require up to eight guy wires with four of these being a mid-set of guy wires. When planning to set up towers, make sure you know who the authority with jurisdiction is in the area you plan on using these towers. Engineering reports, permits, soil engineering, and other items may be required.

Setting Standards

When the rigging working group was formed back in the late 1990s under what was known as the ESTA (Entertainment Standards and Technical Association), the first two issues that were pushed to the forefront were a wire rope ladder standard and the need for truss standards. So the battle over trying to come up with consistency, and making an effort to bring sense out of the chaos of loading figures, safety factors, and design factor began. The main battle cries of 5 to 1 design factors, 10 to 1 design factors, 300%, and 500% still ring in my ears. A factor of design, or a design ratio, is how many times the maximum allowable load the manufacturer expects will be bearable before the structure will fail or become unsafe or unusable.

This factor is sometimes given as a percentage; i.e., 300% means the design should hold three times the recommended load. At least that's what the term *should* mean. But then you start digging into what the manufacturers and engineers really meant by these numbers. Given a choice you would go for a piece with a 500% rating over a piece with a 300% rating, right? But one manufacturer's 300% rating might have meant you would have a structure with a permanent smile or bow in it if you loaded it to three times its recommended weight. The permanent smile or bow means you have permanent deflection. The permanent deflection means that you have probably turned this truss into a recycler's dream. The other manufacturer's 500% rating might have meant that would be the point where you needed to call the recycling company. The main point of this section is that if you have these figures dancing around, make sure you know what they mean.

Several ratios are used by engineers to give these loading figures a margin of safety: deflection shall not be greater than the length of span of the truss measured in inches divided by 160. What does that mean? Well, let's crank some numbers into the equation using a 40' span of truss: 40' is equal to 480" So the measured deflection at the center should be no more than 480/160 = 3".

This is fairly easy to check and does not require anything more than a calculator and tape measure. Say we have a uniformly distributed load on a 40' span of truss. The end chords of the truss are both at 36" off the floor. When we measure the center of the truss we find it is 32" off the ground. Is this truss okay? No, that is a deflection of 4" and it is higher than the ratio given. What should you do? Hang more points? Bridle the points? Remove weight from the truss? Correct answer: d) all of the above, or a combination of the above until you get the deflection to 3" or less.

More recently we have added the repetitive use of the truss into consideration. For example, a truss may be rented from your local production company for a show, which means that truss will be considered as being used repetitively. The loading from this truss will be multiplied by a factor of 0.85. If the truss catalog sheet says 1,000lb uniformly distributed load, this will be reduced to a load of 850lb. The loading will be 1,000lb uniformly distributed load if that same truss is purchased from the manufacturer and permanently installed in a building.

I get asked the following questions a lot: What can we do to protect our truss? What causes damage to truss? Surprisingly, the most common damage is not the overloading of truss but the old theatrical c-clamp. When this clamp is used without a piece of PVC tube to protect the aluminum, a ½" gouge is generally formed by the screw. This damage is then aggravated by a lighting designer that says "I need you to lift that whole fixture up when you focus it." This action starts to saw the main chords in half. To solve this damaging practice, use a half coupler or triggered coupler to locate lighting fixtures or other devices to the truss.

What else damages truss? Forklifts bending or chewing the internal diagonals or the diagonals in the side of the truss are probably the second largest cause of damage. Use care when using a forklift to pick up and place truss. Take care of the truss and it will take care of you. The third cause is allowing someone to drag the truss across the floor by one end. The solution to this problem is always use two people to pick up the truss sections. Invest in truss dollies to help move the truss around. Next is improper rigging from the truss using slings to hold another load underneath the truss without anything like a pipe or tube either in the truss or attached with swivel couplers to keep the sides of the truss from being pulled in towards each other.

Finally, the truss loaded into the truck improperly, allowing objects to rub against the truss or smash into the truss. All of these cause more damage than overloading of the truss.

To prevent the overloading of the truss: follow the loading tables, take a class on rigging, hire professional riggers.

Hanging or Suspending the Truss

How do we get this truss into the air? We use two slings of the appropriate length; from both sides of the truss we choke the bottom chords wrapping the top chords forming a triangle at the top, placing the loose loop of the sling into a shackle. This is done at panel or node points where two or more structural pieces meet. These panel points keep the slings from sliding.

Follow the manufacturer's guidelines when suspending truss from the top or bottom chords. For the bottom chords: choke the bottom chords at a panel point on either side of the truss forming a triangle and attach a shackle at the peak of this triangle.

For the top chords: choke the top chords at a panel point on either side of the truss forming a triangle and attach a shackle at the peak of this triangle. Use the chart below to help you determine the appropriate sling length for the angle to be less than 30 degrees. Notice how the force versus the angle of the sling changes.

Yes, it is possible to pull the top or bottom chords towards each other if you do not keep this in mind.

Picking up a box grid of truss from the bottom chords is not a problem for the center of gravity. Picking up a single span of truss from the bottom chords is not recommended due to it moving the center of gravity of the load, thereby making the load unstable.

You could use a lifting point a device designed to connect both top or bottom chords of the truss and attach to the lifting device. Most lifting points are designed for use on either two top chords or the two bottom chords of a truss span. If you were in a box grid, four trusses connected with corner blocks attaching lifting points to the two bottom chords is possible and good practice. On a single span of truss, however, if you locate the lifting points on the bottom chords you will have a truss trying to do a balancing act. You have moved into a less stable lifting system. Do you move the lifting points to the top chords? Always follow both the truss and lifting point manufacturer's guidelines when you use them.

Innovations in Design

In the 1970s, truss spot operators began being flown from trusses. In the 1980s, moving lights started coming. The first moving lights were huge and placed a moment back into the truss that caused an interesting ride for some of the truss spot operators. A moment would be like a pendulum swinging on the truss, causing it to move in a circular motion or back and forth depending on the lighting cues. And with the addition of these lights came more cables. Control cables to move the fixture around, make the light coming out of it change color, move patterns in and out, change focus, and even rotate the patterns.

To explain my love-hate relationship with cables, let me go back a little bit (cue the dream sequence . . .). The rig I cut my teeth on was 120K pars in Pre-Rigged Truss (PRT), 91″ long, 26″ tall, and 30″ wide. The front of house truss: five sections of PRT with ten Socapex 19-pin connector cables in various lengths taped together in a bundle powering the ten lighting bars, each with six parcans. That meant 60 parcans, each with 1,000-watt lamps. The cables we used were only 1.5mm, or 16awg/18-conductor cable. This cable would generally swag to the rear truss. The rear truss was also five sections of PRT with its own bundle of ten cables. These cables would drop to the dimmers located either upstage left or upstage right depending on where the main power feed was located. I apologize now for my carbon footprint. But back to moving lights, each cable added more and more weight. For a time, I figured with more moving lights the parcans would completely go away. Nope, the rigs just got heavier with a combined weight of moving lights and pars.

Why the size and the width and depth? The pars were on 15″ centers, so when you do the math the lighting bar length worked out to be 90″. The 91″ truss allowed you to contain 90″ lighting bars without bumping into each other.

Another clever development for the lighting truss was the octagonal color frame. This allowed for a tighter focus between instruments than you could have with a square color frame while still making it possible to cut gels into shape with a paper cutter. Has anyone else cut themselves with a razor knife while trying to cut gels for a followspot? Octagonal was a glorious innovation. This frame allows

12 lighting fixtures that are in this pre-rig truss to be focused in many different areas across a stage.

The next innovation in the world of the lighting truss was the LED. The first versions of these looked like a million tiny indicator lights such as the kind you find in a science fiction movie from the 1970s. The first fixtures to use LEDs were box-like contraptions, or simply a par can that had the LEDs shoved in them to replace the lamp. As these fixtures progressed, they became brighter and heavier and developed a fan noise like their moving light kin. But they needed a lot less power: 90–300 watts with similar or greater output than the 1,000-watt pars. This innovation may finally be the death of the parcan. The interesting thing is that now the rigs with the LEDs use less power, but they weigh more than the parcan rigs. And now there are moving LEDs, so they have gained even more weight.

It's a bird! It's a plane! No, it's a video wall! With the move for image magnification came various attempts at masking and using projectors outdoors or in brightly lit convention centers. That worked OK, but to really make a statement you needed huge walls of video monitors to show how important you were. Somewhere back in the 1980s we had a tube-type video display. We commonly called these video walls. Now the LEDs are multicolored and allow resolution limited only by the client's wallet. The current ones are not as heavy as their cathode ray tube (CRT) predecessors, but they occupy huge spaces across the truss. This brings a new set of problems: how do you hang them, and where do you hang them? A bumper or interface from the truss to the LED wall is generally provided. Why some people that own the walls and the bumpers still don't know how much these things weigh is one of the great mysteries of life similar to whether the light in the refrigerator stays on when the door is closed. I have had numerous phone calls similar to the following:

> The truss seems to be deflecting or smiling a lot.
> Really, what is hanging from the truss?
> Well, we have this video wall hanging from five hoists.
> OK, what kind of hoists?
> One ton or two ton. Hang on and let me ask the rigger . . . He says they are two ton.
> Really, so are you saying to me that you are hanging ten tons or around 20,000lb. from this single truss?
> No, the touring group's rider said that it was only 5,000lb.

Before you hang it know how much it weighs. We will all sleep better.

The Future of Truss

So, what is the future of lighting truss? What material will it be made of? Will it continue to be aluminum, or maybe carbon fiber?

Bolted truss connections appear to be here to stay despite the better loading efficiency of a spigot, forked connection, dual forked connection, or the egg connection. In the future will we just have five connections to make? Four of these connections would be the actual chords themselves, and the fifth connection would be a ring circuit connection for the electricity. The control for all our other lights, gadgets, and even the motors would either be wireless or would be controlled by a signal sent down the hot leg of the power input. Is it possible for five simple connections per truss to become reality?

The appropriate question is: when will the next quantum leap in truss design occur? Will it come from new materials or from new customer needs? Whatever it is and whenever it happens, I hope to be around to see it and I hope the end users will say "This is good."

Sources and Credits

ANSI E1.2-2012 Entertainment Technology Design, Manufacture and Use of Aluminum Trusses and Tower.
Chart: Safetysling.com.
Drawings: Angel Hicks and Isaac Cogdill with James Thomas Engineering, Inc.
The Aluminum Design Code 2010.

4

Arena Rigging

ROY BICKEL

Introduction

While it's a bit difficult to pinpoint exactly when arena rigging began, it's a pretty safe bet it started with the touring ice shows somewhere in the early to mid-1960s. By today's standards, the methods used at the time were pretty crude. There was very little truss. Most of the hanging pieces were made of steel pipe and the methods used to hang the pipe, block and tackle, as well as manual chain falls were cumbersome at best. As rock 'n' roll shows became more sophisticated, their crews took this technology and began adapting it for their own use. Instead of block and tackle they started using electric chain hoists. Rudimentary truss replaced pipe to hang lighting and sound equipment. Almost overnight, it seemed, an industry was born.

In those early days the chain hoists were attached directly to the roof beams. This meant that the hoists didn't move and their chains ran up and down. The biggest problem with this arrangement was the hoist weighed at least 130lb and it was no fun to haul 80' or 90' in the air. The other problem was controlling the chain, which was gathered into a bucket up at the hoist. If there was a problem with the chain—and there usually was—then it was up at the ceiling and very hard to reach. The solution to both problems was to turn the hoist upside-down. This meant that the hook end of the chain was raised to the building steel and the hoist was at the floor where it was free to ride up and down on the chain. This solved both the hauling problem and most of the chain issues.

It's been pretty much that way ever since. This is not to say that there have been no technological advances in arena rigging; there have, but the fundamental way we hang a show remains surprisingly the same. The big differences between then and now lie in the size and complexity of the shows. But to understand these differences we should first look at the basics.

The Equipment

A chain hoist is an electrically operated material handling device that is designed to raise and lower objects. It's a not terribly sophisticated machine that, at least

Figure 4.1 Electric
chain hoist

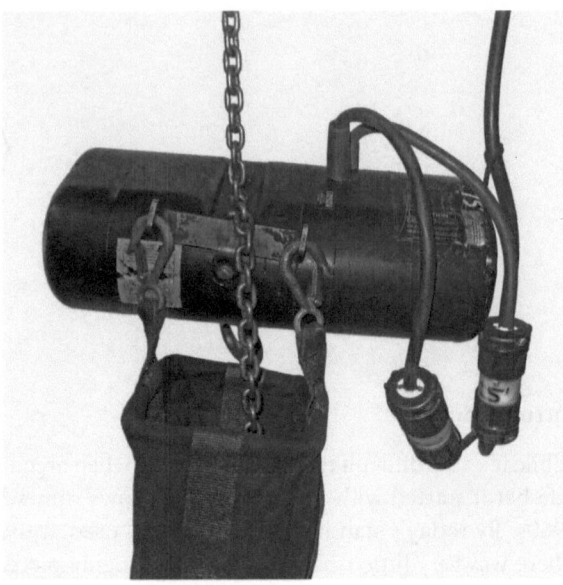

in the basic models, has very few bells and whistles. Standard chain hoists* come
in load ratings ranging from ⅛-ton to 3-ton. Larger capacity hoists are available
but seldom used in the entertainment industry. How high a hoist can lift an
object depends on the make and model of the hoist, the amount of travel
between the upper and lower limit switches, and other factors inside the hoist,
but it is usually around 130′.

The hoist has a hook at the end of the load chain to connect to the building
steel and one on the bottom of its housing to connect to the load being lifted.
A chain bag is hung from the hoist to collect the chain as it passes through the
hoist.

The hoist load chain is most often attached to the building steel by means of
a wire rope sling and screw-pin anchor shackles. The sling is typically wrapped
over the steel beam leaving both ends of the sling hanging below the beam.
When used in this manner, the wire rope sling is called a basket.

Sometimes, when the building is particularly high and the trim height of the
hoist is relatively low, a wire rope sling called a stinger or downtail is placed
between the basket and the chain hook. This is done to reduce the amount of
chain weight the rigger has to pull.

---

*The common practice in the entertainment industry of calling these devices chain motors is
inaccurate. A chain motor is the electric device inside the hoist that turns the shaft allowing
the hoist to lift the chain. The hoist is the entire machine.

Figure 4.2 Roundsling with synthetic core

The load rating of the hoist will determine what size wire rope and shackles to use. For all sizes up to and including 1-ton, a ⅜″ diameter wire rope sling and ⅝″ diameter shackles are used. For 2-ton loads we typically use a ½″ diameter wire rope sling and ¾″ diameter shackles, while 3-ton and higher rigs are custom-designed to meet the specifics of the situation.

The connection of the hoist to the load is normally accomplished with a roundsling. These slings come in two varieties: synthetic core and wire rope core.

The synthetic core sling is made of 100% polyester. The core has filament polyester laid in a circle, creating many loops of the fiber. Polyester is then woven into a shell that fits over the core and protects the core from damage. In the commercial world, the color of the shell identifies its load rating. Not so in the entertainment industry, where the shell color is usually black. For us the load rating is on the label, both in the printed information found there and in the color of the label. A green-rated roundsling is typically used for a 1-ton or smaller rig. Yellow is used for 2-ton.

The wire rope core roundsling is a relatively recent development. It came about when fire marshals began expressing a concern about the relatively low melting point of the polyester in a synthetic sling. (A polyester sling will begin to suffer permanent damage at approximately 200° F.) To resolve this problem, one of the roundsling manufacturers came up with a roundsling made of wire rope. They replaced the polyester core with two lengths of ¹⁄₁₆″ diameter wire rope that are long enough to create approximately 50 loops. The core was then encased in a polyester cover to hold everything together. The result is a roundsling that is just about as flexible as a synthetic roundsling and has a load rating equal to that of a green polyester roundsling. The label comes with a

Figure 4.3
(a) Roundsling
with wire rope core;
(b) Wire rope core
exposed

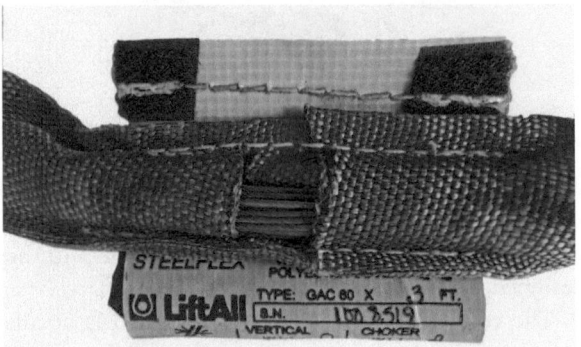

Velcro attachment that allows you to open the sling without damaging it to show the fire marshal the steel wire rope inside.

The roundsling, regardless of what core material is used, is then connected to the hoist with the ubiquitous screw-pin anchor shackle.

Figure 4.4 Screw-pin anchor shackle

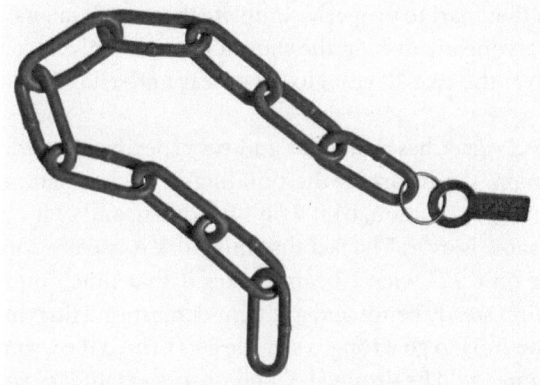

Figure 4.5 Long link
(STAC) chain

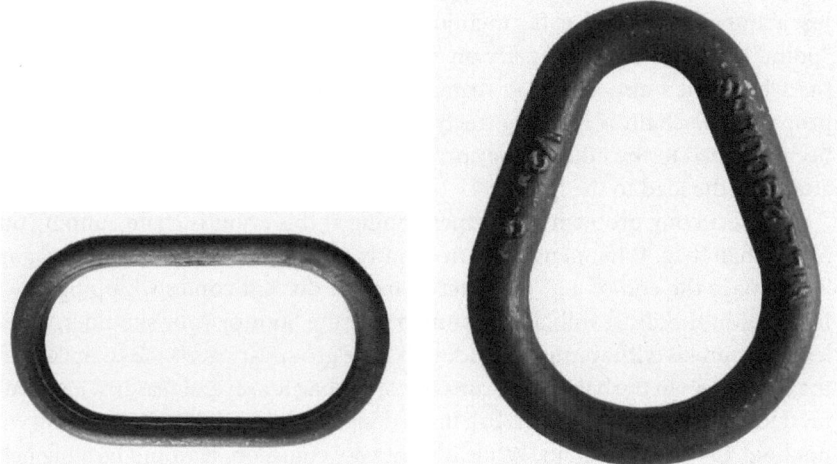

Figure 4.6 (a) Oval ring; (b) Pear ring

Long link chain, sometimes referred to by its brand name STAC, is often used in a hanging point to allow for more precise positioning of the downtail.

There are other hardware devices that are used in this assembly, all of which are used to connect slings and shackles in a hanging point. The most common of these devices are oval and pear rings.

The Riggers

When hanging a show, the most important component are the riggers. Their job requires that they be keenly focused on their work, making sure everything is in order at all times. Failure on their part to properly complete their work can have serious consequences for everyone involved in the show. One need only look at the photos of rig collapses over the past 20 years to get a clear understanding of how serious this work is.

The up-rigger or high-steel rigger has a singular and very specific job to do. They must connect the hanging hardware to the building in such a manner as to ensure that a) it's in the right location, b) it won't fall down, and c) it can be safely removed once the show is over. The fact that this work is usually done 100′ in the air while sitting on a 12″ wide I-beam makes it that much more fun. To be an up-rigger requires steady hands and a patient demeanor. Hurrying through a job may cause something to go wrong; a shackle gets turned the wrong way or, worse yet, a shackle pin could be dropped. Good up-riggers understand that a steady pace and a religious attention to detail make for a safe and fun work day.

The down-rigger, or ground-rigger as they are also called, is responsible for just about everything else. It's their job to make sure the hanging hardware (the "point") has been built properly on the deck before the up-rigger pulls it up to the I-beam. It's also their job to make sure the chain hoist has been set up properly: the chain is running freely through the hoist and the chain bag has been attached to the hoist in the correctly. The down-rigger is also in charge of attaching the load to the hoist.

One recurring problem worth mentioning at this point is chain running out of the chain bag. It happens fairly frequently and is a serious problem. Imagine the damage the end of a ¼″ diameter chain can do as it comes whipping down from 70′ in the air. It will take a chunk out of the floor or your shoulder. It will level a drum kit with wanton abandon. It's the ground-rigger's job to make sure the chain stays in the bag. Chain runs out of the bag for several reasons. First, you need to use the right size bag. A bag that's designed to hold 60′ of ½-ton chain will not hold 75′ of 1-ton chain. While it's not very common, it would be a big help if the chain bags were all color-coded. That would make it easier to match the right size bag with the chain length and hoist capacity. The other cause of chain running out is that the chain isn't seated in the bag properly. When the hoist is floating just off the deck and the rig is set to be raised up to its trim height, the load chain hanging below the hoist should be inside the bag and touching the bottom of that bag. Or, at the very least, it should be hanging two-thirds of the way down into the bag. The weight of the chain will then help keep the chain in the bag when the hoists start up. If the chain isn't far enough into the bag at the beginning, raise the rig only 2′ and stop. Then check all of the bags in the rig to make sure the chain is running properly into the bag. This only takes a few extra minutes. This might annoy those who are worrying about the schedule, but it takes less time to check chain bags then it does to lower the rig to refill a bag.

Ground-riggers don't get much credit but they're the ones who do the lion's share of the work and can make or break a load-in.

Now that we've identified who is doing the work and what they have to work with, we should take a look at how they do it.

The Action

Before any actual rigging starts, the floor must first be marked to show where all the hoists are supposed to go. If it's a touring production then this is usually the job for the venue head rigger and the road rigger. They will measure out the venue and either use chalk to mark the floor or large printed stickers to show where the rigging points are located. The markings not only tell the ground-rigger where to put the hanging point but also what hardware to use when making the hanging point. For example, Figure 4.7 tells the ground-rigger that they are supposed to make up a deadhang (vertical) point that, starting from the top, has a 5' basket, a 20' downtail and uses a ½-ton hoist. The C3 refers to the control channel that's been assigned to it.

How the rigging point is made up on the floor is critical to ensuring it is attached properly and safely on the overhead steel. Figure 4.8 shows the proper location of the basket, work and load shackles, downtail and the rope that is used

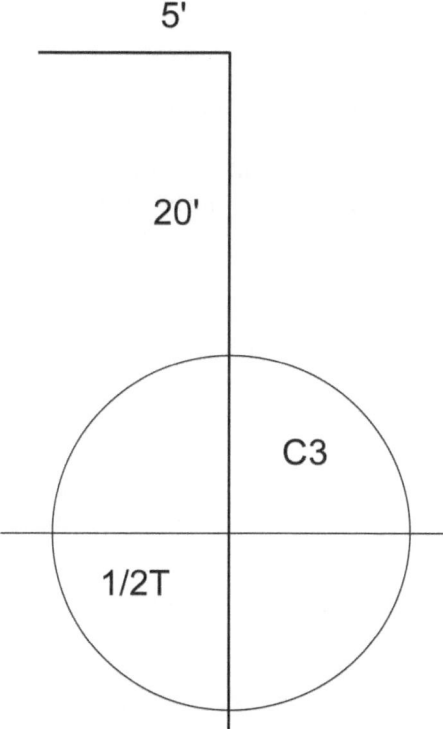

Figure 4.7

Figure 4.8

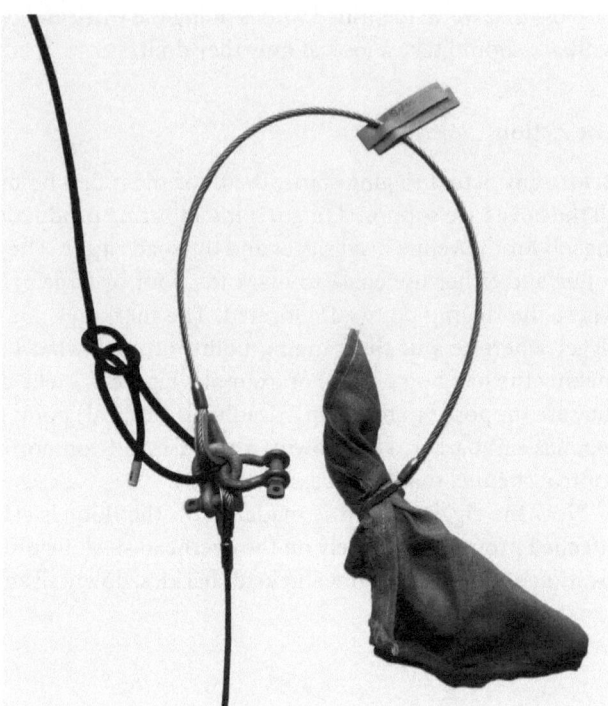

to pull the point up to the steel. The burlap in the eye of the basket is used to pad the beam and protect the basket from damage.

The up-rigger, once they've pulled the hanging point hardware up to the beam, must now go about the business of attaching it to the beam. The steps involved typically look like this:

1 Place the burlap on the beam
2 Lay the basket over the beam on top of the burlap.
3 Take the pin out of the work shackle, connect the work shackle to the loose end of the basket, and put the pin back in to the work shackle.
4 Lower the point until the basket is taking weight around the beam.
5 Make sure all shackles, basket sling eyes, and any other hardware involved in this point are hanging straight. A shackle that is twisted, for example, will become a problem when the rig is loaded and could cause a failure.

Once the point has been put into place it should look like Figure 4.9.

Sometimes a basket is not possible. It may be that the location of the hanging point is critical and the beam flange too wide to use a basket and still have the point hanging in the right location. In this case a choke point is used. A choke point has only one vertical leg that runs down the side of the beam. Care must be taken when using a choke point because there's only one vertical leg and

Figure 4.9

so the capacity of the choke is determined by the diameter of the wire rope sling and the damage created by the beam around which the sling is wrapped. It's also important to remember that a choke point can twist a beam if that beam doesn't have the proper lateral support.

The hardware in a choke point is basically the same except the work shackle is used not to connect the two ends of the basket but to connect one end of the basket to an in-line spot on the other side of the basket. When making up a choke point it's important to make sure the vertical leg isn't deflected or pulled over by the work shackle. If this does happen, the in-line spot on the basket leg will take a significant side load that can cause significant damage to the wire rope.

Then there are those times when the building refuses to cooperate. As you stand on the arena flooring marking out the points you look up and realize that for many of the points you just marked out there's no steel directly overhead. A normal deadhang won't work and so a different way of making the point needs to be found. This is where a bridle is used.

A bridle is an assembly of rigging hardware very similar to a deadhang but instead of having a vertical leg, it has two or more legs at the tip that form a Y

Figure 4.10 Choke style point

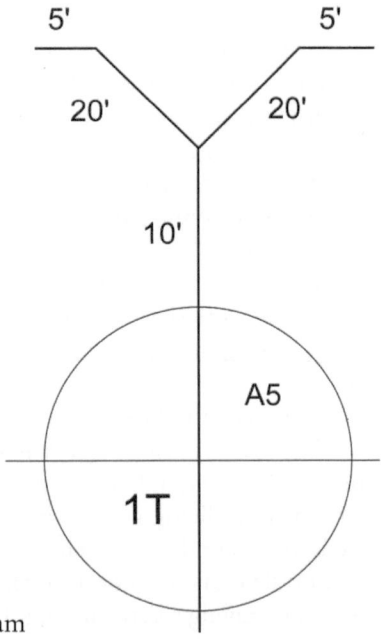

Figure 4.11 Bridle diagram

shape. Figure 4.11 shows a bridle as it's drawn on the floor for the ground rigger to assemble. The assembly has a 5' basket and a 20' sling on each leg.

The purpose of the bridle is to connect the hanging point to the nearest available rigging beam and place the hanging point in the right location. To do this, the length of the bridle legs is adjusted until the point falls where you want it.

Let's assume that the distance between rigging steel overhead is 40'. The point you want to hang needs to be located midway between two rigging beams or, to put it another way, 20' away from any rigging steel. To achieve this, each bridle leg should be approximately equal. The exact length of the bridle is determined by not only the distance between the rigging beams but also the height of the beam and the height of the junction of the bridle legs. To sort this out mathematically you should check out Rocky's chapter in this book (Chapter 1).

You may be asking yourself at this point, why do we rig this way? Why do we wrap wire rope around I-beams with only a bit of burlap padding? The answer is, because it works. The reason it works is because of the size and load rating of the components of a hanging point relative to the loads applied to the hanging point. For example, a 1-ton hanging point uses a ⅜" diameter wire rope sling for the basket. That sling has, in a vertical "pull-to pull" application, a working load limit (WLL) of 1.2 tons. When used as a basket, however, each leg of the sling hangs over the beam, one on each side. Used in this way, each leg will have a WLL of 1.2 tons resulting in a WLL of the basket assembly at 2.4 tons. Considering that we would only ever use a ⅜" diameter basket for a 1-ton point, there is ample capacity in the configuration to compensate for the abuse the wire rope receives when wrapped around the edges of an I-beam. (The burlap padding does help a bit, but not nearly as much as some people hope.)

A 2-ton point has the same geometry but with ½" diameter wire rope. The ½" diameter wire rope sling has a WLL of 2.5, so the basket rating is 5 tons.

When asked why they rig this way, most people respond with "this is the way we've always done it." While true, the reality is that we've rigged this way because the equipment has load capacities considerably higher than the loads they carry. This provides a generally acceptable margin of safety when lifting and moving loads over people's heads. But that margin is shrinking.

There was a time when all chain hoists were simple and uncomplicated. They ran at a single speed, had pushbutton controllers and moved pretty slowly. We used them to lift lighting truss and audio gear into position and hold them there. Little, if anything, moved during the show and the hoists weren't used again until the show was over. A 50–75 hoist show was considered a big show. Well, those days are long gone. Modern day fixed speed hoists run at speeds ranging from 16fpm up to 64fpm. Variable speed hoists may run up to 100fpm. A 200-hoist show is not uncommon. Hoists are no longer used to simply raise equipment into position and hold it there. They now move scenery, lights, video wall—pretty much anything you can imagine—on cue, during the show. Performers are sometimes flown on chain hoists, much to the dismay of the hoist manufacturers. Shows are now complicated enough to require computer control systems to manage hoist operations and rigging cues.

Shows are getting heavier too. Much heavier. The current roof load record for a touring show is right around 240,000lb. It's a safe bet that that record will be history by the time this book is published. Load cells and other load-monitoring devices are frequently used to ensure these loads on complex structures are distributed properly.

In short, it's gotten a bit crazy out there.

The rigging equipment we've been using for the past 50–60 years has proven itself over and over to be safe and reliable. The newer gear coming into the market does so with a higher degree of research and testing and scrutiny. So where is the weakest link in the load path of modern arena rigging? The accidents over the past five years all point to the same thing. People. It's the human component that provides the potential for disaster.

Given the size and complexity of the shows on the road these days, the cavalier attitudes of the past are no longer acceptable. When you consider the sheer quantity of equipment hanging overhead, its weight, and the speed at which it moves, the risks are too high to tolerate any sort of arrogant or careless behavior. It's also become clear that on-the-job training is no longer enough to provide the skills to work as an up- or ground-rigger.

Proper training is now of paramount importance if you wish to reach a higher level of employment. Having a thorough understanding of the equipment's capabilities and limitations along with the manufacturer's specifications and restrictions are the road to a safe, efficient and enjoyable theatre experience, not to mention a better paying career. Fortunately, training has become popular over the past ten years.

We can debate until the cows come home when the number of serious training options in the US began to increase. It seems clear, however, that the rise in interest in training coincided with the introduction of the Entertainment Technician Certification Program (ETCP) in 2005. Whether it was an awakening of an awareness that skill levels had to grow, a fear of being left behind, or a combination of the two, riggers began to understand that the status quo was no longer sufficient.

The ETCP mandate and charter does not allow the organization to conduct training—private individuals and organizations provide that service. What the ETCP has done, through the certification exams, is provide the means to measure the skill level required to be a member of the top level of riggers in North America. This is no small feat and should not be taken lightly. Until the ETCP came along a rigger had only their wits and their résumé to show the world how good they really were. The ETCP has added another layer and the industry is all the better for it.

The arena rigging industry had its beginnings in the circus and the road to the present day has been more than a little interesting. The industry has had its share of successes and failures, but that's to be expected when the way forward leads into unchartered territory. As the industry continues to grow, it also continues to mature. The level of safety awareness today is rising to meet the level of technology found in modern-day shows. One can only hope that this relationship will continue to grow, providing safe and wonderful experiences for all for many years to come.

5

Outdoor Roof Structures

KEITH BOHN

Intro

Welcome to the oldest venue for entertainment, the great outdoors. People have enjoyed the wonderful venue of Mother Nature since they first learned how to dance and tell stories around the campfire. Maybe it is something primitive, but there is some draw and appeal in going to a big festival out in a field, with no one around, limited security, PSAs coming from the stage, inadequate bathrooms, no food, lots of mud, and . . . well, we still like outdoor gigs for some reason.

OK, so less recent than Woodstock and not as far back as the ancients at Stonehenge, Shakespeare's Globe Theatre was open-air and "outdoors" by some definition. But, even in the 17th century they had trouble with the venue not being suitable for the performance, or blunders in the show, when misfired cannon in a performance of *Henry VIII* burned the place down. Let's hope that we have learned from that little trinket of history.

So, what are we talking about here?

There are a lot of different structures used to produce shows and events outdoors. These structures can serve many different purposes and use a variety of designs, materials, and technologies. Things like PA towers, spotlight platforms, mix positions (FOH platforms), freestanding video walls, and the roof covering the stage are all examples of some of the structures you might see in an outdoor show site. These could be made from aluminum, steel, or synthetic materials, or a combination. They might require scaffolding, vertical towers, hydraulics, electric motors, or even cranes to erect.

There are other types of structures used to complete an outdoor event safely that might include portable water closets, tents, and trailers. For the purposes of this chapter, we are going to differentiate between these two categories by identifying that the first list of structures is utilized in the technical production of the show. This isn't to say that those other operational structures are less important, or are not subjected to Mother Nature, it is simply that most trailers or port-o-johns might not present the life safety or structural ambiguities that can exist with a customized temporary structure supporting tens of thousands of pounds overhead.

Throughout the next few pages, the primary focus will be on the over-stage roof structures. However, it is important to realize that most of the same rules apply to PA towers, video walls, etc. Despite all of the different forms and functions of these structures, once they are utilized outdoors, they all are subjected to similar forces and conditions. The challenge is identifying, understanding, and minimizing the risks associated with those conditions.

Inside, Outside—What's the Difference?

Many of the same technology required to produce a show outdoors is also utilized indoors. There are some extraordinary differences, besides the obvious weather. For example, when rigging a show in an indoor venue, most of the time the loads hang from a separate and independent superstructure such as a building. When loads are hanging, the supporting components such as wire rope or hoists are in tension. In most cases, this tensile load is relatively quantifiable with clear direction. In other words, we can identify how much something weighs and we know that the force is in the direction of gravity's pull. The other end of the supporting mechanism simply needs to resist this simple pull.

The complication to this clear direction of gravity indoors is when lateral forces are induced into the hanging objects. (As an aside, there are regularly lateral forces imposed on local components of the superstructure when the rigging configuration requires the use of bridles, but for this discussion, we are focusing on the hanging loads.) Lateral loads on the hanging objects can be created by automation elements, such as tracking video walls and performer flying, or even wind or air movement indoors when loading doors are open, for example. Most times, this lateral force can be resolved with some type of additional rigging specifically designed to resist the direction of this motion. However, since the supporting assemblies are in tension, a lateral load-imposed hanging object may cause the tension cables to be pulled out of vertical, or create a swinging or pendulum effect. And, while this may not be desirable from the perspective of the show, it could be perfectly safe (considering that the anchorage in the superstructure is suitable) in regards to the hanging structure itself. In other words, a tension cable going out of vertical by a few, or even many degrees, may not create an extraordinarily unsafe condition.

In contrast to this, an outdoor structure almost always depends on the main supporting devices to be put into compression. This means that instead of trying to pull something apart, like a cable or chain, the load is trying to press things together. In this situation it is absolutely critical that the direction of compression and how much it is out of perfect alignment is clearly controlled. A tower that is out of vertical by only a couple of degrees can be completely unstable.

You can use a simple drinking straw to further understand this critical tension versus compression concept. To test tension, simply grab on each end and try to pull apart. Odds are pretty good that your fingers are going to slip off long before you pull the straw apart. Now, if you put the straw between your fingers and press together, compressing the straw, it will take very little to get the straw to buckle. This is precisely what we see in hanging (tension) loads, versus ground (compression) supported loads. It also relates to a concept called slenderness, which will dictate size and height limits on towers, but we are getting ahead of ourselves a little.

For now, it is important to understand this key, tensile versus compression concept. If you then toss in the added aspects of wind, rain, and snow that take place outdoors, you can clearly see how different the considerations need to be for outdoors over indoor applications.

This is Easy, Right?

So before we get much further, let's put the complexity of these structures in perspective by comparing the expectation of a temporary outdoor roof structure to that of a brick and mortar permanent building. In the process of erecting a permanent building, an architect completes a design. Then, an engineer validates that design to be suitable for the location and what is required. This includes understanding the soil and foundation, the environment (i.e., wind, snow), and the addition of the live loads, which in most cases are occupancy loads.

Many of the loads the engineer will use are dictated by codes and standards. For example, certain soil types have established bearing capacity, occupancy loads are typically based on available square area, and wind, snow, and seismic are based on the statistical likelihood of occurrence in a given locale. Once these calculations and values are completed, the materials and construction methods can be determined and construction can begin.

During construction, the contractor must ensure that the end result is consistent with the design and engineering documents. If anything is done incorrectly, then the building process could be delayed until the appropriate corrections can be made or modifications are approved. With a permanent building such as this, the entire process could range from six months to a few years to complete. It is also important to note that we haven't talked about building and an electrical inspections that have to take place that are driven by the local municipality and usually required at multiple stages of completion. These inspections throughout the construction process are intended to ensure compliance and safety.

Now let's look at a temporary roof structure. A designer will be involved and an engineer. However, in this case, quite a few of the fixed or known factors in a permanent building are now variable factors for a temporary structure. For example, a roof system may move every few days or weeks. Now the structure must be able to accommodate different soil types, different weather conditions, and a variety of load scenarios (not all shows are the same). Aside from that, the structure must be erected in a fraction of the time of a brick and mortar structure. The allowable setup and dismantle period many times is quantified in hours rather than days or weeks. This means that there is little, if any, time to make corrections or modifications if something isn't exactly right. Despite those differences and variables, the general expectation is that these structures, which can be set up in a very short period of time, are going to be no less safe than the permanent building.

Further to the perception of these structures, let's consider the potential number of players involved in this complex system. Including the aforementioned designer and engineer, a list of who could be involved with a structure might include: designer, engineer, manufacturer, system owner, setup manager, labor provider/supervisor, operator, security, and, in some cases, the promoter. Each entity has a role, in some cases a critical and complex role, that could monumentally affect the reliability of the structure. The long and short of it is that there are lots of people involved, each with a specific job to do, and each likely to be reliant on others to complete their tasks safely. At one time there was the notion that you shouldn't buy a car that was built on a Monday or Friday. But, temporary structures aren't mass-produced machines, and even the slightest misstep along the way at any stage can have catastrophic consequences.

The trap that many people fall into is that they want the temporary structure to be as safe as a building, but they don't consider exactly what is required to achieve those levels of safety. There are different perspectives at play here.

- Attendees and concert goers, i.e., the punter
- Employers/hirers, i.e., promoter
- Providers, i.e., the production company

The attendees assume that the responsible parties producing the event have considered patron safety. And while some common sense should prevail in the case of extreme weather, it typically isn't in the purview of the patron to consider that they are ever in any peril.

Employers and promoters want the safety, and assume that their provider is an "expert" (if they weren't an expert, why would they say they are qualified to do this kind of work?), but may let competitive cost and economics play an important role in their choice of provider. Many times the employer or promoter doesn't know the difference between a qualified provider and unqualified provider. They might put a project or show out for bid and receive a varied response. And, while economics seem to always be part of the equation, employers have to take the time to understand whether the entity they are hiring is truly demonstrating the ability to provide the appropriate equipment and personnel. In other words, you might get what you pay for, and the cheaper provider just might not be the correct decision.

With regard to these production companies or providers, it is expected that they are the experts and have considered all aspects of safety and structural integrity of the structures. However, as mentioned before, they are regularly forced into competitive bid situations. These bid situations can pit a qualified provider against an unqualified provider, creating an uneven competitive advantage to the unqualified provider and forcing the overall quality of the structure to be diminished. At this point, it is not only incumbent upon the previously mentioned employer/promoter to be diligent in their decision, it is also key for the provider to clearly outline all of their qualifications and exactly how their structure is suitable for the show in question.

This suitability is one of the key considerations for these structures and requires careful consideration by all parties involved (other than the punters). What usually happens is a promoter is given a set of requirements from the act or show they have hired. This could include rigging plots, lighting equipment, audio needs, video, etc. It is imperative that, when assembling a bid, the potential provider compares these requirements with the limitations of the structure they intend to use for the show. There must be a clear understanding of all of these loads and requirements and documentation that definitively supports this suitability. Furthermore, the employer/promoter must be diligent enough to ask for this compliance.

Keep the complex nature of temporary structures, as well as the differences to permanent buildings, in mind as you consider these types of structures and, as you read on, note that caution and understanding will provide a better chance for success than a cavalier approach. The amount of respect needed for these types of structures cannot be overstated. The instant you think this is as easy as setting up your tent at Lake Too-Many-Bugs is the precise instant that you invite the potential for disasters and all the lawyers that tend to accompany them.

You Want to Build This Where?

So the real appeal for temporary outdoor structures is that you can produce an event almost anywhere. That is a vast oversimplification, however. The suitability of a site for a temporary outdoor structure is as critical as the structure itself. As discussed previously, there are a number of variables that must be considered to determine this suitability. Some of the more commonly used locations range from farmland (maybe in upstate NY?) to parking lots and stadium venues. Wherever you might choose to hold an event, careful planning must take place far in advance.

One of the most critical, if not *the* most critical, factors in determining site suitability is the foundation on which the structure will be erected. This particular factor involves the foundations on which the towers or support system might rest as well as the anchorage system and guy cables that might be needed. So we have to consider the load-bearing (compression) capacity of the foundations under the supports as well as its tensile suitability for the type of anchors that might be employed on guy cables or other tension components.

For example, looser soils might further compact during the event if there is enough pressure—this can cause towers or supports to sink. Depending on the design of the structure, this can cause myriad negative effects to the integrity and stability of the structure. This might cause the towers to go beyond their acceptable levels of verticality, or could cause the system to shift out of a desired horizontally level condition. In extreme cases, where towers might be in close proximity to each other, a compaction of bearing soils could effectively negate the function of the tower itself. In many systems that employ multiple towers on each side of the stage, the interior towers are supporting the most weight. Effectively losing reliable foundations under these towers can be catastrophic for the entire system.

Further to this are the effects of moisture on the foundations. Some sites are chosen for their naturally occurring amphitheater characteristics. However, this usually means that the stage area is located in an area that is a naturally occurring drainage path (muddy mosh pit anyone?). Certain types of soil, particularly some types of clay, expand and contract based on moisture content. This reaction may not be discernible in very short periods of time, but certainly during a few weeks it is quite possible that simple moisture changes will cause the structure to shift according to the soil reactions.

Solutions for this bearing pressure usually involve spreading the foundation-bearing loads over the appropriate square area. For example, if you know that the maximum load potential down a tower in your particular system is 8,000lb, and the area of the base that is in direct contact with the foundation is 2' square, then you need bearing capacity of at least 2,000lb per square foot ($2' \times 2' = 4'$, $8,000lb \div 4 = 2,000lb$). If your particular soil type is limited to 1,000lb per square foot, then your base area needs to be a minimum of 8' square.

Aside from simply considering different soils, sports venues may have limitations on ground-bearing pressures not related to soils, but related to infrastructure. Some natural turf venues might have irrigation systems beneath the surface that must be considered. There are also synthetic turf systems that are actually built on modular panels. These conditions might necessitate that the maximum load at the base of the tower or support be limited.

Similar instability that is created by soil compaction can also occur with a foundation that simply disappears due to erosion. Once again, the type of soil has a direct impact on this possibility, as could any vegetation. A thick grass with a hardy root system might be less prone to erosion than loose soil in a field.

But wait! What about tensile components like guy cables? This certainly has to be another consideration when reviewing the ground suitability at a site. If guy cables are required, then what type of anchorage is to be used? If earth anchors are used, then the type of anchor must be selected based on the resistance required by the system as well as the type of soil. This will usually dictate the depth and angle of the anchor to sufficiently perform at the required level.

In other words, banging a car axle into the ground isn't usually quantifiable enough to answer these questions; Beamer or Buick, doesn't matter.

In the situation that reliable earth anchors cannot be installed, as is regularly the case in sports venues where the ground surface cannot be broken, a ballasting system is likely to be the solution. Ballast will be discussed more a little later, but suffice it to say that you have to consider the interaction of the different surfaces in both the ground and the ballast type to identify a coefficient of friction that can be used to calculate the required mass. Regardless of the type of anchorage for tensile components, whether it is earth anchors or a ballast system, the notion of erosion potential must also be included in their evaluation.

Additionally, guy cables present another factor related to site suitability. Operationally, where are guy cable and anchors located? External guy cable positions might have an impact on audience areas, production access, emergency egress, or other key operational functions required at an outdoor event. Is security required at each guy cable anchor to prevent someone from trying to climb? Does the guy cable path cross any key areas? Loading dock? Truck or bus access? This starts to touch on the more global planning and communication coordination that is required between all of the key players in the event. Operational planning has to start at site selection.

Wait, Wait, Wait. There's Math Involved?

I suppose the answer to the math question was already revealed in the previous section since we started calculating bearing pressures for foundations. This precise math and calculation must occur on every component throughout the system. Regardless of the type of design, materials or methods of manufacture, every part of an outdoor roof system must be mathematically verified to be appropriate for its function.

In a sense, there are two directions to approach the design and mathematical verification of a structure. The first would be to simply validate the capacity of a system based on the components intended to be used. For example, you own a certain type of truss or structure which you plan to integrate to other components to create the system. For this, you would want to know what it is capable of supporting and in what conditions. This approach should be used on any system that might have been put together with components not designed for this specific purpose. Basically, this approach is used when the question is "how much can this support?"

The second direction to approach the design and verification of the structure is to start with the target objectives that are desired. This would include load capacity in specific places in the system (i.e., PA wings or video walls), as well as operating conditions or locations that might have different wind or environmental requirements. In other words, this is the design approach when the question is "what would it take to accomplish this?"

Regardless of the direction required in your situation, the fact remains that a design professional, usually a licensed professional engineer (PE), must complete the necessary calculations to validate the strength of the system and the components within the system. It is also important to note that, whatever PE you use, it is likely preferred that they are licensed in the location you are building your structure and holding your event.

So, specifically what kind of math are we talking about here? Well, for example, we know that a 1-ton rated hoist is good to support 2,000lb. We know that if we use prefabricated truss components, those items also should have a rated capacity (albeit most of the published ratings are for highly controlled conditions that do *not* include outdoor applications). We also know that commonly used rigging components like shackles, wire rope, and roundslings have rated capacities as well. We even know the capacity of single- and double-wrapped zip ties and gaff tape . . . oh wait, ignore those last two examples. Anyway, many components used in outdoor structures are items that are adapted from existing technology. But, in many cases, as pointed out with the truss, those items haven't been evaluated for outdoor use. So when we get to these outdoor applications, we have to apply the appropriate loads and directions of force to ensure that these components work in this application as well.

Before we get into the specific outdoor loads, let's quickly look at the overhead grid in an outdoor roof structure. Typically, there is a grid that is supported on vertical towers to provide the overhead roof. Since all of the load must ultimately transfer back to the ground, the most efficient path for the load to travel is in a straight line to the closest tower. Using this principle, many outdoor structures utilize main spans of structure between towers on opposite sides of the stage. It is these main spans that must carry all of the load, and their size is dictated by the length of the span and the necessary load capacity.

In the case of a simple four-tower type of system, the structure or truss on the perimeter of the grid effectively carries the load to the towers. Any loads that are placed on intermediate structures within the interior of the grid will have to transfer to these perimeter trusses before getting to the towers. Therefore, it is typically the capacity of these perimeter trusses that dictates the overall capacity of the grid.

However, even within this simple perimeter type of example, the layout of any interior structure and where it intersects with the perimeter structure will have a very direct impact on the usability of the system. For example, if the desire is to simply have some lights on the downstage and upstage spans, and nothing on the interior or sides, then the largest structural spans will simply be those that are carrying that weight. On the other hand, if there is a center span truss running left to right that is carrying weight, not only does that truss have to be sufficient to support the intended load, now the side trusses that are supporting it must be sufficiently sized to carry the intended load and the self-weight of the center truss span. On top of that, this side perimeter truss has to be able to support it in the worst possible location, the center.

So, the easy solution to the above scenario is to add towers to the middle of the side perimeter spans. This would result in that center truss span having a direct load path to the towers. And this indeed, is more efficient and favorable over the simple four-tower perimeter system. It is this same concept that applies to most major cross stage truss spans within an outdoor structure. If you see a system that has three or more towers on each side of the stage, you are likely to see a larger main cross-stage truss connecting directly in line with these towers to provide the greatest and most efficient capacity.

So far we have been talking about "loads" in a very generic sense, but you need to make sure you understand that any evaluation of a system must include all possible loads that the system will have to support. In addition to the presumably easily quantified gravity loads, lateral, dynamic, and environmental loads must also be quantified and evaluated.

Lateral loads are going to be any loads that are trying to pull the system, or a component of the system, sideways. The most obvious of these would be guy cables that might be attached to the system, but you can't forget that any rigging done by way of a bridle is also going to create a lateral load. In fact, most outdoor roof structures prohibit the use of bridles since they could put a lateral load on a component that is insufficient to support it. Furthermore, the use of bridles in an outdoor structure is usually impractical since the structure is already relatively low compared to an indoor arena, so using a bridle for rigging hurts trim height for anything under the structure even more.

One way to get around the use of bridles in an outdoor roof structure is to utilize load beams or intermediate structures that can span between the main support spans of the system. This provides an infinite number of support points that can be rigged as deadhangs rather than bridles, and thereby maximizing the available trim height in the system.

Dynamic loads that are placed on the system must also be evaluated. These would include any automation or acrobatic performance that the system might be supporting. Yes, jumping out of the grid on bungees creates some dynamic force into the structure. Additionally, any simple rigging movements during the load-in must be considered. These conditions of dynamic amplification could range anywhere from a 1.1 to 2.0 increase of the moving load depending on how fast it starts and how fast it stops.

For example, if a truss supporting 1,000lb of lighting is hanging from the outdoor structure on 1-ton hoists that lift at 16fpm, the simple operation of lifting the truss into position will likely result in a higher load than what it weighs statically, or as a non-moving load. The best way to talk about dynamic loads is to put them in terms of a statically equivalent load, which would be the load multiplied by its dynamic amplification factor. One prevalent chain hoist manufacturer has stated that their 16fpm, 1-ton hoist will induce a dynamic amplification factor of 1.25. This means that a 1,000lb load will actually create a 1,250lb reaction into whatever it is supporting it. This 1,250lb would be the static equivalent load just for the lighting. But, in our example we can't forget the weight of the truss (let's say 500lb) and the hoists and rigging (200lb at each end, so 400lb). This equates to a total static load of 1,900lb, and a static equivalent load of 2,375lb. This is a far cry from simply considering 1,000lb of lighting.

Finally, we have to consider the environmental loads that are present on an outdoor structure. This can easily be the most under-estimated of all the loads. For the longest time, a wind speed of 40mph was considered a minimum wind force that a structure must be able to withstand. However, this is a load that has historically been applied to the use of lightweight tents and non-load-bearing systems. Recently, the required values for these types of system have necessarily been increased. The current ANSI E1.21-2013 minimum wind force that a structure must be capable of resisting is 67.5mph based on a three-second gust. This 67.5mph is essentially a 25% reduction of the basic design wind speed in the US of 90mph, and is derived from the applicable code ASCE 37, which covers buildings under construction. It is important to note that this reduced wind speed can only be applied if the structure is intended to be in place less than six weeks, which is the definition of "temporary" according to ANSI E1.21-2013.

Before we look at what that wind load might be at 67.5mph, let's circle back to the support layout of our structure. For that basic four-tower system, we can easily recognize (as a simplification) that each of the towers is likely to carry equal amounts of the load from wind. This considers that the wind load is based upon the square area of surface that will resist the force of wind, such as the entire covered area overhead. Now, when we step up to the six-tower system that utilizes mid-towers on the sides, we have a different scenario. In this configuration, a simplification would be to assume that each of the towers carries equal parts of the load, but this would be incorrect. In this situation, the center truss span, and therefore the mid-stage towers, are carrying not just the center third of the covering, but a full half of that total load. This is because that center span must support everything halfway between itself and the upstage and downstage spans. We call this total area to be supported, the tributary area, as the area that can be attributed to that span. The point in bringing attention to this is simply to ensure that it is understood that any towers or mid-stage spans of truss really could be the weakest link in the system and must be evaluated with the consideration of their location in the system.

OK. Wind. It blows. Wind is calculated as a force against an exposed area of the structure. Most commonly considered is the area that covers the stage, but you can't forget the area of any towers or support structure, as well as equipment like video walls, scenery and audio, and certainly you can't forget backdrops, side walls, and banners (that always seem to show up at the last minute). So how do you know how much force a specific wind speed puts on the structure? Well, there is math for that. The basic formula is simply velocity squared times 0.00256. This gives you a product of pounds per square foot (psf). So, for the "old" and misapplied wind speed of 40mph, the force equivalent is 4.096lb psf

($40^2 \times .00256$). Compare that to the current 67.5mph requirement, which results in a force equivalent of 11.664lb psf ($67.5^2 \times .00256$). Obviously, this is an increase of almost 300% of force with an increase of only 68.75% of wind speed. Figure 5.1 shows this relationship between wind speed and force.

When calculating this wind force, like anything else, it has to be calculated from the worst possible position, so basically the entire wind load is calculated with wind coming from any direction. The charts below, Figures 5.2 through 5.4, are simple grids showing the generalized total wind area for different sized outdoor structures. You can see the force and loads at the various wind speeds that would be present for each sized system. Make a quick note regarding the dramatic difference between the "old" 40mph thinking and the more recent 67.5mph. Also keep in mind that the wind load on the side and back walls is entirely lateral and must be resolved into the towers, grid, or through guy cables; essentially anywhere that these walls are attached, the anchoring structure must be capable of resisting that portion of the lateral load.

There is an allowance in ANSI E1.21-2013 for items that can be removed from a structure within five minutes. This allowance reduces the wind load requirement on these components down to the aforementioned 40mph, but *only* if they can be removed in the five minutes. Many have said that it isn't

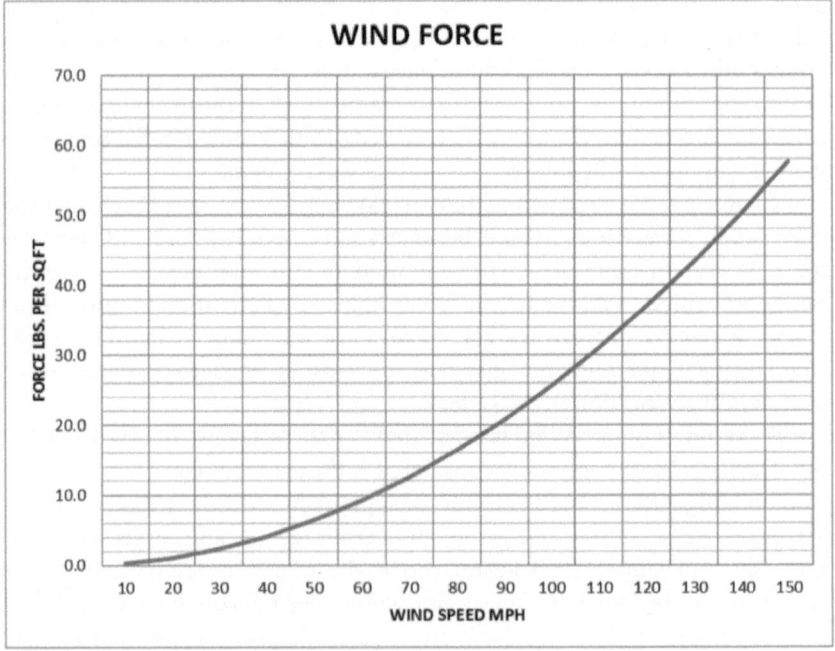

Figure 5.1

For a 40'x40'x30'H structure							
		Overhead		Back Wall		Side Wall	
MPH	PSF	Area (sq. ft.)	Force (lbs.)	Area (sq. ft.)	Force (lbs.)	Area (sq. ft.)	Force (lbs.)
40	4.1	1600	6,554	1200	4,915	1200	4,915
67.5	11.7	1600	18,662	1200	13,997	1200	13,997
90	20.7	1600	33,178	1200	24,883	1200	24,883
120	36.9	1600	58,982	1200	44,237	1200	44,237

Figure 5.2

For a 60'x40'x40'H structure							
		Overhead		Back Wall		Side Wall	
MPH	PSF	Area (sq. ft.)	Force (lbs.)	Area (sq. ft.)	Force (lbs.)	Area (sq. ft.)	Force (lbs.)
40	4.1	2400	9,830	2400	9,830	1600	6,554
67.5	11.7	2400	27,994	2400	27,994	1600	18,662
90	20.7	2400	49,766	2400	49,766	1600	33,178
120	36.9	2400	88,474	2400	88,474	1600	58,982

Figure 5.3

For an 80'x60'x50'H structure							
		Overhead		Back Wall		Side Wall	
MPH	PSF	Area (sq. ft.)	Force (lbs.)	Area (sq. ft.)	Force (lbs.)	Area (sq. ft.)	Force (lbs.)
40	4.1	4800	19,661	4000	16,384	3000	12,288
67.5	11.7	4800	55,987	4000	46,656	3000	34,992
90	20.7	4800	99,533	4000	82,944	3000	62,208
120	36.9	4800	176,947	4000	147,456	3000	110,592

Figure 5.4

practical to move anything in five minutes, and the response to that is simple. If it can't be removed in time, then analyze and design for the full required wind speed. Easy.

Lastly regarding wind and wind area, there have been a couple of myths out there that should be dispelled. First the use of "blow through" materials for side or back walls is a misnomer. Any reasonable engineer will consider these areas as solid surfaces rather than the 70% (or whatever it is) that supposedly "passes through." Part of the reason for this, is that even with a 70% pass-through, what is there will likely billow when wind hits it. Once it billows, the effective openings in the material begin to get smaller as the overall shape of the surface begins to become concave. At a certain point, the material won't be realistically passing anything through it because all of the openings will essentially be closed to the direction of the wind; benefit negated. The other myth is that truss or open towers and structure don't provide as much wind resistance as a solid surface. While this seems to make good sense, the reality is that truss, and truss with round members in particular, can actually be more difficult to quantify for wind than a solid surface. The reason for this is that the round members are going to deflect the wind in a multitude of directions within the truss or tower. This is nearly unquantifiable in realistic terms. So, most engineers default to the 100% solid surface approach in all directions to help compensate for this phenomenon.

Once the capacity and limitations of the structure are known, it is important to now make sure it is suitable for the use. In the case that the structure was designed to the parameters of a particular show, you would hope that this review of suitability wouldn't be too extensive. However, in the case of an existing structure that is used for a variety of events and shows, it is imperative that someone is looking at the requirements of the show and ensuring that they fit within the limitations of the structure to be used.

So, yes. There is a lot of math involved. We haven't even gotten into weld tensile capacity, member buckling, seismic consideration, connection analysis, and so on. It is absolutely imperative that a licensed design professional who is familiar with these types of structures performs a thorough analysis and does all of the math for you.

Badges? We Don't Need No Stinkin' Badges!

Back in the "old" days, no one really worried about making sure that a structure was suitable for its purpose. There was also no clear idea on who would approve the use of a structure or what codes it had to comply with. Thus, 40mph wind became a common benchmark, and it was simply assumed by, well, everybody that if you had the wherewithal to own and build the structure, you must be able to do it safely. This led to the proliferation of many substandard systems being used for the wrong purpose and beyond the purview of any oversight. Sadly, many of those structures still get used, but slowly the industry is becoming more safety conscious.

One of the things that has been a repeated theme throughout this chapter is the need for professional engineering involvement. This input from the licensed smart guys is at multiple stages, including design and suitability. But one more area that is needed is confirmation of compliance with the local codes in the area the structure is being used. Part of the reason for a temporary structure is that it can operate in different locations from one week to the next. However, these different locales are likely to have different prevailing codes and requirements and someone has to make sure the structure meets them all, or that the structure is modified to meet those requirements.

So, where do you find out what these requirements are? Well, that is a pretty complicated question that is without a consistent answer. Part of the reason for this is that the United States is a republic of 50 states, each with their own ability to adopt codes that might be different than another. For example, there are separate building codes for New York and Chicago among others, some states adopt the International Building Code (IBC), or even different years of the IBC. But that is only scratching the surface.

Depending on where you are holding your event, you may not fit into a clear governing authority's jurisdiction, so who do you need to check with to ensure you comply? For example, if you are on city property, or in city limits, you might need to file for a permit with the city building department. However, if you are out of a city and on county property, it might be the county parks and recreation department that is in charge. It is possible that the fire department needs to verify that your structure is safe. These are only a few of the possibilities.

However, regardless of who has to inspect or approve the structure, it is quite likely that they will have little idea of what a temporary entertainment structure like roof system is designed to do. This is where a good relationship with your engineer and some proactive research for your local authority can prove to be very beneficial. The authority having jurisdiction (AHJ) where you are to set up your system needs to be your ally, and the best way to be their ally is to make sure they know that you have covered all the bases. This is most easily done with input from a good engineer who is familiar with the local requirements and can help you convey your compliance to these requirements. The last thing you need is for the AHJ to show up on your show site the day of the event, and be so unfamiliar with the structure and what it does that they are intimidated by it to the point of shutting you down. Take the time to identify who the AHJ is and what you need to do to make them happy, and you will have a better day.

I Have Been Doing This for Years. Doesn't That Count?

There are a lot of myths in the entertainment industry. Many of these are perpetuated with good intentions, but end up being taken out of context over time and pretty soon the original intent has been lost. The notion that just because you have been doing something for years equates to some level of expertise isn't really logical. Maybe you were taught the wrong lessons, but have never been in a situation where the weaknesses in your technique have been exposed.

As an example, it was once argued that aluminum truss rated for a certain load had a great amount of empirical, real-world evidence to show that it was suitable, or even underrated, based on the limited number of incidents that had occurred. But what this argument fails to point out, is that on many of the occasions on which this truss is used it is seldom actually pressed to its limits. In other words, if it was never tested and used to its capacity, then how do you have any real empirical evidence? The same goes for techniques and personal qualifications; if you never learned correctly, but the conditions that you operated within were always relatively safe and controlled, then you have never truly been tested. Many system failures that have occurred were in systems that had been set up the same way for many years, so the operators falsely assumed that what they were doing was fine. But when the time came for the system to be really challenged, it didn't hold up.

Someone claiming to have years of experience as the extent of their qualifications may or may not have a sufficient or clear understanding of the limits of the system they are operating. Once again, this leads to the need for a partnership with a registered design professional to help understand the limits of the structure and understand what can and cannot be done safely. It is important to further understand that what could be done in one configuration may not be completed safely in a different configuration, even under the same structure. This can be due to wind exposure, use of scenery as ballast, different loads in different locations or a variety of other factors. If you think you already know everything, you are already out of your league.

We Are All Set Up. Now We Party Until Load-Out, Right?

Getting the structure set up and the show ready to go underneath it is only part of the process. Managing the structure during use is absolutely critical for a safe event. In 2011, there was an incident at the Indiana State Fair that resulted in quite a few deaths and injuries. One of the follow-up reports to that accident clearly outlined deficiencies in managing the structure and the decisions, or lack of decisions, made to ensure the safety of the event.

An Operations Management Plan (OMP) needs to be created and followed throughout each stage of the use of an outdoor structure. This plan needs to be shared and explained to every single stakeholder in the production of an outdoor event, as each party can have an effect on the performance and outcome of the event. For example, all the parties need to know who the final decision-maker would be in the event of extreme weather, and what the security staff need to do when a weather plan is implemented. These things all need to be clear and objective, which is what a good OMP will provide. If you are trying to decide what to do in an unexpected situation when it happens, then your OMP isn't good enough, and you are likely already in a very tough position.

Some of the things that need to be outlined in the OMP might include:

- daily or hourly system inspections
- shifting conditions to the site
- performance thresholds
- weather plan—what changes need to happen at what weather conditions
- training requirements
- public safety considerations (storm shelter areas)
- venue configuration (access and egress issues)
- venue impact on structure

All of the previously mentioned precautions to ensure the structure is designed and built correctly and meets the proper codes are completely wasted if the structure isn't managed appropriately on site. One thing that is particularly important, and should be clarified at the time the structure is hired or contracted, is a clear understanding on which entity (promoter, structure owner, etc.) is responsible for decisions regarding the structure at any given moment. Someone always has to have clear and unquestioned authority, and it must be very clear on how that authority is to communicate critical decisions at critical times to everyone involved. The real point of an OMP is to make all of these decisions as objective as possible rather than subjective or based on someone's opinion.

Summary

What should be clear and simple at this point is that these types of structures are anything but clear and simple. These are all complex undertakings from start to finish and they deserve the appropriate attention to all details at every step along the way. There are a number of qualified resources available for you to learn more, and if you are in a situation using an outdoor temporary structure where you or the person you are working with doesn't know "why" something is the way it is, seek out an answer that makes sense and ensure that you and your event operate safely.

6

Counterweight Rigging

KAREN BUTLER

History

As near as can be told, theatrical rigging (the suspension of scenery and actors above the performance area) started in Greece. In their plays about gods and heroes, they wanted more spectacle. They searched for better and bigger staging effects to entice the audience and engage them in the tragedy and drama in front of them.

In the "modern" theatre, stagehands utilized the technology of the day to rig scenery. Relying on the rope crafts of the tall-ship sailor, theatre technicians used belaying pins, Manila hemp, wooden grids, and sandbags to counterbalance scenery and lights. Every new show required a skilled crew to spot loft blocks on the grid specific to the incoming show, and to reeve hemp through the loft blocks and head blocks. The load was tied to the hemp, and a crew of stagehands, using block and falls, wire rope sundays, and lots of effort, raised the loads to a working height. Sandbags were attached to the line-sets by means of wire rope sundays or clamps to balance the load, and the scenery was raised and lowered using the line-set and auxiliary lines called haul-downs. It was a poetic but labor-intense way to hang a show.

This continued up into the early 20th century when J.R. Clancy introduced the first counterweight system.

> In 1924, the company took counterweight rigging to the next step, introducing a system Clancy called Manual Counterbalance Rigging, "a method by which scenery could be easily raised and lowered with the least effort and in a way that would, in necessity, be rapid but always sure."
>
> *Theatre Design & Technology*, 2010, p. 61

J.R. Clancy began offering the components to the manual counterweight system in its 1925 catalogue and theatres began installing them. While there are still a few operating hemp houses tucked away, the majority of theatres in North America use counterweight systems evolved from these origins.

Manual Counterweight Rigging

Manually operated counterweight systems are the standard for proscenium fly houses in much of the world. Next to those that utilize rope and sandbags, they are the oldest form of theatrical rigging systems. Simple and direct, they are the backbone of most theatres. While the use of motorized line-set systems has become more common, manual counterweight rigging systems continue to be the essential driving force in theatrical productions; they yield magic via trained people, ropes, sheaves, and sweat.

Counterweight rigging systems are where most theatrical riggers and flymen get their start. They are the classrooms where these individuals are introduced to all the basics of proper operation, applied forces, standards, and to the immense responsibility they hold in their hands. Nothing drives home better the inherent dangers and responsibilities in theatrical rigging than having to wrestle with an out of weight line-set.

In this chapter we are going to look at a brief history of manual counterweight rigging systems, define the components and how they work with each other to create a reliable system, relate the safe operation procedures, and point out some of the warning signs of wear and tear that should never be ignored.

Parts of the System

Blocks and Sheaves

Head Block

Figure 6.1 Head block

The head block is located above the arbor guide tracks. It is a stationary, multi-line sheave that changes the direction of the wire rope lift lines from horizontal coming from the loft blocks, to vertical where they attach to the top of the arbor. It is also grooved in the center for the hand line, also known as the operating line, allowing for operation and control.

The head blocks are installed on substantial I-beams oriented along the side wall of the theatre perpendicular to the proscenium. These beams must be able to withstand vertical, horizontal, and resultant forces. In a single purchase system, the beams and block must be able to withstand the weight of the arbor, the maximum force of the load on the batten, plus the dead load of the head block. In a double purchase system, the beam and block must stand up to twice the vertical load in addition to the maximum lateral load of the batten.

The resultant load is the summation of the batten load and the arbor load, going around the head block. This creates a force on the beam relative to the angle between the lines coming from the loft blocks and the lines going to the arbor. If the angle were zero, the load would be 100%. At 180 degrees it would be 200%. If the angle is close to 90 degrees, the load is 1.41 times the sum of the two loads. This can be calculated using the Law of Sines, or by using multipliers based on the angle between the two lines.

$$F_R = P \times \frac{sine \leq}{\dfrac{sine <}{2}}$$

LOFTBLOCK

Figure 6.2 Underhung loftblock   **Figure 6.3** Upright loftblock

Loftblocks are the individual sheave blocks used to change the direction of the lift-lines between the head block and the batten. They are grooved for the number of lift-lines passing through them. They can be mounted to the grid itself, either overhung or underhung, or they can be mounted to overhead loftblock beams that are often part of a building's roof structure.

IDLER BLOCKS

Figure 6.4 Loftblock with idlers

Idler blocks are small non-weight-bearing sheaves that attach to the side of the loftblocks to keep the longer lift-lines supported on their way to the head block.

TENSION BLOCKS

The tension block, also known as the floor block, is located at the bottom of the T-track below the arbor. It is an adjustable assembly that can be raised to put slack into the operating line for some purposes, and lowered to take slack out of the operating line. When operating the system, it is important to keep the line in tension to prevent twisting of the line and to avoid friction between the operating line and the arbor and other structural parts of the system.

Figure 6.5
Up-floor block

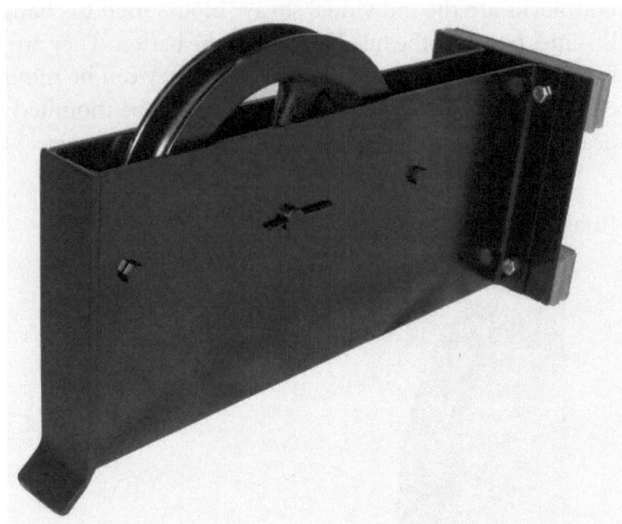

Mule Blocks

A mule block is a sheave block used to reroute wire rope lift lines from positions not in line with the head blocks into alignment with the head block. Typically used for offstage side battens oriented up to downstage.

Figure 6.6 Mule block

The Lines

Wire Rope

The battens in a counterweight system are suspended on steel wire rope that is threaded through the loftblocks and head blocks on the way to the counterweight arbor. There are terminations at both ends. Wire rope is a machine in itself. The way it is constructed, the lay pattern, the core types, and the configuration all play a part in its efficiency. There are three basic components to the machine, wires, strands, and the core.

It is the main artery of strength and support in the system. Just with regular everyday use, with everything in perfect working order, it takes a beating by way of friction, breasting lines, bending and moving back and forth through sheaves.

There are many grades, sizes and classifications of wire rope, and each one is constructed with a purpose in mind; none are constructed specifically for theatre. In counterweight rigging, we first need both flexibility and durability. Most theatrical system applications use a 7 × 19 GAC construction because it has both of these qualities. The GAC stands for "Galvanized Aircraft Cable," the "7 × 19" denotes the construction. There are six bundles of wire rope strands wrapped around a center bundle, with 19 individual steel wire strands in each bundle.

Hand-Line

The hand-line, also known as an operating line, is a fibrous rope used to control the movement of the counterweight arbor. Until the early 1990s the hand-line was made exclusively from organic material, usually manila. Since then synthetic ropes have been designed specifically for this use. The synthetic ropes are typically made of polyester and, in some cases, a secondary fiber called polyolefin. Most theatres use synthetic operating lines, although there are still some using manila.

When operating the system, the flyman pulls on the line closest to the stage to lower the batten and the suspended load. Conversely, the back-line, farthest from the stage is pulled to raise the batten.

The lifeline of the system, the analog control panel if you will, the operating line transmits to the flyman information about the system. The operator might not always be able to see what is happening on stage or in the fly tower, but the feel of the hand-line will tell the operator if something is amiss. There is no motherboard, no RAM, no circuits, no operating program to do the calculations. The system doesn't have a brain of its own; the flymen must use their own.

Other Parts

Lock Rail

Figure 6.7 Lock rail

The lock rail is a heavy steel frame to which the rope locks for each line-set are attached. It is securely attached to the floor of the stage to prevent lifting, and contains some facility for marking each line-set with a changeable identifier.

Rope Lock

A rope lock is a positioning device located on the lock rail. It's meant to be used as a method of maintaining the operating line in a set location. It has a pair of jaws that compress the hand-line to hold it in place. There is a lock ring that is threaded around the hand-line such that when the handle is up (in the locked position), the ring slips over the handle and secures it in the locked position

This is one of the more abused parts of the system. Rope locks are designed as a means of controlling or slowing down an out of balance load; they are not intended to be used as a brake, or to hold a heavily out-of-balance line-set. Rope locks are generally designed to hold no more than 50lb of out-of-balance load; misuse of the rope lock greatly reduce its life span, adversely affect its strength and can lead to a catastrophic failure.

Figure 6.8 Rope lock

Guide System

The guide system is a series of vertical tracks or, in some cases, wire ropes, from stage deck to just under the head block used to guide the arbor and keep its travel smooth and quiet. They also restrict the lateral movement of the counterweight arbors to prevent them from running into one another as they pass. Wire guide systems are generally not used for line-sets that require more than 30′ of travel. They tend to be a little noisier than the other guide systems because the wires allow for movement in the arbor even under normal conditions. T-bar and J-bar systems are the most prominent. They are usually constructed out of 1½ × 1½ × 3⁄16 steel or aluminum that is rigidly fastened to the support wall at regular intervals.

Arbor

An arbor is a rack or carriage used to hold the counterweight. The lift-lines terminate at the top of the arbor, as does the operating line in a single purchase system. Steel rods connect the bottom and top plates of the arbor. The arbors are constructed so as to facilitate the loading and unloading of the weights. They must be constructed to retain the weights even in the case of an unexpected impact. The lift-lines are normally attached to the arbor with a single termination for each line, and then reeved through the head block in a way that keeps

Figure 6.9 Arbor

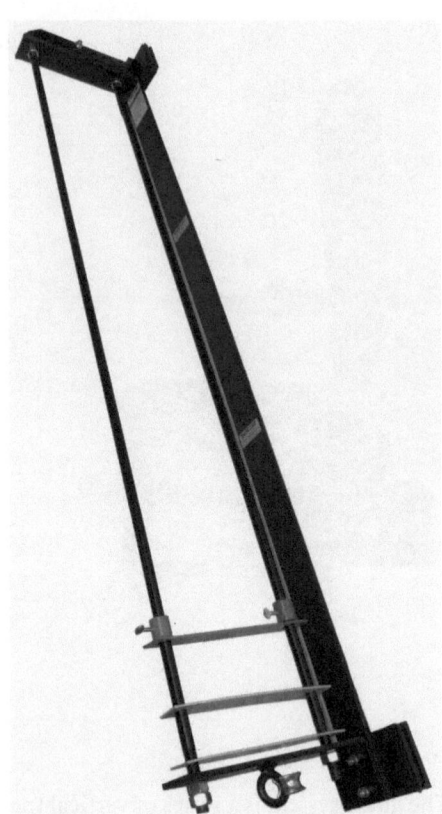

the lines untwisted and free from one another and allows for easy inspection and maintenance.

There are spreader plates located in the arbor, designed to prevent the arbor rods from bending and accidentally releasing the weights in the event of an out-of-balance accident. There should be one spreader plate for every two vertical feet of arbor rod.

Locking collars placed on each arbor rod above the last spreader plate are designed to help prevent the counterweight from being forced out of the arbor in the event of an out of balance accident. There should be one locking collar on each arbor rod.

Batten

Battens, also known as system pipes, are the other side of the balance scale. Battens are attached to the lift lines and suspend scenic pieces, electrics or other equipment above the stage. They can be as simple as pieces of 1½″ Schedule 40 pipe, or they can be an engineered ladder batten or a truss.

Materials used to construct battens must be able to hold at least 30lb per foot of uniformly distributed load and 100lb point loads between two lift-lines. Splicing is a normal method of joining batten sections, but the materials used must be of the same standard as the batten. Threaded couplers should never be used.

Battens, whatever their construction, are beams. They are subject to the same forces as those applied to other beams. As such they are not indestructible. If forces are applied to the battens that compromise their structure then they will fail. It is not uncommon to find battens that are noodle-shaped from abuse. These need to be retired from service.

Once a material has been strained past its yield point and does not return to its original shape, physical and chemical changes render it permanently altered; an alchemy of sorts. Not the transmutation of base metals into gold, but the transformation of a material to a weakened state. It will never have the load bearing properties it did before. Schedule 40 pipe is used in so many different places around the stage house, it seems people believe it to have magical properties—able to be bent, overloaded, and abused and still be in service. Not so.

Terminations and Hardware

Terminations and hardware are the connective tissue that hold the system together. They are comprised of an assortment of hardware elements. They are used to terminate the wire rope lift-lines at both ends in such a way that they can then be attached to the arbors and battens.

Wire rope can be terminated using compression sleeves designed for that use, wire rope clips, or fist grips. There are other pieces of termination hardware used with wire rope, but not in this application.

The terminated ends of the wire rope can be attached to the batten directly or with trim chain and shackles, allowing height adjustments to be made easily. At the arbor the ends are attached to the top of the arbor with shackles or, in some cases, turnbuckles.

A word on hardware; not all hardware is equal—always use hardware designed for the application. There are many pieces on the market at your local big box hardware store that look very similar to the hardware designed for the purpose but rated hardware will be marked as such either on the piece or on the packaging. As the system is only as strong as its weakest link, and as the hardware can be that link, don't jeopardize your safety or the safety of others by using unrated hardware.

Operation

Our industry is unique in that we are allowed to have moving loads over people's heads. The fact that we do so in the dark and in a matter of seconds adds to the potential that something can go wrong. Because of this we must adhere to a higher set of standards, use higher design factors, have stringent rules, and be unwavering in our attention to inspection and maintenance.

There are a number of roles that are filled in the operation of these systems. While there are variations on the specifics due to the circumstance of any individual venue, the responsibilities of each role still need to be fulfilled. The safe and efficient operation of any fly system requires specialized skills. While there is a longstanding tradition of on-the-job training, there is a need for formalized training as well. It is important that no worker be sent to fulfill any task listed below without proper training and oversight to assure that the job is done correctly and safely.

Personnel and Responsibilities

The head flyman is in charge of directing the loading and unloading of the arbors as well as the attachment or removal of loads from the battens. It is important that these activities be under the leadership of one individual in order to prevent accidents due to out-of-balance line-sets.

KNOW YOUR SYSTEM

A flyman may work in a variety of houses with different systems, and must know the general properties of each system. It is important to know the kind of system (single, double purchase, or a combination), the weight limits for each line-set, the heights of the grids, and the vertical travel distance of the line-sets. A flyman needs to be familiar with the counterweight available (full, half, and/or quarter bricks), the position(s) for arbor loading, and the lengths of the battens. Often there is a manufacturer's plate from the installer near the lock rail and/or in the loading gallery area where this information can be found.

LINE-SET SCHEDULE

Prior to (but sometimes during) a load-in, one of the first things the flyman receives is a line-set schedule. This denotes which scenic elements or lighting design elements will be hung on specific line-sets for the incoming show. It should also indicate what each element weighs, its dimensions, whether or not it is tailed down from the system pipe, and where in the house plot it is to be hung. Exactly when this comes to the flyman is often a function of how the space is programmed—some venues allow for preplanning and some must deal with shows with very little to no time for preparation.

Labeling

Labeling the rail and keeping a log book are essential parts of a well-run system. Proper labeling includes a description of what is on the batten, what it weighs, and whether or not the line-set moves during a show. Usually there is a place for a 3 × 5 card to label the line-set on the locking rail. As the only consistent fact about our business is that "things will change," it is necessary to verify the position and weight of each element after the load is hung and balanced. The log book is an important record of system usage as well as a place to memorialize each show's hanging plot.

Loading

The loading and unloading of arbors is one with great potential hazard. Dropping a counterweight brick can set off a dramatic and very dangerous chain of events.

The flyman must assure that the deck crew and loading crew are aware when arbors are being loaded and unloaded. Clear, concise, and consistent communications from the rail are very important.

The steps to safely hang a load on a batten are as follows:

1. Warn the deck crew that a pipe is moving in.
2. Warn all personnel on the loading bridge and grid that the pipe is moving.
3. Bring pipe in to the lowest limit for loading.
4. Secure the hand-line.
5. Attach scenery or electrics.
6. Load arbor with appropriate counterweights; as the entire weight of the elements hung may not immediately transfer to the batten—as in the case of tall flats or heavy drapes—arrangements must be made to control the arbor heavy line-set until all of the weight is suspended from the batten; do not use the rope lock to try and control an out-of-balance line-set!
7. Using snub knots and/or twisting the rear line around the front line four or five times, allow the operator to control an out-of-balance line-set until balance is achieved.
8. Check the weight with the hand-line and adjust accordingly

Calculating and Loading Counterweight

Counterweight can vary in size and weight from house to house. Knowing the weights of the house bricks makes the procedure safer and more efficient. Refer to the line-set schedule for preliminary loading calculations. Once the load is suspended, the weight can be adjusted so as to achieve the desired balance. There are occasions where line-sets that move during a performance may want to be slightly batten- or arbor-heavy.

SETTING TRIM

The low trim or "in trim" is the position relative to the stage where an element needs to be during performance situations. For an electric pipe, this is usually a static position. For a moving piece of scenery this can vary from the stage (as a scenic wall element) or at a prescribed height from the stage (as a chandelier). This position is marked on the operating line with a trim indicator. When using twisted line, the common practice is to use a trim indicator in the form of a piece of colored string, ribbon, or twill tape. Each system will have their own color system, but red for low trim and white for high trim are common. Intermediate and warning trims of different colors may be used when necessary.

The high trim or "out trim" is the highest position the element will travel. This is often the storage position. It is important to mark this with a trim indicator to prevent battens from reaching their ultimate height, which can potentially damage parts of the line-set if not eased into that position.

SHOWTIME OPERATION

During the operation of a performance it is important that all personnel operating the fly system be familiar with the cue sheets, the cue lights, or other cue warning devices as well as the characteristics of the element they are moving. Operators must keep their concentration during these moments and be aware of what is going on not only onstage, but in the line-set itself. An experienced flyman can feel when there is something amiss with the line-set and will know what to do if that occurs.

UNLOADING

The unloading operation is the reverse of the loading operation. Unloading tall pieces that create out-of-balance situations must be planned carefully and executed exactly to the plan. The arbor can be partially unloaded at an intermediate position and the out-of-balance line-set controlled as in the loading process. The deck crew can also assist in bringing the batten to the deck by means of bull-lines thrown over the batten. Capstan winches may also be employed to control unbalanced loads.

WEIGHT-LOADERS

Under the direction of the head flyman, the weight-loaders add and remove the counterweight. Often this job is given to younger, less experienced stagehands. It is vital that they understand the process and the gravity of what they are doing—the safety of the entire crew depends on the proper execution of this task.

Handling 30lb bricks 50′ to 90′ above the heads of your fellow workers, is not to be approached casually. Clear communication from the flyman to the loaders is essential, and the flyman's instructions must be followed to the letter.

The first thing to be aware of is the pipe weight. This is the weight of the system pipe itself and keeps the empty battens in balance when nothing is hung from them. In many houses the pipe weight is marked by painting these bricks red or yellow. These bricks are never to be removed during normal situations.

The next part of the arbor system to understand is the spreader plates. They are a very important safety feature of the arbor as they to ensure that the rods connecting the top and bottom plates of the arbor cannot spread enough to let a brick fall out. This is especially true if there is an unexpected impact, as can occur with a runaway line-set.

There should be enough spreader plates permanently installed to place them every two feet of counterweight. When loading the arbor, the plates must be held up and out of the way. There are a variety of ways this can be achieved: sometimes spring clamps are used, sometimes automotive battery terminal clamps, which grip the arbor rod much better, and sometimes tie-line, as it also poses no overhead danger. When using any kind of clamp, the device must be secured with a line that is tied off to prevent the clamp from being dropped. The spreader plates are set onto the counterweight after every two feet of weight is added to the arbor.

The top spreader plate may have lock screw collars mounted to it. When balance is achieved, these spreaders should be lowered onto the last brick and secured in place with the thumb screws.

DECK CREW

Under the supervision of the head carpenter and/or head electrician, the deck crew is responsible for securing the elements to the battens. In a departmental venue, the carpenters will secure scenic elements and the electricians will hang lights, accessories and cables. The head carpenter oversees the operation of the system as the fly system is ultimately under that department. The head electrician and carpenter must coordinate their efforts to assure that everything suspended on or from the batten is safely secured and within the parameters of the system. Both must assure that the elements are placed in their proper orientation and in accordance with the show's needs.

Taildowns, pipes suspended below the batten, are sometimes used to accommodate unusual loads or trim difficulties. Rigging these would be done by the carpenters. In interdepartmental venues, the work of the department heads is increased as the deck crew is being asked to understand a wider variety of equipment and proper installation procedures and oversight by the department heads is essential.

INSPECTION AND MAINTENANCE

Keeping the system in good working order is of the highest priority. There are simple inspections that should occur with every use, more thorough inspections after every load-out, as well as comprehensive scheduled yearly inspections. Usually very little thought is given to a counterweight system, but as it is a very complicated assembly of parts, inspections must include all pieces of the system. If no one at the venue is trained to inspect the system, then an outside contractor should be brought in once a year to do the annual inspection. In many cases bringing in an outside contractor is a good way to have fresh eyes look at the system. Often this catches things overlooked by the in-house staff who are used to what they see every day and don't always recognize the problem.

LOG BOOK

Keeping a good log book is one of the most important things a flyman can do to assure the system is kept in good working order. Keeping track of what has been loaded and when, and of any problems or accidents that might have occurred, helps to pinpoint problems that need to be addressed either immediately or during yearly maintenance. The entries must be legible and specific in order to direct the maintenance and ensure the system is in good working order. The log book also details the frequency and results of all inspections.

WIRE ROPE

The wire rope in the system does the lion's share of the work and, as this is a heavy-duty job, it is very important to make a close inspection of the wire rope. Inspecting the terminations and wear points around loftblocks and head blocks are good regular inspection tasks.

There are things that will extend the life span of wire rope. First is to be sure the fleet angles are within tolerance. The general recommendation on fleet angles is no more than 1.5 degrees from zero. An easy rule of thumb is that you are allowed one unit of offset for every 40 units of distance.

The formula to find the fleet angle is:

$$\text{angle} = \text{arctangent of } \left(\frac{\text{offset distance}}{\text{measurement distance}} \right)$$

Second is the D:d ratio. This is the ratio between the sheave tread diameter and the diameter of the rope, wire, or fiber. This must be addressed during the initial installation process and is a time-consuming chore to do afterward. Choosing the proper wire rope for the job and the right diameter sheave assures the rope a longer life. The smaller the diameter of the sheave, the sharper the bend; the shaper the bend, the more wear and tear on the wire rope; the more wear and tear, the shorter the lifespan. All reputable manufactures of wire rope and sheaves list a minimum D:d ratio for their products.

During inspections, look for broken strands in the wire rope. While there are some industrial applications that allow for some broken strands on the outer surface, this is never acceptable in theatrical applications. We should also look for bends and kinks in the wire rope. These can indicate a number of problems and require the wire rope be replaced.

Within a counterweight system many things can happen to the wire rope. Shock loads from runaway arbors or suddenly released snags on other line-sets are among the worst. When this occurs the wire rope will "birdcage," looking like it has come unraveled. This symptom is serious and warrants inspection of the entire line-set, not just replacement of the wire rope.

Closer to the batten connection, deformations in the lift-lines can occur because of bridling or the application of stiffeners to keep the batten from rolling. Damage can also occur when line-sets are diverted instead of relocating the loft blocks to position the lift-lines in a desired location.

OPERATION OF SHEAVES

As previously stated, an experienced operator can feel if there is something wrong with a line-set. If there is resistance in the line beyond the normal then inspection is required. An experienced flyman can also hear when there is a problem. Grinding noises or rubbing and chafing sounds are all reasons to inspect the line-set immediately. Generally, the sheaves at the loftblock and head block are a good place to begin when these signs present themselves.

There are times when loftblocks must be moved to create room for other sets to pass or to align with a specific mark on the stage. "Kicking sheaves" is a common practice and these blocks are usually accessible and movable. This changes the fleet angle to the head block and must always be taken into consideration. Excessive fleet angles greatly reduce the life of the wire rope and the sheaves.

Idler blocks are small non-weight-bearing sheaves that attach to the side of the loftblocks that support the longer lines as they traverse the grid on their way to the next block and eventually to the head block.

CONDITION OF TERMINATIONS AND CONNECTIONS

Terminations and connections must be inspected for deformation and proper use. Shackles and turnbuckles should be moused in a way that prevents them from loosening. Thimbles should be inspected to be sure they are not elongated or twisted—one is a sign of overloading or shock, the other is a sign of side-load for which they are not designed. Eyebolts and other threaded devices that have a specified torque rating must be checked occasionally to be sure they have not loosened.

Hand-Line

The system's hand-lines must be inspected for frayed or broken strands, indicating misuse or excessive wear, and abrasions. If there are signs of abrasion, the point of rubbing must be discovered and eliminated.

Deformed lines or overly compressed sections can occur with heavy loading and misuse or maladjustment of the rope locks. Where possible the problem needs to be addressed and the line replaced.

Ropes with an outer jacket and inner core are subject to inner core slippage. This results in "lumpy" rope and gathering of the outer jacket. When this occurs the lines need replacement.

Rope Locks

Rope locks must be properly adjusted so the jaws hold the operating line, but do not subject it to over-compression. The lock must be adjusted so that, when lifted nearly to the locked position, it stays put, but not so tight that a great effort is required to lock it. The keeper rings must be in place and easily placed on and taken off the rope lock handle

Food for Thought

We rely on the good condition of our systems, so they must be maintained in safe working order. Regrettably, the more experience we have with these systems, the easier it is to become complacent about them. It is very easy to start taking shortcuts and skipping steps. Confidence in our own expertise can work against the best interest of the system. Assuming that something out of the ordinary will be noticed because experts operate it is hubris. The more familiar things become, the less they are noticed.

Always remember that as the operator of a manual rigging system you are the finesse; you are the motor control and the speed settings; you are the E-stop and the limit switch; you are the safety measures and the precautions. Never forget that the most important component of any manual rigging system is *you*.

7

Aerialist Rigging

STU COX

Introduction

Aerial arts are some of the most popular elements in live performance today. They can be seen in all aspects of entertainment, ranging from live theatre and special events to tradeshows, corporate events, and even biblical productions in houses of worship. They are fast becoming a form of fitness and recreation much like dance or martial arts. Originating in the circus, aerial arts have evolved into a spectacular art form that amazes, frightens, and mesmerizes an audience.

As the number of aerialists increases, so too does the need for experienced riggers who can combine their rigging skills from other parts of the entertainment industry with an understanding of the dynamic forces involved with this art form.

Aerial Arts

The aerial arts are as varied as they are spectacular. For aerialists, these involve differing degrees of strength and flexibility, as well as specialized aerial equipment. For riggers, it is important to have a working knowledge of the various aerial arts to be able to create safe and appropriate rigging.

Silks

One of the most recognizable acts, aerial silks, also known as fabrics or tissues, involve a long piece of fabric, doubled and suspended at its center, thus giving the performer two long lengths of fabric typically reaching to the ground. Fabrics range in width, thickness, and stretch depending on the performer's experience, size, and routine. Performers use a variety of climbs, wrapped poses, and drops to create routines of striking fixed images and dynamic action. Routines may involve swinging and spinning, as well as dramatic unrolling drops that can create sizeable shock loads. These shock loads create a dynamic "bounce" that can cause system components to move, possibly resulting in side-loaded attachment hardware, such as a carabiner or shackle. The point should be stabilized to avoid this.

Figure 7.1 Silks

Sling

A sling, or hammock, is a suspended loop, most often of fabric, that the performer uses for various sits, hangs, and contortions. These usually involve

Figure 7.2 Sling

swinging and spinning at a fixed height, making them good for venues with low rigging height. Slings can also be made of chain, rope, or net.

Corde Lisse

Figure 7.3 Corde lisse

Corde lisse or rope acts utilize a vertically hanging rope. It may be a braided or twisted rope, sometimes with a fabric sheath. Typically the rope diameter is between 1″ and 1½″ with a spliced eye at the top for rigging. Performers create routines with combinations of wrapped poses and drops.

Spanish Web

Similar to corde lisse, Spanish web takes the rope and adds hand or foot loops. These loops increase the types of grabs, and therefore the number of poses, that the performer can create. Aside from variety, the loops add a more secure hold, enabling the performer to create very dramatic rope spins, often incorporating a second performer at the bottom of the rope, adding tension or spin to the rope as needed.

Figure 7.4 Spanish web

Lyra

The lyra, or aerial hoop, is a circular steel ring suspended in a way that looks like a vertical hula hoop. It can be rigged using one or two points. Performers

Figure 7.5 Lyra

sit, hang, contort, and spin on lyras—typically solo, but sometimes as a duo. Routines do not require much height, but can be raised or lowered before or during the performance to add variety, visibility, or effect.

Trapeze

Trapeze covers an entire family of traditional to modern aerial work. The basic trapeze is a horizontal bar suspended at each end by vertical ropes. Trapeze bars can be metal or wood, and vary in size by performer and routine. Performers use various sits, hangs, poses, and swings in the many styles of trapeze: static, swinging, dance, flying, and the list continues. Some of these involve more than one trapeze bar, and multiple performers. Rigging heights can vary from 12′ to 18′ for static trapeze, and to circus heights for flying trapeze.

Figure 7.6 Trapeze

Straps

Figure 7.7 Straps

The straps, or sometimes known as aerial ribbon, are a pair of straps, typically cotton or nylon, hanging from a spinning point. They may, or may not, have looped ends. Routines most often involve two partners, with a variety of grabs, poses, and spins. Different moves may involve both performers on the ground, one in the air being spun or manipulated by the ground partner, or both spinning in the air at the same time. The height of the straps is often raised and lowered during the act.

There are many other forms of aerial work, often derivations of the ones listed above. These include, but are not limited to: cradle, window, cloud swing, cube, hair hangs, and mouth hangs. Some even involve using body piercings as the attachment points on the performers.

Aerialist Rigging Hardware

The hardware used in aerialist rigging comes mostly out of the rescue and entertainment rigging industries. There are many other types that can be used, but listed here are the most common ones. In all cases, make sure that you are selecting hardware with appropriate load ratings and design factors. Even better is to have hardware with ratings stamped on them, and spec sheets from the manufacturer.

Remember, the hardware used in aerialist rigging is only as good as it is maintained. Proper inspection, maintenance, and removal from service when required are vital safety points.

Beam clamps are available in various styles, and create strong, non-sliding attachment points to the bottom flange of I-beams. Of these different types, only a couple are suitable for aerialist rigging, as they securely capture both sides of the I-beam's lower flange, and have ratings suited to the loads present in aerial acts.

Adjustable Beam Clamps

The jaw-style clamp acts like a vise. When screwed tight, it captures both edges of the I-beam's lower flange. These should only be used for aerial acts that create a downward force, as they are not designed for a dynamic gimbaling motion.

Figure 7.8 Adjustable beam clamp

Cleat Beam Clamps

The other style of clamp places cleats on top of each of the I-beam's lower flanges, that bolt to a piece of channel steel running perpendicular under the beam. A swiveling eye-bolt for rigging attachment is centered on the underside of the channel steel.

Figure 7.9 Cleat beam clamp

Riggers beware: Many other types of hardware and devices fall under the name "beam clamp." These are used for hanging electrical conduit, lighting, threaded rod, or other construction materials, and some only clamp to one edge of the bottom flange. They are not appropriate for aerialist acts.

Synthetic Roundslings

These slings come with and without an internal wire rope core, are incredibly strong, and work well in numerous applications for wrapping beams and trusses of all shapes, sizes, and materials. Care should be taken that they are installed in a way that minimizes the chance of them sliding.

Figure 7.10 Synthetic roundsling

Carabiners

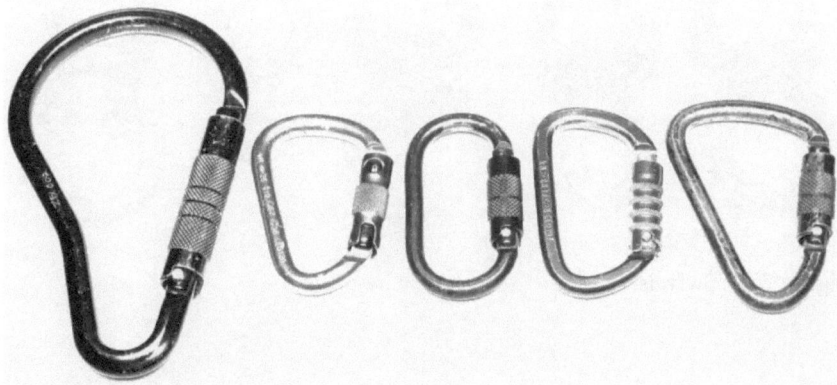

Figure 7.11 Carabiners

Carabiners are connectors commonly used in aerialist rigging. They can be manipulated with one hand and do not have separate pieces that can be dropped or lost. They are manufactured in many styles, with the variations focusing mainly on the gate, shape, and material.

Choose carabiners with appropriate ratings and "auto-locking" gates. The motions involved in aerial performances can easily cause an unlocked gate to open, or a "screw-lock" gate to spin itself into an unlocked position. The shape of the carabiner chosen is determined by how this connection will be used and loaded. Ovals and Ds are for connecting straight-line loads, while the pear-shaped ones give more room for attaching aerial fabric or apparatus on the larger side, but can also function like a shackle, connecting three lines or loads together. The choice of steel versus aluminum is up to the educated user.

Swivels

Swivels are essential in aerialist rigging and performance. These allow the aerialists to safely spin at high speeds, and continue to do so over long periods of time. They also keep the rope and cables used in the rigging and aerial apparatus from becoming overly twisted or unwound during the routines. Make sure the swivels you have chosen have appropriate ratings and are designed for spinning, not simply for rotating gear into alignment.

Figure 7.12 Swivels

Blocks or Pulleys

Blocks and pulleys are used in moveable point rigging, or any time the aerial apparatus is going to be raised or lowered. Whether this is happening before and after the routine or during, blocks should be appropriately chosen for not only their design factor, but also for the size of cable or rope that will travel through them.

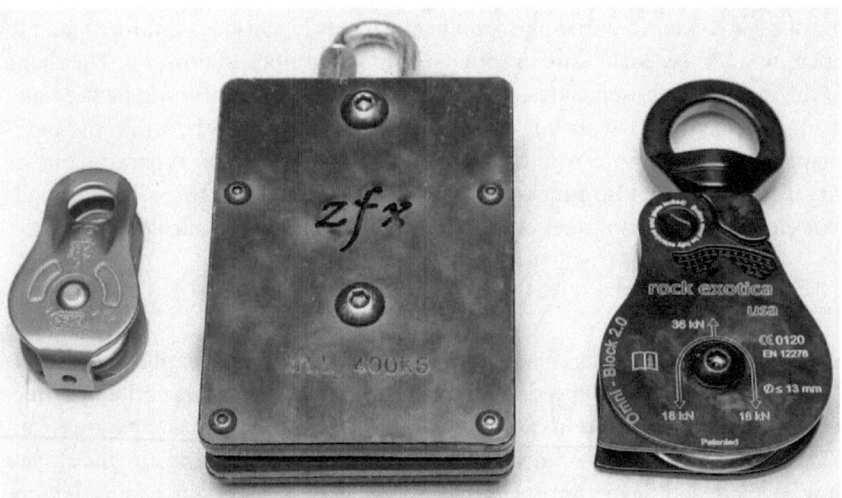

Figure 7.13 Blocks and pulleys

Clew Plates

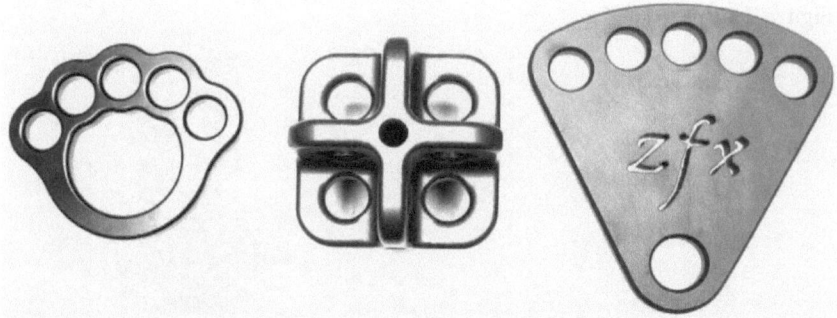

Figure 7.14 Clew plates

Clew plates are used for connecting multiple lines or loads together. They are the rigging equivalent to the stage electrician's twofers or threefers. Typically cut from plate metal, they have multiple holes that give the rigger options for attaching carabiners or other connecting hardware in an organized distribution.

Clew plates are commonly used in straps routines. The two separate straps are attached to the lower outside points of a clew plate, with the top of the clew plate being connected to a swivel.

Rescue 8

Rescue and figure 8s are used for creating the attachment point on an aerial silk. The midpoint of the silk is passed through the large eye and choked over itself making the two hanging lengths used by aerialists. The small eye of the 8 is then connected to a swivel.

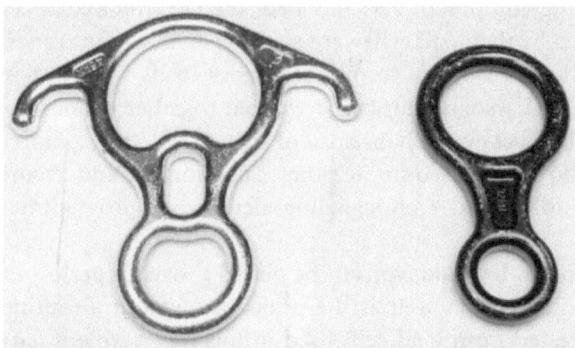

Figure 7.15 Rescue 8

Load Cells

Figure 7.16 Load cell

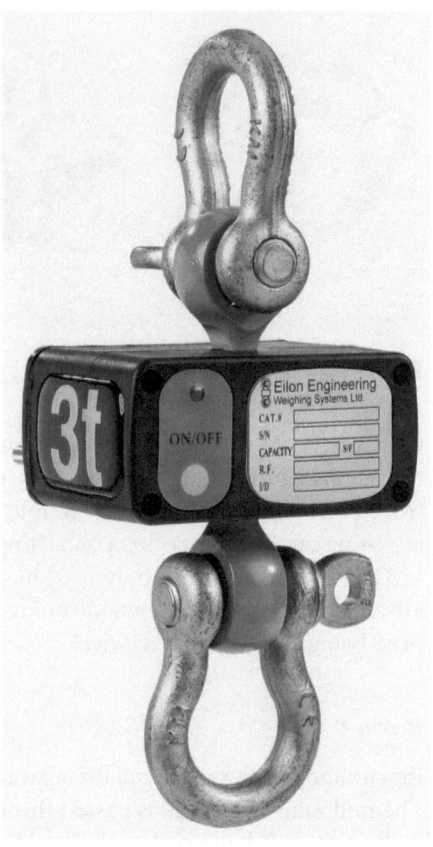

Load cells are a newer technology entering aerialist rigging circles. Strategically placed between the rigging equipment and the load, the electronic load cell monitors the force created by that load. They are available with various ranges and features including digital readouts, recording of peak loads, and wireless connectivity. Some load cell systems can be networked together to monitor entire rigging systems (placed at every chain hoist of a truss grid, for example), transmitting their data to a single computer that can monitor and record the constant changes in force, and even signaling alerts when programmed thresholds are exceeded.

In aerialist rigging, a load cell would typically be placed above the performer to monitor the changes in force, as well as the peaks throughout a routine. Data that have been collected from load cells used in this way have put hard numbers onto the dynamic, and sometimes confusing, forces created during aerial performances.

Anchorage Points and Rigging Points

In order to rig anything in a venue, there must first be some structure capable of supporting the load. This is an anchorage point. Riggers attach the load to a rigging point that ultimately transfers that load to the anchorage point. In some cases these are the same thing and in others there can be multiple layers of rigging or structure that separate them. How direct this load transfer is depends on the venue.

Entertainment venues are designed to offer multiple accessible rigging points that distribute the loads back to various anchorage points within the venue's structure. In arenas, these rigging points take the form of high and low rigging steel, often I-beams and catwalks, but can also include mother trusses—entire truss grids that can be lowered for ground-level rigging, then raised into show positions. Theatres can have counterweight systems and networks of grids and catwalks, providing not only rigging options, but also the ability to easily fly scenery and equipment during a show.

In venues that do not have these conventions, or do not have them in the desired performance location, the rigger must find the anchorage points and create rigging points from those. These could be accessible I-beams, architectural elements, or any number of things. Just because a beam is the only one in the room or "we've used it before" does not mean it is the only rigging option. Get a building manager or engineer involved if ratings are not known or not readily discernible.

In some cases it might be necessary to build from the ground up, literally using ground support truss or stage scaffolding. Location work for events and film often utilize cranes for overhead rigging, and ballast in the form of water containers, large concrete blocks, or even sea containers for anchorage points.

Aerialist Rigging

Aerialist rigging can be divided into two main categories: fixed point, where the aerial apparatus is suspended at height; or moveable point, where the aerial apparatus is attached to a line running through blocks so it can be raised and lowered. Both of these can be broken down into further variations.

Fixed point, in its simplest form, is where the rigging height meets the aerialist's need, and the aerial apparatus connects directly to a single anchorage point. Where more than one rigging point is needed for load distribution or performance placement, a bridled point, or a suspended section of box truss can be used. When the rigging point is too high for the aerialist, the apparatus attachment must be lowered to the desired aerial height. This can be accomplished by adding an extension between the rigging point and the aerial apparatus or, in a more conventionally outfitted venue, by lowering the rigging point. This might be flying in the batten, or lowering the truss.

A typical fixed point aerialist rig would be as follows: starting at the rigging point, a carabiner is connected to the roundsling, beam clamp, mounted eye, or batten clamp. That hanging carabiner is then connected to a swivel, with another carabiner below, making the final connection of the swivel to the aerial apparatus.

Moveable point is when the connection point for the aerial apparatus is attached to the end of a retractable line instead of a rigging point. The line is routed through blocks to some type of lift control. The blocks are attached to the rigging points. The lift control, either a person manually pulling a rope or a mechanical device such as an electric drum hoist, should be placed safely away from the aerialist's performance area, and offstage if during a show. Being able to lift and lower the aerial apparatus can be necessary in many situations.

The aerial performance may involve changes in height as part of the choreography, including raising the aerialist to a specific height, lifting or lowering them during the performance, and returning them to the stage at the end of the act. Moveable point rigging can also be used for safe storage between studio uses or scenes in a show. The convenience of easy access to the connection point for inspecting, swapping, or removing the apparatus cannot be overstated in situations where the use of ladders or lifts is limited by production obstacles such as scenery. It may also be unreasonable for a non-performance venue to have ready access to ladders or lifts at all times.

A simple moveable point aerialist rig would be as follows: starting from the aerial apparatus, its connecting hardware is attached to a line that routes up through a block attached to a rigging point directly over the aerialist. The line continues across through another block and down, attaching to the lift control. This final block functions as a head block and should be attached to a rigging point located directly above what will be controlling the raising and lowering.

Point Stabilization

Whether fixed or moveable point rigging is used, point stabilization should be considered anytime the aerialist rigging involves more than the apparatus connecting directly to an anchorage point. During an aerial performance, the movements of the aerialist may shock the system or cause it to sway. In all but the simplest of fixed point rigging scenarios, these forces are transferred to the entire rigging system, creating a swing that can make it difficult for the performer to maintain balance and control. In some cases the aerialist might collide with other performers or set pieces. The rigging supporting the aerialist may bump into other truss, line-sets, or equipment.

The aerialist rigger's solution to this problem is point stabilization. Stabilizing lines, sometimes referred to as guy lines, are used to immobilize the rigging point from moving with the performer. The number of lines used will depend on the truss, batten, or point you are trying to stabilize. Tensioned wire rope, ratcheting load straps, or metal pipe with cheeseboroughs connect the aerialist rigging to points or structure in the venue.

Venues

Aerial arts are being practiced and performed in just about every type of venue imaginable. From theatre to stadium to outdoor festivals, the aerialist's require-ments are relatively simple: a fixed or movable point at a specific height. Aerialist rigging strategies differ depending on the venue and the aerial act. It is the job of the aerialist rigger to create a safe and efficient rigging plan utilizing the venue's advantages, while avoiding elements less conducive to the aerial perfor-mance or its rigging.

Arenas

These venues usually have a vast expanse of open space, allowing for rigging access with less chance of structural obstructions and point placement limi-tations. Arenas often have steel beams as anchorage points that can support substantial loads. These beams are almost always too high for just a simple fixed point to suffice, and in terms of a plan view, they are rarely located above the desired performance position. To get the proper point placement in an arena, oftentimes it is necessary to bridle between two or more beams.

Consideration must also be given to how the aerialist rigging and perfor-mance will interact with the rest of the production's rigging for lighting, sound, video, etc. Lighting designers tend to be less than pleased to see their lighting instruments swinging along with the aerialists.

Installing the aerialist rigging on separate points or truss from all other production elements of the show will eliminate this interference. Complete moveable point systems, including their control hoists, can be attached to their own trusses. These independent systems can be easily adjusted or inspected without affecting a show's entire truss network.

Point stabilization in arenas can be a challenge due to the long distances, and the often limited options for sturdy points below the overhead beams used for rigging. Consulting the arena's head rigger or engineer can be very helpful in these instances, making sure the stabilizing lines are rigged to suitable points, and will not cause visibility concerns for the production design. Aerialist rigging that blocks audience line of sight or casts shadows on projection screens can be as bad as swinging lighting instruments.

Theatres

Aerialist rigging in a theatre depends on where in the theatre the aerialist will perform, and what the theatre has to offer in terms of rigging points and conventional fly systems. Knowing the aerialist's needs will enable the aerialist rigger to create a rigging plan that complements the show's other production elements and takes best advantage of the theatre's technical capabilities.

Counterweight and motorized batten fly systems can be utilized for aerialist rigging. The aerialist rigger should check with the technical staff for those systems' ratings, and obtain the venue's permission to use them for aerialist rigging. Both fixed and moveable points can be rigged to battens, as long as care is taken to avoid imposing point loads that will bend a batten.

The rigged batten can be flown out to the desired height and secured, so the batten cannot move during the aerial routine. Immobilizing the arbor with safeties to the counterweight system's anchorage points, appropriately locking out the hoist controlling the batten, or attaching the batten to the grid or overhead anchorage points with dead tie lines of chain or wire rope will work.

If the fly system is not in the right location for the aerial performance, or if using a fly system is not an option, the aerialist rigging methods mentioned in the arena section can be used. Most theatres have beams or structural rigging points that can be used as anchorage points.

Many options exist for mounting and securing hoists for moveable point rigging. Catwalks, mid rails, and pin rails are some of the typical theatre structures positioned in places that align neatly with the aerialist rigging. These are very useful for point stabilizing, which is almost always needed in a theatre venue to give the aerialist a solid point at the desired height, and also to keep the aerial act's motion from disturbing the adjacent line-sets that may contain everything from electrics to band shells to projection screens.

Cruise Ships

Aerial acts are very popular on cruise ships. They tend to be permanent, or at least long-running, parts of the shows. The venues for these shows are, like most other places on ships, efficient and economical with their usage of space. The motion of the ship can also be an issue. Everything from scenery to sound equipment is designed and installed in ways that have them organized to fit in among everything else, and to allow them to move only as needed for the show, but not to be swinging or rolling loose as the ship is at sea.

Aerialist rigging is no exception, and is installed in ways that would be considered permanent in other venues. Rigging points are made with eyebolts mounted to the anchorage points. Hoists are mounted to beams and catwalks with bolts.

The lower heights available in most of the compact cruise ship venues are typically well-suited for most aerial work, although many of them are too low for some of the bigger silk routines. Lower height, coupled with the more permanent style of mounting and rigging the equipment, means point stabilizing is rarely needed.

Ballrooms

Ballrooms and their cousins, convention meeting rooms, generally provide good heights for aerial acts, 15' to 20' being common; however, they often have some form of ceiling below the anchorage points. Occasionally rigging grids have been installed above the ceiling to accommodate the more common production needs, such as lighting trusses or projection screens, but usually rigging is limited to the main beams located above the ceiling. These beams can be spread out and, with the low ceiling height, bridling options are limited. The rigging points are often quite a bit above the ceiling, with all sorts of electrical, mechanical, and HVAC in between

In addition, the main beams are usually sprayed with a fireproofing insulation that, when disturbed, breaks away from the beams. A truss section can underhang the ceiling by rigging from two or more main beams, coming down through removed ceiling tiles. Stabilizing lines are necessary to limit sway. Pipe and cheeseboroughs work well if they can be connected from the truss back up to the beams or other solid structure above the ceiling. This will stop insulation from raining down during the performance, and also protect the ceiling from damage.

Ground supported structures can also be used. These boxed-in structures provide a stabilized point and eliminate the need for roof attachments. (They can be supported with grid assists from above, if worried about load limitations.) They provide a rigging option when beams are poorly located, inaccessible, or not rated for the application.

Non-Traditional Venues

Aerialists are performing just about everywhere. While the previous entertainment venues all have their own advantages and disadvantages, they are set up with facilities, staff, and ratings ready for rigging. But many aerial performances are happening in locations just as visually interesting as the acts themselves.

These venues can range from large atriums to studios to outdoor architecture. Aerialist riggers must do extra legwork or consultation to ensure safe rigging options in these circumstances. Engineered drawings and meeting with the engineer may be required to assess and discuss viable load-bearing options. Furnishing the engineer with information as to the type of loads and forces created by the proposed aerial act will help ensure usable rigging specifications.

Smart Aerialist Rigging

Risk assessment and inspections are the "before and after" of smart aerialist rigging. They are the first tools for preventing accidents. Hazards that can be understood before the aerialist rigging plan is developed can be minimized.

Inspecting in a planned and routine manner ensures that everything functions as it should, while at the same time making it easier to identify potentially unsafe scenarios that might develop over time.

For any aerialist rigging show or project, identify the why, who, what, when, and how for both the risk assessment and the inspections. The results from this preparedness and follow-through reduce accidents and instill confidence in the aerialist rigger and rigging. Smart aerialist rigging is better aerialist rigging.

Risk Assessment

WHY

Thorough risk assessment will identify hazards in the aerial routine and rigging. Once known, those risks can be eliminated or minimized. Weak links in the rigging plan can be fixed. If needed, rescue plans can be developed that will respond accordingly and efficiently.

WHO

The aerialist rigger should spearhead the risk assessment process. Involve the aerialist—no one will know the aerial apparatus and routine better. Depending on the complexity of the show, get someone from production involved too. By including these individuals, you may learn of new hazards. They may also be able to provide solutions from their end that are simpler than ones on the aerialist rigging side of things.

WHAT

Include the entire aerial project. Rigging should take into account the aerialist rigging and the venue rigging. Consider the aerial apparatus in general, and the specific piece of equipment being used. For the actual performance, look at the routine, the aerialist, and any operators. Remember show factors, such as apparatus storage, cueing procedures, and communication.

Rescue options are ways of minimizing hazards too, but those rescue procedures should be assessed for risks to the rescue team.

WHEN

Do the first risk assessment before rigging even starts. Consider doing follow-up risk assessments as changes happen. Were substitutions made to the rigging plan during the install? Has there been a change in the show, cast, or crew? Have the needs of the aerialist changed? Is there any proposed new hardware or apparatus?

How

Be thorough. Go in deep, get specific, but leave egos and biases out of it. Get expert help or input when it is needed. Once the risk assessment is complete, get it looked over for a second opinion.

Inspections

Why

Inspections catch things before they cause accidents. Aerialist rigging utilizes many types of hardware and equipment. They all wear over time. Some are complex, and require a sophisticated maintenance regime to retain necessary specifications for safe operation. Mistakes can happen during install or servicing.

The overall production may have elements that require consideration during inspection. Use of haze, fog, and pyro can leave residue on rigging components. Poorly preset scenery can obstruct line of sight, or create a dangerous interference.

A comprehensive and well-planned inspection, performed routinely, is one of the most effective ways for an aerialist rigger to reduce risks and prevent accidents during the show run or for the lifetime of the aerialist rigging.

What

Automatically included on the inspection checklist will be the aerial apparatus, the aerialist rigging components, and any venue rigging affected or utilized for the rigging plan or aerial act. Power sources and any mechanical or automated equipment should be checked.

Refer to the risk assessment. Include the aerialist and any operators. Communication lines should be checked. Special cues should have their test runs. Any parts of the rescue plan—equipment, lifts, or personnel—get inspected too.

Who

The aerialist should always check their apparatus, and be encouraged to inspect the aerialist rigging at least after the completion of install. The aerialist rigger should perform the inspections or train a designated inspector.

When

The aerial apparatus should be inspected before and after every use. The aerialist rigging is inspected at the end of install, then on a scheduled basis depending on the risk assessment. If the aerialist rigging is for a long-running show, or a permanent install, there might be service inspections that involve changing out parts of the rigging.

How

Once the inspection checklist has been agreed upon, organize the inspections so they can be performed expeditiously. Is there a necessary order to steps for the inspection? Consider the accessibility to the rigging. Are lifts needed, or is there an easier way to complete that part of the inspection? Could binoculars be used? Is there a portion that requires video monitoring?

Use checklists during the inspections, and be ready to photograph and document any findings. If any repairs or service are performed, document these as well. All of this is recorded in a logbook. Include the dates, checklists, and inspectors' names.

Thinking Ahead

Aerial arts continue to grow in popularity. Aerialists are taking advantage of the increasing resources for training and practice. Aerialist riggers are now finding opportunities for workshops and education in actual aerialist rigging. New performers and riggers are being drawn to this art form. People on both ends of the rigging are expanding the possibilities. As the world of aerial arts grows and expands, is there a place for aerialist rigging to evolve alongside aerial performances?

Yes. By increasing the understanding of what is physically happening to the rigging during aerial routines, aerialist riggers can devise new methods of rigging, search for alternative equipment options (and even create the need) and contribute to the development of gear designed specifically for aerialist rigging. Comparing data from load cells with video (especially high speed) of aerial routines shows how much force is being created, how it changes throughout the aerial performance, and how it affects the different components in an aerialist rigging system.

The increasing availability of load cells over recent years has enabled aerialists and riggers alike, to see in actual numbers, what is happening. Readings can be taken throughout an entire routine and, when cross-referenced with the routine's video, can create true working explanations of the changing forces and loads in an aerial act. These experiments are being done, and, as the volume of data increases, it will benefit aerial artists and aerialist riggers alike.

Finally

Being an aerialist rigger means being part of the team bringing an aerial act to an audience. The aerialist rigger brings skills and experience to create a safe rigging environment for the aerialist, other production professionals, and the audience. At the same time the aerialist rigger provides the aerialist with a secure and stable point at the height they are prepared to confidently perform.

In aerialist rigging especially, the rigger also brings an attitude that promotes understanding and collaboration between the aerialist, the venue, and the production, while providing professional rigging solutions. In all of this diplomacy, the aerialist rigger must still be ready to confidently say "no" when necessary and back up their decision with knowledge and experience.

Aerialist rigging is combining a rigger's knowledge and experience with a sense of professional responsibility to safely support aerial work while, at the same time, understanding it as a form of art.

8

Performer Flying

JOE MCGEOUGH

Foreword

The field of entertainment technology fascinated me at an early age. I worked on my first show in high school; it was a production of *Fiddler on the Roof* (no performer flying) and I was assigned to the carpenter crew building sets and marveled at the fact that there was a group of wizards known as scenic painters and lighting technicians who could make my crude construction look good. I can remember watching the show and thinking to myself "this is something I would really love to be involved with."

Fortunately I knew the right people in the business and before long I was building stages for rock concerts, setting up lights and loading trucks locally in Portland, Maine. The team aspects of the crew were engaging and before long I was venturing up to the catwalk to assist the tour riggers hanging chain hoists from the roof with steel slings and shackles. The confusion and hectic pace of the work below did not follow me up to the rafters; I found the level of concentration required for high rigging to be very peaceful and calming, and I've never looked back.

Performer flying came when I met Peter Foy in the Ice Capades back in 1978. Like many people who met Peter, I was immediately drawn to both what he did and to the man himself; I knew nothing about his craft but he was very willing to teach me countless things in many areas over the decades that followed. The following is a direct result of that relationship.

History

Stage flying dates back to the time of ancient Greece. Back then the god-like characters such as Hercules would be performing in a drama and find themselves in a near-death predicament when suddenly a device similar to a large crane would appear on stage and lift the performer to safety. It was called "deus ex machina," a Latin term that translates to "god from the machine," and the effect was fitting for the characters portrayed in the shows. The methods are

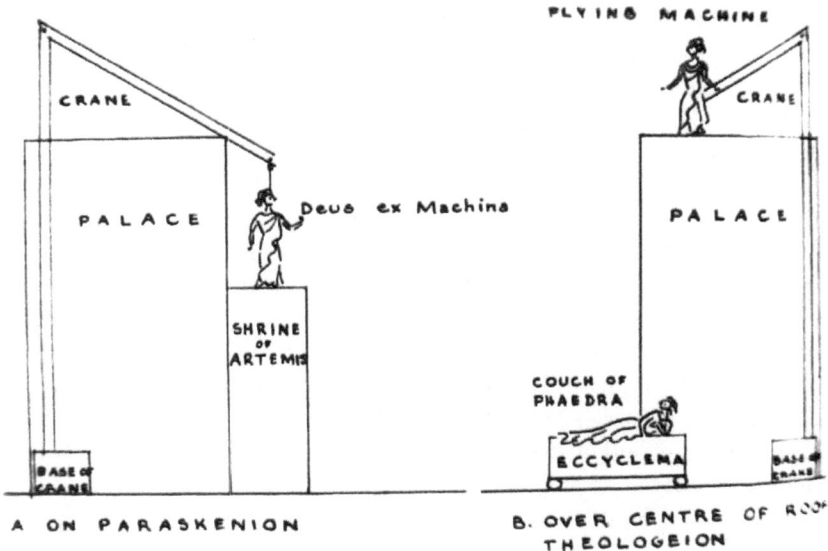

Figure 8.1

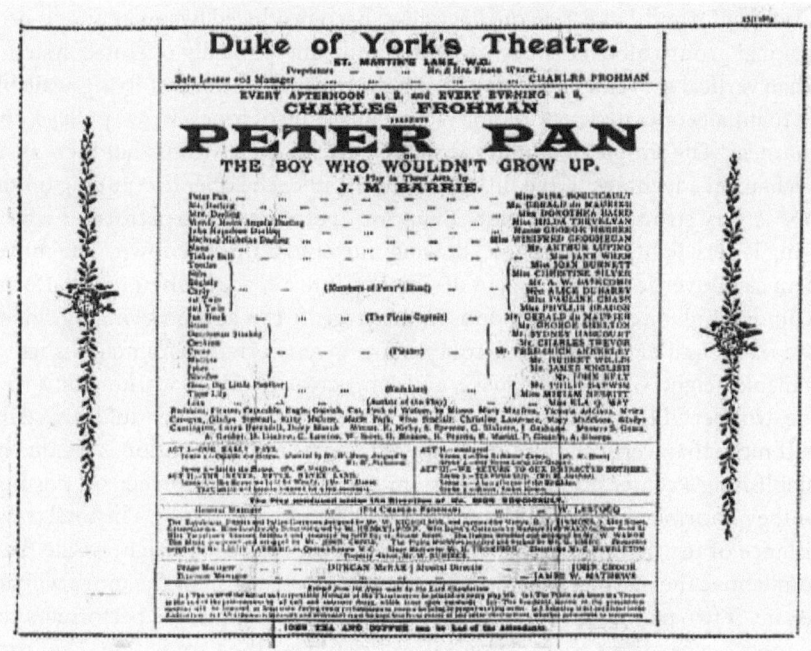

Figure 8.2

considered primitive by today's standards; however, the ingenuity behind this machine places performer flying alongside the trapdoor as one of the oldest special effects in the history of theatre. There was not much innovation in the art of stage flying for centuries until the opening of *Peter Pan* in London, and then—beginning in the 1950s—Peter Foy not only changed the way stage flying was done but introduced many new products and techniques to accomplish the effects.

With the introduction of J.M. Barrie's play *Peter Pan* at the Duke of York Theatre in London in 1904, performer flying took a giant leap. The show was very successful, with much of the excitement generated by Peter Pan and the Darling children flying from their nursery to Neverland and subsequently back to their home in London. The flying effects as staged in the production were so realistic that the show's producers actually had to post warnings for children attending the play that they should not attempt to fly from their bedroom windows; a sort of "we are professionals here, do not try this at home."

Manual Systems

With the success of the show in 1904, thousands of productions of *Peter Pan* (and many other productions that contain performer flying effects, such as *Wizard of Oz* and *Fiddler on the Roof*) have been produced by amateur and professional groups all over the world, often utilizing manually operated systems. When vertical movement occurs, the simplest form of performer flying available is a manually operated pendulum system consisting of ropes, wires, pulleys, and a harness. The simple pendulum is a 1:1 system, which means that there is no mechanical advantage to the lifting medium; either the operator pulling on the rope is very strong, there is more than one operator, or the performer who is flying is very light, allowing free, flowing movement up and down. The movement can be vertical; ascending or descending into a space from or to a platform, behind a wall, or out of a trapdoor in the stage. It can also be swinging side to side or a circular pattern in a cone shape; given certain parameters such as available height, width, floor space, and some creative minds working on a routine, wonderful flying effects can be created with the simple pendulum system.

If more than vertical motion is required, the pendulum motion is created by establishing a center point for the system and then moving the takeoff position for the performer away from the center point by a fixed distance. The total travel distance of the flight should never exceed two-thirds of the height of the fixed point above the performer, otherwise the speed and action of the move will not be safe. If two pendulum systems are used to fly two people, the performers can be moved away from their center takeoff marks so they are within 2' of each other; in this position they could be facing one another or virtually back-to-back. If both performers are lifted together, at the same pace and distance, without providing additional motivation then they will swing away from each

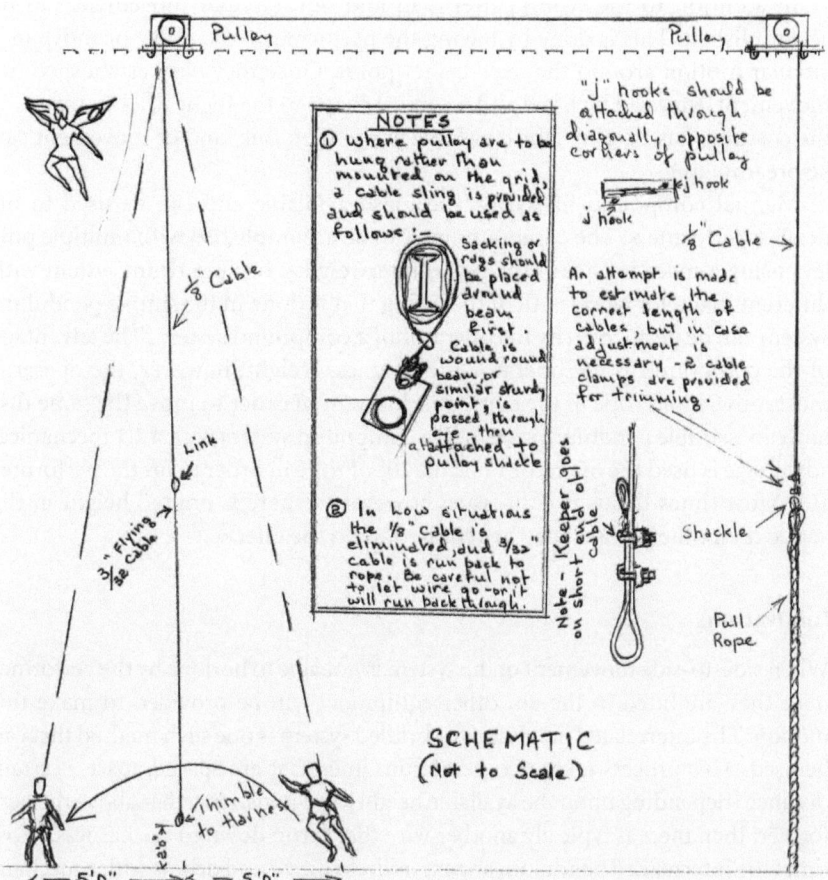

Figure 8.3 A performer's movement is governed by his starting position. If he starts 5′0″ to one side of a mark placed directly beneath pulley and is lifted then he will fly to a point 5′0″ on the other side of the mark, always passing over that mark. If he is landed at the end of the swing, he will have completed a flight of 10′0″. If he is not landed then he will continue to swing to and fro over that path until he is landed. He must always be landed at the end of a swing, trying to catch that moment when he is stationary, before he changes direction. Once this technique has been mastered, the flights can be extended one foot at a time until the maximum desired length of flight is achieved. Note that *total* flight length must not exceed two-thirds of the height the pulley is suspended.

other then back toward one another. If the movement back and forth is not too dynamic, they can join hands and stop in mid-air during the move; they could then be landed together or release and swing, then land.

In addition to the swing patterns in and out, circular movement can be accomplished. This is done by having the performer walk, glide, or move in a circular motion around the fixed center point. Once they have established the movement, they can be lifted and a conical shape to the flight will be created. If the costume has flowing fabric or large wings then this kind of movement can be breathtaking.

Manual compound lift systems are also available and can be used to lift heavier performers. The compounding can be accomplished with multiple pulleys using a mechanical advantage, counterweights, or via a drum system with different diameter wheels. All of the lifting that is done in the simple pendulum system can be performed by the operator of a compound system. The advantage of the compound is the operator is lifting less weight; however, the operator must move more rope in the compound system in order to move the same distance in a simple pendulum system. If a compound system with a 2:1 mechanical advantage is used the operator must lift 20' of rope in order to lift the performer 10'. Most times this is not an issue; however, if there is limited height in the venue then other means may be required to do the effect.

Innovations

When side-to-side movement of the system is not able to be done by the performer once they are lifted in the air, other equipment can be provided to make this motion. The interrelated pendulum or bridled system is one such method that can be used. This process takes two pendulum lines that are spaced apart a certain distance (depending upon the available height) and joined together above the performer; then there is typically another wire that is run down to the harness. Most times the interrelated pendulum system utilizes compound drums with a mechanical advantage of 2:1 or 3:2 allowing the operators to lift and slack their ropes without too much resistance to the load. The distance of the points above, together with the height of the structure and the pace at which the operators pull and slack the operating ropes, will dictate the amount of side-to-side motion available.

The operators of the system are typically close by one another so they can determine who is lifting and who is slacking at just the right moment. (In Figure 8.4 they are placed at opposite sides of the stage for clarity in the diagram.) With both operators lifting together, the performer will lift straight up above the stage; if one operator lifts and the other slacks the line then the performer will move in one direction over the stage. To move the performer in the opposite direction, the operators switch the direction they are pulling or slacking the ropes and the performer will swing in the other direction. To control the swing, one rope is pulled as the other is slacked quickly, which moves the bridle point over the head of the performer stabilizing the swing. Operating this type of system is similar to watching a bell ringer in a church tower; as the line is slackened then suddenly stopped the operator is sometimes lifted off the ground due to the speed and

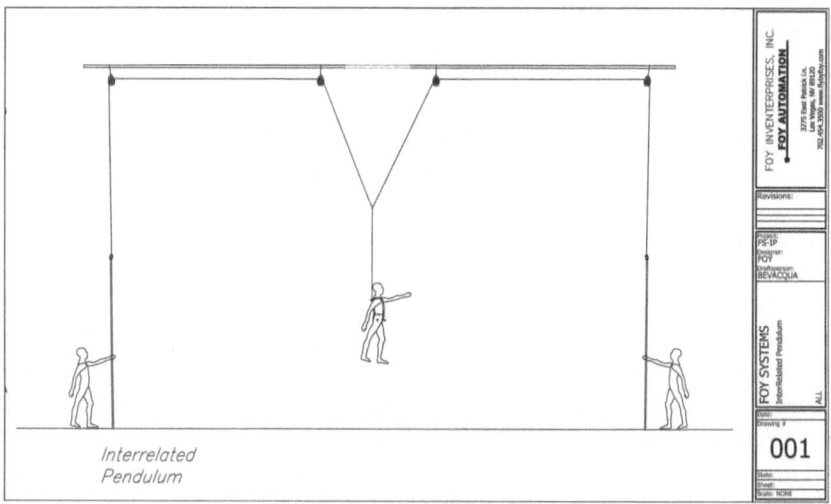

Interrelated
Pendulum

001

Figure 8.4

swing pattern of the performer. Peter Foy utilized this type of system to fly Mary Martin on Broadway in the production of *Peter Pan* in the early 1950s. The flying he did with Mary was—and still is—considered state of the art and extremely dynamic, generating swoops and swings speeds of up to 25mph over the stage. The technique for this system requires a fair amount of experience but, once mastered, can generate fluid and graceful flying effects. It is often referred to as the most difficult but most enjoyable system to operate by technicians.

When the side-to-side movement with an interrelated pendulum system is required but not possible due to height restrictions, a tracking method needs to be employed. The simplest way of lifting and tracking a performer is to use a floating pulley track system. This is a track that has an upper and lower traveling carrier; the upper carrier houses pulleys that allow the lift cable to run over and through to a lower pulley system that "floats" up and down. The upper carrier also contains the termination for the traversing cables or ropes that pull the carrier along the length of the track. The floating pulley track is very similar to a tower crane that is normally operating on the side of a high-rise building; you can see the upper carrier that travels along the arm of the crane and the hook that attaches to the load being hoisted up and down the side of the building. With the lower pulley rigged to the upper carrier with a single run of cable or rope you get a 2:1 advantage, which reduces the lifting load by half. Manually operated floating pulley tracks are very straightforward to set up and operate, the only drawback to them is the lower floating pulley can be in view in extremely low height venues, causing a distraction to the effect. When in view of the audience, the mechanism is like watching a marionette crossbar during a show and is not easily hidden with masking or lighting.

A groundbreaking development by Peter Foy in 1962 was the Track on Track® system. Peter rigged the floating pulley or compensating pulley and placed it inside a separate short carrier track mounted horizontally below a main support long track. Now the pulley that gives you the 2:1 advantage is moving left to right to lift the performer and not up and down, making it much easier to conceal in low height situations. The Track on Track revolutionized performer flying and made it available to all sorts of venues such as multipurpose rooms in schools, tents, outdoor amphitheaters, theatres without grids, and even parade floats! The limiting factor with the Track on Track system is the length of the horizontal carrier; in manual systems this carrier is furnished with a 2:2 compensating pulley system and a 2:1 compound drum, giving you a 2:1 lifting advantage but limiting your amount of distance to raise the performer to the movement of the travel on the 2:2 compensating pulley. In most cases this would be 8' of movement on the pulley or 16' of actual lift.

There are many ways to mount the devices mentioned above and they can be manipulated with human or automated power; all of these systems require precise planning and a good creative outline to be used effectively. It is also important to install the flying equipment in such a way that the audience has a difficult time seeing or hearing it, which gives the element of surprise when the effect actually occurs.

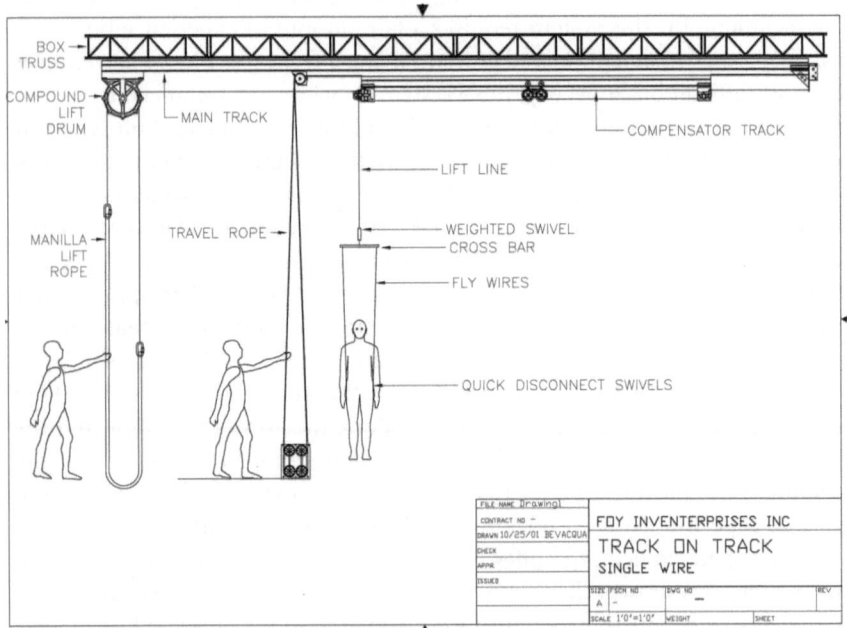

Figure 8.5

Two-dimensional flying systems

The majority of the flying effects that take place in theatre are two-dimensional in nature. Most lighting and scenery used in stage productions is hung or mounted left to right above the stage; this limits the amount of space needed for performer flying and can be quite difficult to secure. In *Peter Pan*, the entrance for Peter is one of the most dramatic in all of entertainment. In order to achieve the upstage to downstage movement (flying toward the audience) there must be an area from 6' to 8' kept clear for the path of the flying wire to move through the window and toward the audience. This is accomplished many times by bargaining with the scenic and lighting designers for that highly desired real estate center stage.

Tinker Bell enters the Darling nursery as a ball of light, first flitting about looking for Peter Pan's shadow; then the windows burst open and Peter shoots through the room, flying high above the window seat and landing softly downstage near the mantle. Even though the audience is expecting Pan's entrance they are quite often taken completely by surprise by the magnitude of it and applaud instantly. This is two-dimensional movement achieved by a very rapid lift and pendulum action that thrusts the performer into the air and toward the audience. This same method is often used for the curtain call, during which Peter is flown over the audience. The pendulum point is fixed over the orchestra pit or first few rows of the audience, the performer walks on stage to take a bow, and in a split second is lifted, rapidly swinging over the audience. The pendulum action swings back and the operator lowers the performer to the stage. In situations where there is low height above the audience, a track system is used to maximize the effect.

The Impact of Hollywood

The next major event that would have an impact on performer flying came with the release of the *Star Wars* movie in 1977. The spectacular visual effects and battle scenes showing actors flying as they wielded their light sabers created countless ideas for the creation of people moving through space in all areas of live entertainment. It seemed as though live events had to compete with cinema in order to give their audiences a "wow factor" that would keep them coming back for more. Corporate events had their CEOs making entrances flying over the audience as if they were astronauts, James Bond, or agents like Tom Cruise from the *Mission: Impossible* movies to the wild applause of their employees; it was as if everyone was sitting in an IMAX theatre waiting for the next big moment to occur, and it was being delivered.

Today it is very likely that if you watch a televised event—be it a commercial, reality or serial program, or live awards show—there is a good chance a performer will be seen flying. This could not be said 25 to 30 years ago; the influx of special effects into the entertainment industry is progressing at a staggering rate. What used to be done with wires, ropes, and pulleys is now being accomplished with sophisticated winches being controlled by high tech control systems. Let's take a look at the evolution of flying equipment and harnesses over the years.

Harnesses

Regardless of what mechanical system is used to fly a performer, the most personal element of the equation is the harness. It's hard to imagine what types of harnesses were used back in the days of ancient Greece. They probably resembled a corset that contained a few bones from animals (like a girdle), but thanks to modern high tech materials such as webbing, synthetic threads, high-density foam padding, and connection hardware, today's harnesses are lightweight, form fitting, and comfortable. The performer must have the correct fit in the harness, almost becoming one with it in order to avoid discomfort when in the air; loose straps or buckles will only cause problems as the rehearsal and run of the show continues. A prime example of "becoming one" with your harness occurred back in September 1984 in San Francisco at the Davies Symphony Hall with a legend of musical theatre.

Peter Foy was sitting in his office back in the summer of 1984 when the phone rang. He picked it up, and before he could say a word a voice came over the line asking, "Peter Foy? This is Peter Pan." Peter Foy hesitated for a minute, then asked, "Mary . . . it this you?"

Although it had been more than 20 years since they'd spoken, it was indeed Mary Martin and she had called to ask Peter to fly her again as Peter Pan at Davies Symphony Hall in San Francisco at a benefit for the Trauma Center where, two years earlier, she and Janet Gaynor had been treated for severe injuries suffered in a car accident. Once Peter got his composure he politely asked if Mary was alright; not from the injuries she had suffered in the accident, but for the request to fly as Peter Pan again at the age of 70! She told him she was determined to do it as long as he was there to make it happen, and so it was that Mary Martin flew over an audience filled with black ties and evening gowns wearing her original costume from Peter Pan on Broadway . . . and that harness. Mary requested the harness be sent to her one month before the show; she wore it diligently for four hours every day to get accustomed to it again, "becoming one" with it. That persistence paid off; she flew magnificently 50' in the air and 100' out over the audience to open *Mary and Friends* in September of 1984; the fairy dust and Peter Foy were there to make sure it all happened perfectly.

Single-wire harnesses that used to be made from leather and roller buckles are now made with the webbing and high tech materials mentioned above and can be set up to pick up the performer in a variety of different positions. The harness can be constructed to lift in a front seated position, leaning backward, side hanging, back vertical, or upside-down. Proper fitting of the harness must include that all buckles are secure, straps and waistbands snug, and any costume must allow for free and clear attachment of the lifting media.

Double-wire harnesses allow the performer to be connected at both hips and, if properly fitted, will give them the ability to fly in the horizontal position, flip somersaults through the suspension media, or assume any position in between. Many performers prefer to hang upside down when in the air waiting for a cue because it relieves the pressure on the waist and between the legs.

While working on the Oscar-winning film *Fantastic Voyage* in the 1960s, Peter Foy developed the multi-position attachment for the double-wire harness. The film called for the actors played by Stephen Boyd and Raquel Welch to save the life of scientist Jan Benes (played by Jean Del Val), who was left comatose with a blood clot in his brain caused by an assassination attempt. They are placed into a submarine and miniaturized to microscopic proportion and injected into the body of Mr. Benes with the task of removing the blood clot. Several obstacles change the course of the mission, but in the end the characters eject from the submarine and swim to the blood clot, remove it and exit Mr. Benes' body via a teardrop in his eye just before they returned to normal size.

The challenge for Peter Foy in *Fantastic Voyage* was to create effects that made the actors appear to be scuba diving. The film's producers were concerned about creating actual scuba scenes due to the costly water tanks or location shots required that were outside the budget for the film. Wires were blended into the background of scenic veins (like vines or seaweed) and the multi-position harness allowed for perfect balance for the actors. This balance enabled the actors to fly horizontally or flip effortlessly as if swimming through water. Peter was

Figure 8.6

able to work with the cast to position them in virtually any direction as they interacted with the set and each other (due to the physique of Ms. Welch, the multi-position harness came in very handy). The end result was a perfect underwater sequence causing one director outside of the project to ask Peter how he staged the swimming sequences in the water. Peter took great pride in the fact that most people who saw the film thought the actors were really swimming; other directors approached him to ask how big the tank of water was to create the effects, and he would just smile and tell them it was all accomplished with wires and harnesses.

Sometimes rapid movement is required, such as actors being ejected from vehicles or blown up in a battle scene. When this type of movement is needed then jerk vests are worn, which create a multitude of hanging options with multiple loops of webbing sewn into the body of the suit. This type of harness allows for higher impact movement due to the fact that harness covers more of the body and the material spreads the load imparted on the body throughout the chest, shoulders, hips, legs, and crotch.

There are also twisting belts or halo harnesses available. These are harnesses that provide the performer with a third axis of motion; they have the ability to twist between the wires like a corkscrew. There are waist, shoulder, and leg straps just like the other harnesses; however, there is also a solid ring around the exterior of the harness that has a bearing surface allowing the twisting action that is generated by the performer. This type of harness requires more rehearsal time with the performer; very subtle body movements are needed in order to make the twisting appear smooth and effortless.

Custom harnesses for special applications are common; the most common harness for special application is the hanging harness. Many productions require a hanging to take place and to make it appear authentic most times a harness is used. These harnesses have a pick point near the top of the shoulder behind the neck; most are somewhat adjustable to allow for perfect placement of the pick point. The noose is cosmetic only and has the support cable or wire run through or just outside of it. The termination on the noose must not be complete. The noose must fall apart if loaded; only the wire must support the performer.

Other custom harnesses can be used for illusions where the performer is lying on a rigid frame and strapped in; this projects the illusion that the performer is in a trance but completely prone or horizontal in space.

The webbing in all harnesses can be padded to allow the performer to hang for several minutes provided the limbs are kept moving and circulation is continuous. Any numbing of the limbs or noticeable discomfort of the performer requires that they are lowered to the ground and disconnected from the system. Before staging any flying sequences, the operator of the flying equipment will always work out a distress signal with the aerial performer. When that signal is given by the aerialist then the operator knows there is a problem and that the performer needs to be lowered to the ground right away.

Costume Interface

The costume and the harness have to be designed together for any production. Many times the harness is integrated directly into the costume and in some instances it is partially attached to the costume. A typical instance of this is on a double wire harness when the performer is carrying out multiple somersaults and the costume has snaps or Velcro to attach it to the harness around the hip area. This is the same location that the lifting media is attached to the harness and by joining the costume in this area it prevents the two from getting snagged in the rotating hardware on the harness.

Flying without harnesses is prevalent today in many circuses and aerial shows; this is seen as flying with apparatus such as silk or tissue acts, straps or webbing acts, and cube or hoop acts. Many times the apparatus is hung in a static position and the acrobat climbs up to it or is lowered to it to perform the act. These acrobats have years of strength training and many times do not wear safety belts. There are of course situations where safety belts are used; it is mandated that performers wear safety belts when performing over other people, both audience and actors.

Automated and 3D Flying

Automated flying systems have their advantages and have become more prevalent in the past 10–15 years. The main reasons for choosing automation over manually operated systems are safety, labor, and repeatability. Winches can be designed to lift much heavier loads and move them safely and efficiently at speeds not possible with manual systems. In a manually operated system the operators have more feel should problems occur; for instance, if a cable jams in a scenic unit while flying a performer the operators are more likely to be stopped then if it were automated. Thresholds for position errors or over-current can be set in an automated system that would fault or trip if a cable jams; however, if the output current is lowered to this threshold it may not be sufficient to lift or swing the performer. In most safety situations when an automated flying system is installed for a show there are spotters put in place with a clear line of sight to the flying performer; the spotters have immediate access to emergency stop stations that can stop all movement if a problem occurs. It is also extremely important the operator of any automated performer flying system have line of sight to the performers being flown; reaction time to a problem is often a key to protecting the aerialist.

Most people have seen TV cameras being flown over an arena or football stadium during a live sporting event; the same type of setup is used for the flying of performers. Computer programs are now available to control multiple winches that operate together to fly performers in three dimensions. These systems are designed to move the performer up, down, left, right, and in circular motions with extreme accuracy. The manual operation of a system such as this would be virtually impossible given most setup and rehearsal times. Now, with the advanced technology, enabling the operator of a flying console the three-dimensional movement described above can be pre-visualized in simulation on a computer then played back in real time once the winches and rigging have been put in place in a specific venue. The savings on rehearsal time using this type of technology are invaluable.

In a recent production of *Aladdin* at an outdoor amphitheater, the director required a magic carpet to fly from the stage over the audience with the two stars of the show performing on the carpet. The dimensions of the amphitheater, scenery, winches, and the rigging were preprogrammed into the control system and the entire flight sequence was run in a pre-visualization program before the system was installed. The show required four winches to lift and move the carpet over the stage and then out over the audience; a wireless revolving unit rotated the carpet as it flew over the house, making the carpet come to life. Pulleys were strategically placed on towers at the borders of the flight path allowing the carpet to move totally over the seating area giving all patrons a close up view of the stars of the show. The wireless rotate winch was located just below the connection point of the four lines that lifted and moved the carpet and was powered by 12-volt batteries that required charging overnight. Its revolve was programmed in the computer to give the audience the best view of the performers as they flew overhead. The performers were tethered to the system wearing harnesses that would not allow them to reach the side of the carpet. As an added layer of safety there were wireless load cells mounted at the connection point above the carpet; these were tied into the E-stop loop of the control system.

Three-dimensional flying requires clear space in order for the wires or ropes to move about the performance area without contact above the performer being flown. This is not always possible in theatres or studios due to the lighting and scenic elements that are installed for most productions. If there is enough space then the result can be magnificent. On Broadway the *Mary Poppins* curtain call totally surprised the audience when she traveled over the stage then proceeded over the orchestra pit toward the balcony; she then lifted 50' up to the ceiling, passing both balconies and nearly touching the people in the front row. It was truly amazing how the roar from the crowd got louder the further she went out toward them then up out of sight. Lighting was a key element to this flying effect as the wires were preset over the audience for the entire show but they were never seen, even when lifting up into the ceiling at the end of the show. This three-dimensional flying effect was accomplished with a two-axis track located over the stage and a pendulum winch located out over the audience; the two systems were connected to create the third dimension of flying out into the auditorium.

Creative Process

About 30 years ago, Peter Foy provided the flying effects for Seattle Operas Ring Cycle segment *Die Walküre*; the set designer placed a mobile tower about 30′ tall on the stage, which housed the female lead in the show, a character named Brünnhilde. She controlled all aspects of the tower, which produced the army of Valkyries as it was moved all around the stage by the many minions below. The director wanted her to fly up and down from the tower showing her prowess as the mighty Valkyrie. We were able to install an electric winch system within the tower structure and with a series of pulleys diverted the lift cables to the top of platform allowing the female lead to be flown up and down the tower. "The Ride of the Valkyries" was truly a spectacular scene for this production and the tower was very similar in design to the crane machine used by the ancient Greeks, albeit a motorized system compared to manual operation.

The director for this production of *Die Walküre* also wanted to do something that had not been done before—he wanted to fly five of the Valkyries on horseback. Peter Foy came up with a brilliant manual system that incorporated the house fly battens and custom revolving carriers designed specifically for the show. The horses were made of fiberglass but were very lifelike both in appearance and size . . . and the female opera singers were very real.

The Valkyries rode their horses high above the stage, floating and turning ever so gracefully. This routine lasted more than 20 minutes and was choreographed with the five horses and female opera singers riding along, they flew together in a complex aerial scene with the stage covered in fog as if flying in the heavens . . . truly a magnificent image during the show.

Figure 8.7

The creative process is what drives the choice of equipment and harnesses for any flying effect. An example of this is a production in Japan called *Nina and the Twelve Months*, a story of a young girl who tragically dies in an accident but when she awakes she is in a magical place with the 12 months of the year as her mentors. As the story progresses, the 12 months of the year take a liking to the young lady and gather to find a way to return her to her parents. They decide the only way to accomplish this is to turn back time: enter the carousel that will fly 13 people. It is a winter scene and the 12 months of the year are placed in a half circle; they will be suspended in single-wire harnesses by 12 cables that will eventually form a 16' diameter circle. The characters are lifted by a single winch with a tapered drum (each character has its own wheel with a slightly different diameter so they can lift at different times but eventually level off at a height of 25'). January lifts first and so on till December is in the air. They level off, all holding hands, then January begins to revolve towards December, who is static until January meets him. Then the 12 months of the year all revolve in a counter-clockwise direction, which represents the turning back of time. As they revolve, a center two-wire system lowers cables into a trapdoor; the young girl is hooked and she rises revolving in the opposite direction of the 12 months of the year. This was designed as a self-contained flying carousel system through the creative need described above; it had to fit in an 8' space center stage, which is why the 12 months begin in a half circle. The system had the ability to fan open or closed; when open it was an 8' half circle; when closed it became a 16' diameter carousel with five moving axis. Lift 12 performers, fan open/closed, revolve 12 performers, lift one performer, and revolve one performer.

Equipment Selection/Engineering

Once the creative process is complete, it is time to select the equipment needed to produce the flying effect and have the system and building structure engineered. This can be a short or a very long time period depending upon what is involved. In most theatre environments the building structure has been designed to accommodate a wide variety of equipment, so it is normally a short engineering phase. Other facilities can be more challenging; stadiums, outdoor venues, cruise ships, and multipurpose rooms sometimes require additional research. The end result has to be that all equipment and personnel are safe and that the equipment can be inspected and maintained on a regular basis.

Today there are many safety, engineering, and rehearsal procedures in place to protect the aerialist, operator, and—in some instances—the audience from any danger that could be introduced by a performer flying system.

All of the flying equipment must be tested prior to shipping and commissioned once it is installed in the venue. This means that not only does the equipment function but it must do so reliably under all conditions. On cruise lines the equipment has to operate under different conditions due to salt air, constant rocking and flexing of the structure, and—in many instances—unstable power.

Risk Assessments

Risk assessments and method statements need to be supplied, pointing out potential hazards and indicating methods to reduce risk. These documents are to be prepared by qualified personnel with full knowledge of the flying effects and the system used to accomplish the flying effects. The fly system installation must also include a rescue scenario for equipment failure up to and including loss of power or catastrophic failure of the system. Training for these types of scenarios is essential and personnel must be signed off. Any new hires must be trained by qualified persons in all aspects of inspections, maintenance, and rescue.

Operator Training

It is best for operator training to take place as the equipment is being installed and during performer training and technical rehearsals. Fitting harnesses on the performers and checking with the performers on the correct fit of the harness must be done prior to any flying effect. All maintenance procedures are covered during installation so the operator will have hands-on experience in trimming cables, making flying wires, attaching ropes and pulleys, assembling track systems, running winches, operating consoles or pendants. During rehearsals the operators of the manual systems are shown the techniques of pulling the ropes, watching the performers on the ground and in the air. If it is not possible to get these procedures in place during installation or technical rehearsals then a proper handover from the systems operator and qualified person must take place for new operators. The timeframe for this is established by the complexity of the flying effects and equipment used to accomplish the flying effects.

For the manual systems, the operator is trained to make long consistent pulls of both the lift rope and the travel rope. Hand-over-hand action is best with the extension of the arm high and finishing at the low waist. The lift rope should be operated so the performer is not exhibiting any stops or jerky movement in the up or down direction. The travel operator must work within the two ropes, feeling for the line to change direction as the performer nears the end of each travel swing. Once the performer is swinging then the operator is trained to cancel the swing by moving the carrier over the performer's head, stopping, slowing down, or controlling the side-to-side movement. The lift and travel operators are trained to communicate with each other so they do not cause any dangerous movements for the performer.

An example of bad fly operator communication occurred during a run of Liberace's show in Las Vegas many years ago. As he was being flown on a manual track system, the lift operator did not pull the rope as the travel operator moved on cue; the travel operator did not notice the lift operator's mistake and kept right on pulling the rope. When this action happened the operators dragged Liberace out of his Rolls-Royce, across the trunk, and onto the stage. After he exited on foot he was heard to say "I don't think that was supposed to happen, and it won't happen again, will it?"

It is essential that the two operators of a manual system communicate constantly so the correct movements occur during the aerial segment.

The programming of the consoles is done exclusively by qualified personnel; operators are trained to make only minor adjustments and to troubleshoot potential issues. Navigation of the system, setup of the equipment on the console, copying and saving files are also provided to the operator. The programming process during rehearsals is the best education for a new operator; observation during this period is the key to learning how to run a console. Cue navigation, fault signaling information, manual jog operation, and study of log files showing how the system logs cue sequences show the operator what to look for in case of system malfunction. Running cues and making sure the operator is aware of when it is essential to stop a cue is a primary function of the training.

Performer training begins with the fitting of the harness and is the most important part of the performer training segment. The harness is fitted first for safety and comfort, then for the functionality when the performer is lifted into the air; a second person must check the harness and the final approval has to be from the performer. As in most things in life, anything can happen—I was involved in a production in San Francisco and an aerialist actually came to the stage during a show to be hooked up . . . without a harness on. Unfortunately it was too late to go back to the dressing room to get the harness; the operator checking them was completely dumbfounded.

Performer Training

Once the harness is fitted; the performer will be hooked up and flown approximately 2′ in the air—this is done out of sight of the audience—in order to make sure the harness is properly fitted. If a single-wire harness is used the performer will be instructed to keep their head up, arch their back, and to utilize their body, arms, and legs to control their rotation. They are also informed to flex their knees when landing so they are not injured. When getting attached to the lifting media, the performer is also taught to receive a tap or physical signal from the person hooking them up to ensure they are actually coupled to the fly system.

For the double-wire harness, the performer is fitted so the harness is snug and comfortable. The pick points on the side of the harness need to be close to the balance point on that particular person's body. Everyone has a slightly different build, so it is important to make sure the balance point is correct; this may require several repositioning fittings on the performer.

Once the double-wire harness is fitted, the performer is hooked up to the system and flown approximately 2′ in the air and taught the correct way to fly horizontally and to generate somersaults. This is done by keeping the back straight and manipulating the arms and legs to distribute the weight of the body on either side of the pick points. A good balance point will allow the performer to make these motions with ease.

Finally, the operator and the performer must have a safety signal that is displayed by the performer while in the air. This signal indicates to the operator that there is a problem with the performer and they need to land immediately.

When executed properly in the context of a show, performer flying can be breathtaking and significantly enhance the telling of the story. It can also be very dangerous if not done by professionals. A good rule of thumb is to always possess the awareness of what is happening at the moment and never fear stopping if something goes wrong; this holds true for the operator or the aerialist (and the author), full stop.

9

Stage Automation

SCOTT FISHER

The What and Why of Automation

What is Automation?

Automation is the mechanization and control of motion tasks or operations that require precision, safety, and repeatability, or are difficult or impossible to accomplish by direct human operation, or all of the above. Automation involves the connection of machinery to the object being moved, controlling that machinery via motion-control computers and safety devices, and providing operators and programmers with the means to alter the parameters of that motion as well as the timing of execution and the coordination with other motions.

Why Automate?

There are several compelling reasons to automate the stage and rigging machinery in a theatre or production. These include:

- **Safety:** Automating machinery provides precisely controlled and highly repeatable motion as well as the ability to stop that motion quickly and safely either by direct operator action or by automatic safety control systems.
- **Manpower:** A production may be operated by fewer stage personnel using automation equipment, allowing smaller theatres and venues the ability to stage larger and more complex productions.
- **Force amplification:** Heavier loads and larger scenery may be manipulated using automation equipment than can be moved by human power and counterweight alone, allowing for the safe application of larger scenery and rigging systems.
- **Consistency:** Regardless of the personnel operating the equipment, the scenic and rigging systems will move in the exact same manner for each performance, providing a more consistent and higher quality experience for both the performers and the audience.

Different Systems

Automation is used worldwide in many venues and applications and, as such, the nomenclature for different aspects of the systems can vary in different locales and countries. For this chapter, we will be using terms common to applications in the United States.

Basics of Motion

Types of Motion

There are two types of motion typically achievable with an automated system. These are linear motion and rotary motion.

LINEAR MOTION

This refers to motion in a straight line, typically horizontal or vertical only. The most common linear motion in most rigging automation systems is vertical, moving a suspended load up and down.

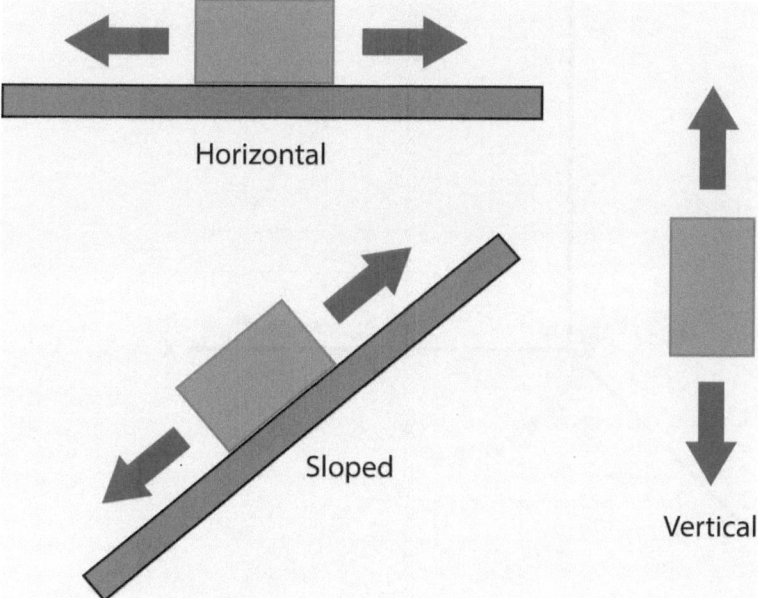

Figure 9.1 Linear motion

Rotary Motion

Rotary motion is motion where an object is set to spin either clockwise or counterclockwise, generally around a vertical axis. The most common rotary motion in stage automation systems is for a stage turntable used to move and present different scenes in a show.

It is also possible to achieve three-dimensional motion of an object through the coordinated use of several automated machines.

Figure 9.2 Rotary motion

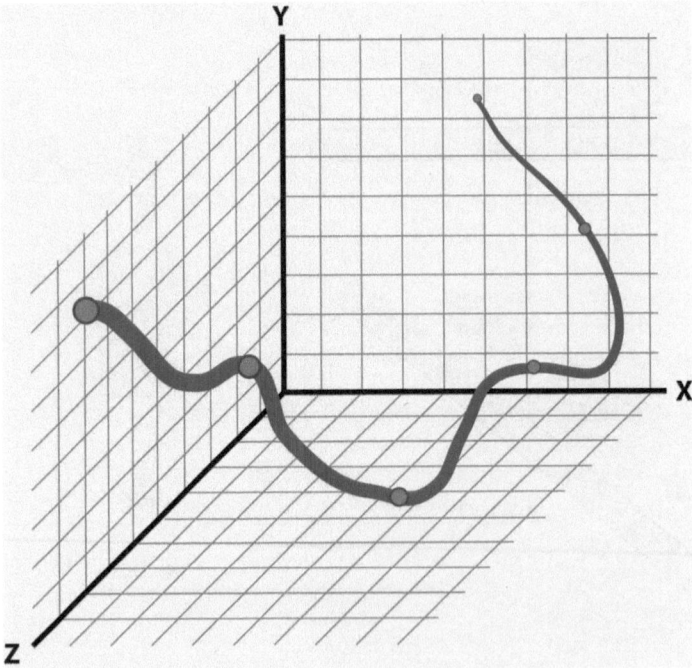

Figure 9.3 Three-dimensional motion

Motion Parameters

The motion of an object is typically defined by several parameters that describe its speed, desired target position, and characteristics of the way it achieves those targets. With these parameters defined by the operator or programmer, an object will move in the desired fashion and its motion can be tailored exactly to the needs of the scene or production. The typical parameters used in defining motion are outlined below:

Position

This refers to the location of the object in space. Position can describe its location at any point in the motion profile, although the most common positions used are its position at the start of the move, and its position at the end of the move (the "target" position). Position can be defined using any unit system desired (feet, meters, inches, etc.).

Velocity

This refers to the speed at which the object is moving, and is defined by its position over time.

The units for velocity are related to units chosen for position and the timescale required. For example, if you chose meters as your position units and seconds as your time units, your velocity would be expressed in meters per second.

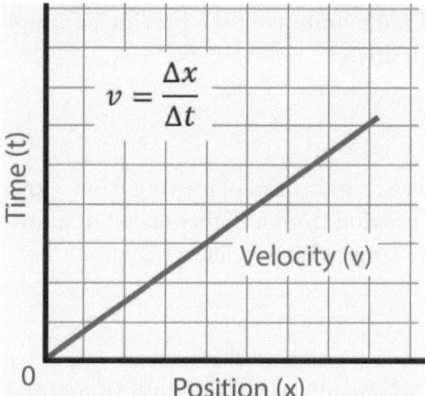

Figure 9.4 Velocity equation and graph

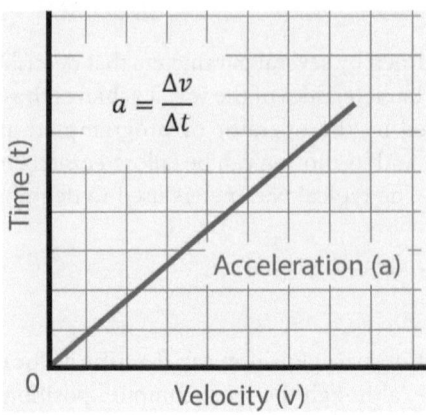

Figure 9.5 Acceleration equation and graph

ACCELERATION

Acceleration is the rate at which the object achieves the target velocity. A higher acceleration rate results in the achievement of the target velocity in a shorter period of time, and a lower rate means the object takes longer to achieve its target speed.

Like velocity, acceleration is a time-based parameter, except in this case it's the rate in the change in velocity over time. Because velocity is expressed in "position units per time unit" (x/t) and acceleration is expressed in "velocity units per time unit" (v/t), the units for acceleration are in "distance unit per time unit^2" ((x/t)/t). If you had chosen the same units as in the previous example, your acceleration would be expressed in m/sec^2.

DECELERATION

Deceleration is the same as acceleration, except instead of moving from a lower speed to a higher speed, the object is moving from a higher speed to a lower speed. Units and definitions remain the same as for acceleration.

JERK

This refers to the rate at which the acceleration or deceleration of an object changes. This is sometimes referred to as the "smoothness" of the acceleration, as it controls the characteristics of how the object changes from one constant velocity to another.

Jerk is also expressed with respect to time, in this case "acceleration units per time unit." Since we've added another "layer" of time for jerk, the units are

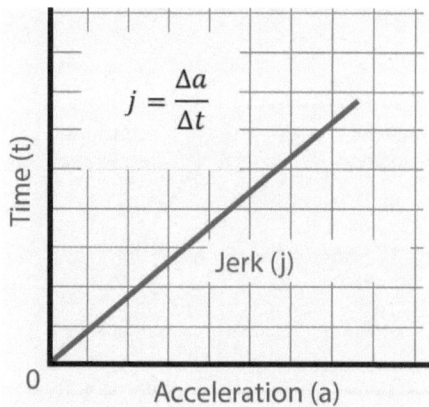

Figure 9.6 Jerk equation and graph

"distance unit per time unit³." Using the meters per second example again, the units for jerk would be expressed as m/sec³.

While this is all useful information for any automation operator or programmer to have, it is not usually necessary to know each of the derivatives with respect to time for motion parameters, as modern automation systems will put these in much simpler terms, often just providing a "0–100%" control for these parameters rather than the absolute numbers. However, when building motion profiles, having a basic understanding of where the numbers come from and what effect a higher or lower value will have on the motion can be invaluable.

Motion Profiles

Application of the parameters described in the previous section will result in a few typical types of motion profiles. Motion profiles are usually described by the shape produced by a graph of the velocity of the motion with respect to time. These profiles are described below:

TRAPEZOIDAL PROFILE

A "trapezoidal" profile is the most common type of profile used for motion control in stage and rigging automation. It consists of the acceleration phase, the constant velocity phase, and the deceleration phase. This is what's most commonly utilized for point-to-point motion commands of the "start here, go this fast, end here" variety.

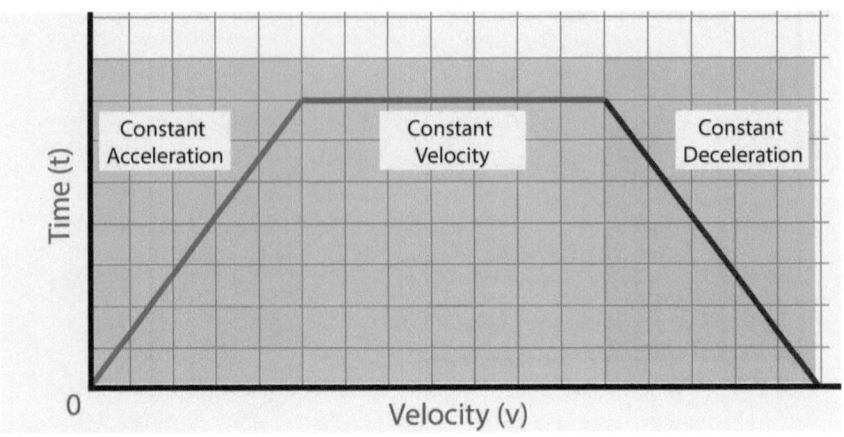

Figure 9.7 Basic trapezoidal motion profile

S-CURVE PROFILE

An S-curve profile is what is produced when more gentle parameters for Jerk are introduced. This basically "rounds off" the corners of a trapezoidal profile, making for more gentle transitions and considerably smoother motion.

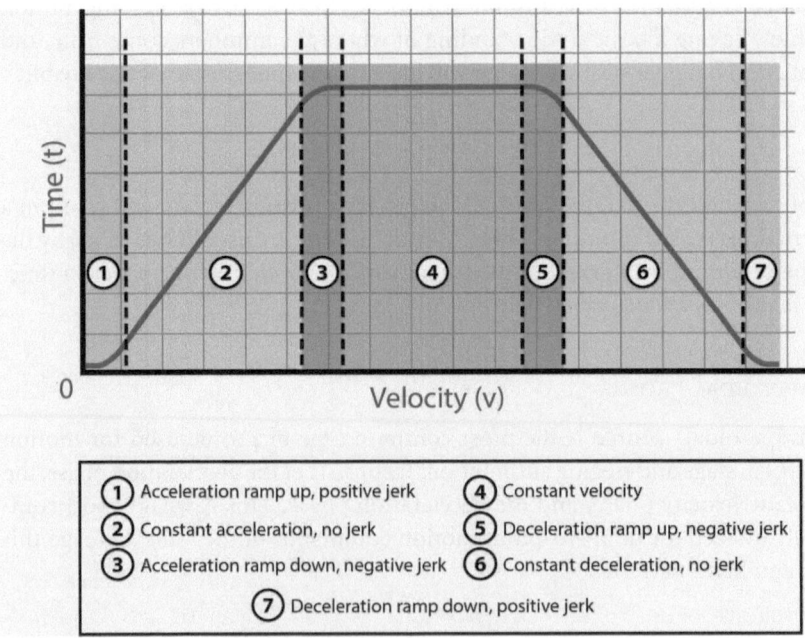

Figure 9.8 Trapezoidal profile with jerk ("S-curve")

In reality, operators who are programming trapezoidal profiles are most likely programming S-curve profiles, as the limitations of the machinery or the built-in parameters of the system will automatically "round off" the corners of the motion without the operator needing to define a specific jerk value. In most applications, the default parameters of the equipment are sufficient and they produce very smooth and precise motion. In higher speed or other unusual applications, the ability for the operator to define the motion parameters more precisely can be critical to the safe and smooth operation of the machinery, and many automation systems provide this ability when necessary.

"STACKED" PROFILES

In many modern automation systems, it is possible to "stack" profiles by providing multiple velocities between the start position and the target position. For example, if you were programming a move that went from a position of zero to 100, you could provide a motion profile like this:

- Accelerate to a velocity of "2.5" and travel to "20"
- At "20," accelerate again to a velocity of "5.1"
- At "50," accelerate up to a velocity of "6.8"
- At "70," decelerate to a velocity of "2.8"
- Decelerate to a stop at "100."

This would produce a motion profile that looks like Figure 9.9

All of the basic aspects of a trapezoidal or S-curve profile still apply to the individual "sections" of the profile, but the programming environment provides

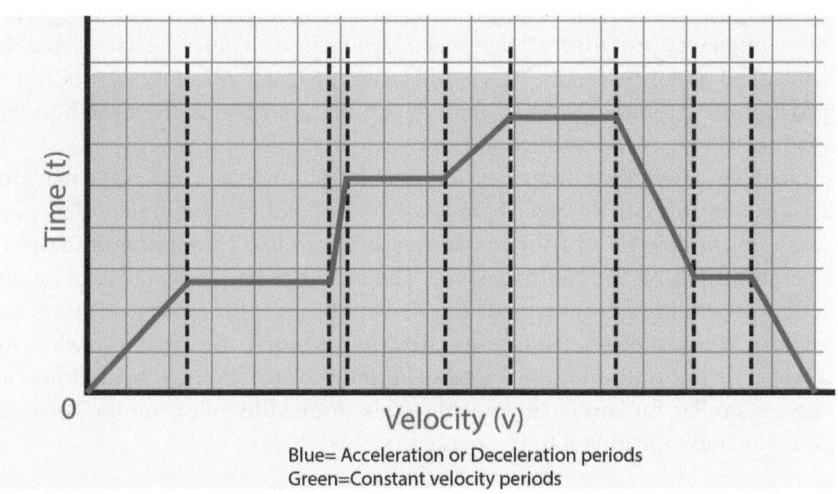

Blue= Acceleration or Deceleration periods
Green=Constant velocity periods

Figure 9.9 Multiple speed profile

the ability to include multiple profiles within a single move, allowing the operator to program more complex motions for the production or application.

Controlling Motion

While automation systems can perform many tasks, the core functionality of stage and rigging automation is the control of the motion of objects, performers, and machinery. Unlike industrial automation, the stage automation environment is typically more fluid and adaptable in order to accommodate the needs of the productions, and modern automation systems provide a variety of methods to initiate and control motion to service those needs.

Manual Control

Manual control is the most basic form of motion control in an automation system. It provides the operator with the means to directly move an object or group of objects, and is most analogous to flymen pulling on ropes in a counterweight system to directly move scenic elements. Typically little or no programming is involved, and the direct physical input of the operator to input devices such as joysticks or pushbuttons is translated directly into object motion through the automation system.

Jogging

The process of moving an object by changing its speed and direction through direct operator input is known as "jogging." This method of motion control is generally used for ad hoc slow speed movement of objects, usually during setup or maintenance procedures. High speed jogging is sometimes used for manually controlled performances or "teach and learn" programming procedures, but as programming tools improve in modern automation systems, these methods are gradually being used less often to improve safety and consistency.

Jogging is typically initiated via an "enable" action that activates the machinery (usually a joystick trigger or similar device) and, once initiated, the speed and motion direction of the machinery is controlled by varying the level of operator input on the control device. The means of level variation can be different depending on the type of device being used—for a joystick input, the angle of stick deflection from center determines speed and direction, while for a wheel or button input, the speed or number of rotations or actuations can serve a similar function. The machinery is stopped by releasing the "enable" function and returning it to its "parked" status.

Control Loops

The "control loop" is the mathematical model and means by which the speed and position of an object is controlled in an automation system. There are typically two types of control loops utilized: "open loop" control, and "closed loop" control.

OPEN LOOP CONTROL

An "open loop" control system is a control system in which the results of applying the command input are not monitored by the system to assure that the desired result is achieved. Open loop systems are very simple and robust, and are useful for basic controllers where precise performance is not necessarily required. An example of an open loop system would be a fan speed controller. A variable voltage is fed to the fan motor controller in order to change its speed, but there is no speed sensor on the motor output to verify that speed has been achieved. If the load on the fan increases or decreases, the speed may also change, as there is no means for the controller to adjust its output to compensate.

In a stage or rigging motion control system, an example of an open loop positioning system would be a chain hoist or winch with a simple position feedback device like a potentiometer and a circuit that shuts off the motion when a certain value is reached. As soon as that circuit sees the targeted position value, it would shut off the motion command, and the machine would come to a stop. However, if the motor continued to "coast" past the position (or "overshoot"), the control loop would have no way to correct for this and assure that it stops at exactly the correct position without overshooting. In many applications, especially those where the machines are moving at slow speeds and the mechanism is constructed such that it's physically unable to overshoot very far, this method of position is good enough, and indeed open loop control can be found in many older motion control systems. However, newer technology and better control methods are available in modern systems, and it is rare that open loop is utilized on these systems.

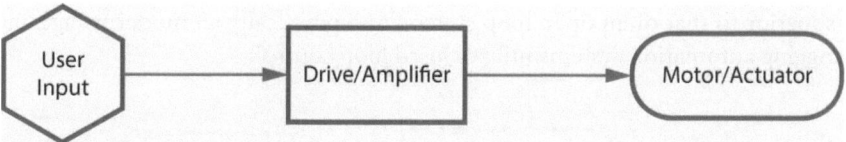

Figure 9.10 Basic open loop diagram

CLOSED LOOP CONTROL

A "closed loop" control system is one in which the process feedback signal (position, speed, force, etc.) is used to modify the motion control command signal and thereby allow the system to compensate for varying loads and to provide much more accurate motion control under a wide variety of conditions. Closed loop systems require more components than open loop to provide feedback, and the circuitry or computing engine is necessarily more complex than an open loop system to provide the precision control characteristics of this method. An everyday example of a closed loop control system is the cruise control in an automobile. The feedback data to the loop is the speed of the car, and the variable motion control signal is the amount of throttle applied to the engine. On level ground, the system will maintain a steady throttle to maintain the preset speed, but if the car encounters a hill or headwind, the system will automatically apply more throttle as necessary to maintain the preset speed. This action will continue on a second-by-second basis as long as the closed loop cruise control system is engaged.

In a stage or rigging control system, the variable that is usually used as the feedback input is the position of the device. If you remember from earlier in this chapter, position is the basis for the basic parameters of motion: Position over time is velocity, velocity over time is acceleration, and so on. Therefore, if you have position information, you can calculate just about everything else you need to move a machine as desired. Position information is generally monitored by the control loop on a schedule in the tens or hundreds of milliseconds; that is, the system checks the position and modifies the output signal that often while the device is in motion. This allows smooth and rapid changes to the motion command signal to compensate for any changes. For example, if a winch is lifting a piece of scenery and six performers step on to it during the motion, this would cause it to slow down in an open loop system due to the additional load. In a closed loop system, the system would detect that additional load within milliseconds and apply a higher control signal level to the machine, causing it to produce more force to counteract the load and thereby produce smooth and steady motion that places the scenery exactly where it's supposed to be throughout the entire range of motion.

For obvious reasons, the performance of a closed loop system is vastly superior to that of an open loop system, and practically all modern stage and rigging automation systems utilize closed loop control.

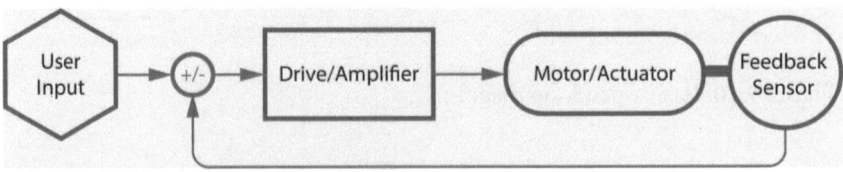

Figure 9.11 Basic closed loop diagram

Variable versus Fixed Speed

With regard to speed control, stage and rigging devices typically fall into one of two categories: fixed speed or variable speed.

FIXED SPEED DEVICES

Fixed speed devices are those where the speed of the machine is set to a fixed value, either by the mechanical characteristics of the device itself, or by the settings of the control system. A common fixed speed device in stage and rigging automation is the chain hoist, where the motor in the hoist simply runs at full speed as soon as the "up" or "down" control is engaged.

The advantage of fixed speed devices is that they're very simple, and typically inexpensive, as they don't require a lot of complex controls to operate. The disadvantage is that they can be difficult to position accurately, as there is no acceleration or deceleration function available to smoothly approach the target position. However, for low-speed applications or applications where either position accuracy or visually appealing motion is not necessary, a fixed speed device can be a great solution, as can be seen by the proliferation of chain hoists throughout the entertainment industry.

VARIABLE SPEED DEVICES

Variable speed devices are those that include controls and mechanisms that allow the speed of the device to be varied between zero (or a very low "creep" speed in the case of some devices) and the full speed that the device is mechanically capable of. Variable speed devices work best with closed loop control systems, and this pairing is the basis for all modern stage and rigging automation systems. A variable speed device on a closed loop control system produces very smooth and visually appealing motion and extremely accurate positioning, which in turn provide for a safer and better-looking show.

The most common example of a variable speed device in stage and rigging automation is the cable drum winch, although there are also variable speed chain hoists available as well as a wide range of hydraulic devices that are provided with variable speed capabilities through the use of proportional or servo controlled hydraulic valves.

Position, Velocity, Torque, Load, and Force

This section provides an overview of some of the key variables that are used in modern stage and rigging automation systems. We will discuss the nature of the parameters themselves, and introduce some of the most common means of measuring those parameters and incorporating them into the control system.

POSITION

"Position" is the physical position of the actuating part of the machine being controlled. For a winch, this is usually the rotational position of the cable drum, which can be extrapolated into how much cable is reeled out, which in turn tells the system where the suspended piece of scenery is located in space. For a hydraulic lift, it would be the amount that the hydraulic cylinder is extended, which would indicate the position of the lift platform. There are many ways to determine position, and many devices and locations for those devices to obtain position information.

FEEDBACK DEVICES

There are two types of position feedback devices typically used in stage and rigging automation: incremental and absolute. An incremental position device only indicates a change in position when it is moving and the increment of that change. Once the system "knows" the starting position, it can use this incremental information to count up or down from there and determine the actual position, but from a "power up" state, the system does need to be provided with a current "real" position. An absolute position device maintains its position information at all times, even without power being applied to the system, and can reliably provide that position information to the system at any time. There are many position devices in use in motion control systems, and several of these are common in stage and rigging automation systems. These include:

- **Limit switches:** A limit switch is an electromechanical on/off switch that engages at a specific and known location. Limit switches are typically actuated either by direct contact with the object being moved by the machine, or by the normal action of the machine or actuator itself. Limit switches are very commonly used for positioning in open loop systems, but are usually used only as 'benchmark' position checks or limit-of-motion indicators on closed loop and variable speed machines.
- **Potentiometers:** Potentiometers are analog variable electrical resistance devices that produce a variable voltage throughout their range of motion

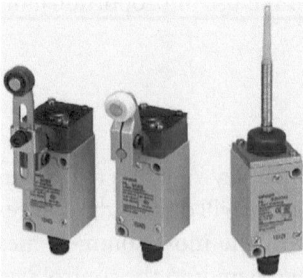

Figure 9.12 Limit Switches

Figure 9.13 Potentiometer   **Figure 9.14** Potentiometer

when a reference voltage is applied to them. For example, if a reference voltage of 10V is applied to the potentiometer, its output would read 5V at the halfway point of its full motion. When a potentiometer is attached to a machine or a load in such a way that its full travel corresponds to the full travel of the machine itself, it can be used as a reliable positioning device. As analog devices, potentiometers can be subject to electrical noise and other interference, which can in turn provide inaccurate position information, but properly applied, they can be very robust and reliable position feedback devices. A potentiometer is an absolute position feedback device.

• **Encoders:** An encoder is a device that provides an incremental on/off pulse output to the system based on the motion applied to its actuating input, which is usually either a rotary shaft or a linear scale. If the information for the encoder scale is known by the control system (i.e., how many pulses it should expect to receive in one turn of a rotary encoder or a specific length of a linear encoder), that information can be extrapolated into actual distance moved as well as current position. Encoders can also be provided with absolute position output, and these devices output a very accurate data that represents a discrete number indicating the position of the encoder shaft or scale. Absolute encoders provide their position from a power-up condition.

Figure 9.15 Encoder   **Figure 9.16** Encoder

- **Other devices:** There are many other position feedback devices that are less commonly utilized in stage and rigging automation, but that you may occasionally encounter. These include laser and sonar rangefinders that bounce a laser or ultrasonic source off of the target object and provide a position number for that object, magnetostrictive devices that report the position of a magnet moving along their length (most commonly seen in hydraulic cylinder positioning), and resolvers or linear variable differential transformers (LVDTs) that provide a variable voltage output through inductive coupling (or basically a transformer with a moveable core). These types of devices are much less common in the entertainment industry, and are usually employed for very specific applications or environments.

Figure 9.17 Linear Variable Differential Transformer

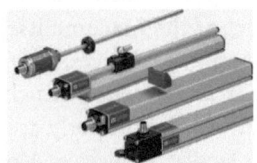

Figure 9.18 Linear Variable Differential Transformer

USING A POSITION CONTROL LOOP

As previously noted, the position control loop is the most common type of closed loop control loop utilized in stage and rigging automation. A position signal is obtained from the position transducer being utilized, fed to the control circuit, and then used to monitor and control speed, acceleration, and jerk as necessary to achieve the desired position of the object being moved.

VELOCITY

Direct velocity control is not as commonly utilized in motion control applications as position control, although it's always there just "under the hood" in most variable speed control systems. Velocity control is generally utilized for devices where velocity is their defining characteristic, such as fans or pumps. For stage automation, devices such as treadmills or turntables sometimes use velocity control rather than position control, but that is typically dependent on the application.

VELOCITY FEEDBACK DEVICES

Velocity feedback devices are typically the same as position feedback devices, but the control circuit receiving the information from the feedback device simply evaluates the signal with respect to time rather than utilizing the signal value directly. For example, if an incremental encoder is used as a velocity feedback device, the circuit will "count" pulses per second, and if the distance a "pulse" represents is known, then the velocity is known and can be used as a control value.

Using a Velocity Control Loop

Our previous example of the cruise control system in a car is an example of a velocity control loop. The speed of travel is the controlling variable, and the amount of throttle is the command variable. A velocity control loop in a stage automation application would work in a similar fashion. For example, if you needed a turntable to move at 1 RPM during a scene, you would set your velocity target to that value. If the turntable load changed due to performers or scenery moving on or off of it, a closed loop velocity control system would detect that change and then alter the control signal to the drive motor to maintain a speed of 1 RPM.

Torque

Controlling torque is one of the options available if a rotational actuator is being used, such as a winch motor. Torque relates directly to force in this case—the harder you twist the winch drum, the harder it pulls on the cable. There are a few special applications where torque control can be important, and these typically fall into one of two categories: constant force control and force limitation. Constant force control is useful for applications where you want to maintain constant pressure or tension on an object, such as stretching fabric or maintaining a steady pressure in a cable. If the load increases on the object, the winch or device will allow the position to change in order to maintain a constant level of force. Force limitation works in a similar fashion, but where variable position and force is available up to a certain limit, after which the actuator will not provide additional force, or will allow position change to avoid exceeding a set torque value.

Torque Feedback Devices

Dedicated torque feedback transducers are specialized devices that are installed in line with the shaft producing the torque. In the case of a winch device, this would typically be in between the gearbox output shaft and the winch drum. These devices provide an absolute value for torque that can be monitored and scaled as necessary by the control system. There are other means of controlling torque through indirect measurement, such as monitoring the amount of electrical current being used by a motor, or by mounting the entire rotating assembly on a plate that can rotate against a load transducer at the end of a fixed-length arm. Any of these methods can provide the necessary data to a closed loop control system that can then control the torque levels.

USING A TORQUE CONTROL LOOP

To utilize a torque control loop, you must first decide if you are using the torque feedback as a maximum limit control or a constant torque control, as these methods utilize different loop variables in the controller. Once the method has been selected, the transducer or other data source output is connected to the control circuit and scaled appropriately. Once these steps have been completed, the machine can be turned on and it will produce or limit torque as appropriate.

Many modern variable speed motor controllers or motion controllers have built-in torque control circuits and control algorithms, so it is unnecessary to build one up from scratch. These are typically well-matched to the devices being utilized, and provide a very accurate and reliable means of torque control.

LOAD AND FORCE

Load and force control are very similar to torque control, and they utilize almost identical devices and methodologies. In fact, in many cases the two can be used interchangeably, especially if rotating equipment is being used. In the case of a winch, knowing the torque on the drum shaft is the same thing as knowing the force at the end of the cable, so either is applicable and useful for force or torque control. In the case of a hydraulic cylinder, load and force are the parameters being measured, as there is no rotational component to the motion.

Like torque control, load and force control are typically used to either maintain a constant level of force, or to provide a maximum limit for force. Load and force control can also be used to simply provide an "overload" alarm level for the operators or riggers to respond to, rather than causing the control system to respond directly.

LOAD AND FORCE FEEDBACK

There are a few common means of measuring and reporting the load on, or force applied by, a machine or actuator. These include:

- **Load cells:** A load cell is a transducer that converts an amount of force on the transducer into a directly proportional electrical output. The strain placed on the device changes the electrical resistance of small wires inside, causing a change in its output voltage. This voltage can be used to produce a very accurate measurement of the load on the device.

Figure 9.19 Load Cell   **Figure 9.20** Load Cell

- **Motor current:** In motorized systems, the amount of current used by the motor is directly related to the amount of load on it. If the load-per-amp value is known, the current value can be scaled into an actual load value. The drawback to this method is that the motor has to be supporting the load at the time of measurement, which can be problematic for static load measurements where the system is actually in an overload condition, or in situations where it is impractical or undesirable to put the load to be measured on the active machinery.
- **Hydraulic pressure:** In a system utilizing a hydraulic cylinder, the force on the cylinder rod is equal to the pressure per unit surface area. In US units, this would be psi or pounds per square inch. If the area of the cylinder piston is known and the pressure is measured, then it's a simple matter of division to come up with the pounds of force on the rod. There are also commercially available load cells that use this method rather than the electrical strain gauge method, where the device is essentially a very small inline hydraulic cylinder equipped with a pressure gauge or transducer.

In all process measurement sensors, it is important to know that your sampling rate is of sufficient frequency to allow the system to react to changes in a timely manner. The sampling rate is the rate at which the sensor acquires and reports the data back to the control system. For example, a sensor with a 100ms sampling rate would return a value to the system ten times per second. Your sampling rate should be a multiple of the amount of time in which a significant change can occur in your system. If you have a piece of machinery that moves at 20′ per second and you are sampling your position five times per second, then that means that you could cover four feet in the time between samples, which is a very significant distance. System and sensor sampling rates are most often set by the manufacturer of your selected system, but in general the sampling rate should be taken into consideration and well-matched to the process and system that you are controlling.

Using a Load Control Loop

Utilizing a load control loop is essentially identical to utilizing a torque control loop. You must first decide if you are using the load feedback as a maximum limit control or a constant load control, as these methods utilize different loop variables in the controller. Once the method has been selected, the transducer or other data source output is connected to the control circuit and scaled appropriately. Once these steps have been completed, the machine can be turned on and it will produce or limit load as appropriate.

Many modern variable speed motor controllers or motion controllers have built-in torque control circuits and control algorithms, so it is unnecessary to build one up from scratch. These are typically well-matched to the devices being utilized, and provide a very accurate and reliable means of torque control.

Synchronized Motion

In many cases, it's necessary to coordinate or synchronize the motion of several machines in a stage or rigging automation system. Sometimes this is just for the visual effect of a scene change or other stage motion, and sometimes it is out of mechanical necessity, where the force of several machines is required to lift a large object or a heavy load. This is one of the trickier problems in automation, and it's important to know the challenges involved and methods for addressing solutions.

Synchronization versus Coordination

While both synchronization and coordination can appear to be similar methods, they are in fact very different. The primary differences between the two are as follows:

- Coordination is a method of controlling multiple machines and objects where command signals are sent to those machines at coordinated times (usually simultaneously), but there is no inter-machine checking or feedback to assure that they stay in coordination with each other. This method simply counts on the accuracy of the control signal timing in reaching the machines and the individual accuracy of the machines to run their motion profiles correctly to maintain coordinated motion. This method is fine for motions where it's not critical that the machines all stay perfectly aligned and timed, and where there are no non-compliant physical links between the machines or loads themselves.
- Synchronization is a method where multiple machines are set in motion and where there is inter-machine communication and control to assure that those machines start, move, and stop as if they were a single device. A typical application of synchronization would be to lift a large rigid structure using several winches or hoists where the failure or uncoordinated motion of one or more of those machines could cause structural issues or failures.

MEANS OF SYNCHRONIZATION

There are several different ways to synchronize the motion of multiple machines, as this is a common application in industrial systems. Some methods involve additional wiring and connections between devices in order to directly coordinate their position feedback signals among one another, and others rely on real time high-speed network communication between the devices to achieve tight synchronization.

While it is beyond the scope of this chapter to go into great detail on the methodologies for synchronization, it is important for the reader to understand the differences between coordination and synchronization, as misapplication of these methods can lead to errors and accidents. First, it is important to be able to recognize when an application requires full synchronization, or if simple coordination is acceptable. Second, you must be able to recognize when you are actually using a synchronized system and when you are not, as some controllers can appear to be synchronized when they are actually just coordinated. A common example is suspending a large truss structure from a number of chain hoists. All of the hoists are set into motion at roughly the same time with a single button push, but there's no guarantee that they're all running at the same rate of speed or all supporting an equal amount of the load. This is simple coordination, not synchronization. Likewise, sending a "go" command simultaneously to a number of machines in a closed loop variable speed control system does not necessarily guarantee that they are all starting, stopping, and moving synchronously. Be sure to familiarize yourself with both the system being utilized and the object(s) being lifted or moved, and be sure you know how the system will react on the failure of a single device in the group or in an emergency stop.

MITIGATION OF FAILURES OR ERRORS

As complex systems, it is inevitable that one or more machines will fail or stop at some point during a coordinated or synchronized move. The system and structures must be prepared to handle this eventuality without compromising safety or the integrity of the machinery and structures. There are a number of ways to go about doing this, both from the physical and mechanical side and from the controls side.

- **Physical mitigation:** Make sure that the structure being lifted or moved can handle a certain amount of localized bending or deflection without compromising its structural integrity. Do not build or lift anything that requires absolutely perfect synchronization. Lifting or pulling devices can also be equipped with slip clutches or other physical load controlling devices that prevent the equipment from being overloaded to the point of failure.
- **Control system mitigation:** The synchronizing control system should have the means of assuring that if one device stops, they all stop well

within the physical "comfort zone" of the object being lifted or moved. Likewise, motion starts should be coordinated across all devices to assure that they move together as a single machine. Load monitoring and control should be employed to assure that no single device takes on more load than it is designed for or that should be applied to any one point on the structure.

Keep in mind that there is always the potential scenario of complete power failure during motion due to lightning strike, infrastructure failure, or other reasons. In this case, it does not matter how capable your control system is, as it can be instantly taken out of the picture. Machines and structures need to be designed to handle this eventuality and stop the move safely, and the control system needs to be able to recover from a power loss situation without causing uncommanded motion or other issues with the system.

Means of Motion

All of the previous information in this chapter dealt largely with the theories and practices behind motion control, but eventually these theories have to be turned into real-world devices and machines in order to make things happen. The following sections describe some of the most common machines and devices in use in stage and rigging motion control today.

Electrical

Electricity is by far the most common means of powering automation devices for stage and rigging. It's clean, easily transmitted, readily available, and there is a wide range of devices available to run on it. Some of the most common are as follows:

- **Chain hoists:** A chain hoist is typically a low-speed lifting device that uses a highly geared motor and chain wheel in an enclosed case to take up or let out chain and lift objects and structures. Chain hoists are commercially available from a number of manufacturers, and they are the workhorses of the touring segment of the entertainment industry.

Figure 9.21 Chain Hoist   **Figure 9.22** Chain Hoist

- **Cable winches:** Cable winches are available from many manufacturers in a large range of speeds, capacities, and configurations. Cable winches are typically much quieter than chain hoists, and are also considerably faster and more accurate in positioning. Most cable winches are used with variable speed closed loop automation systems for control.

Figure 9.23 Cable Winch

- **Linear actuators:** Linear actuators are also known as "electric cylinders," and they are used to produce large amounts of short-distance linear force. The typical arrangement uses a motor to drive a screw assembly, which causes a rod to extend or retract within a housing. These are often seen on trapdoors, platforms, panels, and other applications where high force and short throws are required.

Figure 9.24 Linear Actuator

Fluid Power

Many stage and rigging applications utilize fluid power as a motivation means for the machinery. "Fluid power" includes both hydraulic (oil) and pneumatic (air) powered devices. The main advantage of fluid power is "power density," which is the amount of force that can be produced by a device of a certain size. In the case of hydraulics, the power density is extremely high. A relatively small hydraulic cylinder can produce an enormous amount of force compared to its electrical counterpart.

The tradeoff for this high power density is the amount of infrastructure required to support and operate a fluid power system. Pumps, reservoirs, compressors, piping, valves, accumulators, and other devices are necessary for a complete hydraulic or pneumatic system, and this amount of equipment and the space it requires can sometimes outweigh the advantages of a fluid power system. However, there are cases where nothing else will do, and in those cases fluid power performs very effectively and reliably.

Some fluid power devices commonly seen in stage and rigging automation applications include:

- **Hydraulic cylinders:** Typically used to move or support stage lifts, but also used in many props and specialty scenic pieces.
- **Hydraulic motors and winches:** Hydraulic motors and winches are used in the same stage applications as electrical motors, and several manufacturers have produced hydraulic winch and powered batten systems. While the space savings can be considerable, the disadvantages of potential oil leaks and the large infrastructure requirements for piping and pump systems can outweigh the space and power density benefits.

Motion Commands and Controls: Cueing Systems

In order to be useful in a stage environment, automation devices require a control system to provide an interface for users to program and operate the machinery. Control systems can be as simple as a set of pushbuttons or as advanced as a completely integrated computer-controlled automation system. At a minimum, a modern stage and rigging automation system consists of a networked series of motion controlled machines and a programmable user interface to construct motion sequences for playback and editing. For the purposes of this chapter, we will concentrate on a basic system of this type.

Cueing and Playback Systems

A cueing and playback system communicates with the various machines in an automation system for the purpose of constructing motion events and sequences, saving those events in sequences in a format that facilitates playback during rehearsals or performance, and providing the means to edit, save, and initiate the preprogrammed actions on demand. Systems of this type are available from several manufacturers in the entertainment automation industry and, while each manufacturer provides somewhat different methodologies and features, there are aspects that are common to almost all systems.

Typical System Features and Terminology

A typical system will provide an operator interface that allows manual control of individual devices or groups of devices via a joystick or similar device, "go" and "stop" buttons to initiate preprogrammed motion sequences, an emergency stop button to allow the operator to stop all motion when necessary, and a programming interface to allow the operator to construct motion actions and sequences.

The programmed sequences constructed by the operator are typically referred to as "cues." Cues can consist of one or more motion or action commands and a show can contain multiple cues, usually played back in order during the performance. The following sections will discuss some basic cueing commands and actions and means of playback.

COMPONENTS OF A CUE

CUE STRUCTURE

Cues are generally assembled as sequential lists or trees of motion and action commands. These commands will execute in order to provide the desired sequence of motion on stage when the cue is started by the operator.

TYPES OF COMMANDS

Cue commands generally fall into a few specific categories. Commercial manufacturers provide systems that are capable of executing from a few basic commands up to dozens of complex commands. Most systems will provide command capabilities of at least the following types:

- **Motion commands:** Motion commands provide the means to move a machine from its current position to a target position. Command parameters generally include target position, acceleration, deceleration, and

CUE 4.3	Position Wagons for Scene 2
MOVE	Motor 1 to Position 15ft @ a=2, v=3, d=2
DELAY	3 Seconds
MOVE	Motor 2 to Position 3ft @ a=1, v=5, d=1
MOVE	Motor 3 to Position 3ft @ a=1, v=5, d=1
WAIT	Motor 1 to reach Position 10ft
MOVE	Motor 4 to Position 20ft @ a=5, v=5, d=5

Figure 9.25 Simple motion cue

velocity. Velocity is also sometimes expressed as a function of time (i.e., "move to target position in X time") rather than entering the velocity value or percentage directly.

- **Timing commands:** Timing commands allow the operator to insert pauses or wait periods in the playback sequence. This provides the means to sequence motion starts or other actions without the need to separate those commands into separately executed cues, as well as the means to achieve repeatable and accurate timing in playback of sequences.
- **Sequence commands:** Like timing commands, sequence commands control the playback of motion and action commands within a cue. Sequence commands control the start of subsequent cue actions based on the state of other machines or actions in the system, for instance "wait for curtain to reach 'in' position." Sequence commands can also start other cues from within the currently executing cue, sometimes referred to as an "auto follow."

ASSEMBLING CUES

Using the different types of commands in the proper execution sequence, a system programmer can construct fairly complex motion sequences that execute consistently and reliably from show to show. Interfaces for cue construction are usually list-based, and the actions execute in the sequence that they appear in the list. Generally, back-to-back motion commands in the list (i.e., commands that are not separated by other timing or wait commands) will execute simultaneously, even though they're in a sequential list, as that construction effectively means "do all of these actions at the same time." Some programming systems represent the cue in a nested tree structure to represent the relationships between the commands, and others use a flow chart style, but the most common is the basic command list.

Playback

Playback of cues is usually initiated by the operator by pressing a "go" button of some type on the interface to start the cue at the appropriate time in the production. This is known as "operator initiated." Cues can also be started automatically through the use of various sequence commands and, in theory, an entire show could be constructed where everything plays back automatically and in the correct timing from a single "go" command at the start. However, in practical application, operators and stage managers use their best judgment based on the requirements of the production to construct automated sequences that balance the needs for consistency and ease of execution with the fluid nature and ever-changing nature of live productions.

All systems allow the execution of a single cue by the operator, and some allow the execution of multiple cues from multiple "go" commands, causing more than one cue sequence to be executed live on stage at the same time. Operators should take into account the capabilities of the system that they are operating when formulating their programming and playback strategy for any production.

Motion Safety

Safety is paramount in any motion control system, as stage and rigging automation is one of the few disciplines in technical theatre that produces physical "kinetic" actions on stage. Automation systems are equipped with many safety features and functions to prevent unintended motion or motion outside of normal parameters and to provide a means of stopping all motion quickly and safely in the event of an emergency. The following sections describe some of the basic safety systems and parameters included in modern automation systems.

Emergency Stop and Braking Systems

The most basic and ubiquitous safety measure provided in an automation system is the emergency stop system. This provides the means for the operator or other theatre personnel to initiate a full system stop in the event that personnel or property are in jeopardy due to the motion of the machinery and its attached loads. Emergency stops are typically initiated as one of two types: The "category zero" stop, where power is immediately cut from the machines and all braking systems are immediately engaged; and the "category one" stop, where the control system is allowed to bring the machinery to a rapid but controlled decelerated stop before setting the brakes and removing power. The category one stop is preferred, as it can be considerably more gentle and smooth than a category zero stop, but all machinery and structural systems should be able to handle a category zero stop at any time, as this can also be induced by a complete power failure.

Position Control and Limitation

Another safety measure is the control and limitation of the allowable motion envelope of the machine and the attached load. This is typically accomplished in two ways:

1 Computer controlled limits, where the minimum and maximum position that the operator can enter as a valid target is restricted by the programming system, and the motion of the machinery is also restricted beyond that value during jogging or other manually controlled motion.
2 Electromechanical limits, where the control signal to the machine controller is physically interrupted by a limit switch located at the physical

end of travel. Secondary limit switches that initiate an emergency stop are sometimes located just beyond these limits as well.

Force Control and Limitation

Many machines are equipped with load cells or other means of detecting and measuring the load on the machinery. This feedback can be used to either stop the motion of the machine if a maximum allowable value is reached, or to limit the output of the machine to a maximum force value. This can prevent overloading and damage to the machinery or the attached load.

Machinery Interlocks and Enables

Machinery interlocks can be used to prevent collisions or other motion into dangerous areas. For example, a suspended piece of scenery may be interlocked to the lateral position of a piece of scenery on a stage wagon such that the suspended piece may not be brought to a low enough position to cause contact with the scenery below. Similarly, enable switches and other interlock devices can be provided to stage managers and other personnel to allow a second visual "all clear" check, and the motion of the object being checked will be disallowed unless that enable is constantly activated.

Operation Practices

There are a few key operational practices that will greatly increase safety in and around the stage when using automated machinery. Theatre staff and management should incorporate well-defined and clear safety and operation practices in any facility using automated machinery. Some key items are as follows:

COMMUNICATION AND VISIBILITY

The automation operator must have a clear and constant method of communicating with stage management personnel and other persons monitoring or controlling the actions in the performance space or anywhere automation machinery is being used. The operator should also have a clear line of sight to any devices or objects being moved by the system. This line of sight can be provided via closed circuit video monitoring, but in no circumstances should the operator be operating machinery in a "blind" situation.

AVAILABILITY OF EMERGENCY STOPS/ENABLES

The automation system operator and technicians must assure that emergency stop actuators are available anywhere that motion of machinery or objects is being executed so that qualified personnel in those areas can initiate a stop if any dangerous situations are detected. The emergency stop function should not be delegated to the operator alone at the control station, as communication of the necessity for a stop is not always reliable.

SAFE STATES

A "safe state" for each piece of motion controlled machinery should be determined so that it can be reliably placed in that safe state at any time necessary. Depending on the environment and the type of machinery and load in question, a "safe state" can be anything ranging from the system being in emergency stop mode, to a complete removal of power from the device, to a secondary structural support or stop in the machinery or rigging. "Safe" has a definition, and all operators and maintenance personnel must use the same definition and terminology when describing a device as "safe."

PRACTICE AND REHEARSAL SPEEDS

Initial practice and rehearsal of motion sequences should always be performed at low speeds to assure the motion path and envelope is clear and safe and to provide operators and spotters with more time to initiate a stop if necessary. This is especially important in motion sequences utilizing complex timing or motion interlock commands, or where there is a large amount of machinery in motion. Once the paths and motions have been determined to be free of interference or other issues, speeds can be increased up to the levels to be used in the performance.

NOMENCLATURE AND CLARITY

It is important that all operators, technicians, managers, or other personnel communicating on stage during motion controlled sequences speak the same "language" when giving instructions or information. The methodology and terminology for expressing readiness, clear for motion, directions of motion, desired speeds, stopping and starting motion and cues, and emergency stopping should be known and clear to all personnel, and always used when giving commands or expressing status. Lack of clarity in communication is one of the main contributing causes to accidents, and it is important that it be eliminated from the performance space.

Note: This chapter is intended to familiarize the reader with the basics of mechanical stage automation concepts and systems. Stage automation involves electricity, high pressures, machinery with potential pinch points and other hazards, and overhead loads, and can be dangerous if improperly applied or executed. This chapter is not intended as a "how to" guide or instruction in the construction and commissioning of stage automation equipment, and readers should not take it as such. Just as we would not recommend that readers attempt to forge their own shackles for use in rigging systems, we also do not recommend that readers of this text attempt to fabricate their own automation equipment. It is important to use equipment engineered and manufactured by reputable and qualified suppliers, and to obtain proper training in the installation and use of that equipment prior to operation.

10

The Mechanics of Stage Automation

DAN CULHANE

Winches and hoists have been traced back to the Greeks, who considered them their "deus ex machina" or "god of the machine."

A hoist is a winch that raises and lowers battens, electric sets, scenery, orchestra shells, and almost anything else. In this chapter, we will be using the terms "winch" and "hoist" interchangeably. There are dead haul winches that lift the entire weight of the flown element and there are counterweight assist winches, where the weight of the flown element is partially counterbalanced by a counterweight allowing the hoist to lift only the imbalanced weight of the system. At a minimum, a hoist has one lift line, more often they have two or more. There are three types of dead haul winches: the line shaft winch, the drum winch and the traveling drum, or zero fleet, winch.

Line shaft winches have multiple cable drums connected to each other and the gear reducer with a shaft. A drum winch has a single cable drum with the cable or cables coming off of that one drum.

For line shaft winches the cable comes off of the drum and then attaches to the batten. The cable spools on to the drum from one side; if all of the drums spooled in the same direction the flown object would travel to one side. On line shaft winches, half of the cable drums have a have a left-hand winding and the other half wind to the right. The net effect is that the flown object will not travel to one side as it is flown.

A counterweight assist winch typically uses a chain sprocket attached to the gear reducer. A roller chain is attached to the bottom of the arbor, travels down and around the chain sprocket on the gear reducer, then travels up to the head block. At some point between the chain sprocket and head block, the roller chain terminates to a piece of cable that travels over the head block and attaches to the top of the arbor.

A traveling drum winch is typically a winch in which the drum travels along guides allowing the cable to spool on and off the drum at a rate that equals the cable wrap on the drum. For example, if the cable is wound around the drum at three wraps per inch, the cable drum will travel the distance of one inch in three revolutions.

Figure 10.1 Line shaft winch. Photo courtesy of SECOA.

Figure 10.2 Line shaft winch. Photo courtesy of SECOA.

Figure 10.3 Counterweight assist winch. Photo courtesy of SECOA.

Figure 10.4 This is a 7,750lb capacity traveling line shaft winch for orchestra shell storage. This winch has two gearmotors on a common shaft. Photo courtesy of SECOA.

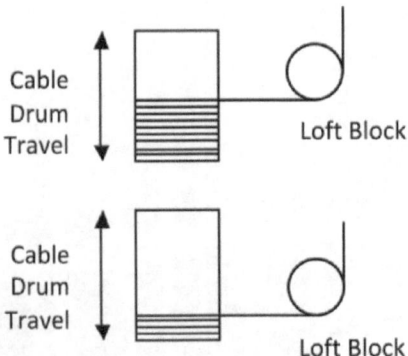

Figure 10.5 Cable drum travels as the cable winds/unwinds off of the drum, maintaining a consistent cable location relative to the loft block

In this manner, the cable leaving the drum will stay in a constant place relative to the building. The advantage of this method is that the fleet angle for the cable from the drum to the loft block remains the same, and the distance between the loft block and the drum can be very small.

Cables terminating at a steel drum will have three dead wraps on the drum before the active wraps start. At a minimum, cable terminations need to hold at least 80% of the cable's strength. In addition to the actual cable termination, the three dead wraps generate friction between the cable and the drum to accomplish the 80%. If the drum has a synthetic surface, additional dead wraps may be necessary to achieve enough friction to obtain 80% of the cable strength.

The diameter of the cable drum is determined by the cable diameter. All cables have a minimum bending radius allowed by the manufacturer. For 7 × 19 small diameter specialty cord, also known as aircraft cable, the minimum bend radius is 26 times the cable diameter.[1] This is known as the D:d ratio; the ratio of the tread diameter (D) of the drum or sheave to the diameter (d) of the wire rope. The smaller the diameter the shorter the service life is for the cable. For ¼" cable the minimum diameter for the drum would be 6½". Example: ¼" × 26 = 6½".

Cable drums used in the theatre have a single layer of cable on a grooved drum. The groove in the drum is a single helix.

Recommended cable drum design:[2]

- The minimum groove radius for new rope: 0.53 to 0.535 × diameter of the wire rope.
- The maximum groove radius: 0.55 to 0.56 × diameter of the wire rope.
- The minimum drum pitch for a single layer of cable: 2.065 × the groove radius.
- The maximum drum pitch for a single layer of cable: 2.18 × the groove radius.
- The drum groove depth: minimum × 0.375 × diameter of the cable.

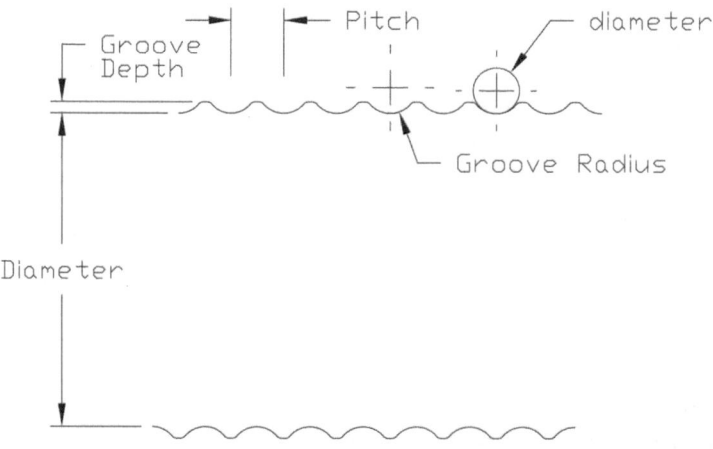

Figure 10.6

The Wire Rope User's Manual published by the Wire Rope Technical Board is a very good source for additional information regarding drum design and cable bending information. When designing cable drums for a system it is often impossible to keep the fleet angle less than two degrees, while keeping the distance between the cable drum and the loft block to a reasonable distance. When the distance needs to be shorter, a larger diameter drum is necessary to maintain the fleet angle at ≤ 2°.

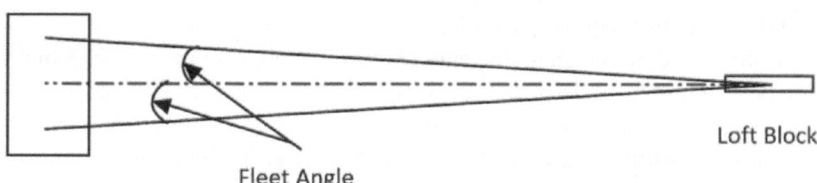

Fleet Angle

Figure 10.7 The longer the travel the greater the distance between the loft block and the cable drum to maintain an acceptable fleet angle of two degrees or less

All winches have a number of parts generally consisting of a motor with a motor brake, a gear box, and a cable drum or chain sprocket. Dead haul hoists have a secondary brake. A necessary part of the winch is a set of limit switches. There are at least four of these, two of which are the normal end of travel limit switches that stop the hoist at the extreme ends of normal travel but allow it to travel in the opposite direction. The other two are over travel, or ultimate limit switches. The limit switches are usually placed just beyond the normal travel limit switches in the event the hoist does not stop when the normal end of the travel limit switches are tripped. The over travel limit switches are wired to a special circuit that will remove power to the hoist and ensure that the hoist stops. No further movement of that hoist is allowed until the limit switch is reset.

Rigging in the 21st century began in earnest with the advent of the packaged hoist, brought to market by Hoffend and Sons. They designed and mass produced an integrated hoist and control system to a market receptive to that concept. Mass production and standardization of features was key to producing a system that was affordable. Many safety features were integrated into the hoists and controls, such as load-sensing, which informed the controls that the intended load would stop the hoist when the batten snagged on something that made it too heavy or landed on something that overly lightened the load. They made a passive drive-through brake that would hold the load should there be a motor brake failure. The controls have intuitive graphical touchscreen displays to select features. The operator can combine hoists together into groups that can be assigned, or individual hoists into queues. The graphic display shows the operator the process precisely as the queue is played out. In the packaged hoist market there are numerous competing manufacturers. The dominant players are: Daktronics Vortek line of hoists, J.R. Clancy's Power Lift line of hoists, and Electronic Theater Controls (ETC) with their Prodigy line of hoists. The packaged hoist is a self-contained zero fleet winch.

Winches designed for the theatre need to conform to many standards. At a minimum, ANSI E1.6-1, Entertainment Technology-Powered Hoist Systems must be referenced and followed.

Most winches today use a paired motor and gear box integrated into one unit which is called a "gearmotor." One of the many advantages to this unit is that the motor and gearbox are designed to work together and are built as a single unit that does not require assembly.

Motors used on winches are generally three-phase type motors with brakes. The type of motor used is totally enclosed and fan-cooled, designated as TEFC. This type of motor does not have openings, which facilitates keeping dirt and dust out of sensitive areas. The fan is essential for keeping the motor cool and, mounted on the end of the motor opposite to the gear reducer, it blows air over the outside of the motor frame. Normally the fan is plastic and is relatively light in weight. Most motors that use a reversing contactor use a high inertia fan, which is heavy cast iron. When a motor starts and stops with a regular fan it does so very abruptly, which can jar the load. This abrupt start and stop is hard on lamp filaments and other moving lights. Using a cast iron fan adds weight to the rotor of the motor which creates higher inertia and slows down the motor wind-up and wind-down, allowing for smoother starting and stopping. The high inertia fan is the "Z" option for both NORD Gear Corporation and SEW Eurodrive Corporation motors. If a motor is going to be controlled with a variable frequency drive (VFD), the motor does not need the high inertia fan because the VFD is capable of producing ramped starts and stops in the motor.

Single-phase motors are not used for hoisting winches; they are used in low horsepower industrial applications, such as fans or on farms.[3] Single-phase motors are generally not instant-reversing. This means that when started in one direction, if the motor is reversed without stopping the motor rotation, the single-phase motor will continue rotating in the same direction. This can be a problem when, in an emergency, it is necessary to back away from a problem quickly. A more expensive instant reversing single-phase motor is available.

A brake motor comes as a pre-connected package. The brake is a spring-set, stop and hold type. When power is removed from the brake, the brake sets and holds the load. On most winches using a reversing contactor, the power for the brake is taken directly from the power sent to the motor. There is a rapid reaction option to activate the brake. This is advantageous for hoists, as it sets the brake very quickly after the motor power is turned off. If the motor is using the high inertia fan, the rapid reaction option for brake activation is not used to stop the motor, since the high inertia fan is cancelled out by the rapid reaction brake activation. On winches with a variable frequency drive, the brake power must be switched separately from the power provided to the motor. There is a double motor brake option from both NORD and SEW.

A motor is comprised of a couple of parts. The rotor controls the rotation while the stator, which is stationary, contains a set of motor windings. Three-phase induction motors are high-efficiency, have a high starting torque, a relatively low current draw, and are available in various speeds. The speed of the motor comes from the number of poles wound into the stator. The following table gives the approximate RPM speed at rated load for 60hz. The most common three-phase motor is a four-pole motor.

No load speed can be determined by the following formula: [5]

$$RPM = \frac{Frequency_{(Hertz)} \times 120}{Number\ of\ Poles.}$$

Motors are rated by horsepower (HP). 1 HP = 745.7 watts.
= 33,000 ft-lb$_f$/minute.
= 396,000 in-lb$_f$/minute.

Table 10.1

Number of Motor Poles	Motor RPM[4] at 60 hz.
2 pole	3,450
4 pole	1,725
6 pole	1,140
8 pole	850

The 745.7 watts is typically rounded off to 746 watts. Since no motor can operate at 100% efficiency, 80% efficiency is a reasonable approximation to use, until the actual motor efficiency is obtained. The other 20% is used in creating heat or friction. A one HP motor that is 80% efficient will use 932.5 watts of energy.

$$\frac{746 \text{ watts}}{.8 \ (80\%)} = 932.5 \text{ watts}$$

Horsepower can be calculated if the torque is known using the following formula:[6]

$$HP = \frac{\text{Torque}_{(lb-in)} \times \text{RPM}}{63,000.}$$

By rearranging the above formula torque can be broken out:

$$\text{Torque (lb-in)} = \frac{HP \times 63,000}{RPM.}$$

The motor service factor (SF) is the amount of continuous overload capacity the motor can operate without damage or overload, provided the other design parameters such as frequency, rated voltage and ambient temperature remain within norms.[7]

Example: A 1½ HP motor with a service factor of 1.10 SF can operate at 1.65 HP without overheating or other damage if the frequency, rated voltage and ambient temperature are within norms.

1.5 HP × 1.10 SF=1.65 HP

All motors need thermal or overload protection. This prevents motors from getting too hot and causing a fire. Overload protection devices protect the motor from both excessive current and temperature.

Gear reducers take the RPM and torque input provided by the motor and convert it to a lower RPM and correspondingly higher output torque. Gear reducers are also defined by the number of stages they have. A single-stage gear reducer reduces the input RPM to the output RPM in one set of gears or one stage. A two-stage gear reducer has two steps or stages to go from the input shaft to the output shaft. All gear systems have gaps or clearance between the moving parts—this amount of movement is called backlash. Backlash can be significant enough to cause repeatability problems in winch systems, particularly when the encoder for the winch is connected to the end of the motor and the piece of scenery on the other end is hanging by cables. Generally, the more gear stages involved, the larger amount of backlash in the system. While unavoidable, the effects of backlash in the gear system must be kept to a minimum for the winch system to provide consistent results every time.

Figure 10.8 Helical gears and helical gearmotor. Photo courtesy of NORD Gear Corporation.

Figure 10.9 Helical gears and parallel shaft gearmotor. Photo courtesy of NORD Gear Corporation.

Figure 10.10 Helical bevel gears and helical bevel gearmotor. Photo courtesy of NORD Gear Corporation.

Figure 10.11 Helical worm gears and helical worm gearmotor. Photo courtesy of NORD Gear Corporation.

There are four basic types of gear reducers on the market today: helical gears, parallel shaft helical gears, helical bevel gears, and helical worm gears. The latter three types of gear reducers are often used in the theatre industry. Both parallel shaft and helical bevel gear reducers are very efficient in transmitting power from the motor to the output shaft. The helical worm gear reducer is inefficient in its transmission of power from the motor to the output shaft. The helical worm design is the most widely used gear reducer type, offering long service life, overload and shock tolerance.

Efficient gear reducers are used on most of today's packaged winch systems. They are also used on most fire curtain motor systems as they are easy to back-drive, which is the reverse of the normal operation. The output shaft is used to turn the input shaft. Many fire curtain release systems release the brake on the motor, using the weight of the fire curtain to turn the output shaft of the gear reducer, which turns the motor. A hydraulic dampener attached to the motor controls the descent speed of the fire curtain.

An inefficient gear reducer design does, however, have advantages in the theatre world. A helical worm gear reducer with a gear ratio greater than 60:1 statistically cannot be back-driven. This means that a load on the output shaft will remain stationary even if the motor brake is released. Most line shaft, drum, and counterweight assist winches use this type of gear reducer because of this feature. Helical worm gear reducers are a simple design that is very cost-effective to produce. This design lends itself to the lower output RPM and higher gear ratios used in the theatre industry.

Gear reducers are filled with oil and are vented because the oil will expand as the reducer is used. The oil helps to keep the reducer cool. The reducer will have a breather vent and a drain plug. While it may look like as if they can be mounted or oriented in any position, it is very important to be certain that the vent is at the top and the drain at the bottom when the reducer is mounted in its final position. The amount of oil with which the reducer is filled varies with the orientation that is used. The orientation is also important regarding the bearings that are installed into the reducer. The manufacturer will install different bearings depending on whether the bearing is below or above the oil level in the reducer. Thus it is important to order a gear reducer for the specific orientation for which it will be used.

Gear reducers also have their own service factor, which is defined by the American Gear Manufacturers Association (AGMA). AGMA adjusts a reducer's ratings relative to the individual load characteristics of the reducer. AGMA's ratings are based upon time duration. For winches used up to three hours per day a 1.00 service factor is recommended. For winches used for between three and ten hours per day the service factor is 1.25. For winches used for more than ten hours per day the service factor is 1.5.[8]

There are a number of considerations in the selection of a gearmotor. These include service factor, speed, horsepower, ratio, physical size, location, orientation, and temperature, and the cost of the unit.[9] All gearmotor suppliers are more than willing to assist you in determining the best gearmotor for your situation. SEW Eurodrive, a leading gearmotor manufacturer, has a number of programs online and available for download that will help you select the best gearmotor for your situation. Their online tool is called "PT Pilot" and is widely used by most rigging suppliers to insure that the gearmotor has been sized properly. PT Pilot performs the necessary calculations to properly size a gearmotor and provides spare parts lists, documentation and downloadable CAD details for the gearmotor selected. SEW also has a standalone program called "Workbench," which will perform the necessary calculations to size a gearmotor.[10] The software will also identify product-specific documents such as spare parts lists and data sheets, and will provide downloadable CAD drawings. SEW has many online tutorials and documentation to facilitate the maintenance of your gearmotor.[11] NORD Gear, another leading gearmotor manufacturer, has extensive online documentation and downloadable CAD drawings as well.[12] NORD has the better written documentation for sizing a gearmotor, but at this time does not have an online calculation tool for sizing a gearmotor. As mentioned previously, all gearmotor manufacturers will accommodate you in sizing a gearmotor to meet your specific requirements. They will walk you through the sizing calculations to make certain the gearmotor will work for your application.

There are a number of steps that need to be taken as part of selecting the proper gearmotor for a specific application. Sizing a gearmotor can be an iterative process. The design and calculations may have to be repeated several times as you work through the process before realizing a successful winch design. This is especially true the first few times the design and calculations are performed.

The very first thing in the design of a winch is to determine its capacity. This will also define many things such as the quantity and size of the lift cable(s). For a counterweight assist winch the capacity will determine the size and makeup of the chain used. For a line shaft winch, the cable size will determine the minimum diameter for the cable drums. It is important to know the size of the drum or sprocket. This determines the amount of torque that the gearmotor needs to generate. The size of the cable drum or sprocket is necessary to determine the speed in rotations per minute (RPM) for the gear reducer. It is important to know the motion profile for the winch as this determines the size of the gearmotor. A gearmotor that is moving a load at a fixed speed of 20' per minute will be much smaller than a gearmotor moving the same load at a fixed speed of 100' per minute. It is important to know the minimum amount of time needed to accelerate the load to its top speed. Accelerating a load from zero to 20' per minute in half a second will be a different value than accelerating the same load from zero to 20' per minute in ten seconds.

The following are some steps necessary in determining the proper gearmotor.[13]

1 Determine the speed or gear ratio of the reducer.
2 Determine the required torque.
3 Determine a service factor or service classification.
4 Select the gearmotor.
5 Determine the required mounting position.
6 Add options to the gearmotor.
7 Perform the required checks:

- Overhung load.
- Thrust load.
- Thermal considerations.

The speed of the gear reducer is usually expressed in rotations per minute (RPM) and can be determined using the following formula:[14]

$$RPM = \frac{FPM \times 3.820}{Dia}$$

Where:

RPM = rotations per minute
FPM = feet per minute
Dia = pitch diameter in inches of the drum or chain sprocket

Next, determine the required torque. Most gearmotor manufacturers express their output values in pound-inches (lb-in) of torque. It is necessary to work in the same value as it is very easy to confuse the different values. We will be using lb-in of values. Another common value used is pound-feet (lb-ft) of torque.

Determining output torque of the gearmotor can be an involved process requiring a number of steps.

Initially, determine inertia calculations for all parts of the hoist components beyond the gearmotor. Inertia is typically represented by the letter "J" in these formulas. Each piece of shaft, drum and sprocket will need to be calculated separately.

For a solid cylinder the inertia formula will be:[15]

$$J = \frac{(r^4 \times L \times \rho)}{C_2}$$

Where:

J = inertia in lb-in-s^2
r = radius of cylinder in inches
L = length of cylinder in inches
ρ = specific gravity of the cylinder material

Steel = 0.284 lb/in³
Aluminum = 0.0955 lb/in³
C_2 = conversion factor 246.0

For a hollow cylinder the inertia formula will be:[16]

$$J = \frac{W}{g} \times r^2$$

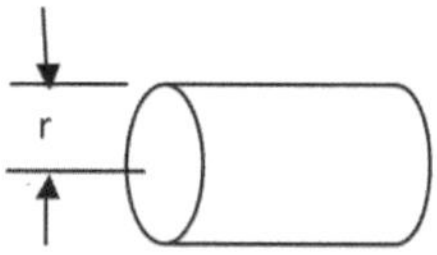

Figure 10.13

Where:

J = inertia in lb-in-s²
W = weight of cylinder in lb
g = gravitational acceleration 386 in/s²
r = radius of cylinder in inches

Now the load torque is calculated. The force in the following formula is the lifting capacity of the hoist.[17]

$$T_L = F \times r$$

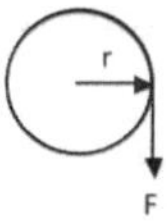

Figure 10.14

Where:

T_L = torque in lb-in
F = force perpendicular to the rotated shaft
r = radius to where the force F is applied in inches

The motion profile of the motor must be understood, as accelerating a weight will take more energy than it would take to maintain constant speed for that weight. The less time it takes to accelerate a weight from a dead stop to operating speed, the more torque from the motor is required. A typical motion profile graph would look like Figure 10.15.

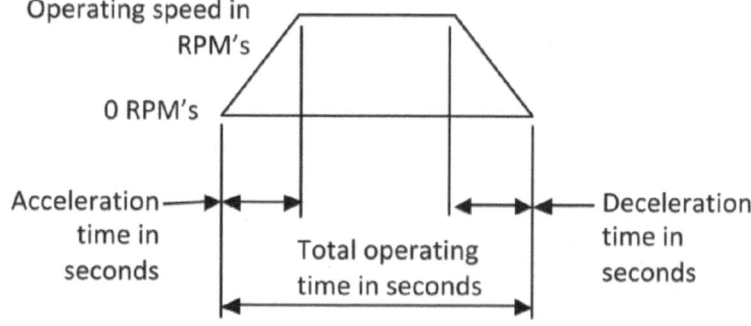

Figure 10.15

A typical acceleration time for a constant speed hoist to accelerate from zero to 20' per minute will be in the ½-second range.

The inertias, J, calculated previously can be converted to torques using the following formula. This calculation needs to be done for all of the inertias previously calculated.[18]

$$T = \frac{J \times RPM}{9.55 \times t}$$

Where:

T = torque in lb-in
J = inertia in lb-in-s^2
RPM = rotations per minute
t = acceleration time in seconds

Next the root mean square (RMS) torque is calculated using the above information. In this formula all of the above torques are compiled. This includes the load torque (T_L) value calculated previously.[19]

$$T_{RMS} = \sqrt{\frac{T_1^2 \times t_1 + T_2^2 \times t_2 + T_n^2 \times n}{t_1 + t_2 + t_n}}$$

Where:

T_{RMS} = torque root means squared in lb-in
T_1, T_2, T_n = Torque 1 through n in lb-in
t_1, t_2, t_n = times 1 through n in seconds

The T_{RMS} value is the minimum torque output from the gearmotor that is required to move the load given the amount of lifted weight, shafting, drums and time parameters.

The next step is to multiply the T_{RMS} value with a design factor. A typical design factor in the order of 1.25 is not uncommon. A formula multiplying the T_{RMS} value with a design factor would be the following.[20]

$$T_M = T_{RMS} \times D_f$$

Where:

T_M = required torque in lb-in
T_{RMS} = torque root means squared in lb-in
D_f = design factor

The T_M value is the output torque of the gearmotor along with the RPM required and can be found in the gearmotor tables of the manufacturer's catalog.

It is very rare that both of these exact values are available in the catalog. When selecting a gearmotor it is important to choose a torque value that is greater than the T_M value calculated. The associated RPM of the selected gearmotor may also be slightly faster or slower than the ideal RPMs calculated. It will be necessary to use the RPMs from the selected gearmotor and reverse calculate how fast the hoist will be traveling. The following formula will yield the feet per minute speed of the hoist.[21]

$$FPM = Dia \times RPM \times 0.2618$$

Where:

FPM = feet per minute
Dia = pitch diameter of the drum or chain sprocket in inches
RPM = rotations per minute

After a gearmotor has been selected several checks must be performed to be certain the gearmotor will perform satisfactorily.

An overhung load is a force that is applied at right angles to the shaft beyond the shaft's outermost bearing. Sprockets and drums will create an overhung load. The amount of this load will vary depending upon its location on the shaft. Overhung loads exceeding the capacity indicated in the gearmotor selection table will cause premature bearing failure. Overhung loads given in the tables are applied at the midpoint of the shaft and are without thrust loads to the shaft. If there is a combination of overhung and thrust loads on the shaft, consult the manufacturer to determine if the loads are acceptable for the reducer.[22]

An overhung load can be calculated in the following formula:[23]

$$F_{OHL} = \frac{2 \times T \times f_z}{d_o}$$

Where:

F_{OHL} = calculated overhung load (OHL) on the gear reducer shaft
T = load torque on the shaft in lb-in
d_o = pitch diameter of overhung component in inches
f_z = power transmission component factor from table below

Table 10.2

Transmission Component	Factor f_z	Notes
Chain sprocket	1.40	13 teeth or less
Chain sprocket	1.20	14 to 20 teeth
Chain sprocket	1.00	21 teeth or more
Cable drum	1.00	

After calculating F_{OHL}, compare it to the overhung capacity found in the motor selection tables.

If $F_{OHL} \leq F_R$ for the output shaft the reducer is fine. If $F_{OHL} > F_R$ then the reducer either needs larger bearings that will meet or exceed the F_{OHL} or a larger reducer must be selected.

If the overhung load is not in the center of the shaft, consult the manufacturer of the reducer to calculate the overhung load and shaft strength to ensure acceptability.

Thrust loads, or axial loads, are loads that are directed toward or away from the gear reducer along the axis of the shaft. Output shaft thrust loads are commonly designated as F_a and are listed in the gearmotor selection table. Thrust loads cannot exceed the value indicated in the table; if it does, a larger gear reducer must be selected. If there is a combination of overhung and thrust loads on the shaft, consult the manufacturer to determine if the loads are acceptable for the reducer.[24]

Consult the gearmotor manufacturer if the gearmotor will be running constantly and has two or more of the following: is in an environment where the temperature is greater than 86° F; has a motor oriented either up or down; has a motor speed greater than 1,800 RPM. This usually is not applicable for the theatre industry as most hoists are not run constantly.

As mentioned earlier in the chapter, most manufacturers use gearmotor sizing software that will do most of the calculations and document the results. The motor sizing software is good but there are a couple of calculations that should be performed outside of the software for input into the software. The software does not calculate the inertia for the sprockets, shafting and drums. Fortunately calculating the inertia for drums, shafting, and sprockets is not very difficult. The software will guide you to a selected set of acceptable gearmotors that saves having to page through the entire catalog trying to find an acceptable gearmotor. Most gearmotor catalogs are over an inch thick and are mostly tables of information.

When the software has selected a gearmotor, all of the checks listed above must be performed to ensure that the published values are not exceeded.

Using PT Pilot to calculate a 2,000lb cable drum speaker hoist, Figures 10.16 and 10.17 are a printout of the calculations generated. The first page is mostly information that is contained in the software and includes: the lifted load, speed, drum pitch diameter, additional inertia (the calculated inertia for the drum, side plates and shafting), the lifting efficiency for the cable on the drum, operating criteria (the motion profile), brake torque, and high inertia fan. The second page has the results from the software calculated for both up and down motion including: calculated static HP, actual speed, acceleration time, starting time, and braking distance. The lower part of the page has the selected gear motor model, the HP, gear ratio, output speed in RPM, output torque, nameplate service factor, reducer efficiency, and brake torque.

12/12/13 PT Pilot

SEW-EURODRIVE

PT Pilot® Calculator Report

Company
Project Name Center Speaker Hoist
Created By
Date 12/12/2013

Crane Hoist - Cycling Without Inverter Page 1

Input:	Imperial		Metric	
Maximum Weight/Mass - Up	2000	lb	907	kg
Maximum Weight/Mass - Down	2000	lb	907	kg
Counterweight Mass	0	lb	0	kg
Load Velocity	20.0	ft/min	0.102	m/s
Counterweight Velocity	--	ft/min	--	m/s
Reeving Ratio	1		1	
Drum or Sprocket Diameter	12.94	in	328.7	mm
Additional Inertia	14.008	lb-ft^2	0.59	kgm^2
Minimum Acceleration	--	ft/s^2	--	m/s^2
Minimum Deceleration	0.00	ft/s^2	0.00	m/s^2
Lifting Type	Steel Cable & Drum			
Lifting Efficiency	0.93			
Hours Operation Per Day	8			
Run Time	146.00	sec		
Cycle Time	150.00	sec		
Desired Starts/Hr	24.0			
Desired Travels/Hr (1 Travel = 1 Up + 1 Down)	12.0			
Desired Nameplate S.F.	1.00			
Motor Poles / Frequency	All / 60	Hz		
Brake?	Yes			
Brake Torque Desired	100%			
Rapid Reaction?	No			
Include High Inertia Z-Fan?	Yes			
Class H Insulation plus Forced Cooling Fan?	No			

v4.ptpilot.com/PTPilotNET40/CalcReportABHO.aspx

Figure 10.16

12/12/13   PT Pilot

SEW-EURODRIVE

PT Pilot® Calculator Report

Company   **Date**   12/12/2013
Project Name   Center Speaker Hoist
Created By

Crane Hoist - Cycling Without Inverter   Page 2

	Imperial			Metric		
Results:	**Up**	**Down**		**Up**	**Down**	
Calculated Static Power	1.77	-0.75	HP	1.32	-0.56	kW
Actual Speed (Velocity)	20.5	20.5	fpm	0.104	0.104	m/s
Acceleration Rate	0.31	0.49	ft/s²	0.09	0.15	m/s²
Starting Time	1.10	0.71	sec	1.10	0.71	sec
Load Torque @ Starting (Stat + Dyn)	20014	-8198	lb-in	2261	-926	Nm
Load Torque After Starting (Static)	13914	-12034	lb-in	1572	-1360	Nm
Permissible Starts per Hour	14.7	79.2		14.7	79.2	
Braking Rate (Deceleration)	0.41	0.30	ft/s²	0.13	0.09	m/s²
Braking Time	0.82	1.19	sec	0.82	1.19	sec
Braking Distance	1.94	2.73	in	49.3	69.4	mm
Braking Accuracy (+/-)	0.23	0.33	in	5.9	8.3	mm
Braking Work (Mechanical)	1090	1661	lb-ft	1478	2253	J
Load/Motor Inertia	0.00	0.00		0.00	0.00	

Brake Service Life (Average) till readjustment	3695	hrs
Cyclic Duration Factor	97	%
SEW Recommended Nameplate S.F.	0.95	
Permissible Travels/Hour	12	
(1 Travel = 1 Move Up + 1 Move Down)		

Selection:

Model	SA87DRE90L4BE2Z	
Power	2	HP
Ratio	288.00	
Output Speed	6	rpm
Average Accelerating Torque	218.5	lb-in
Total Motor Inertia	2.4881	lb-ft²
Output Torque	15000	lb-in
Nameplate Service Factor	1.35	
Reducer Efficiency	0.700	
Brake Torque	14.70	lb-ft

Comments

Figure 10.17

Endnotes

1 MacWhyte Wire Rope Company, Catalog G-18 (Kenosha, WI) p. 170.

2 Unirope, Ltd., Wire Rope, www.unirope.com/wireropes/pdfs/wr_technical_info.pdf, p. 11 (December 15, 2013).

3 Leeson Electric Corp., Basic Training Industrial-Duty Electric Motors, Gearmotors & Commercial-Duty Gear Reducers, AC & DC Drives (Grafton, WI: Leeson Electric Corp.,1999), p. 24.

4 Leeson Electric Corp., p. 25.

5 Leeson Electric Corp., p. 24.

6 Leeson Electric Corp., p. 25.

7 Leeson Electric Corp., p. 26.

8 NORD Gear Corp., Constant Speed Drives, Simple Reliable Efficient Drive Systems 100200001/ 01.11(Waunakee, WI: NORD Gear Corp.), p. A68.

9 NORD Gear Corp., p. A11.

10 SEW Eurodrive Corp., SEW Workbench www.seweurodrive.com/produkt/electronic-catalog-system-engineering-sew-workbench.htm (December 14, 2013).

11 SEW Eurodrive Corp., Documentation www.seweurodrive.com/support/index.php (December 14, 2013).

12 NORD Gear Corp., Documentation www5.nord.com/cms/en/documentation/dop_documen-tation.jsp (December 14, 2013).

13 NORD Gear Corp., p. A11.

14 NORD Gear Corp., Formula Card (Waunakee, WI: NORD Gear Corp.).

15 Kollmorgen Corp., Application Sizing Guide, Servomotor Selection www.kollmorgen.com/en-us/search/?searchText=application+sizing, p. 3 (December 14, 2013).

16 Kollmorgen Corp., p. 3.

17 Kollmorgen Corp., p. 5.

18 Kollmorgen Corp., p. 6.

19 Kollmorgen Corp., p. 6.

20 Oriental Motor Corp., Technical Reference www.orientalmotor.com/products/pdfs/2012-2013/ G/usa_tech_calculation.pdf, p. G-4.

21 NORD Gear Corp., Formula Card (Waunakee, WI: NORD Gear Corp.).

22 NORD Gear Corp., p. A12.

23 NORD Gear Corp., p. A52.

24 NORD Gear Corp., p. A52.

11

Training in the 21st Century (US version)

EDDIE RAYMOND

We live in an age where most information is available to us at the click of a mouse or keyboard. We also live and work in a world where technological changes have created opportunities to do more things and to jump into doing them sooner. This is a great thing and has allowed unprecedented growth and change in the entertainment technologies sector. It does, however, often impart the feeling that we know what we need to move to the next level. So, given all this access to information and opportunity, in a constantly evolving workplace, what are we doing to ensure that our technicians are properly trained and vetted?

Let me begin this chapter on training with a notion about learning: learning is linear. Over the span of a lifetime there's a beginning, several middles and, occasionally, an end to the learning curve. In the entertainment industries we often think that once we learn the basics we can skip those middle steps and jump right in at an advanced level. However, in doing so, we are denying the complexity of the crafts as well as dishonoring the real experts in the field. This is as much because the crafts in the entertainment industries are "avocational" long before they are vocational, as we have lost the tradition of progressive learning in our industry. We do these jobs because they are interesting and sometimes fun and we don't want to deny ourselves the experience (or the money) just because we have yet to learn the nuances.

There is a metric taught in coaching clinics that explains the levels of learning in a succinct way. There are, in this metric, four levels of competency:

1 Unconsciously incompetent—you don't even know what you don't know.
2 Consciously incompetent—you are aware of your lack of knowledge.
3 Consciously competent—you know stuff, but you have to think about it.
4 Unconsciously competent—you can perform certain tasks without thinking about it.

For example, a person who has never rigged professionally or who comes from a sport climbing background may not know to tie a bowline with a long loop to send a basket and bridle to a high rigger. He's unconsciously incompetent.

Once he's been told to use a bowline, but before he learns how, he's consciously incompetent.

After some practice, he can tie a bowline and knows when to do so most of the time. He's consciously competent.

After rigging for a few years, he no longer has to think about how to tie a bowline or when to use it. He is unconsciously competent.

This is a useful tool when talking with students about the learning curve and progressive learning. It isn't pejorative, but it does illustrate the point in an accessible way.

Another way to think of this is in terms of the foundations necessary to learn certain skills. Jean Piaget was a Swiss developmental psychologist and philosopher. In his theory of cognitive development, Piaget defined learning as mental processes dependent on biological maturation and environmental experience. He posited that an individual was incapable of "owning" a new concept until, through experience and exposure, they "owned" (had a working grasp of) the concepts on which it was built. Until a child had reached a certain biological maturity and been exposed to ideas from which they learned, they could not learn above the level of that experience.

A child who doesn't own the concept of "more and less" can't learn to count. They can repeat the sequence of numbers, but they will not understand the values of those numbers until they own the concept of "more and less" on which it depends. He referred to this as conservation, the idea that things have a value and that the redistribution of things doesn't change their inherent value.

In the same sense, learning to be a rigger (or an electrician or an automation technician) relies on ownership of the concepts that precede the new skill you are learning. You may be able to imitate what someone else has done in building a bridle or hanging scenery on a batten, but if you don't understand the reasons why it was done in a particular method or order, eventually you will make a mistake, and this could have serious consequences. We must be exposed to a number of variations in our crafts in order to own the concepts on which those crafts are based.

Learning is a different experience for each of us. Adult learners in particular are very much set in their ways of learning. Depending on who you read, there are variations on how those ways are defined. I favor David Kolb's theory that there are four primary styles of learning and that each of us operates with a combination of all four. They are as follows:

- **Convergers:** These are people who learn best by abstract conceptualization (AC)—understanding an idea by reading about it or listening to an explanation of it—and active experimentation—doing it. They are good at making practical applications of ideas and using deductive reasoning to solve problems.
- **Divergers:** Those who tend toward concrete experience (CE)—learning by doing while being coached—and reflective observation—watching

someone else demonstrate a skill. They are imaginative and are good at coming up with ideas and seeing things from different perspectives.

- **Assimilators:** These are characterized by those who learn by reflective observation (RO) *and* abstract conceptualization. They are capable of creating theoretical models by means of inductive reasoning.
- **Accommodators:** These use concrete experience (CE) and active experimentation, trial and error to learn a skill. They are good at actively engaging with the world and actually doing things instead of merely reading about and studying them.

Few of us learn purely by using one of the four learning styles. Each of us learns through a combination of these styles, although we usually have one or two that work best for us. There are ways to approximate an individual's learning preferences and illustrations that graphically demonstrate the results. Below is one such exercise.

Exercise Instructions

There are four sets of four descriptions listed in this inventory. Mark the words in each row that are most like you, second most like you, third most like you, and least like you. Put a four (4) next to the description that is *most* like you, a three (3) next to the description that is second most like you, a two (2) next to the description that is third most like you, and a one (1) next to the description that is *least* like you (4 = most like you; 1 = least like you). Be sure to assign a different rank number to each of the four words in each row; do not make ties.

EXAMPLE

happy 4   fast 2   angry 1   careful 3

Some people find it easiest to decide first which word best describes them (4 happy) and then to decide the word that is least like them (1 angry). Then you can give a three to that word in the remaining pair that is most like you (3 careful) and a two to the word that is left over (2 fast).

The person "described" above shows a preference toward learning through active experimentation and concrete experience over reflective observation. This also demonstrates that this person will do poorly if the only option for learning a concept is abstract conceptualization (reading about it). Notice, however, that there is some possibility for learning in each style. The primary point here is not that we have to test each student to determine their strongest learning style, but rather that we cover all four styles whenever possible.

To use a simple example of how this strategy might be employed in an effective training, think about how we teach a stagehand to tie a bowline. There are those who could read or listen to a description of how to tie the knot and figure

EXERCISE

A	Feeling ____	Watching ____	Thinking ____	Doing ____
B	Concrete ____	Reflective ____	Abstract ____	Active ____
C	Experience ____	Observation ____	Conceptualization ____	Experimentation ____
D	Practical Application ____	Modeling Another's Behavior ____	Reading ____	Trial and Error ____
	TOTAL CE ____	**TOTAL RO** ____	**TOTAL AC** ____	**TOTAL AE** ____

Figure 11.1

it out from that alone. Many could learn the knot by watching someone else tie it. Most can learn the knot by being coached step-by-step through the process. Still others might simply look at the knot already tied and figure it out on their own.

However, if in a single class the students are taught using all four learning styles, then the possibility of success would be much higher. In this scenario they would be:

1 handed a sheet of paper with an illustration of the knot and an explanation of its uses and properties;
2 presented with a demonstration of how the knot is tied;
3 followed by a coaching session where they practice tying the knot;
4 finally, left to practice on their own.

If you subscribe to this idea, then any effective training must include opportunities for each of these styles of learning. Being exposed to a variety of

LEARNING STYLES GRAPH

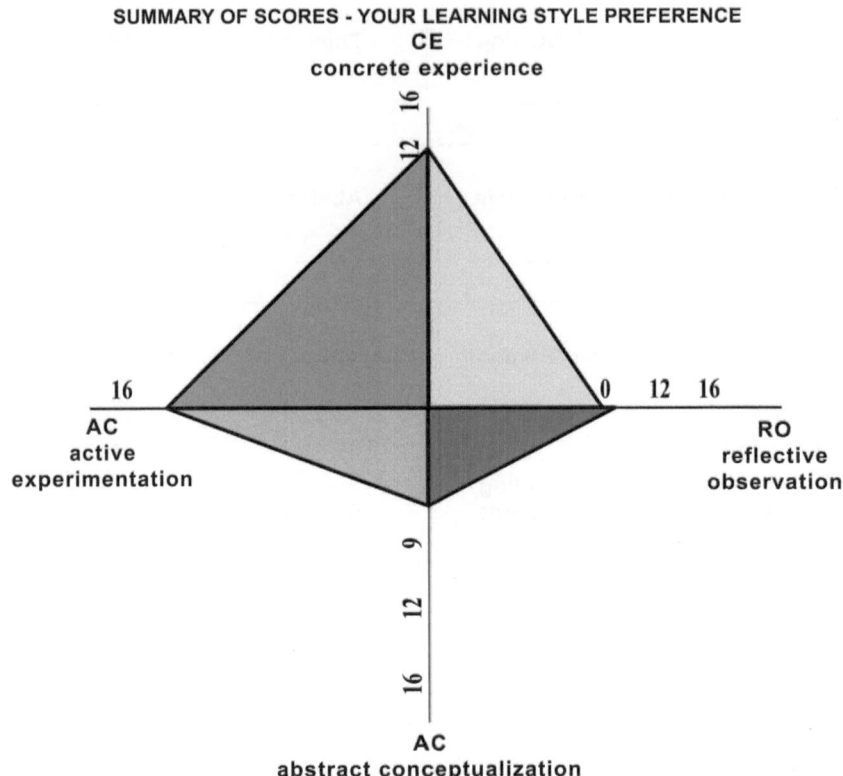

**Place 4 dots corresponding to each of your scores for each learning style.
Then connect the dots. This will show you your learning style preference.**

Figure 11.2

experiences over a prolonged period of time provides us with opportunities to learn in our most effective style or styles. But even with the best teaching strategy there are no shortcuts; skills can't be learned overnight.

OK, enough psychobabble for now. Let's look a little at how the craft of rigging has evolved and where we are in terms of training.

Theatrical rigging goes back at least to the Greeks. Granted, the materials they used were only vaguely similar to what we have at our disposal now, but the concept and problem-solving matrix was the same—gravity makes things fall down, rigging prevents things from falling down, at least when you don't want them to.

In the early days of "modern" theatre, the rigging systems were based on the practices used in the tall ships that sailed the oceans. While we may think that this was a crude way to do theatre, the skills learned by the sailors employed as riggers and flymen in the old hemp houses were learned under the most harrowing and crucial of circumstances. If they screwed up the rigging on a three-masted schooner in a squall or hurricane, people died—and they themselves were among the dead. It was imperative that they understood the entire system of ropes, pulleys, block and falls, and belaying methods in order to survive. Recruiting them to take charge of the rigging in theatres was a logical choice. Their survival proved their skill and worth, and they brought with them a sense of professionalism and competence.

Today there are only a handful of hemp houses left in the country. While you'll not likely ever work in one, learning how things are done in a hemp house has great residual value as you move through your career as a rigger. If you can, go spend some time talking to an old-timer about how it's done. If nothing else, the stories are worth the effort. Telling stories is part of the rich heritage of stagehands around the world.

In the first part of the twentieth century, counterweight rigging systems began to be developed. While these systems were possibly less flexible than the old hemp systems, they were certainly more efficient for the majority of theatrical applications. Fixed line-sets with single operating lines and counterweight arbors with uniformly weighted "bricks" were easier to load, balance, and use than any five-line hemp set with a sandbag and a haul-down line. Load-ins and load-outs were streamlined and more shows hit the road. Just as learning from the old hemp house flyman is valuable, learning to properly use, inspect, and maintain a counterweight system is equally valuable. The lessons are inherent to all rigging.

In the late twentieth century, counterweight systems began a slow decline in popularity as new motorized systems came on the market. We are now in the midst of what will likely be a decade-long transition from counterweight to computer-controlled mechanical systems.

In addition to fixed rigging systems, theatres more and more utilize rigging systems that were developed for touring concert and arena shows by pioneers in the industry such as Roy Bickel and Rocky Paulson. These systems have evolved from older and more traditional elements of rigging to highly specialized systems using engineered truss systems, motorized chain hoists and drum winches with computer controls, load sensors and a variety of other specialized hardware.

The variety of rigging tools in the theatre has never been greater. Many houses use combinations of counterweight systems, mechanical and arena style rigging, and even hemp systems to hang and run a show. Many elements of modern scenery have made this a necessity. For one thing, scenery is heavier. Automation systems that work above the stage require stability that is not always possible with counterweight sets. Lighting units are heavier and more complicated than ever before. Projection systems require specialized rigging as well.

The modern theatre rigger must be familiar with all of the possibilities in order to be at the top of their game.

Which brings us back to training and learning. Remember the basic precept that learning is linear? In a world where a practitioner of rigging in a modern theatre must know at least three and hopefully four systems of rigging, there is a need to develop curricula that ensures that as a young person learns the ropes (pun intended) there are no holes in that learning that may cause problems down the road. So where are these curricula coming from today? Good question!

In past decades, young stagehands often learned from older, more experienced craftsmen (or parents) and were told to keep their eyes open, their mouths shut, and their hands busy. This "mentor" model served the industry well for a very long time, but as the technologies evolved, became more complicated, and the stakes grew higher (both in dollars and in risk) "this is how I have always done it" became a less viable answer to technological problems. The value of mentoring has been eroded, even though there is great value in learning the oral history of how and why the industry evolved from someone who was there.

Universities, with a few notable exceptions, are by design not intended to be craft schools. In the first place, their purpose is not necessarily to develop "blue collar" practitioners of the crafts, but to develop designers, technical directors, and managers. Due to this, and budget constraints, many don't expose the majority of their theatre students to the craft side of the industry in a studied manner. All are required to participate in some craft activities, but not so as to thoroughly learn the crafts. They do their stint in the shop or on the stage to support a production and then go back to their chosen discipline. Which is not to say that a university education for a theatre technician is a bad thing. A person who graduates with an MFA in theatre, who is willing to put the time in to learn the crafts, brings a more developed sense of the big picture and can be a great asset in the collaborative world that is theatre.

The International Alliance of Theatrical Stage Employees (IATSE) has taken a long time to embrace training from the top down. However, under the current leadership, many new training and education initiatives have taken root in the past four years. Many individual locals have developed programs to train young stagehands so as to meet the demands of the workplace and to protect their members and their jurisdictions. Building on an already strong workforce and traditions of old school apprenticeships or learning from the elders, these locals have utilized a number of resources to bring training to their locals. Training Trust Funds have sprung up around the Alliance, and now IATSE itself has a robust Training Trust Fund to aid locals in this endeavor. Each local has developed its own method and criteria, and there is a nascent but coordinated effort to define the skills needed in each craft and to standardize training. Individual trainers are being identified and put into a database so that locals can find the training they need to meet their needs.

As the stakes have risen and it has become more important to know exactly who is being trusted with the work at hand, the industry has embarked on efforts to introduce certification to some of the skill sectors. In North America under the PLASA banner, the Entertainment Technician Certification Program (ETCP) has launched three certifications—Entertainment Rigger Theatrical, Entertainment Rigger Arena, and Entertainment Electrician—so that employers can have some assurance that the people they hire to run crews, analyze jobs, and practice the higher levels of the craft have been vetted in a systematic way that is recognized across the industry here. With more than 1,500 certifications issued to date, this effort is proving to be useful and popular.

In turn, the demand for training so that technicians can study, practice, and eventually qualify to become certified technicians has increased greatly since the introduction of ETCP in 2005. That resources are being developed to deliver the much-needed training is a great sign that the craft is being taken more seriously. Individuals, unions, equipment manufacturers, and production companies are all jumping in to answer the call for more and better training.

What's missing at this point is an agreed curriculum that lays out a progressive order of training from how to tie a knot to how to calculate complicated loads, that includes everything in between and builds skills based on the concept of linear learning. Additionally, there is yet to be developed a way to train instructors in the best methods for adult education (yes, I know, not all stagehands are adults) and to ensure the best success of the majority of students they teach. Many stagehands are far removed from the classroom and training methods must take that into account. We must accommodate all styles of learning.

The Canadian Human Resources Council has addressed the first of these missing elements with the development of a Chart of Competencies for a variety of theatrical jobs, including riggers. This is an attempt to determine what a person must know to be a competent craftsperson, what resources are in place to impart that knowledge and skill, and where there are unmet needs in training resources. I submit that there is a need in the US as well as worldwide to further develop such charts and to develop resources that parallel those findings. It is in our best interest to begin to develop the learning curve, the lessons to achieve that learning and the people to deliver those lessons.

In looking at the requirements of technicians in today's quickly evolving and financially restrictive world, there are other skills equally important to the future success of individuals that need attention. Communication skills are among the least recognized and developed of these. It is often thought that people either are or are not good communicators. While this may be true at face value, the skills employed by good communicators are learnable. Learning how to translate an idea into words, how to assess your audience and how to evaluate your success as a communicator are all things that can be taught and learned.

Learning how to motivate a crew and/or fellow workers is another skill that can be learned. It is most important in the entertainment industries where the hours are often long and the information frequently slim. Working with motivated people can make the difference not only between a good or a bad day at work, but also in the success and safety of a project.

Learning to be a good communicator and to motivate those around you to a high level of performance creates efficiencies and a safer working environment for all involved. These skills make clear the expectations of those in charge and ensure that everyone is aware of the procedures, expectations, and resources of a work site. Especially for lead positions, these are essential skills that can and should be taught.

The entertainment industry has come of age recently and acknowledged that what we do is complicated. Being a professional entertainment technician requires recognizable skills gained through experience and training, and the stakes are high. A concern for the safety of our workers and the talent and audiences we work around has been more in focus of late than ever before.

The response to this acknowledgement has led to a great enthusiasm for more and better training. In order to make this a reality, we need to break the crafts down into bite-size pieces, recognize the progression of skills needed so as to train technicians on a strong foundation, and use diverse training methods that play to the strengths of those learning the crafts. We are blessed with a small handful of great trainers. Now we need to develop a small army of new trainers who understand what is at stake and the best methods to meet our goals.

The future is bright in this regard. The changes in just the past decade or so are evidence that there is a movement afoot that can yield a better trained and more skillful workforce in the entertainment industries. However, there is a great deal of work to be done before we get there and we all need to be a part of that process.

12

Training in the 21st Century (UK version)

CHRIS HIGGS

The phrase 'a steep learning curve' is generally thought of as referring to something that's difficult to learn. In fact, when plotted against time and experience – on the x and y axes, respectively – a steep curve actually more accurately represents a quick process. However, the learning curve for a given task generally levels out once the basics are accomplished, allowing someone to achieve more in a given time. The entertainment industry is usually very fast to react to advances, almost always in response to production or creative demands. Therefore it's odd that rigging has been around for so long, but in many ways, until recently, has progressed fairly slowly relative to other disciplines in the industry.

This learning curve is not so different from the entertainment rigging industry itself! In the UK, rigging has been a stand-alone trade for about 25 years. Once the province of concert lighting crews, the development of touring lighting systems eventually required dedicated crewmembers to take care of the rigging. A twentieth century master carpenter was required, someone with a fairly unique set of skills combining engineering, carpentry, mechanical handling and a keen understanding of the business of show.

This was all quite new as a blend of skills and activities, so nobody knew what worked or had any preconceptions; there were definitely no codes of practice . . .

Structures were required to support first lights, then flown sound systems and scenic pieces. Frequently those structures were suspended on ropes, then wire ropes and chain motors – we'd done that for years over stages, but now they started to encroach on public areas, the auditoria, exhibition hall floors and sports spaces. The principles first employed to get some front lighting on a stage in a hall were being used to transform warehouses and ballrooms into theatres and studios. Sports venues were being used to stage concerts and, before we knew it, using chain motors was as common as using line-sets in theatres had been for 100 years. 'Industrial theatre' (corporates) and the rise of business communications had great influence and effect. Large numbers of theatre technicians were attracted to the even brighter lights and larger budgets of the corporate events. Production houses and the wide range of support services – including rigging and staging – blossomed.

People are people, of course, and some technicians were also climbers. The mind-set of technician blended with climber meant that very few places in a venue were inaccessible. Couple that attitude with the evolution of access machines and industrial fall protection systems and the world was our oyster. We could, and did, hang sh*t anywhere.

The pressures of a touring schedule required speed and efficiency in rigging. Climbing is faster and cheaper than using access machinery. Producers and production managers love a 'can do' attitude – and the smaller budget required. After all, if they can save renting access equipment then that's a bonus, and the skill and courage of the crew meant anything was possible.

The entertainment industry rigger had been born and was often an essential part of a project. Expectations of riggers and from riggers were almost as big as the egos involved.

Riggers were frequently known to and trusted by production managers as someone who had the right attitude (who wouldn't say no). Whether that individual actually understood what he was doing with the equipment was often secondary; if the show went up and nobody got hurt then it was a success. Trust accounts for a great deal when there is risk involved – financial or otherwise – but frankly, reputation was usually the thing nobody wanted to risk. You're only as good as your last job.

Productions got bigger, more riggers were needed; this spawned bespoke rigging companies and venue rigging departments. Soon the 'freelance rigger' emerged, someone who would probably work in every capacity in an average year. So the independent 'head rigger', the rigging company employee, the house rigger and the freelance rigger all became common job descriptions and functions.

Something puzzling is that, if training is so important, why have so few of the riggers we see at assessment had any training? There are very few courses available, but the vast majority have learned what they know on the job. The proportion is increasing slightly but the vast majority of both company and freelance riggers who have also had some formal training are few and far between. It's possible that training is seen as either too expensive or not necessary. Achieving the National Rigging Certificate without attending a training course is perhaps testament to the fact that there are other ways of learning than attending a course. I firmly believe in an apprenticeship route, but it is expensive for employers and in the UK there are many hoops to jump through to obtain funding, presumably to maintain quality and discourage using such schemes to provide cheap labour. It seems that as a society we are pushed down the road of being trained to a standard simply to gain employment. The most worrying reason is the thought that because an individual is already being paid for work as a rigger, that individual believes their competence to be sufficient; they already know all that they need to know, so why do they need any training? As we find at assessment, frequently very experienced and adroit people have no understanding of the fundamentals of statics, for example, or the correct use of fall protection equipment. Riggers don't need to be engineers but they do require a solid grasp of how forces work, what the law says and how manufacturers designed their equipment to be used.

One of the driving forces for me was that I wanted to know things and often couldn't find anyone who could answer my questions about forces, factors of safety and the like. Of course, the more you look, the more you find and it becomes a realisation that a lot of the information was not understood by the industry at all and there was a lot of job protection going on. This is where a Bertrand Russell quotation applies very well: 'Fools and fanatics are always so certain of themselves and wiser people so full of doubts.' Without suggesting that there are too many fools or fanatics involved with rigging, the principle is probably right. 'Ignorance is bliss' would be another way of putting it.

It is a question, then, of who to ask and how to check their competence. In time a network of engineers, suppliers, manufacturers and like-minded colleagues can compile information. Nobody has the whole picture and probably nobody ever will have, but what is happening to arena rigging, now at least, is a more formal approach. Theatre still lags a little way behind in the UK because the landscape is very different.

In the early 1990s, the flagship venue rigging teams were sent to get their powered access training and to the CM 'Motor School'. That was because that was virtually all that was available. The head rigger of the largest exhibition venue (and a great mentor) very wisely sent his riggers on a five-day (industrial) mechanical handling course. Not completely relevant but a very sensible move because it was a recognised, well-established course with the support of large construction companies and there was an element of assessment involved in the certification.

Otherwise there was very little; a couple of theatrical flying courses, a few lifting equipment inspection courses but that was about it.

People learn rigging by doing. They also copy what they assume is correct or appropriate. Sometimes they are right. Often they need little more than confirmation that they are doing the right things. However, the extra layer of quality and competence comes from knowing why.

People need to touch and feel the materials they will be using. Call it a rehearsal. Remove production pressures and the feeling of being scrutinised by peers.

The law in most parts of the world is exactly the same – don't hurt people. Plan work, make sure you are acting reasonably. Be prepared to defend yourself if it all goes wrong. Understand your competence and its limits, know your trade and communicate. Using problem-solving in training (which is really the definition of rigging anyway) will usually level the field of play in terms of age, experience, gender and attitude. Throw people in at the deep end but in a controlled environment where they know they are safe and can't damage anything.

Clearly there are common items in rigging inventories across the globe – the Lodestar, the 20.5" truss, the polyester roundsling and the ⅝" bow shackle. There are many variations in chain clutches (grabs), rigging steels, rings and links, but the really important thing is to ensure the people attending understand what the item is for, how it works and why. This is when the knack to getting people to remember stuff comes in. People like a real life story, and frequently there will be a part of the story that everyone can relate to. Maybe that's why people with experience in the field can deliver the information better than anyone else, even if they are not 'educators' by training. They have probably got a story for every occasion and can deliver it with authority and confidence. It also begs the question about an industry so self-absorbed that there is seemingly no funding or organisation to train trainers. These people are the key to the future.

It's all well and good having found the next generation to employ, but the industry also needs people who can properly deliver the information and skills to them, on the job and externally.

Our people are resourceful and imaginative; a demonstration is often all they need before being given the resources and time to try out the new information. They don't want to sit behind a desk for hours; they get fidgety when grouped around a flightcase for five minutes. They need to be given the tools and the time to figure it out. If they struggle or if it is clearly outside their comfort zone then intervention is called for in proportion to the issue at hand. What people don't need, at least at the beginning of the process, is longwinded theory, legal and safety stuff. That can come later when they see the relevance of it. Things that aren't relevant are the things they won't use or can't practice.

Training on offer in the UK has nothing to do with the National Rigging Certificate (NRC) and is completely unregulated. The providers aren't scrutinised and there is a blind belief that the information being delivered is kosher; the fact that someone is offering training must mean they are competent. Is that information current and valid? There will be a need to accredit training providers soon. It is often the case that content is based on opinion and experience rather than solid facts, especially regarding legislation and standards.

The fundamental part of designing training is that the employer needs to identify what their staff need, not the training provider. Too often 'you know what they need' is the reply when a brief is requested. One way to combat the management's woeful lack of knowledge is to ask the employer to confirm that the syllabus is appropriate to their needs. At the conclusion of the course, the management often assume attendees will be expert on their return. Employers will often say they want as much from the training as possible – 'cover everything'.

The commercial aspects of training are not healthy. We are in an environment where training is still seen as a luxury because the people ordering it probably had none themselves and they have managed perfectly well without it during their careers. If you want to 'learn everything' then take an engineering degree – but that costs even more. Time is money. It is unlikely that the numbers of people requiring rigging training in our industry are ever going to be enough to allow rigging training to be anything more than a short-term course. Taking time out from work is always difficult, especially for freelance workers. Rigging by its nature needs practice in a real workplace situation, so the ideal method of delivery is the apprenticeship style. This requires commitment from the individual and the employer and needs to be structured so each obtains benefit without either side suffering from lack of involvement. It doesn't exclude short courses, they are complementary and can be repeated if necessary. What it does ensure is that the apprentice gets practice and supervision as well as seeing how they contribute to the production or service. However, by definition they are expensive and the employer has to be able to see the outcome as being of benefit. The desire to commit to any long-term arrangement doesn't seem that common these days and young people frequently aspire to be producers, not technicians.

Two fatalities at Earls Court, one in 1999 and another in 2000, and a serious fall in 2004 started to shake things up. At the time people were complaining that riggers were too inexperienced and lacking the basic skills. It was felt in some quarters that much of the industry was working beyond its competence at that time.

A small group of people had already worked together in the late 1990s on a UK project to design a rigging qualification. This was supported by a college and a trade association but was ultimately a commercial scheme. Education is big business. The college registered the course but the lack of numbers was its downfall. There just aren't enough people who need the qualification to justify the expense of delivering the training.

We regrouped and approached several trade associations before finally deciding to work with PLASA on what turned into the NRC. We were determined to produce a nationally recognised qualification that could hold its head up rather than something perceived as a commercially driven exercise or an old boys' network. It had to be transparent (riggers being a sceptical breed at the best of times) and it had to be relevant. The last thing we wanted was something that was imposed by a well-meaning but misinformed government department.

It became apparent very quickly this was going to be expensive and the basic plan was that PLASA would blaze the trail, develop the rigging qualifications and use the learning curve to develop qualifications in the future for other sectors of the industry such as temporary electrical installation.

There were a number of factors that influenced our thinking with the qualification. Probably the foremost in our minds was that we hadn't had any training. It therefore seemed inappropriate to demand that the qualification had to be the result of any training. We wanted the next generation of riggers to be better informed and better skilled than we had been. Most of what our generation learned was by trial and error, through practice good and bad. People who had broken the ice previously then handed down the knowledge. The idea of training riggers would steepen the learning curve – reduce the hours needed to really know something or to be able to perform a task. What it couldn't do was teach people behaviours or how to communicate. Most importantly, it would never be able to teach that most important factor in our business – attitude.

The overarching principle was that, by establishing benchmarks, because we assess against National Occupational Standards, we were producing a known quantity that could be used by employers when recruiting or putting teams together and by venues when vetting contractors working on their premises.

The National Rigging Certificate (NRC) is the only entertainment rigging qualification in the UK and it is starting to gain acceptance in several European countries. It is not just about certification. The NRC is about the recognition of skills that are vital to the rigging industry, which are often overlooked.

The qualification is for life, but in order to allow the industry to self-regulate and ensure riggers are meeting current practices the ID Card expires after five years. This renewal will only involve assessment if there are any significant changes to legislation, techniques or equipment during that time. It does not mean the whole assessment process has to be gone through again. On achievement, the candidate receives a certificate for the qualification and can then apply for the NRC ID Card.

Assessment takes into account knowledge, skill and experience in equal measure – so simply being good at climbing or physics is not enough – it is the combination of the essential skills that make up competence. The body of knowledge required is contained within a comprehensive NRC handbook and candidates are sent this when they register with PLASA Qualifications.

The assessment is based not on the assessor's opinion of a candidate's performance or knowledge, but on National Occupational Standards (knowledge and performance standards set out by the industry). These act as the benchmark by which the assessors make their decisions. Assessors are monitored by qualified internal verifiers based at the assessment centres, who themselves are verified by the awarding body, making it a three-tier quality assurance process.

The qualification is available at two levels. Level 2 is for riggers working independently, not fully supervised. Level 3 is the qualification for someone who supervises riggers. These levels represent the minimum competence required for rigging and are just as applicable to the theatre, concert touring, event, film and TV – anywhere entertainment or event rigging is carried out.

This is holistic assessment and builds a picture of an individual's competence overall, not necessarily requiring outstanding ability in any or all areas of competence, simply meeting the National Occupational Standards at the required standard is sufficient.

At Level 2, practical assessment covers the methods generally used in entertainment rigging. A written test examines underpinning knowledge in combination with verbal questioning. Training records and qualifications further support the candidate's evidence of competence. Importantly, previous experience is demonstrated by verified witness testimonies from employers and submission of rigging plots. After 12 months of appropriate rigging experience candidates are eligible to be assessed at Level 3, should they so wish. Centre assessment involves carrying out a risk assessment and sitting a written paper, submission of paper evidence including risk assessments, method statements, rigging plots, qualifications and training records. Following achievement at the centre, Level 3 candidates also undergo an on-site assessment that must include supervising riggers working at height.

The NRC assessment is entirely separate from training. This was a key feature in its development. We wanted to get away from the idea of doing training and then being assessed at the end of the week, especially by the same person that delivered the training. The NRC rules prevent a trainer from assessing someone they've trained and there has to be a 12-month gap between training and assessment at a facility where the candidate's training was carried out.

For the same reasons, training providers may not advertise their courses as preparing for the NRC, simply that training 'may provide some of the knowledge required at assessment for the NRC at Level 2 or 3' to prevent 'cramming' for assessment. The scheme assesses experience, knowledge and understanding over time rather than being an examination.

The benefits of the NRC have been enormous. It has brought all the major players together, standardised many working practices and generally raised the standards of safety and workmanship. It has created a series of benchmarks and established a credible and mature industry body. The dangers are also many. Managers and venue people often don't understand rigging, so certification simply means to them that the rigger is qualified to do anything that has to do with rigging rather than it being a prescribed set of skills and standards at a particular level of responsibility.

There will always be riggers who satisfied the assessor they were competent but who choose not to use good practice on site. The similarities with driving are the best analogy. We all have to pass our driving test to obtain a licence but look at the standards of driving. Some people display exemplary skills and attitudes, some frighten us to death, but the majority of us have licences.

Can you imagine what it would be like if none of us were tested? A colleague has an expression that sums it up. He says 'just turn up and be average'. Meaning that employers don't need superstars. A good employer will have planned the work and will direct people on what to do and how to do it, the way it's been designed and, importantly, in the way it has been budgeted.

These days many companies are looking to buy in skilled labour on short contracts rather than train in-house because of the cost. It could be argued this is short-sighted, but this industry is notorious for 'just in time' philosophies. Lead times are short, sometimes to the point of lunacy. Promoters know people want the work and use that as leverage to obtain the best value for money. As an employer, you need the flexibility to be able to service those contracts and that will often mean using freelance labour and requiring them to perform a mix of tasks for an intense period. Very difficult to plan and to budget for if you've also an obligation to supervise young and inexperienced apprentices. You need experienced, versatile, resourceful people. Those people are now 45 years old, and there is a dearth of young riggers in the UK. The fascinating thing is that of the many people who have attended training, very few have ended up as riggers. It's always been a contention that providing knowledge and teaching rigging and work at height skills actually puts people off rigging. It is almost a necessary safety valve, to leave people in no doubt that rigging is not an easy option, it's not a job for people who don't want responsibility and it is not glamorous. The reason that someone wants to be a rigger is something none of us can put our finger on. It's almost a calling. On the other side of the coin, people being satisfactorily employed as riggers and being paid as such seem to believe they know all they need to know and don't need any training. We can always learn more, and as a responsible person should be aware of any conscious incompetence and commit to improve that to at least the point of conscious competence.

Certification is one way forward – employers and venues can use it as a benchmark and our experience in the UK is that is has formalised many of the methods we use, it has promoted safe systems of work and it has really sharpened our understanding of fall protection. It is alarming to see how many very experienced riggers coming for assessment bring PPE that is well past its sell-by date and don't know how to use it correctly. That in itself proves that a rigger will spend many hundreds of pounds on gear and work with it on site with nobody picking up on the shortcomings, including employers, co-workers or, alarmingly, venue management – another symptom of a lack of training.

Now the NRC exists, there is a minimum requirement to understand fall protection. Nothing complex, it's very basic stuff but it provides a benchmark and, for the venues, it means that, by demanding proof of NRC qualification, there should be no reason why the rigger shouldn't be competent to carry out the tasks the candidate was assessed against. It is therefore very important that employers know what the essential knowledge and skills are at each level.

Another important factor to recognise is that, like a driving licence test, certification isn't going to stop accidents or make everything safe. What it should do is mean that accidents are less likely because of the increase in awareness, better knowledge and that the people working have all reached at least a basic standard when they were tested. This is when the whole issue of currency or retesting rears its head. Should we be retested? The thinking seems to be that if you are working you will be using the skills and increasing your knowledge, therefore you will keep the skills honed. However, it may be that most people benefit from the threat of a retest every few years to keep their attention on the bigger picture. You can use some skills more than others; while you doubtless remember how to pedal the bike, you may need some practice to ride it without injuring yourself or others. This clearly entails cost and, given the current climate, we are probably a way off being in that position at the present.

In summary, we are in a period of consolidation. The industry has woken up and started to invest in its own future. We have instigated our own certification, won some battles with the safety authorities and in the process started and maintained healthy dialogue with them. We have a number of effective trade associations, we have regular seminars and even a conference where ideas are shared and discussed. We have a number of the 'original' riggers starting to plough back what they have learned. Their apprentices, now close to middle age, are running facilities and companies. It should only get better. Time will tell. We're still on a learning curve.

13

Working Safely at Height
A Common Sense Approach to the Challenges of Rigging in the Entertainment Industry

BILL SAPSIS

Rigging in the entertainment rigging industry got its start in the early 1960s on the touring ice shows. There was little concern for safety in those days and it took about 30 years, and some highly publicized accidents, to get the industry to start adopting a culture that included personal safety. Fortunately, that culture has now taken hold and safety equipment is becoming more commonplace around the country. The trick now is to not only keep this trend moving forward but also ensure that the right equipment is being used, and used properly.

Fortunately, we get some help from folks in high places:

- The Occupational Safety and Health Administration (OSHA) sets the rules for worker safety in the United States.
- The Canadian Centre for Occupational Health and Safety (CCOHS) does the same for Canada.
- The American National Standards Institute (ANSI) provides a framework for trade associations to develop voluntary standards that assist in the protection of workers within their respective industries.

Out of these organizations and government agencies have grown rules, regulations, standards, white papers, articles, and all manner of information that comprises our collective safety awareness. It is the goal of this chapter to distill this information into a digestible size and make it easier for all of us to understand what it takes to stay safe while working overhead.

There are three common terms—fall protection, fall restraint and fall arrest—that are used on an interchangeable basis within the rigging industry. They shouldn't be. They each have their own place in our industry and it's important to understand the distinctions between them.

Fall protection is what you do to eliminate a fall hazard. This should be the first action taken after identifying a fall hazard in the workplace. Taking a fall is no fun, so the best response to a fall hazard is to simply remove the possibility of a fall. A railing placed on the edge of a platform, for example, eliminates the fall hazard and no further action is necessary.

Fall restraint is used when you can't eliminate the hazard but you still want to avoid the falling part. An example would be running a rope or cable from the technician to the back of the platform that is long enough to allow the technician to move around and complete their work but not long enough to let them fall over the edge.

Fall arrest is what you do when neither fall protection nor fall restraint are possible. What you are left with is arresting (stopping) the person's fall before they hit the deck or other obstruction. Arresting a fall is a complicated and dangerous business and most safety experts feel that if you are using fall arrest to resolve a hazardous condition, you have failed to address the problem properly. Most safety experts, however, have not worked an arena concert load-in.

The issue with fall protection and fall restraint is they are situational and site specific and it's very difficult to cover all the potential scenarios and permutations in any sort of detail. Fall arrest, on the other hand, lends itself to general discussions and is something we can discuss in more global terms and still cover the bases.

First, the Rules

OSHA provides the rules, which can be found in 29 CFR Parts 1910 & 1926. However, these rules are not designed specifically for the entertainment industry and many times we are left to our own devices to sort out how to stay safe while we work.

First there's the 6' rule. This rule states that if you have the ability to fall 6' or more, you must be protected from injury or death. Using fall protection to eliminate the hazard would be ideal, as would fall restraint. When these don't work, fall arrest is required. This means that you have the ability to fall but the safety equipment will stop (arrest) your fall before you hit the deck or some other object.

It's important to remember that the 6' rule requires you to be protected from hitting anything that can hurt you. If you are on a catwalk 60' above the stage and the catwalk doesn't have a railing, you need to protect yourself from hitting the deck should you fall off the catwalk. However, if there's a plaster ceiling 3' below the catwalk, then you have to protect yourself from hitting that ceiling. Obviously, if your fall is arrested before you hit the ceiling then it follows that you cannot hit the stage floor either. The point is, take nothing for granted. Evaluate the entire situation before determining the best course of action.

More Rules

All fall arrest equipment, when used by one person, must maintain a minimum tensile strength of 5,000lb (22.2kN).* Tensile strength means that the equipment could fail at 5,000lb but not at a force less than that. Not only must each component maintain this strength level but also the way each component is joined to the other parts of the assembly.

Then there's the one-time-use rule. Should you take a fall and survive, all of the safety equipment that was involved in arresting your fall must be removed from service. Most of the equipment—the harness, lanyard and lifeline, for example—should be discarded. Other elements of an arrest system, a self-retracting lifeline, for example, can be sent back to the manufacturer for repair and recertification. The reason for the one-time-use rule lies in the math of arresting a fall. Lots and lots of testing, combined with real world experience, has shown us that it takes, on average, about 3,000–3,300lb of force to arrest a human free-falling 6'. Knowing that the equipment is designed to resist failure up to 5,000lb, you are well protected the first time you fall. However, if you continue to use the same equipment and you have the misfortune to fall a second time, your equipment may no longer have the available strength to arrest that fall.

We could eliminate this rule by simply making the fall arrest equipment stronger. We don't do that for the simple reason that in making the equipment stronger it would become so big and bulky you'd look like the Pillsbury Dough Boy and you wouldn't be able to work. As a result, you wouldn't wear the gear at all. The one-time-use rule exists to help ensure that you actually do wear the gear.

Some of the other rules appear, on the surface, to be a bit mundane, but they're just as important as the more exciting ones.

First and foremost, your employer is responsible for providing you with the safety equipment. Not just fall arrest equipment, but all safety equipment. Please note that I said *providing* you with the safety equipment. I did not say they could sell you the equipment or make you purchase it on your own. You do not have to pay for the safety equipment that you need to conduct the job you've been hired to do.

The employer is responsible for proper training in the use of the equipment. Handing you a harness while saying "here, put this on" is *not* training. Make sure you know exactly how to wear and use the equipment you are being given. Never forget that it takes at least 3,000lb of force to arrest a fall and you don't want that

---

* kN (kiloNewtons) are the scientific (and appropriate) way to measure force. Riggers tend to not be appropriate about many things and measuring force is one of them. For reasons beyond the scope of this book, we'll continue to measure force in pounds.

force impacting on your body in the wrong place because you weren't wearing the gear properly.

The employer is also responsible for keeping the equipment safe by storing it in an appropriate location when it is not in use. The floor of the loading bridge or catwalk is not an appropriate location. A storage locker in the shop is.

Finally, the employer is responsible for maintaining records of the equipment and training. This is a government program, after all.

Your job, gentle reader, is to wear the equipment whenever it's required and abide by the training you have received. You may think this sort of thing would go without saying but roughly 50% of the fatalities from falling in the entertainment industry over the past 25 years have involved a technician who was either wearing the proper fall arrest equipment but did not use that equipment or was wearing the wrong equipment. Do not follow in their footsteps.

A cautionary note: in other industries it's relatively simple to identify the employer as typically they're the one signing the paychecks. That's not always the case with us. When your paycheck comes from a payroll service in Des Moines, the employer may be a little more difficult to identify. Ambiguities and confusion with things like identifying the employer will create problems for you should a legal issue be raised and you end up in court. Lawyers love to play in loopholes but it's in your best interests to keep those holes plugged as best as possible.

The Gear

Safety harness. Fall arrest harness. Full-body harness. Five-point harness. These are the most common terms used for the primary defense against becoming a brown spot on the floor with a chalk outline. No matter what you call it, a fall arrest harness must meet specific specifications for it to work properly and be approved for use. Fortunately, these specifications make sense and do exactly what we need them to do to protect ourselves.

It's practically impossible to find a fall arrest harness in the US that doesn't meet the OSHA specifications but, just to be sure, a quick review is in order.

When using a harness that claims to be approved for fall arrest, first check the label. If the harness simply says that it meets OSHA requirements but does not add a reference to a particular ANSI code or OSHA regulation number then it does *not* meet OSHA requirements. The label must identify a specific standard that the harness has been built to meet. There are a number of standards that are acceptable. An example is ANSI Z359.1-2007. So remember: no standard number on the label means that the device is not approved.

The harness must be a full-body style harness with leg and shoulder straps. In a suspender-style harness where the shoulder straps come straight down to your waist, there must also be a chest strap connecting the two shoulder straps. This is to ensure that you cannot fall out of the harness should you end up

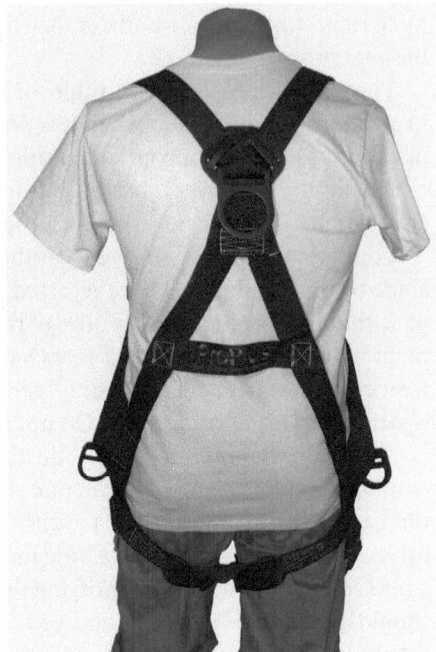

Figure 13.1 Basic harness, front   **Figure 13.2** Basic harness, back

hanging upside down. The design of crossover style harnesses inherently solves this problem.

Please note that body belts and climbing harnesses are not permitted, nor are they safe. Because the tie-off point on this equipment is down around your belly button, the arresting force is transmitted through your stomach and over to your lower back and it is very easy, and quite likely, that you will break your back should you fall when wearing a body-belt or climbing harness. If that's not enough of a deterrent for you, there is a very good chance that once you stop bouncing around after the fall you will be left hanging upside down. That position is extremely uncomfortable but will be resolved in just a few moments. Because neither a body-belt or climbing harness has shoulder straps it is altogether likely that, once you stop bouncing around, you will fall out of the harness.

Once you've established that the harness you have is OK to use, the next thing is to make sure it fits you properly. Just as wearing the wrong type of harness can kill you, so too can a harness that doesn't fit properly. Some points to remember:

- The harness should be snug around your body but should not restrict your movements.
- The leg straps should be pulled up into your groin, snug against your leg and away from any personal bits you don't want damaged. Loose or

poorly placed leg straps will do serious and possibly permanent damage to technicians of either gender. You don't want to think about the consequences should you ignore this advice. Make sure your harness fits properly.

- The base of the dorsal ring (where the harness webbing weaves through the dorsal ring) should be even with the centerline of your shoulder blades.
- If your harness has a sternum (front) ring it should be located between your belly button and the bottom of your rib cage.

A cautionary note: if you wear a crossover-style harness, make sure that the point where the shoulder straps cross on your chest is low enough and far enough apart so they do not press against your neck. If these straps are not positioned properly and you do have a fall, you may compress your carotid arteries and restrict blood flow to the brain. It can take just a few seconds for brain damage to occur if the brain isn't getting enough blood. (Still want to be a rigger?)

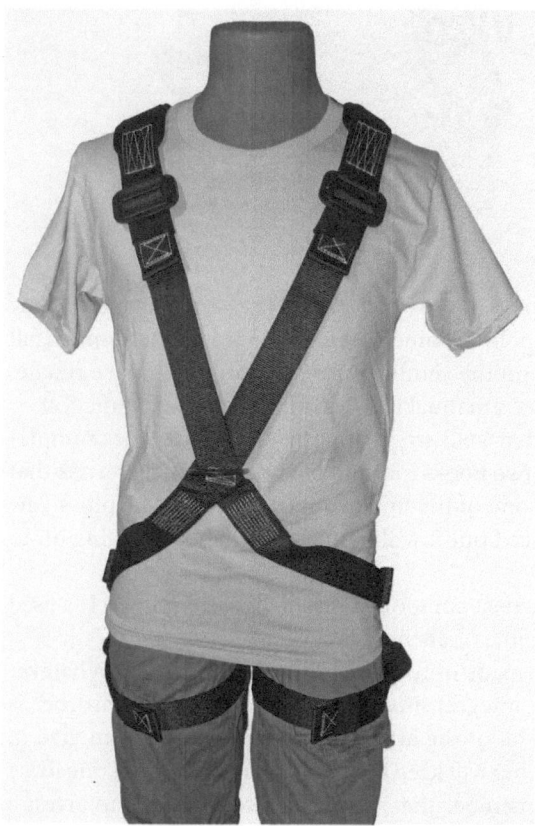

Figure 13.3 Crossover harness, front

Figure 13.4
Crossover harness,
back

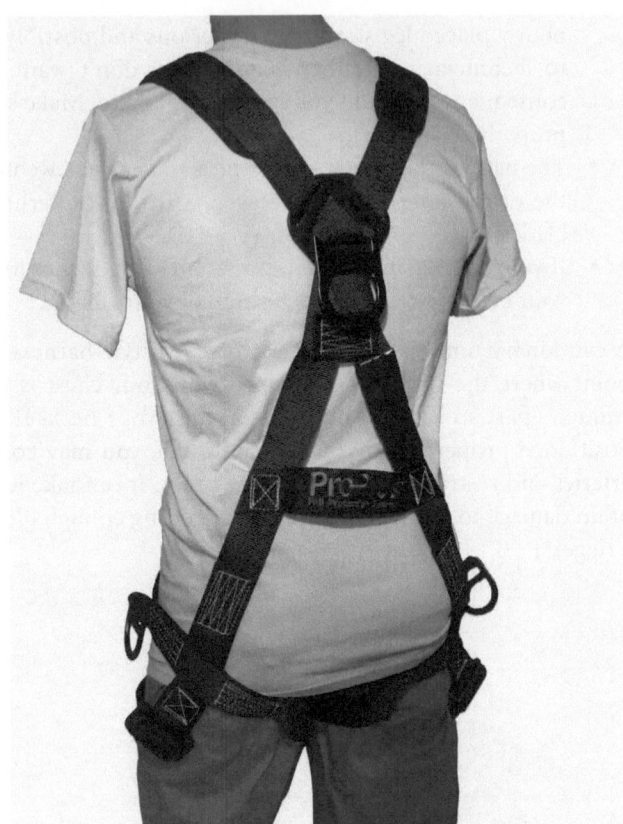

Your harness may have other rings, loops and hooks on it. More than likely, they are *not* fall arrest connection points. Some rings are for work positioning (fall restraint) and others, usually on the shoulders, are for confined space rescue. You should consult your owner's manual to determine each item's function.

Only one hook is permitted in a fall arrest ring (the dorsal ring, for example) at a time. Should you fall with two hooks mounted in one ring there is a risk that the arresting forces will push one of the hooks out of the ring. Murphy's Law says that if a ring does get forced out it will be the one you were relying on to protect you.

A lanyard is commonly the next component in a fall arrest system. It's used to connect the harness to a lifeline or anchorage point.

A fall arrest lanyard can be made of rope, webbing, or steel cable. Whatever the material, it must have an integral shock absorber. The shock absorber is your BFF and should receive all of the attention and respect you can give it. When you fall, the shock absorber will keep the lion's share of the arresting force from reaching your body. Remember the 3,000lb of force required to arrest a

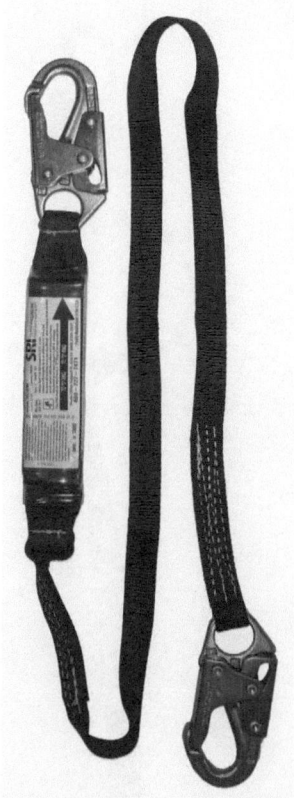

Figure 13.5 Single-leg lanyard

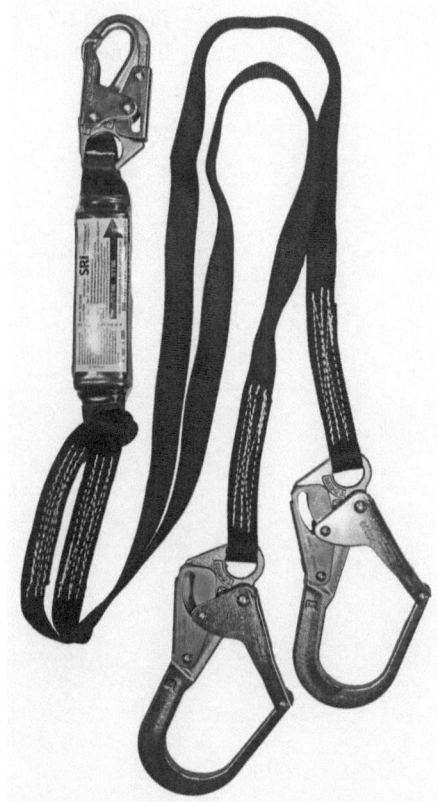

Figure 13.6 Double-leg lanyard

fall? The shock absorber on your lanyard can reduce that amount to less than 1,000lb.

Many riggers tend to think that lanyards are available in only 6′ lengths. The truth is that lanyards are available in almost any length up to and including 6′. The key is to have the right length for the job. Remember, the shorter the lanyard the shorter the fall. The shorter the fall, the less force required to arrest that fall. Less force means less pain for you. If you only need a 4′ lanyard you should only use a 4′ lanyard. No point in taking a fall that's 2′ longer than necessary, right?

It's also important to remember that as it deploys, a shock absorber will add up to 3′-6″ to the overall length of the fall. If you are standing on an 8′ tall platform, a 6′ long lanyard isn't going to do you much good. Yet another reason for using a lanyard of the appropriate length.

Self-retracting lifelines (SRL) are typically used for protection in ladder applications. They are reliable and extremely efficient, usually activating before the worker has fallen 1′.

Figure 13.7 Self-retracting lifeline (SRL)

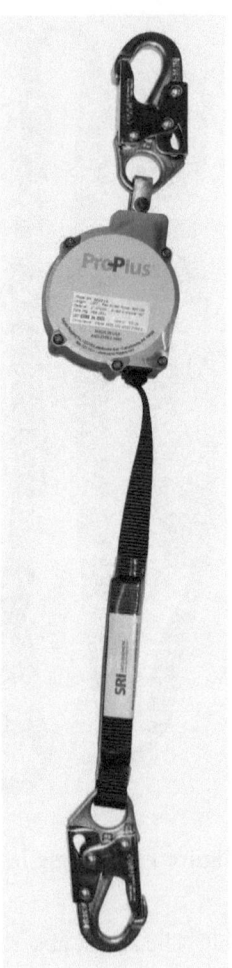

They function pretty much the same way as the seatbelt on your car. They use cable or webbing to connect the worker to the device and, when that cable or webbing is pulled out quickly enough (about 5′ per second), their centrifugal braking mechanism engages and locks up the line, thereby arresting the worker's fall. There's no subtlety in this action. The line runs free until the speed is reached and then . . . instant stop. This is why it's so important to use these devices properly (something the entertainment industry hasn't quite figured out yet).

For any number of reasons, none of them good, SRLs are typically placed in the wrong location. This is particularly true when it comes to truss-mounted wire rope ladders. It's important to understand that an improperly placed SRL may lead to a catastrophic failure in the safety system.

Here are two rules that will help you avoid said failure:

1 An SRL must be above the technician at all times.
2 The SRL must be located such that, after a fall, the technician can reach whatever it was they were climbing when they fell.

An SRL locking mechanism can only work when it senses the line being pulled out at approximately 5′ per second. If the technician is situated above the SRL when they fall, the SRL won't know it until the technician has reached a spot lower than the device and the safety cable starts to pull out. This could mean a significant fall distance. As there is no shock absorption in a standard SRL, any fall greater that 6–12″ is going to really hurt and could cause serious injuries.

The same effect might happen when the SRL is located too far away, laterally, from the ladder the technician is climbing. In addition to the extended fall distance, there's also the possibility of the technician hitting something as they swing back and forth, eventually coming to a stop under the SRL. The end result is an injured worker who cannot rescue themselves because they can't reach the ladder. You now have to go get them down. Where's the fun in that?

The recent development of smaller, lighter SRLs has led to an increase in their use, especially in unorthodox locations.

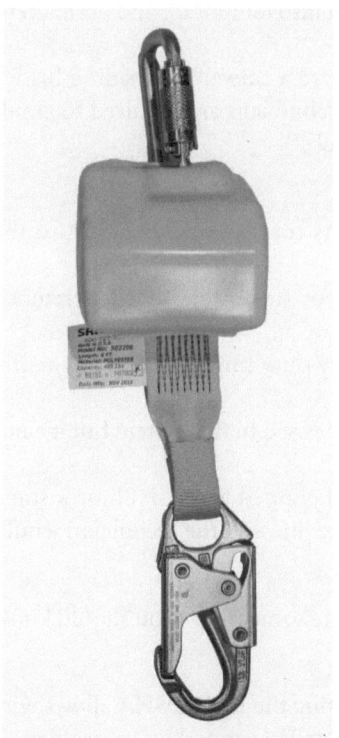

Figure 13.8 Mini-SRL

Of particular interest is on horizontal lifelines (HLL). Slap an appropriately rated cable trolley on the HLL, hang a small SRL from the trolley and you get a safety line that not only moves along the catwalk or truss with you but also engages fast enough to actually eliminate the fall. This is great, especially in places where falling, arrested or not, is a really bad idea. A counterweight loading bridge, for example, typically doesn't have an OSHA approved railing on its offstage side. Even if it did, loading counterweight in and out of an arbor usually requires you to lean over or through the rail, placing your center of gravity outside of the railing. A lifeline is clearly needed here but a 6′ lanyard will allow the technician to fall far enough to land on top of nearby counterweight arbors. Ouch! If they use a shorter lanyard then they may not be able to reach the arbor. Enter the SRL. It is robust enough to allow the technician to complete their work and sensitive enough to engage fast enough to stop the technician from hitting an arbor top.

The final safety system on the agenda is the horizontal lifeline (HLL) itself. There are so many variables when designing, installing, and using HLLs that many safety officers simply won't allow them in their venues. Many engineers refuse to design them. However, riggers, electricians, audio crew, and other entertainment technicians still have to walk on truss, catwalks, and loading bridges on occasion, so solutions need to be found. To help make our HLL discussion a little less confusing, let's break it down into temporary and permanent HLL installations.

A permanent HLL is typically mounted above a catwalk or loading bridge that either has had its railing removed or the technicians are required to climb over or through the railings to complete their job.

The components of a HLL will include:

- Anchorage points. These are the brackets that connect the HLL to the building structure.
- The lifeline itself. This will be a length of steel cable that is stretched between the two anchorage points.
- Tensioning device. This is usually a heavy duty turnbuckle permanently mounted in-line with the HLL.
- A heavy-duty shock absorber is sometimes used in this system but it's not required.
- The technicians can either use a lanyard clipped to the HLL or a small SRL/trolley assembly can be placed on the line and the technician would clip directly into the SRL.

Regardless of how you attach to the HLL there are some things you should know about the installation:

- Do not use wire rope clips when terminating the HLL. OSHA allows wire rope clips as long as they are sized and installed properly. The problem is that wire rope clips require annual maintenance or they could loosen over time. Let's face it, no one ever does this maintenance so it's pointless to use clips. Use a compression sleeve instead. A compression sleeve provides a 100% efficiency rating for the wire rope termination and requires no maintenance after installation. Ever.

- When tensioning the cable please keep in mind that you are not installing a tightrope for a circus act. There is no need for the wire rope to be tightened to a point where it is exerting a significant force on the anchorage points. A small dip in the cable is fine.

- Do not use ¼″ Ø wire rope. At 7,000lb tensile strength, the wire rope has an OK rating, but this is a permanent installation that will probably not be inspected or maintained for many years. Over time you could lose enough rating in the wire rope that you might run into a problem. Spring for the extra $0.15 a foot and go with a larger diameter wire rope.

- To shock absorb or not to shock absorb? It all depends on the situation. Using a shock absorber will increase the fall distance and in some cases that might mean the difference between a worker hitting something below, or not. You will have to evaluate your situation and decide what works best. Keep in mind the two mantras of fall arrest: limit the amount of force that is applied to the workers' body and make sure the worker cannot hit anything when they fall.

Temporary HLL are trickier. A lot trickier. The most common place you're likely to find a temporary HLL is on a lighting truss. The HLL is there because someone will eventually have to be on that truss to focus a light, run a followspot, or maybe conduct an emergency repair on a light fixture. In any case, the dynamics of the situation are complex.

The elements of a temporary HLL are the line, the anchorage points, a tensioning device and a shock absorber. Usually the line is a synthetic rope. A ⅜″ Ø

Figure 13.9 Temporary horizontal lifeline (HLL)

double braid is the most common. The tensioning device is an adjustable rope grab; one that allows you to pull the slack out of the line and locks in place so the line cannot run backward through it. Shock absorption is a must on a temporary line and the most common shock absorbers look and work just like the ones found on most lanyards.

Finding appropriate anchorage points on a truss is difficult. Keeping in mind that the minimum tensile strength of 5,000lb must be maintained throughout the system, where on a stock piece of truss can you tie off the line? There isn't a single point on a truss that has that kind of rating. The best solution is to attach the line to the truss in multiple locations. One popular method calls for round-slings to attach the HLL to the bottom chords of the truss in four places at each end of the HLL. That gives you a total of eight connection points with which to absorb the shock of someone falling off the truss.

The other thing to remember is that the synthetic line will stretch a lot more that wire rope and that will add to the fall distance. There's also a shock absorber in place that will also increase the length of the fall. When using an HLL on a truss please make sure you have at least 20′ of clear space below the truss to make sure the person falling off the truss doesn't hit anything under the truss.

Given the multitude of horizontal lifeline and truss configurations coupled with the enormous forces that can be developed when arresting a fall from one, great care should be exercised when working with a horizontal safety line. You should always have a professional engineer work out the details, including the calculation of the anticipated forces, before attempting to install one of these systems.

While we are on the subject of horizontal lifelines on truss, we should discuss a common problem faced by riggers and electricians who, as part of their job, have to climb a wire rope ladder to the truss and then walk along the truss to reach their final destination.

This business of climbing a ladder and then walking on a truss involves two separate safety systems. An SRL is usually used for the ladder climb and an HLL with a lanyard is used on the truss. The SRL clips directly to the harness but what do you do with the lanyard? You can't clip it into the dorsal ring because the SRL is there already and you're only allowed one hook in a ring. Carrying the lanyard with you and clipping in when you reach the truss requires you to first unclip the SRL and then clip in the lanyard. The dorsal ring is not all that easy to reach and doing all this while sitting on a truss 43′ in the air isn't the best idea you've ever had. The solution lies with the harness.

Some harnesses come with an additional dorsal ring, usually on an 8–9″ web extension.

The secondary dorsal ring is for the SRL and your primary dorsal ring takes the lanyard. Here's how it works:

1 Connect the lanyard hook to the standard dorsal ring on your harness. Put the other end of the lanyard in your pocket, clip it to a convenience ring on your harness or let it hang free.

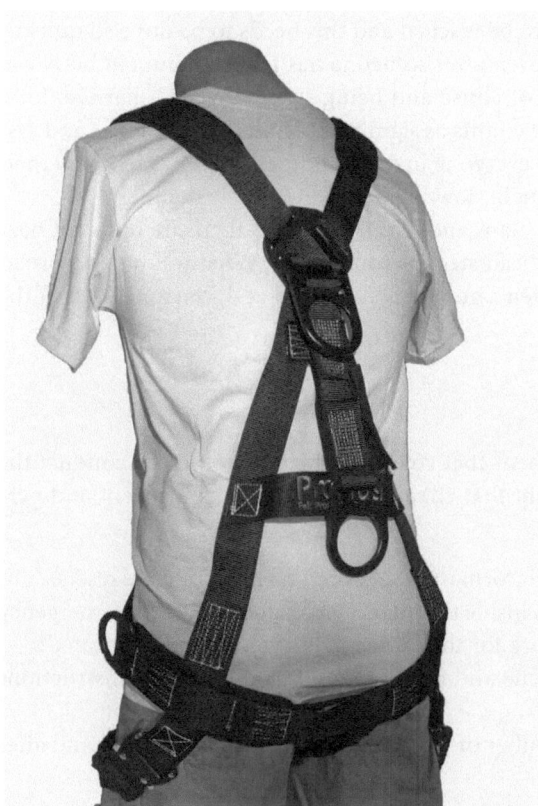

Figure 13.10 Harness with additional dorsal ring

2 Connect the SRL hook to the secondary dorsal ring with the web extension.
3 Climb the wire rope ladder.
4 When you are sitting comfortably on the truss, take the lanyard hook out of your pocket or wherever you stored it and connect it to the HLL.
5 Disconnect the SRL from the secondary ring (it's easy to reach because it's on that web extension) and walk out on the truss.

Once you are finished your work then you reverse the steps and make your way safely down to the deck.

Rescue

Which brings us to the last, and arguably the most important, topic in this chapter. Rescue.

Having safety systems in place for all of your people who work at height is a wonderful thing. But there's more to it than just arresting the fall. Once the worker has fallen they need to be rescued and this needs to be done so quickly. Many things can go wrong even after someone has fallen. A human body can only take a certain amount of abuse and being suspended in a harness for a relatively long period of time counts as significant abuse. Rescue plans and systems have to be in place and everyone involved in the rescue has to know their role if the victim is to be brought down in a safe and timely manner.

There are as many rescue plans and requirements as there are different hazardous locations in the overhead steel of your venue. What follows is a broad outline of what is needed when a good day turns bad and someone falls off the high steel.

The Rescue Plan

This plan is a written document that covers everything needed to conduct the rescue. It is a living document that changes as situations, equipment and personnel change. It contains:

- the names and contact information for everyone needed in the rescue (this includes not only those inside the building but also those in the emergency response units who work for the municipality where you are located);
- the location of the rescue and first aid equipment as well as instructions for their use;
- the roles that each member of the rescue team must play during and after the rescue;
- the documentation on the training of the staff and the outside organizations (EMTs, high-angle rescue team, etc.) who will be involved in the rescue.

The Training

Everyone who works backstage must be trained in the rescue plan. They should be trained in as many of the roles as possible because you won't know who will be in the building when the accident happens. This should be hands-on training. A classroom setting can only go so far in preparing people for the real thing. Set up scenarios based on the requirements of your venue and run drills until everyone involved can take on any role in the plan. Time the drills. Real-time drills can reduce the rescue time by 50–75% even after just two or three drills.

The Rescue Kit

The contents of the kit will be determined by the needs of your venue. Have extra pieces of hardware. In a real-time adrenaline-pumping accident situation, many things will go wrong. Carabiners and shackles will get dropped. Rope will get tangled. Make sure you have enough gear to cover as many possibilities as you can. Then get a little more.

The Roles

It's easy for panic to set in once an accident takes place. Proper training and rehearsals help reduce that panic. Knowing what is expected of you and your coworkers helps reduce the panic level even more. In an ideal situation and assuming an assisted rescue, the rescue team consists of five people.

1 The leader: This person is in charge. They keep an eye on everything and everyone else.
2 The announcer: This is the person who alerts the venue staff and the outside organizations that an accident has happened and maintains contact with these people.
3 The rescuer: This is the person who goes up to the victim with the rescue kit and sets up the rescue equipment.
4 The floor person: This person handles the actual rescue. Their role will depend on the type of equipment being used but it's their job to make sure the victim is lowered to the deck safely.
5 The communicator: This person has the job of staying in constant contact with the victim. The communicator keeps the victim informed of the progress of the rescue and what the victim can do to help promote that progress. The communicator will also continually assess the victim's physical and mental condition and alert the leader of any changes in those conditions, no matter how minor.

Ideal situations seldom occur, however, and it's incumbent on the staff to make sure everyone knows and understands all of the roles. Murphy's Law says that an accident will happen only at the worst possible time and in the worst possible location. Proper planning and training will cover anything that Mr. Murphy throws in your path.

While all rescues are different, there are a couple of rules that you can follow no matter what the situation.

1 Never work alone when working overhead. There should be at least one other person in the immediate vicinity that can begin the rescue process. If you are working above the stage and the only other person in the building is working in the box office with the door closed and the radio on, that person is not going to be of any help to you. The second person should have clear line of sight and hearing.
2 Always lower the victim to the floor, even if the catwalk they just fell from is only 5′ away. The goal is to get the victim safely on the floor. You don't want to have to carry them across the catwalk, down the ladder, around

the spiral staircase and down five flights of steps. Any injuries they may have sustained in the fall will only be exacerbated by a protracted trip through the venue.

3 Once the victim is on the stage do not attempt to diagnose their problems unless you are trained to do so. A medical person of some stripe—be they EMT, paramedic or something else—should have been notified as part of the rescue plan and should already be on site. Let them do their job.

4 Do not let the victim convince you that they're OK. Do not let them wander off to shake it off or go home. Internal injuries are not an uncommon result of a fall, whether the victim feels them or not. Keep the victim calm until the medical personnel have made their assessment and then follow their instructions.

Summary

In an industry that seems to change faster than the weather, it's nice to know that there is at least one constant in this business. Gravity works exactly as it always has and it doesn't look to be changing any time soon. Defeating gravity, or at least amicably cohabitating, need not be the dangerous occupation that it once was. But it remains a challenge. And who doesn't like a challenge now and then?

Staying safe while working at height requires skill, training, experience, respect, and a decent dose of common sense. Fortunately, it's no longer a case of "Climb up there and hang this hoist. Don't fall." We now have safety gear to help protect us and, yes, on occasion, save us from ourselves. Whatever the intent, the result can be the same. Work hard. Enjoy it and come back and do it again tomorrow.

Happy rigging.

14

Medical Issues in Fall Arrest/Rescue
Orthostatic Intolerance, Harness Trauma, and Fall Arrest/Rescue-Related Medical Issues

CARLA D. RICHTERS

And now for something completely different . . .

This chapter addresses the point when the human/rigging interface goes wrong. None of us likes to think about this. However, quickly recognizing that something bad is happening and having a plan to do something about it greatly increases the chances of surviving the event.

An unarrested fall is an ugly thing. The ballistics of a body's fluid post fall is more than nine meters, and is described as the splash zone for a reason. Even after a professional decontamination, small dried droplets will be found for years. Think glitter.

The discussion in this chapter assumes an arrested fall. If you are not wearing a harness in conjunction with a properly designed fall arrest system and with a crew well versed in the rescue then this chapter cannot help you. Once, when walking past an open road box I noticed a photo of one of the most adorable children ever. On the bottom of the photo was this caption: "Dakota says, 'Wear your harness!'" Enough said.

"Orthostatic" means to stand upright. The definitions vary as to whether the "static" part of the word includes "stillness" as part of the definition. There are some animals that stand on their hind legs to browse or as a form of display. As a rule, these animals do not walk upright. That particular party trick is a relatively new development on the human evolutionary timeline. We are, as a species, still working on it. The second word "intolerance" speaks to exactly that. There are situations when we physically do not cope very well.

You may have heard of other terms used for orthostatic intolerance such as orthostatic insufficiency or orthostatic hypotension. They all mean basically the same thing. The heart cannot pump enough blood to maintain the blood supply to the brain. Similarly a search for "parade ground faint" on YouTube will result in hundreds of video hits for soldiers, sailors, and marching band members lopping over. The same reaction occurs in the costume shop when actors lock their knees during a long fitting. The blood flow back to the heart is restricted and blood pools in their lower legs. They are encouraged to walk around periodically and to sit down if they feel lightheaded. This helps to redistribute the blood from the lower legs; it gets the blood back to the brain, and keeps the actors from taking a nosedive.

From the brain's perspective, if it can get the body more horizontal, the heart will be able to reestablish that blood flow to vital organs. The solution, then, is to pass out and fall down. Your body solves the blood flow problem and you come to.

Harness or suspension trauma describes the same grouping of signs and symptoms with blood pooling in the legs until there isn't enough blood available to supply the brain with oxygen and nutrients. Except in the case of harness trauma, the bodies aren't allowed to fall down and self-correct in that they cannot become horizontal, because they are being held upright by the harness.

Almost everyone has experienced the brief lightheadedness of standing up too quickly, rising past their available blood pressure. Bending at the waist helps to increase the blood flow to the brain. This brief lightheadedness and vague sense of nausea are the harbingers of orthostatic intolerance.

Let's take a moment here and talk a little about the implications of orthostatic intolerance. Plain and simple: it can kill you. Even after the fact, we don't understand all of the ramifications of this brisk return of circulation to the tissues or reperfusion syndrome.

We are a group of people making our living by using our brains and strength to alter our world. We use physics to make magic for the audience. We are often pushing our own laws of physics working in stressful situations, against a time deadline, in need of just that one tweak before the end of the call, the start of rehearsal, the well-earned break. Orthostatic intolerance is physics as well. It can sneak up on you, or it can hit you like a hammer. We need to pay attention to the signals and to understand and respect our physical reaction to even a relatively simple arrested fall.

The onset of symptoms happens quickly and unpredictably. There is no marker to predict who is more at risk, or how long it will take to become symptomatic. The recommended rescue time is ten minutes, fall to floor. Since symptoms can begin much sooner, you may very well be working at an extreme disadvantage.

These symptoms cascade, occurring in no particular order, tumbling over each other, and may include:

- cool clammy skin
- shortness of breath
- loss of judgment
- dizziness
- nausea/vomiting
- confusion leading to panic
- syncope (fainting)

which may lead to:

- obstructed airway
- seizures
- cardiac arrest
- death.

While there are no predictable signposts for the development of orthostatic intolerance, other than the fall, there are several factors that can contribute to its development.

Any situation that leaves you dehydrated can, and will, contribute. Possibilities include some medications, working in the heat, untreated hyperglycemia, vomiting or diarrhea, and hangovers. If you stay hydrated then there is less of a possibility for the relative hypervolemia of orthostatic insufficiency. Health care professionals recommend keeping your body's fluid volume at appropriate levels. This helps ensure there's enough blood volume to deal with minor insufficiencies.

What is Happening

Short answer: the body is going into shock.

Long answer: Shock, aka hypo-perfusion, is the circulatory system's inability to provide nutrient-rich blood to the tissues. In order to compensate for any sort of shock you need a number of things: enough oxygen in the atmosphere, intact and functioning organs, and the ability to offload the nutrients on the cellular level.

During an orthostatic intolerance event the body is lacking the circulating blood volume. Fluid volume is drawn down as the blood pools in the legs. The veins in your legs expand to collect the pooled fluid rather than push the blood past the mechanical impedance of compressed vasculature and gravity without the assistance of the efficient muscle pump of the legs.

When the body is under stress, the sympathetic nervous system swings into gear. It produces adrenaline, and this kicks in the "fight or flight" reaction. Push the body too far down this response path and it moves into uncompensated shock, which is when you see the signs of panic that are the classic hallmarks of the onset of suspension trauma.

But, wait a minute! No one said anything about blood loss . . .

Not yet anyway.

Anatomy and physiology

Anatomy is how something is built. Physiology is how it works, and why. When you are looking at something going wrong, it helps to understand how that something is put together and how it's supposed to work. This is true regardless of whether it's a rigging system or the body. The below will give us a basic working vocabulary so we speak the same language; we'll know what we're talking about. Here's the basic list:

Orthostatic: to stand upright on the back legs

Intolerance: (in the context of this chapter) the heart and lungs cannot keep up with the demand placed on them—they don't tolerate or support a particular position

Harness trauma: the constellation of signs and symptoms surrounding hanging vertically after an arrested fall.

Symptomatic: showing or feeling the effects of a physical process. Later on in this chapter there is a list of symptoms of orthostatic intolerance; in this chapter "symptomatic" refers to showing or feeling the effects of orthostatic intolerance.

Hypervolemia: "hyper" means over or above; "volemia" refers to the volume of fluid in the system. In this chapter, "hypervolemia" refers to there being too much fluid in the vascular system.

Hypovolemia: "hypo" means under or below; "volemia" refers to the volume of fluid in the system. In this chapter, "hypovolemia" refers to there being too little fluid in the vascular system.

Vascular system: the arteries, arterioles, capillaries, venules, and veins that carry blood.

Hypo-perfusion: "hypo" means under or below; "perfusion" refers to supplying the body's tissues with oxygen and nutrients. Hypo-perfusion is the inability of the bodily systems to supply the body. It is the definition of shock.

Pulmonary system: "pulmonary" refers to the lungs, so the pulmonary system is its structures and functions.

Venous system: part of the vascular system. This refers specifically the pipes that carry blood back to the heart and lungs. Blood in the veins has delivered its oxygen and then picked up the waste products from the tissues.

Cardiovascular: the heart and the veins and arteries.

Cardiopulmonary: how the heart and lungs interact.

Central nervous system: the brain and the spinal cord.

Several structures require a little more in-depth discussion.

The veins are the body's storm drain, returning blood laden with waste products to the waste treatment plants of the liver, kidneys, and lungs. They also serve as the body's municipal water tank. The vein's smooth muscle walls expand to accommodate a relatively large volume of blood for when the body would have the need. The walls can then contract to decrease the size of the container. This increases the *relative* blood volume and raises the blood pressure. This is called a compensation mechanism, and is designed to solve a lot of problems.

Arteries move oxygen-rich blood away from the heart and are protectively located deeper in the muscle. Veins are closer to the surface. Arteries are held open by the systolic blood pressure of the pumping heart. Back flow in the veins is managed by a series of valves in the outer extremities and something called the "muscle pump" of the larger muscles, particularly in the legs. This particular function *can* be lost if you are unable to use those muscles when suspended in a rescue harness.

The central nervous system requires a constant supply of oxygen and glucose. Denied these things it turns into a cranky toddler, throwing itself to the ground in a fit of temper.

Now for the physiology. Hypovolemia, in this case, is not caused by the traditional culprit, blood loss. Here, the available blood volume is reduced as it pools in the lower extremities. A combination of three factors contributes to this: gravity, a lack of efficient muscle pumping action, and the compression of the greater veins from the body's weight on the harness straps ganging together and reducing the available blood volume.

As the blood remains pooled in the lower extremities several things occur:

- The heart continues to pump available blood into the lower extremities, but it is not efficiently returned. It will continue to do this until there simply isn't enough blood to pump.
- The business of tissue nourishment continues. The body continues to offload oxygen from the hemoglobin and add the waste products back into the mix. This will become very important. These waste products have a lower pH. So, a sort of acidotic stew is cooked up in the pooled blood. With no chance of moving this blood back to the lungs, things begin to get messy.

After the arrested fall victim is on the deck you will want to consider these facts very carefully, as that increased blood flow back to the heart becomes a tsunami of acidotic sludge.

As the brain is deprived of its blood supply things start to deteriorate. It sends several warning signals out along the central nervous system. Combined with the first symptom, lack of judgment, bad choices are made: the drunk frat boy/stagehand interface; the rigger who simply slips off the steel and decides not to call for help.

Everyone who has ever walked into a bar understands that consciousness is a continuum. We can go from rational humans to talking like pirates after several adult beverages. In fact, the emergency services have a way to codify this progression called the Glasgow Coma Scale. As in the bar and with the toddler, the first thing to go is your judgment. Things then move down the list of symptoms from there.

Accidents Happen

Unscheduled. Unplanned. Often in a "jelly-side down" sort of way.

That's why they are called accidents.

This is *not* going to be a first aid class. It will, however, speak to some logistical planning. Part of that plan should be to have solid CPR and automated external defibrillator (AED) training in place. If first aid training is available, please include that. Many things can happen.

Similarly, much of the training around accidents is their prevention, and where to find and file the paperwork once they do happen. Please include at least one additional step that adds a list of where to find the first aid kits, where the cell phone dead zones are, and where the ambulance should respond if the accident is in the steel or in the front of house. You might think about the things you'd want other people to know if you were the one that was hurt. Now, I don't think you can expect your ground-rigger to be a trauma surgeon, but it couldn't hurt to ask.

When an accident happens around you or you happen upon one, there is the possibility of you being the first on the scene. This is different than being a first responder. The goal in a "first-on-scene" situation is to make it somebody else's problem (SEP) as quickly as possible. Don't waste any of the platinum ten minutes on the time between the discovery and accessing the 911-system. Make sure you are safe, and then call 911.

Please don't miss the bit in the last paragraph about "make sure you are safe." Don't become part of the problem.

Let your EMS providers take on the liability of assessment, treatment, and— if necessary—transportation. Do not throw yourself into your vehicle and make a run for the hospital. The ambulance personnel have the training and equipment, and are paying the insurance premiums for it. Let them handle the patient. The fall arrest victim will receive care the moment they make patient contact. And that care will continue until the patient is transferred to definitive care. That's best practice for this situation.

Sometimes teaching people to recognize that something bad has happened is the most difficult concept. We don't want to be seen as the weak member of the herd. And then there is the whole denial thing. When hurt, many people will isolate themselves. You might see them go into the bathroom with their severed finger or refuse to be seen by the EMTs after an arrested fall. Please keep your index of suspicion high. Get help.

The rule of thumb is: ten minutes "fall-to-floor." Own it. Love it. Practice it. Really. Those in emergency services call this process scenario-based training. It creates providers who move through situations and can think past basic logistics. In the theatre we call it "rehearsing." It's brilliant for creating a crew that can handle these situations and feel confident in their abilities.

After the fall arrest victim is on the deck, if the EMS crew is not on-site, your next question should be "what is going to kill them first?" The standard wisdom heard around the arrested fall is "don't *let* them lie down." This is assuming, of course, that they are able to listen and follow directions. Placing the fall arrested on a chair could lose you points in the style competition when they pass out and fall off the chair.

If they are not talking to you, but they are breathing and there are signs of life, put them in a "Barcalounger" position: butt and heels on the deck. Knees bent. Hips flexed in an approximately 30–45-degree angle. This position will help keep the heart from being overwhelmed with acidotic blood that will cause electrical malfunctions, as well as slow down the wave of pooled blood returning back to the heart, which creates structural stretch and possible physical damage. It is also very easy to turn their head to the side should they start to vomit, and to get them flat to start CPR should you need to. Apply the pads for the AED now if you haven't already done so. Watch their pulse and responsiveness.

If they are talking to you but they're not making much sense, or if they are combative, make sure that the ambulance is on its way, and keep track of the patient. Do not let them wander off. It sounds silly, but patient searches are not fun. Whatever you do, don't let them go in a restroom by themselves. Prying someone out of a bathroom after they've passed out makes for a very bad day.

If they are talking to you and claim they're fine, and they seem just fine, this is when you need a policy for the situation. This policy should be written in conjunction with the medical control officer for the nearest hospital, the rescue squad responsible for the venue, the employer, union rep, and anyone else who should weigh in. It should include language regarding the circumstances when employees need to be seen by medical professionals. The parameters of these circumstances need to be clear and within local prehospital protocols.

There is something called "reperfusion syndrome" that can occur an hour or more after an event. Orthostatic insufficiency, reperfusion syndrome, or even positional rhabdomyolysis (which would require an entire other chapter on renal failure to cover fully) do not care if you are the crusty codger, the call steward, or the largest donor's son, so please do not let personality or seniority dictate survival. Get everyone to definitive medical care.

There is also "the index of suspicion." Somewhat along the lines of "just because you're paranoid." Be alert for a second patient. The stress of the accident and rescue can exacerbate heart problems or other chronic conditions. Consider also whether the fall might have had a medical component. Did they pass out and then fall? Or vice versa? Voice any concerns regarding these possibilities to the emergency responders.

It helps to understand your employer's concerns and requirements. Realize that many times the only written rule is how the accident must be reported. Be active in crafting something more proactive, supportive, and effective. Prove to the office that you're interested in creating a safe work environment.

Besides the paperwork, the last part of the fall arrest event is transitioning your friend's care to the EMS crew that has come to treat them. The ambulance crew will ask for demographic information: name, address, and date of birth. They will want to know if their patient has any allergies to medications, or any health issues. It helps if someone on your crew has this information. A good recommendation is for every person on the crew to keep it in their cell phone or wallet.

In the best of all possible worlds, someone should have thought to keep track of the time that things happened, but that is "Advanced Emergency Scenarios 204" (or a knee-jerk reaction for the stage manager). Time seems to expand and contract at will in an emergency situation, and one person with a watch, a pencil, and a clipboard can make all the difference.

Send their wallet, cell phone, glasses, jacket, and—believe it or not—shoes with the patient to the hospital. If there is more than one possibility, find out where the ambulance crew will take them.

Finally, make sure that the event is documented and reported appropriately. Plan a formal review that is not simply a search for the guilty. While painful, these are incredibly rich teachable moments. Don't let them slip away without serious consideration.

I'll Take "Hodgepodge" for a Thousand

Risk management has made a distressing move toward risk removal. There are venues that do not have first aid kits because of the liability involved in the treatment of small wounds. Orthostatic intolerance isn't a splinter. We need to work to ensure that appropriate training is funded and maintained.

Take your training seriously and be proactive. It helps to put together proposals, budgets, and training calendars on our own that fit into our regularly scheduled lives. This will help ensure that the investment in your health and well-being is a good one.

Encourage attendance and a positive attitude with a professional demeanor. One of the most discouraging questions for the instructor/trainer is "how long is this going to take?" Remember that if you are the one that's hurt, you will need others to take care of you. Hopefully they stayed through the end of class.

There will need to be recertification too. Keep that in mind, both in the budget and on the calendar. Keep concise records, and consider mentioning completed training and continuing education in annual performance reviews.

Like your fall arrest systems, other safety equipment needs to be inspected and updated periodically. There will be expiration dates. Having someone come through and ask for a couple of aspirin, only to find an outdated bottle looks careless. Because it *is* careless.

Consider training with your local rescue squad or fire department. The CPR and AED instructors need to teach classes in order to maintain their certification. Bringing them in and developing a rapport with your colleagues is helpful and just plain fun. You haven't lived until you've been roped down in the Stokes basket from the grid.

In a recent "meeting of the pickup trucks," EMS and theatre people were astonished at the knowledge both groups had of high-angle rescue. The EMS people didn't quite realize that those riggers worked at high-angle rescue every day of their work life. There was newfound respect on both sides over pizza at the end of the day. That simple understanding and camaraderie goes a long way to smoothing the path during a high-stress rescue.

Many times, nothing is done at an accident scene because people are afraid of doing the wrong thing, or being sued. Look into your local Good Samaritan laws, as each state is different. While Good Samaritan laws will not stop you from being sued, they will, however, support and encourage the basic human responsibility for each other.

If you are a certified and licensed emergency or health care worker, know and respect your personal scope of practice. Good Samaritan laws are particularly tricky for you. Do not, I repeat, *do not* exceed your scope of practice. Let people know what your scope of practice is so they don't think you should be cracking a chest, or anything else they saw on television last week.

ICE stands for In Case of Emergency. Stop right now. Put this book down and check your phone's contacts list. Please put in a phone number for your emergency contact person. First responders are trained to look at your cell phone for the number under ICE. "Mom" could very well be someone in a condo in Boca. Check this number on a regular basis, like you check your smoke detectors and change the batteries. Remember that after a bad breakup, an emergency worker calling your ICE number may be told "let them die, it's what they would have wanted." Make sure your emergency contact person has access to your basic medical information and your DNR if you have one. Oh, and please sign your donor card.

Index